Computational Physics Using C

Efficient Programming with Ease

John W. Fattaruso

Southern Methodist University, Dallas, Texas, USA

AF794360

WILEY

Copyright © 2026 by John Wiley & Sons, Inc. All rights reserved, including rights for text and data mining and training of artificial intelligence technologies or similar technologies.

Published by John Wiley & Sons, Inc., Hoboken, New Jersey.

No part of this publication may be reproduced, stored in a retrieval system, or transmitted in any form or by any means, electronic, mechanical, photocopying, recording, scanning, or otherwise, except as permitted under Section 107 or 108 of the 1976 United States Copyright Act, without either the prior written permission of the Publisher, or authorization through payment of the appropriate per-copy fee to the Copyright Clearance Center, Inc., 222 Rosewood Drive, Danvers, MA 01923, (978) 750-8400, fax (978) 750-4470, or on the web at www.copyright.com. Requests to the Publisher for permission should be addressed to the Permissions Department, John Wiley & Sons, Inc., 111 River Street, Hoboken, NJ 07030, (201) 748-6011, fax (201) 748-6008, or online at http://www.wiley.com/go/permission.

The manufacturer's authorized representative according to the EU General Product Safety Regulation is Wiley-VCH GmbH, Boschstr. 12, 69469 Weinheim, Germany, e-mail: Product_Safety@wiley.com.

Trademarks: Wiley and the Wiley logo are trademarks or registered trademarks of John Wiley & Sons, Inc. and/or its affiliates in the United States and other countries and may not be used without written permission. All other trademarks are the property of their respective owners. John Wiley & Sons, Inc. is not associated with any product or vendor mentioned in this book.

Limit of Liability/Disclaimer of Warranty: While the publisher and the authors have used their best efforts in preparing this work, including a review of the content of the work, neither the publisher nor the authors make any representations or warranties with respect to the accuracy or completeness of the contents of this work and specifically disclaim all warranties, including without limitation any implied warranties of merchantability or fitness for a particular purpose. No warranty may be created or extended by sales representatives, written sales materials or promotional statements for this work. The fact that an organization, website, or product is referred to in this work as a citation and/or potential source of further information does not mean that the publisher and authors endorse the information or services the organization, website, or product may provide or recommendations it may make. This work is sold with the understanding that the publisher is not engaged in rendering professional services. The advice and strategies contained herein may not be suitable for your situation. You should consult with a specialist where appropriate. Further, readers should be aware that websites listed in this work may have changed or disappeared between when this work was written and when it is read. Neither the publisher nor authors shall be liable for any loss of profit or any other commercial damages, including but not limited to special, incidental, consequential, or other damages.

For general information on our other products and services or for technical support, please contact our Customer Care Department within the United States at (800) 762-2974, outside the United States at (317) 572-3993 or fax (317) 572-4002.

Wiley also publishes its books in a variety of electronic formats. Some content that appears in print may not be available in electronic formats. For more information about Wiley products, visit our web site at www.wiley.com.

Library of Congress Cataloging-in-Publication Data has been applied for:

Paperback: 9781394318537
ePDF: 9781394318551
ePub: 9781394318544

Cover Design: Wiley
Cover Image: Simulated chaotic motion of the double pendulum by John W. Fattaruso

To Marsha

Contents

Preface

This text is intended to be a reference for a single-semester course in Computational Physics for upper-division undergraduate students or first-year graduate students in Physics. It is written assuming the student has had a foundation of physics classes in undergraduate classical mechanics, basic quantum mechanics, and undergraduate electromagnetism. There is quite a bit of computer programming presented in these pages, with the intention that students should be prepared to contribute to the growing capability of the physics community to seek solutions to more and more complex problems by making use of computing hardware in the most efficient manner possible. The programming content in this text is comparable to that of typical first-year classes for Computer Science majors. After the first three chapters of introductory material, the text is organized into chapters around a progression of numerical analysis algorithms, also bringing a study of the basic mathematics of numerical analysis into the discussion. In each of the chapters from Chapters 4 through 14, a particular analysis algorithm is highlighted, and several physics problems are solved that are proposed as useful prototype solutions. The sequence of algorithms has been chosen so that prerequisite numerical methods are covered in prior chapters. Example problems left as exercises for the student are collected at the end of each chapter, but these can and certainly should be supplemented with other numerical problems from the instructor's experience. There should be a sufficient number of chapters from which an instructor may choose a subset to focus on as time and interest dictate.

Contemporary computational physics is tackling problems of impressive scope, from molecular to galactic simulations. For computations on this scale to produce meaningful results and to complete execution both within a researcher's lifetime and while consuming a minimum of computing power, the coding of the problem into a programming language must make the most efficient use of computing hardware as is possible. Chapter 1 will lay out the basic plan for the programming system described in this text. Standard C programming is exclusively used, and emphasis is placed on using C language features that minimize the programming time needed for code development as well as the hardware machine cycles needed for execution. It is assumed that the student has at least some introductory programming skills in C, C++, Java, or another structured language. If not, there are plentiful resources available online to bring the student up to speed. There are excerpts of C code the student will find on the pages of this text, but primarily the code written for all the example problems presented here is to be downloaded from the Wiley Student Companion website.

Before launching into numerical analysis with C code, however, the fundamental limits to the precision of numerical algorithms are first addressed in Chapter 2. This material is essential to ensure that all our efforts in designing and writing computer code will generate meaningful results for the physics problem being studied.

The use of inefficient interpreted languages such as Python, Tcl, or Perl is avoided here. It is assumed that as soon as a programming tool is developed to solve a particular problem, there will next be a need to put that solution inside a programming loop to step some parameter over a range, or embed that solution in a root finding search, or use that solution in a function embedded in a differential equation solution, or some other numerical algorithm at an upper hierarchical level. Indeed, this tendency is illustrated several times in various chapters that follow. For larger and larger computation tasks in hierarchical levels, attending to computational efficiency at every level is essential. Several examples are given demonstrating the very significant execution time savings possible when working with compiled and optimized C code relative to an interpreted language such as Python.

For students not yet confident in making use of pointers and structures in C code, two language features that are key for writing efficient and modular code, Chapter 3 provides a detailed background. Much of this chapter could be skipped by students who are already accomplished C programmers. Basic facility with using the `gcc` C compiler and using command line shells to execute code is also assumed. This universal compilation tool is native to the Linux and MacOS environments, and it may also be easily run under Microsoft Windows by using the powerful MSYS2 package (https://www.msys2.org/), which sets up an emulated Linux work environment on a Windows system. No particular coding environment is assumed, but the use of any of the many available powerful C language coding editors that help the programmer find and correct coding errors immediately at edit time, such as the cross-platform Microsoft Visual Studio Code editor (https://code.visual studio.com/), is highly recommended. The `gnuplot` plotting tool (http://www.gnuplot.info/) is demonstrated throughout the text to render and present graphical results, and the student should have basic capability with this powerful, cross-platform tool that runs under Linux and Windows. Again, plentiful online resources are available to learn this nearly universal tool. Besides laptop and desktop machines running Linux or Windows, even a simple "single-board" computer system, such as the popular Raspberry Pi running the Linux operating system, makes an effective and low-cost platform for running the development environment of the VS Code editor, the `gcc` compiler, the `gnuplot` plotting tool, the `ffmpeg` animation rendering tool, and the `gdb` debugger for the coding system in this text.

We use a programming language such as C to describe in a very precise manner exactly how the computational procedure we have designed to solve a particular problem is to be carried out with optimum efficiency. The greatest possible flexibility in compiling and executing our numerical analysis programs over a variety of physical situations and parameters is likewise afforded by learning the Linux and MSYS2 `bash` command line language for executing tools and programs in a terminal window. As we appreciate the versatility of a plotting tool such as `gnuplot`, we will learn that its command structure is essentially another type of programming language, one that describes in detail how we wish the numeric results of some computation to be rendered and presented graphically. Examples of running text-based commands and developing facility in these languages in a terminal window are provided throughout this text. In a contemporary graphical window display system with wide monitor screens, users typically extend the width of terminal windows to well beyond the number of columns available in the enclosed example boxes used in this printed text. The convention used in this text is to allow the narrow coding boxes to wrap lines that do not fit within their limited widths to the next line below and to indicate a wrapped line with a hooked "→" symbol. These longer lines will typically fit completely and naturally on one line of a displayed terminal window.

Another goal of this text is to develop the powerful possibilities open to us by making use of the C code that accomplishes our computation tasks to also programmatically generate the `gnuplot` code for subsequently graphically presenting its own computational results. Developing this capability fully will let us use `gnuplot` to iteratively render a sequence of graphical plot images that will form the frames of an animated video result of our simulations. Rendering animated graphics is an effective way to visualize and communicate many physical processes, and we will make good use of this capability in chapters where equations of motion are simulated. The animated media associated with some of the example simulations here may also be downloaded from the Wiley Student Companion website. The text that explains the details of this animated graphics approach is generally left to the last section of a chapter, and may be omitted if time does not permit covering this topic.

One common trend in contemporary Computer Science is to design our code in a hierarchical structure of modular and encapsulated programming objects, described in an object-oriented language such as C++. This is not an approach taken in this text. The author's experience is that more often than not, the language constructs designed to accommodate many layers of object class inheritance can get in the way of intensely numerical code, and application of these powerful object-oriented languages is better left to problems more appropriate for polymorphism and class inheritance. Moreover, the language structures in C++ can involve sacrificing execution efficiency at run time. Calling polymorphic functions, handling exceptions, and automatically reclaiming allocated memory can involve hidden execution overhead. Therefore, standard C for procedural code is used here, but at the same time making heavy use of structures and pointers to form the code as a system of versatile and reusable modules.

The predictions that validate the laws of physics are ultimately rooted in experimental data, and the numerical results from realistic experiments are always subject to uncertainty within practical sources of error. For all physical sciences, there are well-defined guidelines for the number of significant digits that may be used to represent numerical data within a known precision limit. Numerical data in this textbook are generally displayed or stored in computer system files using eight significant digits, as generated by the standard C library output and input routines. It should be understood, however, that this arbitrary choice of precision is only for demonstration purposes, and that the actual number of significant digits from practical programs that are displayed for publication should always be chosen by the programmer to be only those justified by the precision of the data that form the program input.

The use of standard SI units should be assumed throughout many of the programming examples and solutions presented in this text. One exception is the occasional use of more convenient units for arguments expected to be typed on the command line of some programs. For example, for programs modeling behavior at a quantum mechanical scale, the student will encounter code written to allow units such as electron volts and nanometers on the command line, but then with the quantities immediately translated to the corresponding SI unit for internal calculations. Other problems in this text may be approached in a general numerical way with quantities expressed in arbitrary units when appropriate.

The library of structures and functions performing numerical analysis algorithms in this book will likely not be the ultimate library that students would use for future research projects. Those are more likely to be well-developed and thoroughly optimized packages already available. The libraries for this text serve as simple examples to get started and modify, to learn the basics of the algorithms, to understand the consequences of choices for algorithmic parameters, and to learn to choose one algorithm from among many.

A common theme among the physics problems presented in this text is numerically exploring the gradual transition between discrete physical systems with a few elements and continuous systems analyzed as a continuum of infinitesimal elements. Examples are the transition between a few coupled harmonic oscillators and a longitudinal wave propagation medium, a few quantum potential wells and a lattice of solid state wells in a crystal structure, a few colliding particles and an ideal gas, and others. These will hopefully lend more physical insight to the students.

Finally, it is necessary to keep in mind when studying the physical sciences with a computational emphasis that our goal is always to enhance our physical and mathematical understanding, not to try to replace it. The availability of fast and powerful computation hardware at times can lead us into the temptation to approach a problem with a battery of simulations that have not been thoughtfully planned or interpreted. The chapters that follow hopefully will lead the student in the direction of using computation as a companion to physics understanding.

This textbook, as is usually the case, grew out of a rich collaboration with others. The author is indebted to Gary Boyd, Thomas Coan, William McElgin, Fredrick Olness, Randall Scalise, Stephen Sekula, and Ryszard Stroynowski for their indispensable assistance. The author is further indebted to Marsha Dekan for her meticulous proofreading of the manuscript.

About the Companion Website

Computational Physics Using C: Efficient Programming with Ease is accompanied by a companion website:

http://www.wiley.com/go/CompPhysUsingC

The website includes:

- CPSolutions
- CPExamples

1

Introduction

1.1 What Is Computational Physics?

Outside the CERN Physics Laboratory in Geneva, Switzerland, is an intriguing metallic sculpture, shown in Figures 1.1 and 1.2. One surface is engraved with a historical progression of many of the salient theoretical equations of physics, such as those attributed to Kepler, Newton, Maxwell, Schrödinger, Einstein, and others. As we study physics, we apply equations that describe these fundamental mathematical relationships of the universe to particular physical situations, seek solutions to the equations, and test the conformity of those solutions to observations.

Newton's equations of motion applied to two bodies in space give solutions that mathematically describe the elliptical orbits of planets around the sun in our solar system. Solving Maxwell's equations applied to various geometries of conductors and insulators allows the design of electromagnetic communication technologies. Applying Schrödinger's equation to a straightforward spring potential predicts the observed quantized energy levels for quantum particles.

Physics textbooks typically emphasize these sorts of physical arrangements, where the resulting equations may be solved, or at least treated within a reasonable approximation, by analytic mathematical transformations. This is always appealing, as a solution expression can often give us a wealth of understanding of the interdependence of physical parameters. However, in general, there are far more physical scenarios that yield equations with no meaningful analytic solutions than those that may be manipulated into neat relationships. Systems of equations with transcendental nonlinearities, high-order polynomials, or systems of equations of too high an order can frustrate the search for analytic solutions.

For example, most quantum mechanics textbooks solve Schrödinger's equation to find allowed wave function solutions for a particle bound in an idealized rectangular potential well with infinitely high walls [1, Chap. 6; 2, Chap. 7; 3, Chap. 4; 4, Chap. 1; 5, Chap. 5]. The allowed energy levels for a particle of mass m bound in a well of width a can be derived, and these energy levels will follow the ideal relation:

$$E_n = \frac{\pi^2 \hbar^2 n^2}{2ma^2} \tag{1.1}$$

at the quantum level n. From this relation, we can see the profound result that the particle has a nonzero energy at the lowest possible energy state corresponding to $n = 1$, and that successively higher energy levels are quadratically related to the quantum level. However, a potential well with infinitely high walls implies an unbound amount of binding force being applied to the particle in an infinitesimal space, and only approximates real physical environments. We will visit bound particles with more realistic binding forces progressively in Sections 5.11, 7.7, and 13.6 as we develop more numerical tools.

As another example, we will first analyze the magnetic fields of the Helmholtz coil in Section 4.4. This simple geometric arrangement of two circular coils carrying electric current has long been used experimentally to create a region of approximately uniform magnetic fields in a laboratory setting. Many textbooks in electromagnetism derive an analytic integration of the Biot–Savart law that predicts the field only along the axis of the structure [6, Chap. 2; 7, Chap. 4; 8, Chap. 8]. We can find a relation for the magnetic field at the center point of the geometry with circular paths of radius R carrying a current I to be

$$B = \frac{\mu_0 I}{\left(\frac{5}{4}\right)^{\frac{3}{2}} R} \tag{1.2}$$

Computational Physics Using C: Efficient Programming with Ease, First Edition. John W. Fattaruso. © 2026 John Wiley & Sons, Inc.
All rights reserved, including rights for text and data mining and training of artificial intelligence technologies or similar technologies.
Published 2026 by John Wiley & Sons, Inc. Companion Website: http://www.wiley.com/go/CompPhysUsingC

Figure 1.1 Sculpture outside the CERN Physics Laboratory in Geneva.

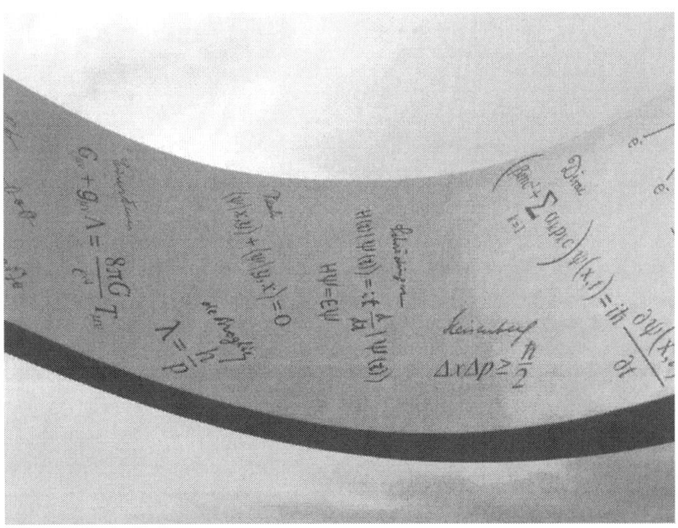

Figure 1.2 Closeup view of the sculpture outside the CERN Physics Laboratory in Geneva.

While inspection of this relation provides some important understanding, such as the dependence of the field strength at this one point directly on the current and inversely on the first power of the radius dimension, the important ability to predict the field at all other points in space around the central axis is dependent on numerical tools. We will visit solutions of successive scope in Sections 4.4, 10.6, and 11.4.

Without the capability to apply numerical techniques and find solutions beyond those that are analytically tractable, we have limited capability to analyze entire areas of rich physical behavior. Numerical solutions open the door to describing the trajectories of multiple bodies under gravitational interaction, or the amazing interactions of nonlinear oscillators, or the chaotic behavior of the double pendulum. Indeed, the whole field of study of chaotic dynamics only became practical in the 1960s with the advent of affordable and powerful scientific computation.

Another emphasis of this text is exploring and visualizing the physical behavior that lies on the boundary of what we consider discrete particles or elements and an ideal, continuous distribution of a medium as the number of physical elements in a system increases to large counts. For example, using these numerical tools, we are able to visualize the transition between discrete coupled spring-mass oscillators (Section 6.5) and a continuous medium supporting a longitudinal wave (Section 14.5), collision trajectories between discrete spherical balls (Section 4.8) and the distribution of particle speeds in an ideal gas (Sections 8.11), the allowed energy states in discrete quantum potential wells and the energy bands and forbidden regions of a solid-state crystal lattice (Section 13.6), and others.

As we are guided into numerical solutions of many interesting physical systems, we must then rely on the centuries of development of numerical algorithms [9–13]. Many of these techniques are attributed to early natural philosophers such as Newton, Lagrange, and Gauss. These attributions reveal that numerical solutions have been a part of the science now known as "Physics" since seminal times.

The execution of these numerical algorithms will almost invariably make use of contemporary computation hardware. Effective and efficient use of computing hardware will then naturally rely on many decades of development of computing languages and software techniques.

So in summary, Computational Physics may be seen as the intersection of three fields of study typically isolated in three separate departments in universities: Physics, Computer Science, and Mathematics, symbolically pictured in Figure 1.3.

Among the many approaches possible for the coding phase of solving physics problems computationally, we must pay attention to the efficiency of the software tools, with the intention of minimizing two universal equations of economics:

$$\begin{aligned} t &= Money \\ E &= Money \end{aligned} \tag{1.3}$$

In the sense of computation for the physical sciences, t, representing time in this equation, may refer to program development time, program execution time, or both. E represents the electrical energy that must be expended in electronic hardware to accomplish the required computation. Computational tasks are growing more and more complex and extensive,

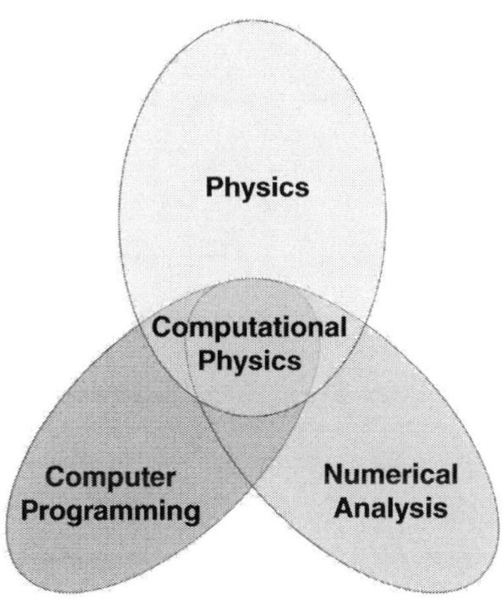

Figure 1.3 Computational Physics is a field at the intersection of three disciplines.

with a significant portion of the world's electric power generation now being dedicated to supply data centers and computing farms. Accordingly, this text will, for the most part, emphasize the minimization of execution time and computational power.

Moreover, we may take caution from the famous quote attributed to numerical analyst Carl-Erik Fröberg, referring to the rapid development of computational capability in the twentieth century: "Never in the history of mankind has it been possible to produce so many wrong answers so quickly!" By developing systematic and efficient methods of coding numerical solutions, we may gain assurance that our code development efforts are rewarded with meaningful solutions to the intended physical problem.

1.2 Modularizing and Reusing Code

Figure 1.4 is another symbolic plot, not from any careful scientific study, but rather just summarizing general observations over time. The vertical axis of this plot is the number of coding bugs included in every 100 lines of newly written computer code, versus time on the horizontal axis. (For the purpose of this text, we will consider a programming bug to be any coding error that can cause numerical errors, beyond fundamental roundoff errors, in the program results, or in a complete program failure.) When we first sit down to learn programming, perhaps as part of a school class or independent study, there are so many details to learn at one time that it seems as though we introduce a bug in every line of code. The details are quickly learned, however, as we slide down the learning curve. Whenever we tackle the adventure of learning a new programming language, at first again we may struggle with bugs introduced in our new code, but these quickly decay down. So to avoid the danger of Fröberg's warning above, one lesson we might take away is that to eliminate coding bugs, we should never actually write any new code! Obviously if we took this guidance literally, it would mean never solving any new problems. However, the closer we approximate this absurd limit, the more likely our numerical results will be meaningful and correct, and involve a minimum of development time. So our task is to develop and maintain modular techniques for coding our programs that allow us to reuse as much previously written and debugged code as possible in each new program.

Figure 1.5 shows perhaps a helpful image by which we can conceive the overall structure of our program code. This image shows a hierarchical structure of interacting modules of data and coded instructions. Each box symbolizes some module of instruction code or data representing a physical entity. By symbolically "plugging in" modules into others, we construct a software system that accomplishes our computation task. We develop a coding system so that for each of our new programming projects, we will make use of as many previously written and debugged coding modules as possible. For the new code we must write and debug, we will conform to a modular model as well, so that the new modules in turn may be reused for another problem.

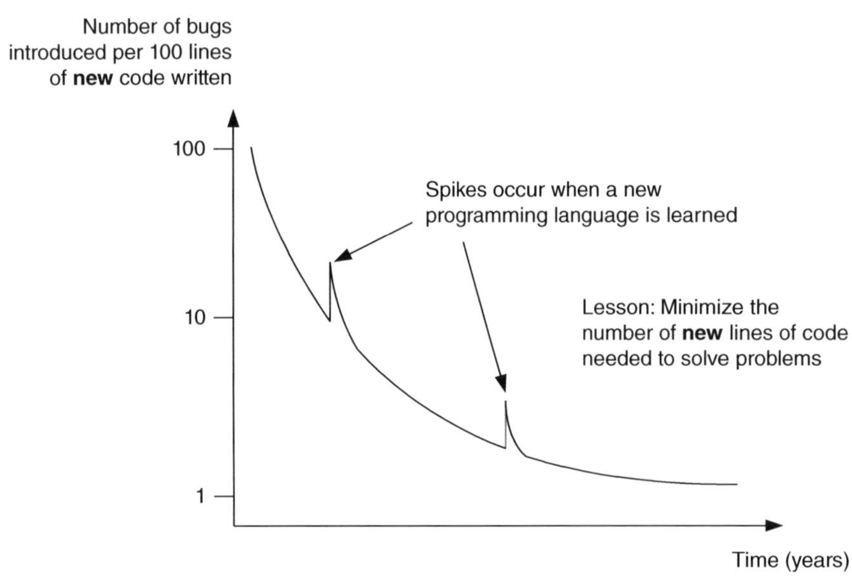

Figure 1.4 Learning curve for coding and avoiding bugs over the long term.

Each program module is coded in the C language in a `.c` source code file, and a `.h` header code file summarizes enough knowledge of the interface of each tool to be able to "plug" the object modules together and interchange data.

Table 1.1 shows how this modular technique is demonstrated in the example code solutions in this text. For each of the main programs listed in the left column that are selected from those we will be examining in the following chapters as solving some physics problem, the right column lists modular library tools residing in the CPLibrary directory that we will slowly develop as we study various numerical analysis algorithms in each chapter. Each of the modular library tools listed will be linked to the corresponding main program by compiling them all together. The component capability provided by each library module is "plugged into" the function hierarchy to complete the overall solution. This general scheme of collecting code modules that are commonly used for a variety of tasks minimizes the need to write new code that is subject to bugs or logic errors.

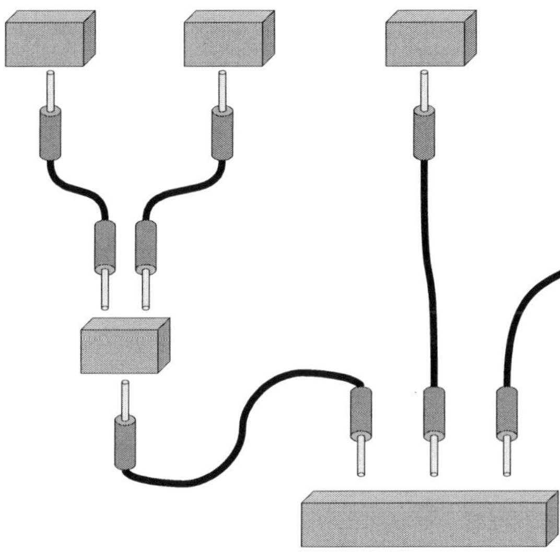

Figure 1.5 Structuring a large-scale program into interacting modules.

Table 1.1 Selected example textbook programs and their required tools linked from the CPLibrary.

Example program	Linked library code required
`damped_oscillator.c`	`step.c array.c`
`two_pool_balls.c`	`collide_balls.c ball.c`
`wave_packet.c`	`step_complex.c array_complex.c`
`quantum_abruptstep_packet.c`	`step_complex.c step.c`
	`array_complex.c array.c`
`efieldvectors.c`	`point_charges.c array.c`
`catenary.c`	`root.c`
`orbit_step.c`	`root.c step.c array.c`
`quantum_rectwell_roots.c`	`step.c step_roots.c root.c array.c`
`coupled_osc3_step.c`	`lineq.c step.c array.c`
`wheatstone.c`	`lineq.c array.c`
`strings2.c`	`sys_root.c lineq.c array.c`

(Continued)

Table 1.1 (Continued)

Example program	Linked library code required
quantum_2rectwell_root.c	sys_root.c lineq.c step.c array.c
walk.c	random.c array.c
gas.c	collide_balls.c random.c ball.c
interpolate_step.c	interpolate.c array.c data_file.c
spline_step.c	cubic_spline.c lineq.c data_file.c array.c
lprecord.c	interpolate.c step.c array.c random.c
integrate_data_file.c	integrate.c data_file.c array.c
integrate_gaussian.c	integrate.c step.c array.c
helmholtz.c	current_loops.c integrate.c array.c
helmholtz_axis_minvar.c	minimum.c step.c array.c
fit_maxwell_boltzmann.c	minimum.c step.c data_file.c array.c
pendulum.c	diffeq.c array.c
ballistic_range_step.c	diffeq.c interpolate.c step.c array.c
ballistic_range_solve.c	diffeq.c interpolate.c step_roots.c root.c array.c
ballistic_range_max.c	diffeq.c interpolate.c minimum.c array.c
efieldlines.c	diffeq.c point_charges.c array.c
quantum_linbarrier_packet.c	diffeq_complex.c step.c array_complex.c array.c
quantum_rectwell_target_roots.c	step_roots.c root.c diffeq.c step.c array.c
wave1.c	diffeq_implicit.c lineq.c array.c
gravitation_implicit.c	diffeq_implicit.c lineq.c array.c

It is assumed that the student may wish to prepare any number of work directories or folders in a file system where programs solving various physics problems are edited and executed. For example, the student may wish to create a separate work directory for the examples and exercises in each chapter. All the library tools we will develop are bundled together in one archive file named `CPLibrary.zip`, available for download from the Student Companion website. All the `#include` lines at the beginning of each example solution program and example `gcc` C compiler command lines presented in this text are written assuming that the source code files from the `CPLibrary.zip` archive have all been placed in a subdirectory of the current work directory named `CPLibrary`. However, the example solutions and command lines may be edited to reflect whatever other directory organization is desired. For example, one directory named `CPLibrary` may be placed at some fixed location in the student's file system and referred to on the `gcc` command lines. In that case, the `#include` lines at the beginning of each example program solution that refer to library routine header files would have to be edited to locate that library directory.

1.3 Introduction to Computational Efficiency

As computational physics simulations grow to be more and more complex in nature and wider in scope, it is important to study software structures and techniques to make as efficient use of the computational hardware on which our analysis code will be executing as possible. In this section, we briefly look at some details of computer hardware to appreciate the choices made in the software design of this text.

An example of a computer structure known as the *von Neumann* architecture is shown in Figure 1.6. This can be considered an architecture that most contemporary processors are either physically constructed from, or at least virtually appear to function as. There are many enhancements in computer hardware design that have greatly improved the performance of hardware beyond that possible by strictly following this architecture, such as cache memory, pipelined instruction execution, and multicore designs. However, we can consider this figure to be an example of a basic functional model.

In computer hardware, all values of data numbers or memory addresses are stored as patterns of 0 and 1 bits in the binary numbering system. At the heart of the architecture is the Arithmetic and Logic Unit, or ALU, which is the digital logic block capable of arithmetic and logical operations between the numerical quantities A and B represented in binary, producing a numerical result at its output. The functions of each of the digital hardware blocks in our model of Figure 1.6 that operate on these binary representations are described in Table 1.2.

Let us take, for example, a case where an arithmetic expression such as $x = x + y$ written in some programming language is to be evaluated and executed on the hardware. Let's assume that the binary bits representing the numerical value of the variable x are stored in one of the general purpose registers in the GPreg block, but the value of the variable y is stored at a particular address in memory. At other memory addresses, the programming language compiler has stored hardware instructions that code the machine operations that form the program. For evaluating the arithmetic expression $x = x + y$, the machine operation is to add the contents of a memory location to the contents of a register and then store the sum back into the data register. So an example sequence of events that must be executed in the hardware for the software expression $x = x + y$ to run is:

- Transfer the memory address of the next hardware instruction from the program counter (PC) register to the memory address register (MAR)
- Initiate a memory read at the address in the MAR, loading the memory data at that address into the memory data register (MDR)

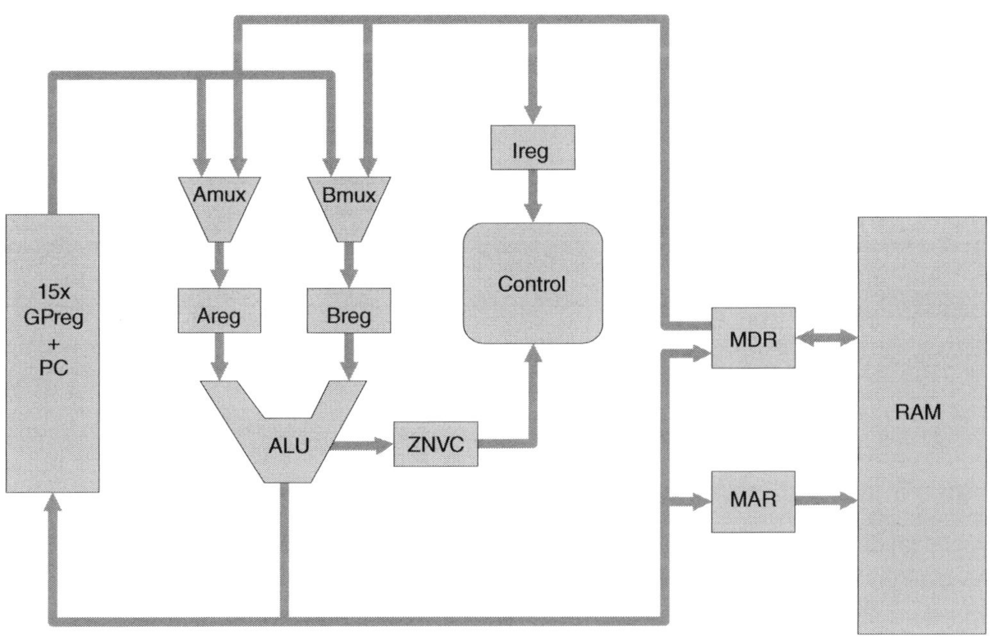

Figure 1.6 Block diagram of a basic central processing unit (CPU).

Table 1.2 Computation functions of the hardware blocks in a basic central processing unit.

Block	Description
Areg	Register for A operand to ALU
Breg	Register for B operand to ALU
Amux	Multiplexer to select source of A operand
Bmux	Multiplexer to select source of B operand
ALU	Arithmetic and Logic Unit – performs arithmetic and bitwise logical operations with A and B
GPreg	General-purpose data storage register (15)
PC	Program counter – holds the address of next instruction
Ireg	Instruction register – holds currently executing instruction code
Control	Instruction decoder and control signal generator
ZNVC	Flag status register – zero, negative, overflow, carry
MDR	Memory data register – holds data read from or written to RAM
MAR	Memory address register – holds address of RAM access
RAM	Random access memory

- Transfer the MDR contents, which is the binary code for the "add" instruction, into the Ireg and initiate the instruction decode
- Transfer the contents of the PC register into the Areg and the constant 1 into the Breg
- Initiate an "add" operation in the ALU, computing the memory address of the next instruction to be executed, storing the result back into the PC
- Transfer the memory address of y embedded in the instruction stored in the Ireg to the MAR
- Initiate a memory read at the address in the MAR, loading the memory data at that address into the MDR
- Transfer the x register value into the A register through the Amux and the MDR into the B register through the Bmux
- Initiate an "add" operation in the ALU, forming the sum of the Areg and the Breg, storing the ALU output back into the x register

At this point, the hardware is ready to execute the next instruction in memory at the address now contained in the PC, and this process is repeated for each hardware instruction that the language compiler has prepared. We can see that the variety of arithmetic operations that may be performed by hardware in typical processors is actually not much wider than that for a crude four-function push button calculator, except many orders of magnitude faster. With this fine granularity of digital operations in the processor hardware, all other arithmetic and function evaluation, such as computing trigonometric or exponential functions, must be accomplished with software.

We can see that what we might consider a "step," in a mathematical and computational sense, is actually composed of a number of more atomic steps at the hardware level, where binary information is transferred between the various blocks and then operated on inside each block. It should be understood that each of these hardware steps takes place at a limited rate determined by the "cycle time" of the particular processor the code is executing on, and requires a certain expenditure of electrical energy. (However, in a contemporary pipelined processor, many of these cycles may actually take place simultaneously.) The actual cycle time is determined by competing engineering and thermodynamic realities: The manufacturing cost of the particular semiconductor circuit technology purchased, and the level of electrical power consumed by the hardware and the rate of heat energy removed.

Besides the monetary investment of purchasing computing hardware and software, the two primary costs of the computation of a solution to a given problem are the time investment needed to wait for the computing tools to come to a solution and the electrical power needed to supply the hardware. For simple problems, the speed of modern hardware may only require the user to wait milliseconds, and the program development time dominates. Although the cycle time of contemporary high-performance processors is typically on the order of a nanosecond, for a very complex problem, a physical simulation or solution may involve an enormous number of cycles. For extensive simulations, for example, of galactic formation or molecular behavior, computation time may involve hours, days, months, or years, and cost many kilowatt-hours of generated power. In many cases, the ability to make the most efficient use of computation hardware may determine thousands of hours of expensive salaries that are being burned while waiting for a large-scale program to finish and report its results. In general, we tend to simulate problems of larger and larger complexity until the cost becomes prohibitive, as governed by Equation (1.3).

Even physical simulations of limited complexity can benefit tremendously with attention paid to computational efficiency, when those tasks are to be run on smaller scale hardware of limited speed and battery power. The technology for small, economical processors is currently proliferating, and the throughput and power performance of simulations that are a part of interactive applications on dedicated small processors can be critical. So as we study how to code the algorithms of computational physics for best efficiency, we are basically answering the question: "For any given processor, on its own scale of capability, what is the most amount of physics we can solve for the least time and energy?"

The design of the C language, and of the compilers we use to write C code, basically allows the programmer to come close to directly accessing and controlling the function of the various blocks in Figure 1.6, but without dealing with any details of the hardware directly. Indeed, this was the motivation of the historic development of the C programming language, as applied to the need to absolutely optimize execution efficiency of operating systems. In particular, the inclusion of the "pointer" variable types in C allows the programmer to manipulate the registers in the processor that hold memory addresses most efficiently. It is significant that the kernel of the Linux operating system and the fundamental `glibc` library are coded almost entirely in the C language.

We have seen in the example sequence of operations above that at times the ALU is used to calculate memory addresses in addition to numerical data. In the example above, the ALU is used to increment the one register dedicated to holding the memory address of the next hardware instruction to be executed, the PC register. After an instruction is fetched from memory, that memory address must be incremented to point to the next instruction in the program so that the program execution process may iterate. Similarly, other registers may be used to hold memory addresses of numerical data variables forming computation operands. There are often aggregations of data elements in consecutive memory addresses; for example, when data is held in arrays or structures, and we rely on one of the registers to hold the address of some initial and subsequent place in that aggregation. To read or write each successive data element from or to memory, the machine must use the ALU to increment or decrement the address to that of the next element, transfer the address into the MAR, and perform a read or write cycle using the MDR for memory data for each element. In the C language, we can directly manipulate pointer variables, which are symbolic names for those registers or memory locations that hold a memory address where data is located. Rather than programming in languages where memory addresses are hidden from us, coding in C with pointers gives the programmer direct control and efficient processing of these memory addresses.

Benchmarking particular processors for the time required to accomplish certain particular computation tasks is an entire research field unto itself, but we can look at a very simple example to get general guidelines on efficiency. Consider a simple C program for evaluating an eighth-order polynomial.

$$y(x) = a_8x^8 + a_7x^7 + a_6x^6 + a_5x^5 + a_4x^4 + a_3x^3 + a_2x^2 + a_1x + a_0 \tag{1.4}$$

In general, the ALU unit in the processor design of Figure 1.6 will not be capable of raising a particular value to some arbitrary integer power. This operation must be coded in software. Our first coding approach, seeing the expression of Equation (1.4), may be to evaluate the power terms in a very straightforward term-by-term manner, such as

$$\begin{aligned} y(x) = {} & (a_8 \cdot x \cdot x \cdot x \cdot x \cdot x \cdot x \cdot x \cdot x) + (a_7 \cdot x \cdot x \cdot x \cdot x \cdot x \cdot x \cdot x) \\ & + (a_6 \cdot x \cdot x \cdot x \cdot x \cdot x \cdot x) + (a_5 \cdot x \cdot x \cdot x \cdot x \cdot x) \\ & + (a_4 \cdot x \cdot x \cdot x \cdot x) + (a_3 \cdot x \cdot x \cdot x) + (a_2 \cdot x \cdot x) + (a_1 \cdot x) + a_0 \end{aligned} \tag{1.5}$$

Counting up the number of multiplication and addition operations that we are required to cycle through the ALU of the architecture of Figure 1.6 for a general nth order polynomial, we would need n additions and a total of $0 + 1 + 2 + \cdots + (n-1) + n = \frac{n(n+1)}{2}$ multiplications. Therefore, the number of required machine cycles would grow as the *square* of the order of the polynomial being evaluated. This is known as an *order 2* computation method. In the coding example above, it ends up with a rather inefficient use of our computing hardware, given that with a small amount of algebraic reordering, we can reduce the machine load to only an *order 1* method. Consider evaluating the polynomial with Horner's method, in which each term is nested inside the term for the next lowest order [10, Chap. 1], as in

$$y(x) = a_0 + x(a_1 + x(a_2 + x(a_3 + x(a_4 + x(a_5 + x(a_6 + x(a_7 + x \cdot a_8))))))) \tag{1.6}$$

With Horner's method, the various powers to which the argument x is raised are realized by how many nested terms a particular multiplication of x is contained in. A great increase in computation efficiency is possible simply by algebraic reordering. Now we require only n additions and n multiplications through the ALU of our processor, giving a computation time that grows only *linearly* with the order of the polynomial.

Let us make a rather contrived assumption that, for a particular computation task, this polynomial must be evaluated at 10 000 000 different values of its argument x. The first part of the program in Listing 1.1 shows how the straightforward term-by-term method for evaluating polynomials may be coded in C. The second part iterates through the same 10 000 000 evaluations, but uses the nested term method instead. Both code sections are surrounded by calls to the Linux system function `times()`, which records the current state of a timer the operating system maintains, measuring the elapsed processing time of any program. By subtracting the time values, we can measure the processing time consumed by the two code sections and report the measurements to the output stream.

```
// File: poly_times.c
#include <stdlib.h>
#include <stdio.h>
#include <unistd.h>
#include <sys/times.h>

int main(int argc, char **argv) {
  double x,y,t;
  double a[9] = {9.0,-8.0,7.0,-6.0,5.0,-4.0,3.0,-2.0,1.0};
  int i,j;
  struct tms cpu1,cpu2;

// Evaluate test polynomial by the term-by-term method
  printf("10,000,000 evaluations of an 8th order polynomial "
         "term-by-term\n");
// Record CPU time at beginning of calculation
  times(&cpu1);
  x = 9999999.0;
  while (x > -0.1) {
    y = a[0];
    i = 1;
    while (i < 9) {
      t = a[i];
      j = 0;
      while (j < i) {
        t = t * x;
        j++;
      }
      y = y + t;
      i++;
    }
    x = x - 1.0;
  }
// Record CPU time at finish of calculation
  times(&cpu2);
// Calculate elapsed time by the difference of the time stamps,
// using the system setting of clock time tics per second
  printf("Elapsed user time = %.3f seconds,\n",
         ((double) cpu2.tms_utime - cpu1.tms_utime) /
         ((double) sysconf(_SC_CLK_TCK)));
  printf("y=%f\n",y);

// Evaluate test polynomial by the nested term method
  printf("10,000,000 evaluations of an 8th order polynomial "
         "with nested terms\n");
// Record CPU time at beginning of calculation
```

```
  times(&cpu1);
  x = 9999999.0;
  while (x > -0.1) {
    y = a[8];
    i = 7;
    while (i >= 0) {
      y = a[i] + (x * y);
      i--;
    }
    x = x - 1.0;
  }
// Record CPU time at finish of calculation
  times(&cpu2);
// Calculate elapsed time by the difference of the time stamps,
// using the system setting of clock time tics per second
  printf("Elapsed user time = %.3f seconds,\n",
         ((double) cpu2.tms_utime - cpu1.tms_utime) /
         ((double) sysconf(_SC_CLK_TCK)));
  printf("y=%f\n",y);
  exit(0);
}
```

Listing 1.1 Test C program evaluating a polynomial by both the term-by-term and nested methods

Compiling and running this program shows an execution time of about 2.08 seconds for the term-by-term methods, as opposed to only 0.28 seconds for the nested method, on a typical desktop processor running in a Linux environment:

```
$ ./poly_times
10,000,000 evaluations of an 8th order polynomial term-by-term
Elapsed user time = 2.080 seconds,
y=9.000000
10,000,000 evaluations of an 8th order polynomial with nested terms
Elapsed user time = 0.280 seconds,
y=9.000000
```

This simple but important demonstration makes it clear that significant processing efficiency can be gained, in this case by more than a factor of seven, by attending to careful choices of algorithms.

Code written in interpreted languages such as Python, Perl, Tcl, Mathematica, Matlab, etc. must be processed by an interpreter program before any actual computation is done. There is a layer of computer code that decodes the text characters in a program file one at a time to assemble instructions on what to do next, and that decoding process alone requires many machine cycles and significant energy. Before the computation blocks in Figure 1.6 can get around to actually computing the quantities involved in your numerical algorithm, first they are occupied with the task of reading the text of the interpreter language statement and decoding it character by character to interpret what computation you want done. Compiled languages perform that text interpretation process up front, compiling source language statements into a series of binary bytes that the computer hardware can interpret directly, and deciding the sequence of computation steps to take in a minimum of machine cycles.

When small-scale computation tasks are coded into an interpreted language, such as Python or Tcl, often with contemporary high-performance computing hardware, the task can be completed before the user's finger has been removed from the "Return" key on the keyboard. This can give us the false impression that these languages are suitable for large-scale tasks as well. To see the run-time penalty incurred by using an interpreted language, let's look ahead into Section 4.7 for an example, the computation task of predicting a two-dimensional diffraction pattern rendered by a single circular aperture with the light from two separated sources passing through. The light intensity seen on an observation screen is shown in Figure 4.24, and significant diffraction artifacts can be noted. Running the code that will be developed in Chapter 4 for this problem through the Linux `time` command reports that the simulation code written efficiently in C and optimized by the compiler consumes 4.602 seconds of processor time on a representative contemporary processor. We can recode this problem in Python, and the reader may see the source code in the file diffraction2D_step_2src_flat.py from the Student Companion website. Running this code through the Python interpreter, which generates the same plot as in Figure 4.24, costs us a very large run-time penalty, by over a factor of 31:

```
$ time python3 diffraction2D_step_2src_flat.py 1 0.2 20 20 -3 0 3 0 -6 6 0.2 -6 6 0.2 >|
    ↪ diffraction2D_step_2src.dat
real     2m25.177s
user     2m25.170s
sys      0m0.004s
```

For an even more stark demonstration, we can look ahead a bit further into Section 8.4. There we will demonstrate a straightforward example application of the Monte Carlo technique to estimate the value of π. Because the accuracy of the Monte Carlo method applied to a two-dimensional geometry is generally dependent on the square root of the number of pseudorandom samples we use, let's test the example program presented in Chapter 8 with 100 000 000 samples. On a typical processor, the optimized C code in that chapter requires 0.753 seconds of central processing unit (CPU) time to arrive at an estimated value of π. Again recoding the simulation code in Python, the resulting test program can be found on the download website as pi_MonteCarlo_flat.py. Running this code through the Python interpreter, and again using the Linux time utility to record processor usage, we see a run time penalty of a factor of *almost 100*:

```
$ time python3 pi_MonteCarlo_flat.py 100000000 1
3.1416672
real     1m15.238s
user     1m15.235s
sys      0m0.004s
```

The difference in time penalties for an interpreted language between the two computation tasks above can be explained by the makeup of the arithmetic needed for the two applications. In the diffraction case, significant time is expected to be consumed by computing trigonometric functions, which can be implemented with special efficient packaged code in the Python interpreter. However, the code to estimate π is dependent on straight algebraic expressions, each one of which must be slowly interpreted.

The simplistic benchmarking exercises above are certainly taking the use of an interpreted programming language to an absurd extreme, where the very lowest level of computation tasks is coded in the most inefficient language. This is not to say that interpreted languages have no place in scientific computation. The structure of typical programs written to solve equations for physics problems on a large scale is a hierarchical layering of iterative loops. Perhaps the outermost loop in the hierarchy will iterate over all the molecules in a simulated gas, or all the stars in a simulated galaxy. The next inner loop may search for the roots of a nonlinear equation, as in Chapter 5, or converge on a minimum of a function, as in Chapter 11, or develop a sampled solution to a differential equation as in Chapters 12, 13, and 14. Perhaps the innermost code level involves the evaluation of a polynomial or other function. The ease of development and testing afforded by interpreted languages does allow time savings during the code *development* phase. But they should only be applied to the *outer* computation tasks in a hierarchical computation structure. This ensures that the time penalty due to the continuous interpretation of the program text is negligible with respect to the computation load of *inner* tasks that are coded in an efficient compiled language. Contemporary Python and Tcl implementations offer the capability of linking to low-level code modules. But that still leaves the task of ensuring that low-level functions are written in a programming language that allows the timed cycles of the computer hardware to be used as directly as possible on the actual computation task at hand. Many pre-packaged low-level routines may be available, but unfortunately may not fill all needs, or may be opaque to needed modifications and enhancements.

Since there are a variety of inner-level tasks that may be needed across a gamut of physics problems, a function that is at a lower level in one application may be a higher level function for another. In Chapter 3 we will study efficient coding methods in the C language, including how the language is designed to allow multiple levels of code to interact efficiently. So all this leads us to the approach taken in this textbook, which is to develop all the fundamental solution methods in the most universally efficient way, compiled from C source code. This approach can potentially save orders of magnitude factors of computation time and energy. These considerations may appear to be irrelevant for simple problems that require only milliseconds of execution time. But, in general, whenever we have the capability to solve one elemental problem, the need arises to nest that solution within more and more complex solutions in our search for physical understanding, and carefully attending to the details of computation efficiency can end up saving large expense. We will revisit this issue in Sections 13.4 and 13.6 with some further examples of numerical tasks requiring substantial processing time.

1.4 Exercises

1.1 **(a)** Study the manual pages or other descriptions for the Linux commands `ls`, `cd`, `less`, `cat`, `wc`, `tail`, `source`, `echo`, and `grep`.

(b) Download the archive file `drill.zip` from the Student Companion website and unpack it in a work directory. This will create a `drill` subdirectory.

(c) Working in a Linux or MSYS2 command shell window, display the contents of the two files `readme1.txt` and `readme2.txt` in the `drill` directory, and follow the instructions for two "treasure hunts" down into various lower subdirectories. Submit the text you find at the end of each hunt as the two solutions.

1.2 In order to develop flexibility at the command line for inspecting and storing the numerical output of the programs developed in this textbook, this exercise demonstrates the use of the powerful output "redirection" capability built into the Linux and MSYS2 command shells. Study the redirection facility by executing and explaining the following commands. First, download and study the source code file `redirect.c` from the Student Companion website. In this program, strings of example text characters are sent to both the standard output stream and the standard error stream of the program process with variations of the `printf()` C library function.

(a) Download and read through the example `redirect.c` source code. Note that in this source code file we may also see the use of the implied concatenation of adjacent character string constants in the C language. This is a convenient facility used throughout the example source code in this textbook for wrapping lines of code containing long output strings.

(b) Open a command terminal window in the directory containing the C source code file `redirect.c`. Compile this source code file into the binary executable file `redirect` with the `gcc` C compiler by running the command

```
$ gcc -Wall -Wshadow -O2 -o redirect redirect.c -lm
```

(c) Execute the compiled program with

```
$ ./redirect
```

and note the text merged from both output streams displayed on the terminal window.

(d) Now redirect only the text sent to the standard output stream into a file by running

```
$ ./redirect > redirect.txt
```

and note the text displayed on the terminal window. Display the contents of the newly written file `redirect.txt` by running

```
$ cat redirect.txt
```

and note the text displayed on the terminal window.

(e) Now append more redirected standard output text into the existing text file by rerunning the command

```
$ ./redirect >> redirect.txt
```

any number of times. Note the text displayed on the terminal window for each run. Check the new contents of the text file with the `cat` command and note the text displayed on the terminal window.

(f) Now rerun the program with the original command line

```
$ ./redirect > redirect.txt
```

Depending on how your "noclobber" command shell option is configured, this command will either overwrite the contents of `redirect.txt` or will make no file change and report an error message. If enabled, this protection mechanism is intended to prohibit the user from accidentally "clobbering" an important existing file. Check the contents of the text file with the `cat` command.

(g) Now explicitly tell the command interpreter to overwrite the existing text file by running

```
$ ./redirect >| redirect.txt
```

Check the contents of the text file with the `cat` command. This style of output stream redirection is demonstrated extensively with the example programs in this textbook.

(h) Now pipe the standard output stream of the `redirect` process into the standard input of the Linux pager tool `less` by running

```
$ ./redirect | less
```

and note what text lines are visible in the `less` tool and which are routed to the terminal window after the `less` tool is terminated.

2

Precision Limits of Numerical Computation and Algorithms

With what numerical precision may we expect to perform our simulations of physical systems? What precision is required for a given problem to be solved satisfactorily?

In Section 13.4, we will examine numerical solutions to the spatial differential equation that is to be solved to draw classical field lines for electrostatic and magnetostatic fields. An example of a simple electrostatic dipole is shown in Figure 13.14. The intention of generating the graphical image of this figure is simply for a visualization of the general pattern formed by the field. All that is required of the solution is to provide enough precision to properly fill in pixels in an image, and therefore, the spatial differential equation need not be solved with any finer precision than one part in a few hundred.

Compare this with the precision required for being in the proper place and time for the best photography of the total solar eclipse shown in Figure 2.1. The gravitational equations of motion must be solved to a precision that allows the placement of an observer within a few city blocks in a gravitational system that spans 93 million miles. This level of precision is many orders of magnitude smaller than that needed for the field line problem.

So to guide our choice of algorithms and algorithm parameters that will afford us the required solution precision, in this chapter we will explore the fundamental limits on computation precision imposed by processing hardware and analysis algorithms.

Figure 2.1 Total solar eclipse as seen in Madras, Oregon, 21 August 2017.

Computational Physics Using C: Efficient Programming with Ease, First Edition. John W. Fattaruso. © 2026 John Wiley & Sons, Inc.
All rights reserved, including rights for text and data mining and training of artificial intelligence technologies or similar technologies.
Published 2026 by John Wiley & Sons, Inc. Companion Website: http://www.wiley.com/go/CompPhysUsingC

2.1 Computer Numerical Representation

Figure 2.2 shows the layout in a typical computer random access memory (RAM) where data and variable values are stored in the process of running a program. In order to make use of the same computer architecture to process information for both text and numerical applications, the smallest addressable aggregation of binary bits is designed to be one 8-bit byte, the minimum size capable of storing a character of text. Larger aggregations of binary bits needed to store large numerical values in the binary encoding system will span several addressed locations in memory, although to maximize processing speed, blocks of byte addresses are actually read and written simultaneously by the hardware.

When numerical variables or constants in a C program are allocated storage space in RAM to hold their binary numerical values, consecutive bytes are allocated as shown in Figure 2.3 for four example variable types: `char`, `int`, `float`, and `double`. The sizes shown are not strictly fixed by the C language standard, but these are typical sizes prepared by most compilers for typical computing hardware.

Internally in the processor hardware, all numerical data are stored and operated on with a radix 2, or binary, numbering system. We are accustomed to entering numbers into and reading output from software tools in our everyday decimal, or radix 10, system, and the standard C library provides library routines to translate between the two radices.

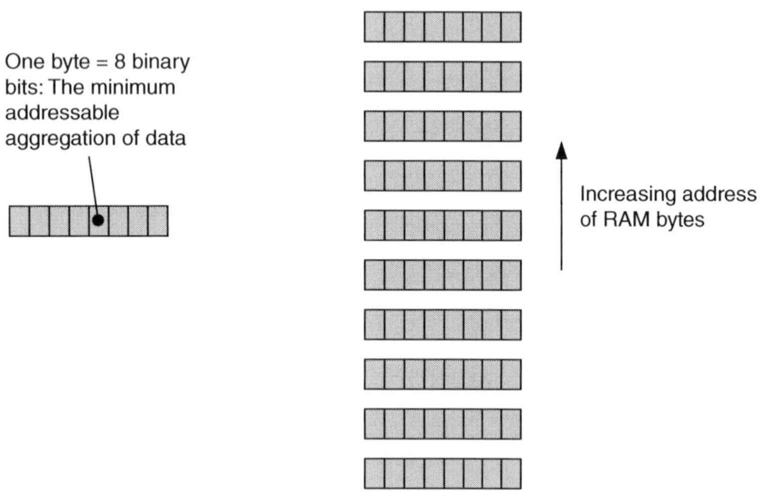

Figure 2.2 Layout of addressable bytes in computer random access memory (RAM).

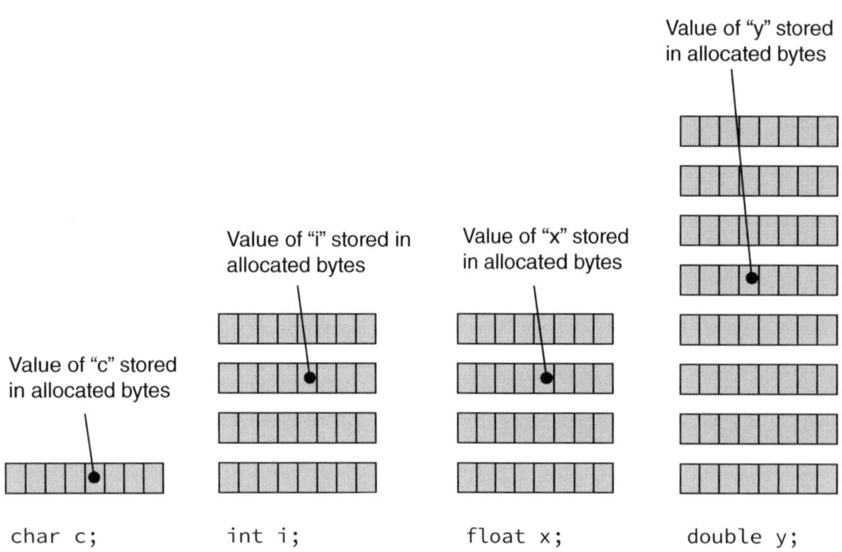

Figure 2.3 Addressable bytes for four example variable allocations.

The diagrams describing how the binary bits in the 32-bit integer and IEEE standard single- and double-precision floating point numbers are mapped to the numerical representation are shown in Figures 2.4, 2.6, and 2.7. For the 32-bit integer, a cluster of four consecutive bytes at four consecutive hardware memory addresses are assembled into one 32-bit binary number. The figure shows the order of assembly for the so-called "little-endian" class of processor. In this class, the "little", or least significant, end of the 32-bit integer comes from the byte at the lowest memory address of the cluster, but the opposite arrangement of "big-endian" is also implemented in some processors.

In the binary numbering system, the two possible values for each bit b_i are either 0 or 1. The integer numerical value associated with a particular assemblage of bits is computed as the sum of products of each binary bit with weights of each bit position, and these weights are successive powers of 2. If a variable is declared as an "unsigned" integer in a C program, the C compiler will prepare code for the hardware to interpret the binary bits according to the sum

$$I = b_{31} \cdot 2^{31} + b_{30} \cdot 2^{30} + b_{29} \cdot 2^{29} + \cdots + b_2 \cdot 2^2 + b_1 \cdot 2^1 + b_0 \cdot 2^0 \tag{2.1}$$

Of course, since the bit values b_i may only be 0 or 1, this is a sum that simply includes or excludes the binary bit position weights.

In order to represent both positive and negative integers, we may also declare an integer variable as "signed" (the default for an integer declaration in C). To encode signed integers, most processors employ the "twos-complement" system of bit position weights, and the represented integer value is computed as

$$I = -b_{31} \cdot 2^{31} + b_{30} \cdot 2^{30} + b_{29} \cdot 2^{29} + \cdots + b_2 \cdot 2^2 + b_1 \cdot 2^1 + b_0 \cdot 2^0 \tag{2.2}$$

For 32 binary bits, the total number of possible combinations the bits may assume is 2^{32}. The encoding scheme of Equation (2.2) maps signed values onto the integer number line for each of the 2^{32} binary combinations as shown in Figure 2.5. Starting from the case of all b_i bits being zero, this combination maps to an integer value of 0. As bits of higher and higher significance are turned to 1 in the binary counting sequence, larger and larger integer numbers are represented, until we reach the maximum possible integer value of $2^{31} - 1$, or 2147483647 decimal. This configuration corresponds to the highest order bit being 0 and the 31 remaining bits 1. Incrementing this binary combination once more results in the integer representation wrapping around to the least possible negative number, $-2^{31} = -2147483648$ decimal, for the bit configurations of the highest order bit being 1 and all the 31 remaining 0. Then as these bits are incremented further, we move through the possible negative integers on the number line until all bits are 1, and the corresponding decimal integer is -1. One last increment completes the cycle through the 2^{32} combinations possible with 32 bits, and we arrive back at an integer value of 0.

In the decimal number system that we use for everyday computation, there is the device of a "decimal point," a marker that divides decimal digits between those representing integral values and those representing fractional values. The corresponding mark in the binary system is called the "binary point." Of course, for these integer values, there is an implied binary point position just to the right of the least significant bit.

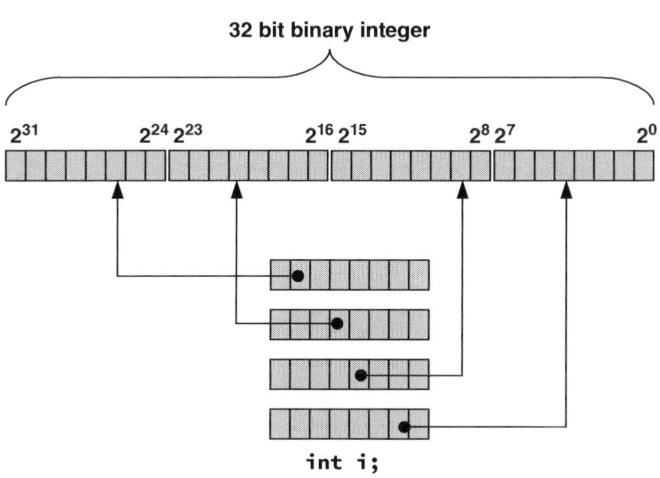

Figure 2.4 Bit map for 32-bit signed integer numbers.

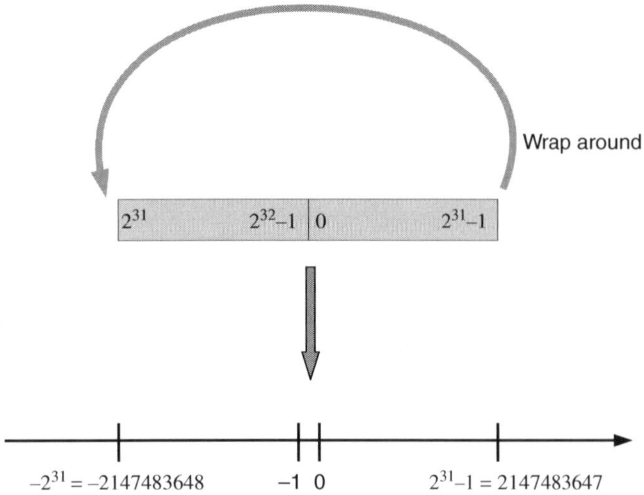

Figure 2.5 Mapping 32-bit signed integer numbers to the integer number line.

When we declare a variable in a C program as having an integer type, either signed or unsigned, its value will be stored in binary bits in a processor register or memory location according to an integer mapping. This type of variable is essential for quantities that represent a count of discrete objects, a discrete index into an array of variables or physical objects in a collection, a count of iterations to make in an iterative loop, and other needs, taking on only integral values. When we aim our programming application at modeling physical systems, we need variables that represent an approximation to physical quantities whose values can be in the continuum of numbers on the real number line. For that purpose, various systems of encoding "floating point" numbers have been defined. These systems encode a position of the binary point that is allowed to "float" over a wide range of placements between binary bits, allowing numbers ranging from very large down to very small fractional values to be represented. Most contemporary processors conform to the IEEE-754 standard for "single-precision" and "double-precision" floating point representations.

Floating point representation can basically be seen as a binary equivalent to the "scientific notation," we conventionally use in the physical sciences. Typically we write numbers such as

$$1.23 \times 10^9 \quad \text{and} \quad -9.876 \times 10^{-6}$$

with a sequence of three elements: first the presence or absence of a leading "–" sign, then the decimal "significant digits," and finally a multiplication by a power of 10. We intend this notation to represent the numbers

$$1230000000 \quad \text{and} \quad -0.000009876$$

where the true decimal point position has "floated" to the right by nine digits and to the left by six digits, respectively. By this notation, we can represent numbers with long strings of zeroes in a compact form. The same idea applies to computer floating-point representation, except that the base number that is raised by an exponent is 2 rather than 10, and we float the binary point by a number of binary bits instead of decimal digits.

The standard bit interpretation for the 32-bit single-precision floating-point representation is shown in Figure 2.6. This is a more elaborate arrangement than that for integer numbers, but includes three fields of bits. One field, of 24 bits of length and called the "significand," encodes the significant bits to the right of an implied binary point. Another 8-bit field encodes the position of the floating binary point. A third field, actually just a single bit, encodes the sign of the number. The floating-point number x represented by the bit patterns in Figure 2.6 is assembled according to the "scientific notion" of Equation (2.3).

$$x = (-1)^s \times q \times 2^e \tag{2.3}$$

The factor q is the value associated with a leading and "hidden" 1 bit, followed by the implied binary point, followed by the 23 stored bits that lie to the right of the point.

$$q = 1 + b_{-1} \cdot 2^{-1} + b_{-2} \cdot 2^{-2} + b_{-3} \cdot 2^{-3} + \cdots + b_{-21} \cdot 2^{-21} + b_{-22} \cdot 2^{-22} + b_{-23} \cdot 2^{-23} \tag{2.4}$$

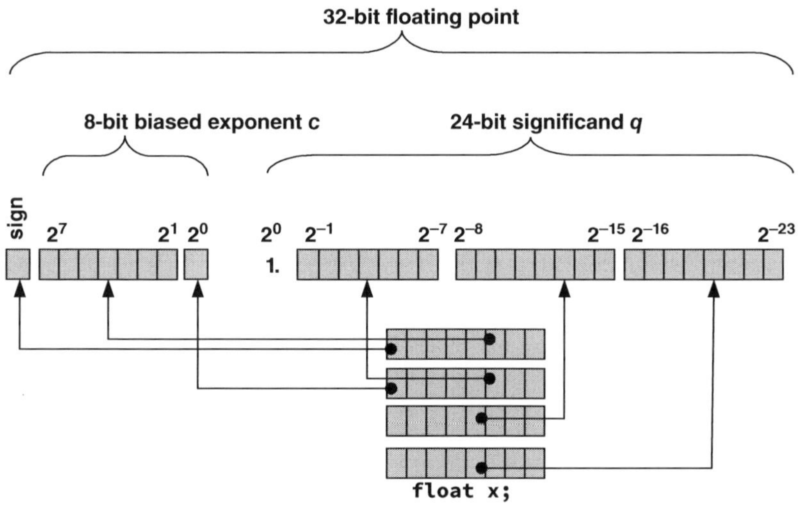

Figure 2.6 Bit map for standard single-precision floating-point numbers.

The factor q can therefore be 1 at a minimum and slightly less than 2 at a maximum. The standard assumes that the floating-point number will always be stored in a normalized form, with one significant bit to the left of the binary point that is always 1, so there is no need to actually take up memory storage space with a bit we know is always 1. For arithmetic operations, such as addition or multiplication, this implied bit is filled in with its 1 value in the Arithmetic and Logic Unit (ALU) of Figure 1.6. But for storage, the bit is stripped off and remains hidden. This convention effectively adds an additional bit of precision for a fixed 21 bits of storage. The exponent e, to which the base of 2 is raised in this binary form of scientific notation, is the difference $c - 127$, where c is the binary integer value stored in the biased exponent field. Finally, the sign bit indicates a positive floating-point number for 0, and negative for 1. Arithmetically, we can express this as the factor $(-1)^s$ in Equation (2.3), where s is the single sign bit stored.

The IEEE standard for a 64-bit double-precision floating-point number is similar, except the significand and exponent fields are expanded with more bits, as shown in Figure 2.7. Fifty-three bits are allocated to the significand and 11 to the biased exponent. The significand is evaluated in this expanded format as

$$q = 1 + b_{-1} \cdot 2^{-1} + b_{-2} \cdot 2^{-2} + b_{-3} \cdot 2^{-3} + \cdots + b_{-50} \cdot 2^{-50} + b_{-51} \cdot 2^{-51} + b_{-52} \cdot 2^{-52} \tag{2.5}$$

The expression guiding the assembly of the resulting floating-point number is the same as in Equation (2.3), except now the exponent e to which 2 is raised is $c - 1023$, where c is the integer biased exponent stored in the binary bit field. The "hidden" 1 bit is again used to add an extra bit of precision.

The careful reader may be wondering how we represent the number zero with this standard, since a hidden 1 bit is always implied in the significand. An exception must be made to the interpretation standard of Equation (2.3). If all bits in both the exponent and significand are 0, then the number is interpreted by the arithmetic hardware as the exceptional value of zero. The largest possible number that can be assigned to a variable with the double-precision format is with the biased exponent field set to 2046, and all significand bits set to 1. This representation works out to be an equivalent decimal number of about $\pm 1.8 \times 10^{308}$. Similarly, the smallest possible number is found with the exponent field set to 1, and only the hidden bit being nonzero in the significand. This representation works out to be an equivalent decimal number of about $\pm 2.2 \times 10^{-308}$. These are enormously large and small numbers, and it would be a rare computation for physical systems that would require any number outside this range. For example, expressing the approximate known size of the universe of 9.1×10^{10} light years in the absurd units of Angstroms would need only a number as large as 8.6×10^{36}.

Sections 2.5 will make clear the numeric benefits of the 64-bit double-precision standard, relative to the 32-bit single-precision standard, for reduced roundoff error. Although the memory storage requirements are doubled using the double-precision format, most physical computation can be performed with roundoff error levels insignificant relative to other error sources, such as truncation error introduced in Section 2.4. The IEEE double-precision standard represents just one of a large number of possible ways to map the 2^{64} possible bit combinations to numerical values on the real number line, as shown in Figure 2.8, but its allotment of bits between significand and exponent is a reasonable compromise between roundoff error and total numeric range.

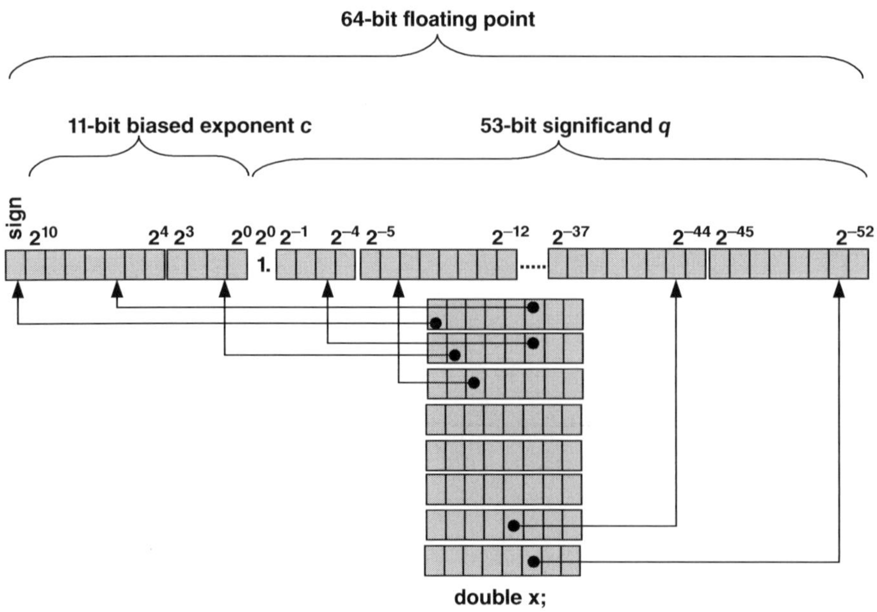

Figure 2.7 Bit map for standard double-precision floating-point numbers.

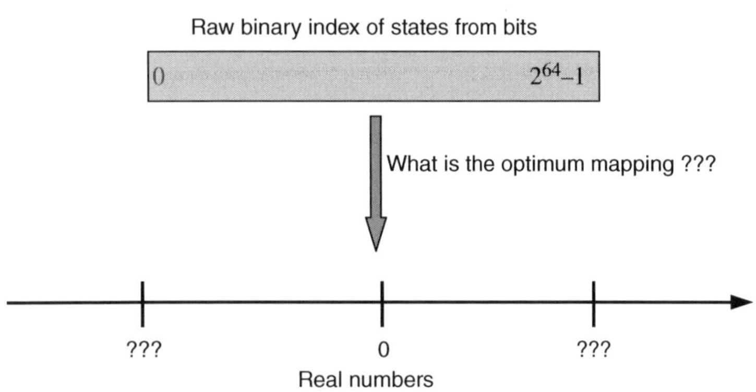

Figure 2.8 Mapping 64-bit floating-point numbers to the real number line.

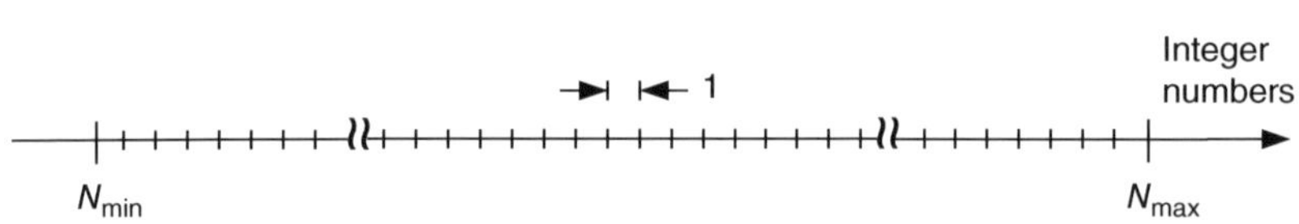

Figure 2.9 Density of 32-bit signed integer numbers on the integer number line.

For the 32-bit integer format of Figure 2.4, the density of numerical values on the integer number line that may be represented is constant, between the range limits established by the format, as shown in Figure 2.9.

However, for the floating-point formats, the density of numbers on the real number line that can be represented is nonuniform. Figure 2.10 shows the various regimes of densities on the real number line for the double-precision format. When the exponent field in the floating-point number is small, the density is high on the real number line, as each possible combination of the significand bits represents very finely separated numbers. When the exponent field is large, the possible density is coarse. For numbers within the span of 1 to 2, the density is about 2×10^{-16} between numbers. This compares to a density of about 1×10^{-7} between numbers within the span of 1 to 2 for the single-precision standard. Since the number of significand bits increases from 24 to 53 from single- to double-precision, the additional 29 bits will shrink the spacing between represented numbers by a factor of $2^{29} \approx 5 \times 10^8$.

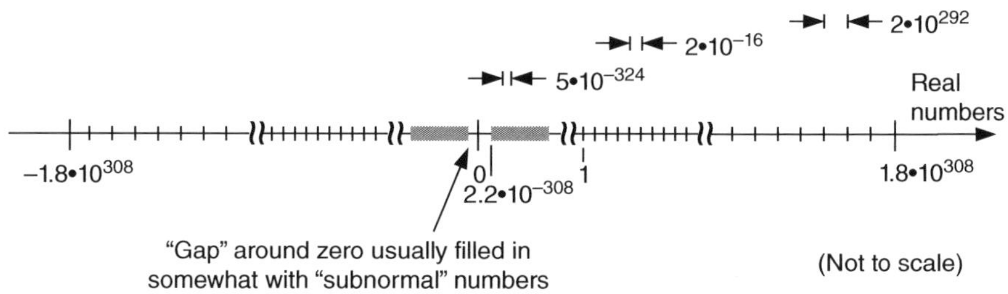

Figure 2.10 Nonuniform density of 64-bit floating-point numbers on the real number line.

Table 2.1 Example encodings in the double-precision floating-point standard.

Decimal	**Double-precision binary**		
		Biased	
	Sign	**Exponent**	**Significand**
1.0	0	01111111111	(1).00000000000000000000000000 00000000000000000000000000
2.0	0	10000000000	(1).00000000000000000000000000 00000000000000000000000000
0.5	0	01111111110	(1).00000000000000000000000000 00000000000000000000000000
−1.0	1	01111111111	(1).00000000000000000000000000 00000000000000000000000000
10.0	0	10000000010	(1).01000000000000000000000000 00000000000000000000000000
−0.625	1	01111111110	(1).01000000000000000000000000 00000000000000000000000000
0.1	0	01111111011	(1).10011001100110011001100110 01100110011001100110011010
1.0000 0000 0000 0002	0	01111111111	(1).00000000000000000000000000 00000000000000000000000001

We can look at several binary encodings of some example decimal numbers in Table 2.1 to gain more insight into how this coding scheme functions.

Let us start with the simple basic decimal number 1.0, with the encoding in the double-precision standard shown in the top row. Since this is a positive number, the sign bit is 0. The significand is simply 1.0, making use of the implied 1 hidden bit and the 52 0 bits actually stored for this value. Then the true exponent is 1023 subtracted from the biased exponent field. The bits in this field represent an unsigned binary integer of 1023, so the true exponent is simply 0. Following the formulation in Equation (2.3) results in a value of 1.0.

In subsequent rows in the table, the encoding of other example values are listed. The value 2.0 is almost identical to 1.0, except that the biased exponent field is the number 1024. This results in a true exponent of 1, and a total value of 2.0. Similarly, the number 0.5 is also encoded with a significand of 1.0 and a true exponent of $1022 - 1023 = -1$. In fact, any number that is a power of two, either positive or negative, will be encoded as a floating-point value with all zeroes in the stored significand and simply different integer values in the biased exponent field.

Two examples of numbers that are not simple powers of two follow in the table. The number 10 is encoded with a significand of 1.25 in binary and a true exponent of 3. This gives a value of $1.25 \times 2^3 = 1.25 \times 8 = 10$. The number -0.625 is similarly encoded with a significand of 1.25 (and a sign bit of 1) but now with a true exponent of -1, decoding to $-1.25 \times 2^{-1} = -1.25 \times 0.5 = -0.625$. The next example highlights the roundoff error that is intrinsic to the encoding of constants that seem to be convenient, round numbers such as 0.1. There is no exact binary representation of this number within a finite number of fractional bits in the significand. This limitation is similar to that seen when expressing a fraction such as $\frac{1}{3}$ in decimal. The representation would require an infinite number of decimal digits in the repeating decimal $0.333333 \cdots$. In a similar manner, the decimal number 0.1 would require an infinite number of binary bits in a repeating pattern of $0.0001100110011001100 \cdots$ that can be seen in the table. In a practical ALU with a finite number of bits this pattern must be terminated at some point and rounded to the closest represented number. The decimal number actually equivalent to the bits in the table entry is 0.100 000 000 000 000 01. So for many numbers that we would typically code in our C programs, there are already numerical errors committed simply by the process of the C compiler translating them into a fixed number of binary bits for the hardware to operate on.

The last row in Table 2.1 shows the decimal number just greater than 1.0 that results from changing the least significant bit of the significand field from a 0 to 1. This decimal number is $1 + 2^{-52} \approx 1 + 2.2 \times 10^{-16}$. As will be described in more detail in Figure 2.2, this establishes the resolution available in the double-precision standard of about 2.2×10^{-16} for numbers in the vicinity of unity.

If even 64 bits of floating-point precision are insufficient for a particular task, leaving still unacceptable levels of roundoff errors, most C compilers targeting common processors also allow a variable to be of the type `long double`, which uses an 80-bit format, with 64 bits of significand, 15 bits exponent, and one bit sign.

2.2 Roundoff Errors

Consider a simple example computation performed in the decimal system, such as the product of two numbers $(2.34 \times 10^{12})(9.87 \times 10^{-7})$, each with three significant digits. The product is 2.30958×10^6, with six significant digits needed to exactly express its value. Suppose that we are working with hardware that may store decimal numbers with at most three significant digits. The arithmetic hardware would then choose the three-digit number that best approximates the true product. This approximation process is called rounding, and in this example, it would store away the product as 2.31×10^6 by "rounding up" the digits 0958 to 1000.

This is the same sort of process by which binary numbers that are the results of computations in an ALU are rounded so they may be stored in registers or memory with a limited number of bits. There are different strategies for rounding that may offer some statistical advantage of lower errors in different classes of computations, but the typical default strategy is the "round to even" method, which will round up a 1 bit that is past the last bit that can be stored when it results in an even number. This rounding method is explored further in Exercise 2.3.

In Figure 2.11, we can see graphically the results of two example computations, perhaps a multiplication or division, that happen to land on the real number line between two allowed numbers that may be represented by a standard floating-point format. The hardware will make an adjustment to the true result, rounding to the closest hash mark on the number line, effectively adding or subtracting a small amount and generating a small roundoff error in the computation. Since there is no reason to expect a true result to end up in any particular place on the line relative to the hash marks, roundoff errors are in general similar to uniformly distributed pseudorandom errors. The errors may range from minus to plus half the distance between representable numbers, and this distance varies along the real number line according to the exponent value in the floating-point format.

Let's suppose we start from a nominal value of 1.0 in the double-precision format, represented with all significand bits at 0. Now imagine incrementing the number by tiny amounts until the least significant bit just turns to 1. Since we assume that the smallest bit will be rounded up to 1 when the change in the number reaches half the bit value, we can derive what is called the "machine ϵ" of about $\epsilon = 2^{-53} \approx 1.1 \times 10^{-16}$ for the double-precision standard. The corresponding metric for the single-precision standard is $\epsilon = 2^{-24} \approx 6.0 \times 10^{-8}$. For whatever floating-point format we are using, results of computations may be displaced by a pseudorandom roundoff error that will range from $-\epsilon$ to $+\epsilon$.

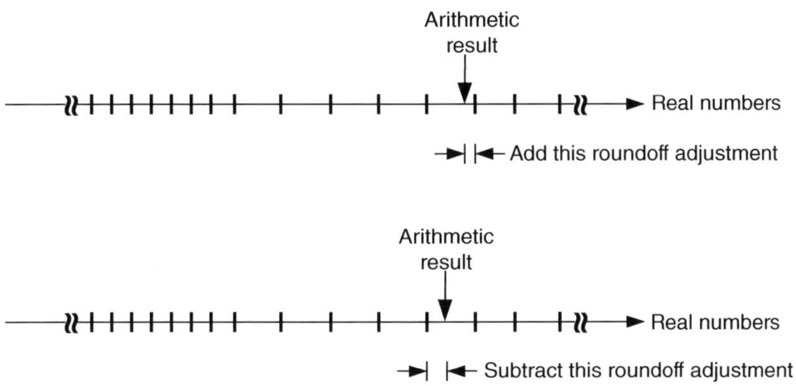

Figure 2.11 Results of arithmetic operations rounded to the nearest allowed floating-point number.

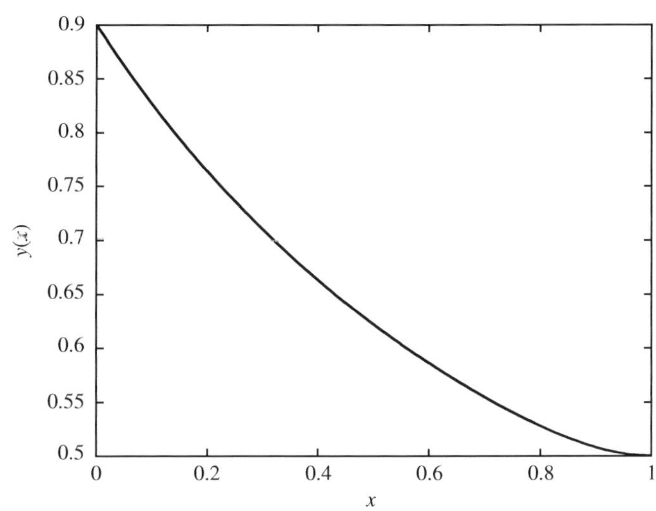

Figure 2.12 Example polynomial value to study floating-point roundoff error.

Let us revisit the numerical problem of evaluating an eighth-order polynomial we used as an example of computational efficiency in Section 1.3. Using the efficient Horner's method of evaluation, we can easily see the introduction of pseudorandom roundoff errors even with this minimum amount of computation.

Using some example coefficient values, the polynomial we will evaluate is

$$y(x) = 0.1x^8 - 0.2x^7 + 0.3x^6 - 0.4x^5 + 0.5x^4 - 0.6x^3 + 0.7x^2 - 0.8x + 0.9 \tag{2.6}$$

Over an example range of x values of 0 to 1, a graph of the value of this polynomial looks like Figure 2.12.

So the values of the polynomial are on the order of unity, which would mean that the pseudorandom roundoff displacements of quantities in the process of the evaluation would be on the order of 10^{-7} for single-precision arithmetic and on the order of 10^{-16} for double-precision arithmetic.

Consider the example C program, which performs the same evaluation for each value of x with three different floating-point formats: single, double, and long double precision:

```
// File: poly_roundoff.c
#include <stdlib.h>
#include <stdio.h>
#include <math.h>

int main(int argc, char **argv) {
  long double xld,yld,sumld,eyf,eyd,esumf,esumd;
  double xd,yd,sumd;
  float xf,yf,sumf;
  int i,state;
// Assign polynomial coefficients for long double, double, and float
```

```
  long double ald[9] =
    {0.9L,-0.8L,0.7L,-0.6L,0.5L,-0.4L,0.3L,-0.2L,0.1L};
  double ad[9] = {0.9,-0.8,0.7,-0.6,0.5,-0.4,0.3,-0.2,0.1};
  float af[9] = {0.9F,-0.8F,0.7F,-0.6F,0.5F,-0.4F,0.3F,-0.2F,0.1F};

// Initialize polynomial sum to measure accumulation of roundoff
// errors for long double, double, and float
  sumld = 0.0;
  sumd = 0.0;
  sumf = 0.0;
// Initialize polynomial argument for long double, double, and float
  xld = 0.0;
  xd = 0.0;
  xf = 0.0;
// Initialize the state machine that determines the alternating sign
// with which each polynomial contributes to the summation
  state = 0;
// Iterate over the range of polynomial arguments
  while (xd < 1.001) {
// Evaluate the polynomial at the current argument value for long
// double, double, and float with the nested method
    yld = ald[8];
    yd = ad[8];
    yf = af[8];
    i = 7;
    while (i >= 0) {
      yld = ald[i] + (xld * yld);
      yd = ad[i] + (xd * yd);
      yf = af[i] + (xf * yf);
      i--;
    }
// Calculate error for a single polynomial evaluation of double and
// float relative to long double
    eyd = (long double) yd - yld;
    eyf = (long double) yf - yld;
// Contribute current polynomial evaluation to the accumulating sum
// according to the alternating sign of the current state
    if (state == 0) {
      sumld += yld;
      sumd += yd;
      sumf += yf;
    }
    else {
      sumld -= yld;
      sumd -= yd;
      sumf -= yf;
    }
    state = state ^ 1;
// Calculate error for the accumulated sum of double and float
// relative to long double
    esumd = (long double) sumd - sumld;
    esumf = (long double) sumf - sumld;
    printf("%.2f %.8g %.6Lg %.6Lg %.6Lg %.6Lg\n",xd,yd,eyf,eyd,
           esumf,esumd);
// Increment polynomial argument for long double, double, and float
    xld += 0.01;
    xd += 0.01;
    xf += 0.01;
  }
  exit(0);
}
```

Listing 2.1 Test program for roundoff errors from simple polynomial evaluations

For the purpose of this visualization, we'll consider the `long double` version of the arithmetic to be accurate, and compute the errors between both the single- and double-precision numbers relative to the `long double` numbers. (Of course, there are still roundoff errors for the `long double` computations. But with 11 more bits available for the significand, these roundoff errors are more than three orders of magnitude smaller, and we can assume for these plots that they are insignificant.) The errors using the single- and double-precision formats are plotted in Figures 2.13a,b.

In these plots, we can visualize the pseudorandom influence of roundoff error at the expected order of magnitude for the two floating-point formats. So as we design our numerical programs and imagine our algorithms generating our desired analysis result, we must always keep in mind that working with practical computing machinery we will only be able to generate approximate solutions that are always displaced from the true mathematical solution by these pseudorandom errors.

Figure 2.11 shows a roundoff of one arithmetic calculation, and Figures 2.13a,b show the error that accumulates with several calculations needed to evaluate a polynomial, but any numerical algorithm will generally require a large number of operations to produce some analysis result. Since the roundoff error from each operation may be seen as an addition of a pseudorandom error term, roundoff error will typically accumulate throughout an extensive algorithm in a manner similar to a random walk, and may reach quite large magnitudes. This process is statistically complicated due to the variation of the density of the floating-point numbers on the real number line as the magnitude of the numbers involved varies. We can visualize an example of roundoff accumulation with a simple simulation. The example program in Listing 2.1 also accumulates a sum of successive polynomial evaluations, but with alternating signs to keep the magnitude of the numbers approximately constant. Figures 2.14a,b show the cumulative error of single- and double-precision arithmetic in this long-term sum. We can see that the random walk can in some cases, carry the accumulating error to levels much larger than the order of a machine ϵ.

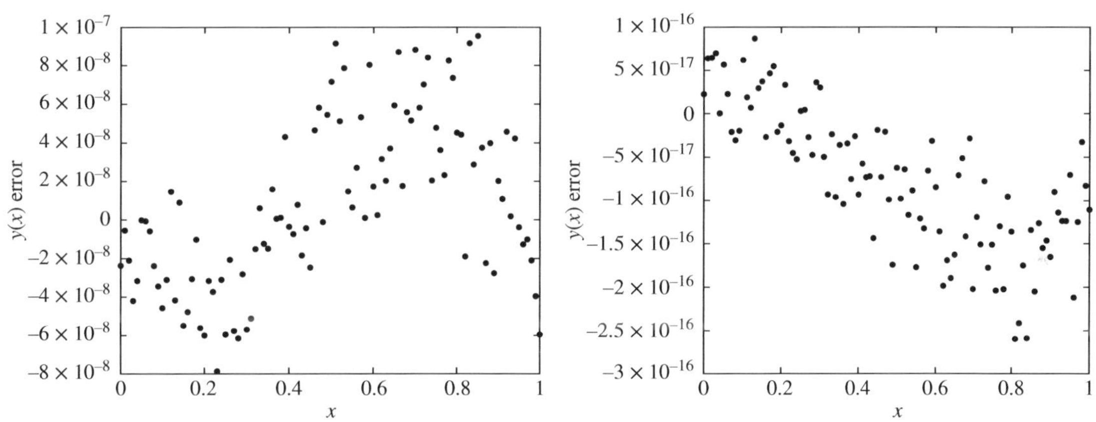

(a) Computation with single-precision floating point **(b)** Computation with double-precision floating point

Figure 2.13 Roundoff error introduced by the evaluation of an eighth-order polynomial over a step of argument values.

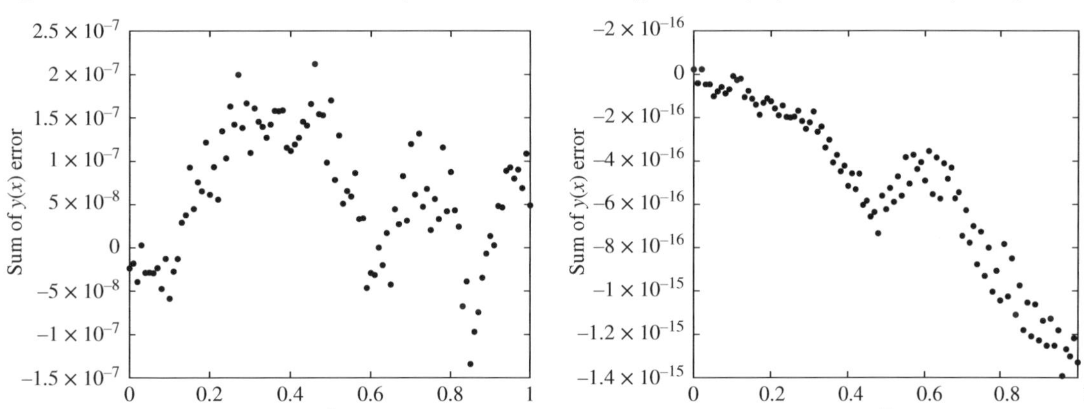

(a) Computation with single-precision floating point **(b)** Computation with double-precision floating point

Figure 2.14 Roundoff error introduced in the cumulative sum of polynomial values over a step of argument values.

We will revisit the topic of roundoff errors in Section 3.3 and see how it affects the details of how we must write our numeric C code.

2.3 Loss of Precision Errors

Let's examine another example decimal arithmetic operation to introduce another source of computation errors. Consider an example six-digit decimal subtraction problem 1.23456 − 1.23321. Each of the two operands are specified to a six-digit precision, but the difference of 1.35×10^{-3} is only known to three digits. Therefore, relative to the full six-digit capability of the arithmetic system being used, three significant digits of precision have been lost to the unfortunate choice of subtracting two operands very close in magnitude in a computation and storage system limited to six digits.

In another example, consider the subtraction $1.23456 - 1.23456 \times 10^{-3}$. The result, rounded to six significant digits, is accurate at 1.23333, but the displacement from 1.23456 is only 1.23×10^{-3}. This means that the significance of the subtraction of 1.23456×10^{-3} term has lost three of its six digits of precision, and this loss may easily emerge with any subsequent subtraction operations in the computation process.

The errors committed in these types of operations fall under the general classification of "loss of precision" errors. They are mirrored by errors of a comparable nature in the binary arithmetic systems of computer hardware, and are again the consequence of the limitations of a finite number of decimal digits or binary bits. As an example of the loss of precision observable with the double-precision floating-point standard, consider the following segment of the `precision_double1.c` program:

```
double t,x,y,z;
int i;

x = 1.0;
t = 0.1;
for (i = 0; i < 20; i++) {
  y = x + t;
  z = y - x;
  printf("i=%2d z=%.16f\n",i,z);
  x *= 8.0;
}
```

Although this is a somewhat contrived example for the purposes of illustration, note that the `double` variable x is initialized to 1.0, and is multiplied by 8.0 at the end of each iteration of the `for` loop indexed by `i`. This means that in each iteration, the value of `x` will be 2^{3i} for `i` ranging from 0 to 19. The `double` variable `t` is set to 0.1 and never changed. (Recall from Section 2.1 that the decimal value 0.1 cannot be represented exactly in a finite number of binary bits.) At the beginning of each iteration, we add `t` to `x` and store the sum in the variable y. Then we subtract x from y and store the difference in the variable `z`. Ideally, we would like to see a repetitive output of the value 0.1 for the value of `z` over all the iterations. However, when executed, this program produces the output:

```
i= 0 z=0.1000000000000001
i= 1 z=0.0999999999999996
i= 2 z=0.0999999999999943
i= 3 z=0.1000000000000227
i= 4 z=0.1000000000003638
i= 5 z=0.0999999999985448
i= 6 z=0.0999999999767169
i= 7 z=0.1000000000931323
i= 8 z=0.1000000014901161
i= 9 z=0.0999999940395355
i=10 z=0.0999999046325684
i=11 z=0.1000003814697266
i=12 z=0.1000061035156250
i=13 z=0.0999755859375000
i=14 z=0.0996093750000000
i=15 z=0.1015625000000000
i=16 z=0.1250000000000000
i=17 z=0.0000000000000000
i=18 z=0.0000000000000000
i=19 z=0.0000000000000000
```

As the iterations proceed and the value of x takes on higher powers of 2, we can see that the number of true significant digits with which the ideal difference of 0.1 is represented gradually degrades, with both positive and negative roundoff errors, until the significance of the difference is completely lost at $i = 17$. Even at $i = 16$, when $x = 2^{48}$, there is only one significant bit remaining in the representation of the difference of 0.1, since the displayed value of 0.125 is actually 2^{-3}. That one remaining significant bit occupied the last possible 2^{-52} position in the significand of y.

We will see the effect of these errors again in Exercise 2.2 and in Section 2.5.

2.4 Taylor's Theorem

Starting with Chapter 4, each chapter of this text will examine a series of numerical algorithms needed for solving a particular class of physics problems. Many of these algorithms, for example, numerical root finding, integration, interpolation, and differential equation solution, are all based on the mathematical foundation provided by the Taylor theorem.

Equation (2.7) shows the basic expression of the Taylor theorem. Here we assume we know the value of a function $f(x)$ at some anchor value of its argument x, and we may write that the value of the function evaluated at some argument a step h away from that anchor value can be expressed as a series of terms dependent on the derivatives of $f(x)$ of increasing order, plus an error term:

$$\begin{aligned} f(x+h) &= f(x) + f^{(1)}h + \frac{f^{(2)}}{2}h^2 + \frac{f^{(3)}}{6}h^3 + \frac{f^{(4)}}{24}h^4 + \cdots + \frac{f^{(n)}}{n!}h^n + E_{n+1} \\ &= f(x) + \sum_{k=1}^{n} \frac{f^{(k)}}{k!} \cdot h^k + E_{n+1} \end{aligned} \tag{2.7}$$

where the error term E_{n+1} is given by

$$E_{n+1} = \frac{f^{(n+1)}(\xi)}{(n+1)!} h^{(n+1)} \tag{2.8}$$

The error term E_{n+1} is referred to as the "truncation error", or the error we commit when truncating the Taylor series at some practical limit of n terms. The evaluation of the $(n+1)$th-order derivative function in the error term is performed at some value ξ in the function argument space. According to Taylor's theorem, in an assertion that traces back to the Fundamental Theorem of Calculus, this value ξ is necessarily bracketed between x and $x + h$.

When algorithms for numerical analysis that involve functions are derived, in general, we will be dealing with function values sampled at discrete values in the function's argument space, at some step distance away from a previously known value. We will be approximating the function with a fixed number of terms in the Taylor series, and the number of terms we include will affect the size of the inevitable error we commit in this approximation process. As we select the order of the algorithm to use for a particular problem, taking more terms in the equivalent Taylor series will involve more computational effort and longer computation time, but for reduced error levels. We will examine this tradeoff in more detail in Section 2.5.

2.5 Truncation Errors

In this section, we will assume we are faced with some numerical computation task involving a mathematical function, which is to be analyzed over some range of its argument variables. Perhaps we will need to compute the integral of the function, or we are interpolating a polynomial function between sparse data points, or we are solving for a function that is a solution of a differential equation, or other typical analysis tasks involving functions. Since we are attacking this task numerically, we will end up with samples of the actual mathematical function evaluated at certain discrete values of the argument, as illustrated in Figure 2.15. In general, the numerical analysis algorithm will be based on a Taylor series approximation to the function involved.

If we assume the sampling is done over a grid of uniformly spaced values, then the grid is specified by its initial and final values and the step size h. Let's examine the factors that influence the selection of an appropriate algorithm, and what is often the most important parameter belonging to the numerical algorithm itself; namely, the step size h.

Consider a simple first-order method to estimate the derivative of a function $f(x)$, derived from truncating the Taylor series of a function (Equation (2.7)) at the first term:

$$\frac{df(x)}{dx} \approx \frac{f(x+h) - f(x)}{h} \tag{2.9}$$

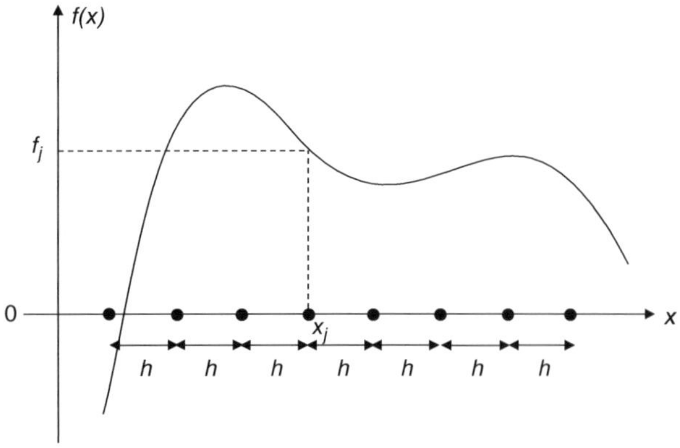

Figure 2.15 Sampling a function's value on a uniform grid of stepped argument values.

for some step size h. This is usually called a "one-sided" estimate of the derivative, since the function $f(x)$ is evaluated, besides at the nominal argument x, only at a step h higher than the nominal value of x. The following test program varies the step size over several orders of magnitude to visualize the tradeoff between roundoff and truncation errors as the step size varies. The reference function chosen to differentiate in these examples is simply the exponential function $f(x) = e^x$, chosen because this function requires an infinite number of Taylor series terms to express. Evaluated at a sample argument value of $x = 2$, the first derivative is, of course, e^2.

```c
// File: precision_deriv1.c
#include <stdio.h>
#include <stdlib.h>
#include <math.h>

// Estimation of derivative of f(x) = exp(x) at x=2



int main(int argc,char *argv[]) {
  double x,dfdx,h,error;
  int i;

  x = 2.0;
  h = 0.1;
  for (i = 0; i < 1000; i++) {
// Estimate derivative from simple first order one-sided ratio
    dfdx = (exp(x+h) - exp(x)) / h;
// Test for complete loss of precision from
// roundoff error and terminate
    if (dfdx == 0.0) break;
// Compute error from known derivative function
    error = fabs(exp(x) - dfdx);
    printf("%.8e %.8e\n",h,error);
// Reduce step size by a factor of 0.1 ^ 0.1 to test 10
// points per decade with logarithmic spacing
    h *= 0.7943282347242815;
  }
  exit(0);
}
```

Listing 2.2 First test program for tradeoff between roundoff and truncation errors

For small step size, loss of precision and roundoff errors dominate, since the numerator in Equation (2.9) becomes the difference between two numbers very close in value. For large step size, the truncation error dominates, since the error term in the Taylor series Equation (2.7) becomes large for a function with sharp curvature. This characteristic is plotted in the upper curve in Figure 2.16.

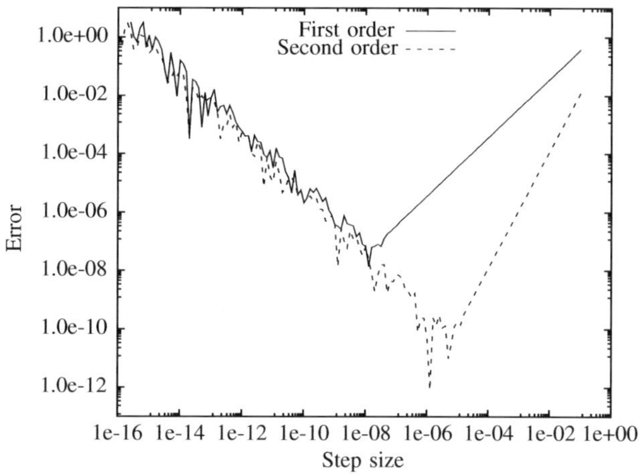

Figure 2.16 Tradeoff of roundoff and truncation error versus step size for first- and second-order estimates of the first derivative.

Now consider truncating the Taylor series after one additional term.

$$\frac{df(x)}{dx} \approx \frac{f(x+h) - f(x-h)}{2h} \tag{2.10}$$

Rewriting the test program to use this second-order estimate:

```c
// File: precision_deriv2.c
#include <stdio.h>
#include <stdlib.h>
#include <math.h>

// Estimation of derivative of f(x) = exp(x) at x=2

int main(int argc,char *argv[]) {
  double x,dfdx,h,error;
  int i;

  x = 2.0;
  h = 0.1;
  for (i = 0; i < 1000; i++) {
// Estimate derivative from second order two-sided ratio
    dfdx = (exp(x+h) - exp(x-h)) / (2.0 * h);
// Test for complete loss of precision from
// roundoff error and terminate
    if (dfdx == 0.0) break;
// Compute error from known derivative function
    error = fabs(exp(x) - dfdx);
    printf("%.8e %.8e\n",h,error);
// Reduce step size by a factor of 0.1 ^ 0.1 to test 10
// points per decade with logarithmic spacing
    h *= 0.7943282347242815;
  }
  exit(0);
}
```

Listing 2.3 Second test program for tradeoff between roundoff and truncation errors

This is usually called a "two-sided" estimate of the derivative, since the function $f(x)$ is evaluated at both a step h higher and a step h lower than the nominal value of x. We will make use of this double-sided derivative estimate in Section 6.6 to find an electric field vector as the gradient of an electrostatic potential, and again in Section 7.6 in algorithms that solve systems of nonlinear equations. The lower curve in Figure 2.16 of this characteristic shows a slope that reduces the truncation error at double the rate. With the higher order algorithm, the truncation error is reduced to much lower levels with a larger step size. But again, as the step size is reduced further, the floating-point roundoff error begins to dominate. Sampling two points in the program output on each curve; for example, step sizes of 10^{-2} and 10^{-3}:

Step size	First-order error	Second-order error
10^{-2}	$3.70687399 \times 10^{-2}$	$1.23151551 \times 10^{-4}$
10^{-3}	$3.69575987 \times 10^{-3}$	$1.23150873 \times 10^{-6}$

We can see that for the first-order estimate, the error decreases by one order of magnitude as the step size decreases by one order of magnitude, but for the second-order estimate, the error decreases by two orders of magnitude as the step size decreases by one order of magnitude. This is consistent with the power that the error term is raised to in the Taylor series of Equation (2.7).

Now compare three curves of the error of the double-sided, second-order derivative estimate computed with three levels of floating-point precision. The three curves in Figure 2.17 are for computing with the standard C types 32-bit `float`, 64-bit `double`, and 80-bit `long double`.

As a generalization, all numerical analysis algorithms that are based on the Taylor series of a function will follow some similar characteristics as shown in these two plots. We will also see a similar characteristic in Figure 10.3 when we study truncation error in numerical integration in Section 10.1, in Figure 12.9 when we study explicit methods for solving differential equations in Section 12.2, and in Figure 14.19 when we study implicit differential equation algorithms in Section 14.2. All of these numerical methods are examples of other algorithms that are similarly based on the Taylor series. Algorithms that encompass more terms in the Taylor series before truncation will have an error dependence that falls with higher power as the step size is reduced. But at some point, the reduction in step size will reach a limit imposed by whatever floating-point precision is used, and roundoff and loss of precision errors will dominate. The greater the number of significand bits in our floating-point representation, the lower is the floor of roundoff error we will encounter. Since Figures 2.16, 2.17, 10.3, and 12.9 are plots of numerical *error*, and since the universal goal of numerical analysis is to approach analytical results as closely as possible, we must ensure that the algorithm order, step sizes, and floating-point standard representing our numerical data are selected appropriately to operate as far down toward the minimum of these curves as is needed for the application.

Moreover, these plots basically characterize the errors committed by our algorithms over only one step. Another factor to consider for algorithm selection is that if a small step size is required for a given algorithm to achieve a given error level, then by the necessity there will be more steps required to cover some needed range of the function argument variable, as in Figure 2.15. Taking a very large number of steps of any algorithm leaves us susceptible to greater accumulated roundoff and loss of precision error at the point the total function span is completed, as illustrated in Section 2.2.

We may also derive a simple approximation to the second derivative by the subtraction of two Taylor series approximations to the function.

$$\frac{d^2f(x)}{dx^2} \approx \frac{f(x-h) - 2f(x) + f(x+h)}{h^2} \tag{2.11}$$

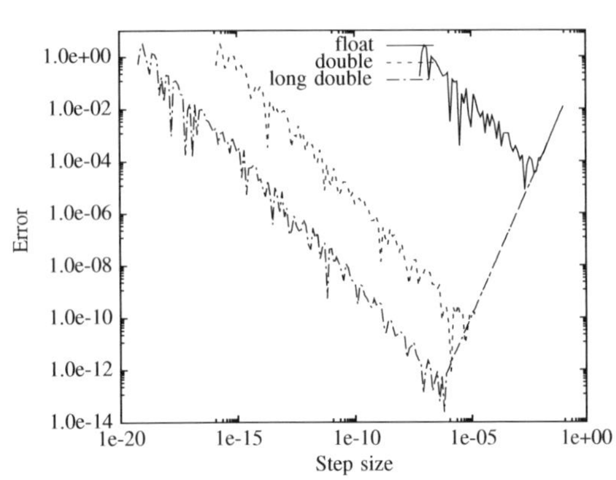

Figure 2.17 Tradeoff of roundoff and truncation error versus step size for different floating-point precision.

The second derivative of the reference function $f(x) = e^x$, evaluated at the sample argument value of $x = 2$ is also e^2. Rewriting the test program to use this second-order estimate of the second derivative:

```
// File: precision_2ndderiv1.c
#include <stdio.h>
#include <stdlib.h>
#include <math.h>

// Estimation of second derivative of f(x) = exp(x) at x=2

int main(int argc,char *argv[]) {
  double x,d2fdx2,h,error;
  int i;

  x = 2.0;
  h = 0.1;
  for (i = 0; i < 1000; i++) {
// Estimate second derivative from second order two-sided ratio
    d2fdx2 = (exp(x+h) - 2.0*exp(x) + exp(x-h)) / (h * h);
// Test for complete loss of precision from
// roundoff error and terminate
    if (d2fdx2 == 0.0) break;
// Compute error from known derivative function
    error = fabs(exp(x) - d2fdx2);
    printf("%.8e %.8e\n",h,error);
// Reduce step size by a factor of 0.1 ^ 0.1 to test 10
// points per decade with logarithmic spacing
    h *= 0.7943282347242815;
  }
  exit(0);
}
```

Listing 2.4 Third test program for tradeoff between roundoff and truncation errors

Again sampling two points in the program output on the curve; for example, step sizes of 10^{-2} and 10^{-3}:

Step size	Second-derivative error
10^{-2}	$6.15756593 \times 10^{-5}$
10^{-3}	$6.16004198 \times 10^{-7}$

Again we can see that the error decreases by two orders of magnitude as the step size decreases by one order of magnitude, verifying that Equation (2.11) is a second-order estimate of the second derivative. We will make use of this estimate to numerically solve the Laplace equation later in Section 6.6, and again in Section 14.5 where we examine numerical solutions of the wave equation.

We posed the question at the beginning of this chapter of what level of precision we may expect from the numerical analysis solutions we will study in this text. In general, as we have seen in this example of numerical differentiation, there are parameters in our algorithms that directly control the precision we will obtain, with greater precision to be had at the expense of greater computation effort. For the algorithms based on the Taylor series, the algorithm order and the step size set the error level. For other algorithms such as iterative root finding, we will directly set the tolerance with which we will converge on the target root with other parameters. However, we must always keep in mind that there is a floor level of error established by floating-point roundoff and loss of precision, below which the error cannot be reduced by algorithm parameter adjustment, as shown in the figures above. Therefore, selection of an appropriate floating-point format in our C code can also be a part of controlling precision.

In practice, if the error level for a given problem is particularly critical, we may explore the behavior of our algorithms by testing solutions at different step sizes or tolerance settings, and see where we lie on the typical error curves shown in Figures 2.16 through 2.18. Starting with the double-precision floating-point variables in our C coding is almost always a

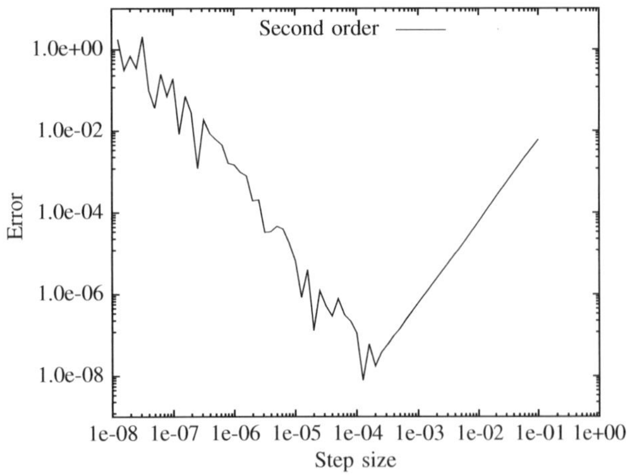

Figure 2.18 Tradeoff of roundoff and truncation error versus step size for second-order estimate of the second derivative.

good tradeoff of roundoff error versus storage space requirements for typical problems in the physical sciences. The double-precision format is also the default for various elements in C code, such as function arguments. The example code in the chapters that follow are written with parameters that give solution precision appropriate for most purposes, and they may be adjusted as needed.

2.6 Introduction to Numerical C Programming

This text will assume that the reader has some basic skills for programming in a structured language such as C, C++, or Java and is comfortable writing code for straightforward algorithms. The student should be confident of writing code with arithmetic expressions, iterative loops, conditional if-then-else blocks, and functions with arguments and return values. There are several introductory numerical C programming exercises at the end of this chapter, and these may be used in preparation for starting a more detailed study in the next chapter of coding techniques that we will use throughout this text.

As an instructive prototype C program that will illustrate many important coding details for the numerical programmer, consider Borwein's algorithm for an iterative method of calculating the value of π that converges very quickly. Each iteration at least doubles the number of significant digits in the estimate of π relative to the prior iteration. We index each successive iteration with an index value i, and in each iteration we calculate three variable values a_{i+1}, b_{i+1}, and p_{i+1} for the next iteration from the current values a_i, b_i, and p_i. The variable p_{i+1} is the next estimate of π, and we may observe the very rapid convergence of the successive values of p_i in this algorithm. We begin by initializing

$$\begin{aligned} a_0 &= \sqrt{2} \\ b_0 &= 0 \\ p_0 &= 2 + \sqrt{2} \end{aligned} \tag{2.12}$$

Then in each iteration we calculate

$$\begin{aligned} t &= \sqrt{a_i} \\ a_{i+1} &= \frac{t + \frac{1}{t}}{2} \\ b_{i+1} &= \frac{t(1 + b_i)}{a_i + b_i} \\ p_{i+1} &= \frac{p_i b_{i+1}(1 + a_{i+1})}{1 + b_{i+1}} \end{aligned} \tag{2.13}$$

Listing 2.5 shows an example C program to execute Borwein's algorithm. This program is also available for download from the Student Companion website as `progex_pi.c`.

```
1  // File: progex_pi.c
2  #include <stdio.h>
3  #include <stdlib.h>
4  #include <math.h>
5
6  int main(int argc,char *argv[]) {
7    int i;
8    double ai,aip1,bi,bip1,pi,pip1,t;
9
10 // Initialize a,b, and p
11   ai = sqrt(2.0);
12   bi = 0.0;
13   pi = 2.0 + sqrt(2.0);
14 // Perform 5 iterations
15   for (i = 1; i <= 5; i++) {
16 // Calculate temporary variable with square root of a
17     t = sqrt(ai);
18 // Calculate the next iteration a and b values
19     aip1 = 0.5 * (t + 1.0 / t);
20     bip1 = t * (1.0 + bi) / (ai + bi);
21 // Calculate and output the next estimate of pi
22     pip1 = pi * bip1 * (1.0 + aip1) / (1.0 + bip1);
23     printf("Iteration %d: Estimate of pi = %.15f\n",i,pip1);
24 // Update the current values of a,b, and p
25     ai = aip1;
26     bi = bip1;
27     pi = pip1;
28   }
29   exit(0);
30 }
```

Listing 2.5 Example program for numerical programming in C

Lines 2 through 4 include system library "header files" that are necessary to call the standard C library functions typically needed in numerical programs. In line 6 we declare the start of the main program block to be executed by the operating system. In lines 7 and 8 we declare the variables needed by the algorithm, either of the integer or double-precision floating-point type illustrated in Figure 2.3. Lines 10 through 28 are the numerical coding implementation of Equations (2.12) and (2.13). Integer variable *i* indexes the loop iterations governed by the `for` statement in line 15. The next estimate of the value of π is sent to the standard output stream from the program by the call to the C library function `printf()` in line 23. This stream of text characters generated by a program will be displayed on the command terminal window by the operating system by default, but much use will be made in future example programs in this text by commanding the operating system to "redirect" output characters to be stored in data files instead. The formatting string passed as the first argument to `printf()` provides a very flexible and versatile specification of the conversion of variable values to text characters, and the student should gain facility with this library function. Finally, the program calls the C library function `exit()` to terminate and return control to the operating system.

In a command terminal window on the Linux and MSYS2 environments, we can compile and run this example program with the dialog

```
1 $ gcc -Wall -Wshadow -O2 -o progex_pi progex_pi.c -lm
2 $ ./progex_pi
3 Iteration 1: Estimate of pi = 3.142606753941623
4 Iteration 2: Estimate of pi = 3.141592660966045
5 Iteration 3: Estimate of pi = 3.141592653589794
6 Iteration 4: Estimate of pi = 3.141592653589794
7 Iteration 5: Estimate of pi = 3.141592653589794
```

In line 1, we invoke the `gcc` C compiler with several command line options. A good choice is to include the options `-Wall` and `-Wshadow` for the compiler. These options set the "warning" level for the reports from the compiler so that students who are developing C programming skills will receive all possible warnings on coding issues that may result in program bugs. The `-O2` option sets the optimization level of the compilation process. An optimization level of 2 is a good compromise choice that can often yield a factor of two or three reduction of execution times of typical numerical programs. Examples of the methods of improvements in computation efficiency. The `gcc` compiler can automatically apply various methods to improve computation efficiency.

Examples of these methods are the transient storage of variables in registers instead of in memory and the temporary storage of the values of "common subexpressions" in the source code. These expressions are arithmetic evaluations that may be common to more than one C language statement in a region of code. The compiler will arrange for this expression to be evaluated only once and its value then stored in a temporary variable to be referenced in all the subsequent statements. It should be mentioned, however, that if a programmer encounters a coding bug that requires the use of the `gdb` or other debugger to track down, the `-O2` option should temporarily be changed to `-g` during the debugging process. Sometimes compiler optimization can reorder the sequence of the hardware execution of elements of the source code in ways that may confuse the debugger and make tracking a bug down to a particular line of code unreliable. The `-g` option commands the compiler to include important debugging symbols in the resulting executable file.

The name of the output binary executable file that the compiler will generate from the C source code files follows the `-o` option on the `gcc` command line. All input C source code files are listed on the command line next, although with this simple example only one source file is required. Finally the `-lm` option informs the compiler that the standard system mathematical library will provide the needed mathematical functions. Future examples in this text will show more elaborate examples of invoking the `gcc` compiler with multiple source code files to be linked with the library of numerical tools that we will gradually develop.

Finally, in line 2 of the command dialog, we execute our program and see the resulting stream of output characters reporting the convergence of the algorithm. The convergence rate is so rapid that only three iterations are actually required for the sequence to report the value of π within the full double precision. Note that we must include the characters `./` before the executable file name in order to specify to the operating system that the executable file is to be found in the current working directory. This is necessary for the Linux system, unless we make arrangements to locate our compiled executable files in a directory that has already been specified to be in the command search path.

This example program may be copied and modified for the programming exercises below.

2.7 Exercises

2.1 Following the examples in Table 2.1, write the bit encoding in the IEEE standard double-precision format for the following decimal numbers:

(a) 3

(b) −23.5

(c) 0.01953125

2.2 Consider the following segment of the `precision_double2.c` program:

```
double x,y,z;
int i;

x = 1.0;
y = 0.125;
for (i = 1;i <= 20; i++) {
  z = x + y;
  printf("i=%2d z=%.16f\n",i,z);
  y *= 0.125;
}
```

When executed, this program produces the output:

```
i= 1 z=1.1250000000000000
i= 2 z=1.0156250000000000
i= 3 z=1.0019531250000000
i= 4 z=1.0002441406250000
i= 5 z=1.0000305175781250
i= 6 z=1.0000038146972656
i= 7 z=1.0000004768371582
```

```
i= 8 z=1.0000000596046448
i= 9 z=1.0000000074505806
i=10 z=1.0000000009313226
i=11 z=1.0000000001164153
i=12 z=1.0000000000145519
i=13 z=1.0000000000018190
i=14 z=1.0000000000002274
i=15 z=1.0000000000000284
i=16 z=1.0000000000000036
i=17 z=1.0000000000000004
i=18 z=1.0000000000000000
i=19 z=1.0000000000000000
i=20 z=1.0000000000000000
```

(a) What is the last output line where the ideal value of z is computed with full 16-digit precision? Explain this observation knowing the details of the double-precision floating-point standard.

(b) At what output line is the arithmetic displacement from the addition operator using double-precision arithmetic totally lost? Explain this observation.

2.3 Consider the following segment of the `precision_float1.c` program:

```
float x,y,z;
int i;

x = 16777216.0;
y = 0.0;
for (i = 0;i < 6; i++) {
  z = x + y;
  printf("x=%.3f y=%.3f z=x+y=%.3f\n",x,y,z);
  y = y + 1.0;
}
```

Note that the floating-point variables x, y, and z have been declared as single-precision variables, and that $16777216 = 2^{24}$. When executed on a particular processor with a particular rounding policy, the output stream from this code segment is:

```
x=16777216.000 y=0.000 z=x+y=16777216.000
x=16777216.000 y=1.000 z=x+y=16777216.000
x=16777216.000 y=2.000 z=x+y=16777218.000
x=16777216.000 y=3.000 z=x+y=16777220.000
x=16777216.000 y=4.000 z=x+y=16777220.000
x=16777216.000 y=5.000 z=x+y=16777220.000
```

Explain the results seen in the rightmost column for each of the six lines.

2.4 We learned in elementary mathematics classes about the commutative property of addition. In ideal mathematics, the order in which we sum quantities should have no effect on the value of the sum. To test this property with floating-point arithmetic with a finite number of binary bits:

(a) Download and study the program `progex_sum.c` from the Student Companion website. Note that the program assigns the four numerical values 0.6, 0.7, 0.8, and 0.9, to the double-precision floating-point variables a, b, c, and d in ascending numerical order. The values in those variables are summed in some order into the variable `sum`, the value of which is then displayed in the standard output stream.

(b) Find the ideal sum of the four numerical values.

(c) Compile and run this program with the example commands

```
$ gcc -Wall -Wshadow -O2 -o progex_sum progex_sum.c -lm
$ ./progex_sum
```

Note the small roundoff error in the reported sum.

(d) Edit the `progex_sum.c` source code to assign the same numerical values to the variables *a*, *b*, *c*, and *d*, but now in descending numerical order. That is, assign 0.9 to *a*, 0.8 to *b*, etc. Recompile and rerun the program with

```
$ gcc -Wall -Wshadow -O2 -o progex_sum progex_sum.c -lm
$ ./progex_sum
```

Note the roundoff error resulting from this version of the code. Is the commutative property of addition apparently obeyed by floating-point arithmetic?

(e) Discover another set of four numerical values that numerically violate the commutative property.

2.5 **(a)** Write a C program `progex_fibonacci.c` to attempt to send the first 100 values of the integer Fibonacci sequence to the standard output stream, starting from 0. Use `int` variables to store the sequence values, and the `%d` output format for calls to `printf()` C library function.

(b) Compile and run your program above. Verify the first several integers in the sequence with the expected values. Explain the sign of the integer values displayed later in the sequence with respect to Figure 2.5.

2.6 The limit of the expression $\left(1+\frac{1}{x}\right)^x$ as $x \to \infty$ must be the base of the natural logarithm *e*.[10, Chap. 1] Test this expression by writing a numerical program `progex_e.c`. To see the limiting sequence, calculate and output the value of this expression for 10 values of *x*: 8^1, 8^2, 8^3,...,8^{10}. Also call the C library function `exp(1.0)` and display its return value.

2.7 Central to the relationships in special relativity is the Lorentz factor

$$\gamma = \frac{1}{\sqrt{1-\left(\frac{v}{c}\right)^2}}$$

where *v* is the relative velocity of the frames of reference, *c* is the speed of light, and γ is the factor by which coordinates are transformed. As the relative velocity approaches the speed of light, as in the case of a high-energy particle in an accelerator, the denominator decreases to a small value. However, if this expression is used to compute the Lorentz factor in a straightforward manner, we can spot the potential for roundoff error resulting from the subtraction of two numbers very close to each other in the denominator.

(a) Download and study the starting program `precision_lorentz.c` from the Student Companion website. The code is primarily an iterative loop that evaluates the Lorentz factor for four particular values of the ratio $\left(\frac{v}{c}\right)$ as the velocity approaches the speed of light. Note the careful use of both the `double` and the `long-double` data types for variables, constants, and output formats to compute the Lorentz factor γ twice in each iteration, once with each precision level. We will consider the computation with the `long-double` data type to be the reference value of γ, and a fractional error for the `double` data type computation relative to this reference is calculated and sent to the standard output stream. Compile and run this program, and observe the level of roundoff errors reported. You should see

```
$ gcc -Wall -Wshadow -O2 -o precision_lorentz precision_lorentz.c -lm
$ ./precision_lorentz
9.0000e-01 1.48299611e-16
9.9000e-01 6.72226228e-16
9.9900e-01 6.66075764e-15
9.9990e-01 1.80642373e-13
```

The output table shows that on the first two lines the relative errors in γ are the minimum of those expected with double-precision arithmetic, on the order of 10^{-16}. However, as the velocity increases, the last line reports a significantly higher error level by three orders of magnitude.

(b) Modify the `precision_lorentz.c` program to extend the approach of the relative velocity toward the speed of light. Use 12 discrete values of the $\left(\frac{v}{c}\right)$ ratio, starting with 0.9 and 0.99 as before, but now continuing to 0.999999999999. Discover the level of roundoff errors now encountered.

2.8 Write a program `progex_fahrenheit.c` to optimize a conversion formula from degrees Celsius to the Fahrenheit temperature scale that is suitable for quick conversions in our heads. The exact conversion formula is $°F = °C \cdot \frac{9}{5} + 32$. However, $\frac{9}{5} = 1.8 \approx 2$. How close would we be if we just used an approximate formula formula $°F = °C \cdot 2 + x$ for some optimum value of *x* instead? Write a program that steps through candidate values of *x* with an increment of 0.5 and calculates the overall conversion error for each value. Define the overall formula error as a sum of squared errors between the approximate converted value of °F and the actual converted value of °F over a range of temperatures typically found in weather forecasts, that is,

$$\text{error} = \sum_{°C=c_1}^{°C=c_2} (°F_{est} - °F_{act})^2$$

where c_1 and c_2 are the lower and upper summation limits in °C that correspond to actual of 40 and 90, respectively, and the conversions in the sum should be performed with a step of 0.1°C. Send a table of the candidate values of x and the corresponding error to the standard output so that we may find the value of x that minimizes the overall conversion error by inspection.

3

C Programming Details

In this chapter, we will continue our discussion of the details of the C programming language from Section 2.6, especially focusing on efficient and reusable numerical code. Let us take off from there and examine in detail those features in C that let us manage numerical data characterizing physical systems in a way that makes the most efficient use of conventional computing hardware.

3.1 Structures and Pointers

In Section 1.2, we introduced the idea of encapsulating and modularizing the data and code in our programs to make the most efficient use of our development and run time. In this section, we will see in detail how the C language makes available the constructs of structures and defined variable types to greatly aid in our data encapsulation. Then in Section 1.3, we looked at the motivation in hardware for the power of the C language to allow efficient control over the address locations of computation data held in memory. So also in this section, we will examine the details of C programming that exercise this control with pointer variables. The encapsulation of data into structures, the definition of the structure as a new data type, and then the reference to items of that new data type with pointers to an area in memory where they are stored are all C language features that work together neatly.

3.1.1 Pointers

A practical processing system requires a very large amount of memory where the values of numerical quantities for computation can be stored. The hardware performing the computation is able to distinguish each area of memory by associating a unique numerical address for each byte of memory. But programming language compilers provide symbolic naming mechanisms that allow us to refer to variable values stored at a location in memory by some name that is meaningful in our algorithm, such as `mass`, or `m`. The C language facility for pointers allows the programmer further detailed control over memory addressing. A good image for understanding the meaning and use of pointers for a memory system is Figure 3.1.

This image is an array of postal boxes on a wall at a post office. Each box has a capacity to store some numerical value as its contents, and each box is also labeled with a numerical address, presumably a field in the postal address used for a sender to identify a particular destination box. A data variable is a box that holds some integer or floating-point number used in computation. However, a pointer variable is a box in which the contents are an address value of another box, rather than a computation number. The box that the pointer variable points to may contain a numerical value for computation, or it may be another pointer variable to point to another box.

What purpose do pointer variables serve in a C program? In order to code the instructions in the general structure of programs as illustrated in Figure 1.5, we will be arranging the code as a hierarchical structure of the C main program and C functions, as illustrated in Figure 3.2. Execution control is passed from the operating system, along with the text strings representing the command-line arguments that the program was run with. The main program may call various C functions in the process of its computational tasks, which may call other C functions, and so on.

Each function in the hierarchy will need arguments and data to operate on. The C language is a "call by value" language, meaning that the literal values of function arguments are passed to functions. This allows functions to call themselves in a recursive manner if the computation algorithm can be efficiently expressed in that form. We can greatly decrease the

Computational Physics Using C: Efficient Programming with Ease, First Edition. John W. Fattaruso. © 2026 John Wiley & Sons, Inc.
All rights reserved, including rights for text and data mining and training of artificial intelligence technologies or similar technologies.
Published 2026 by John Wiley & Sons, Inc. Companion Website: http://www.wiley.com/go/CompPhysUsingC

Figure 3.1 Image for memory storage at addresses.

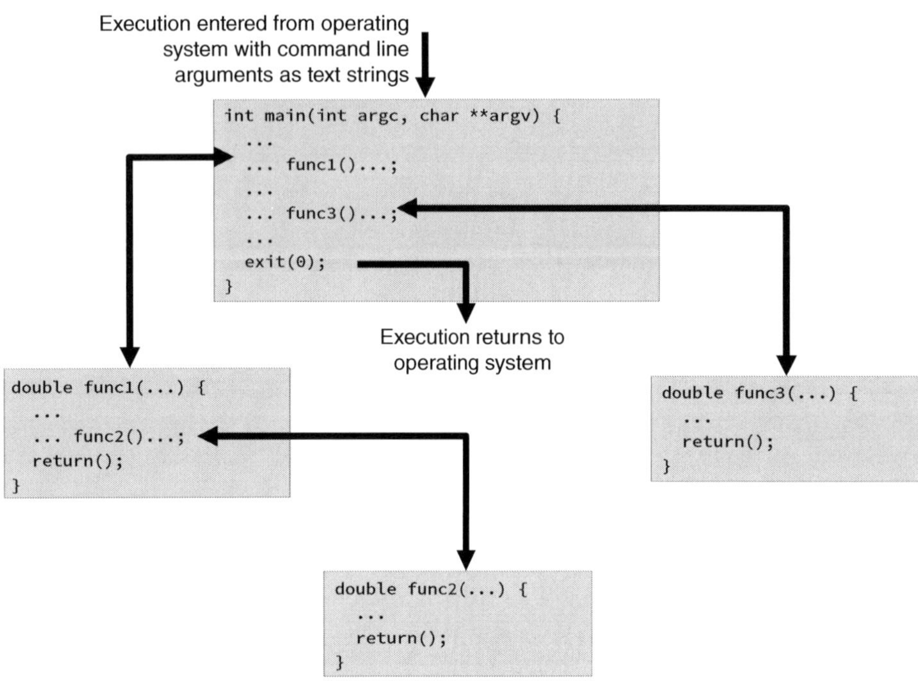

Figure 3.2 Hierarchical arrangement of program execution modules.

binary information that must be passed to each function by aggregating large collections of data into C language structures, and then simply passing a pointer to the location in memory where that structure is stored to a function as an argument. The function then "dereferences" the pointer argument to access the data members needed for its task. Dereferencing a pointer is analogous to a sort of treasure hunt where we open one postal box to extract an address stored there, and then follow that address to a different box to open that door and either read or write data there. Moreover, we allow the code in each function to store the results of its computational task directly back into memory, where it can be used in subsequent computation, using the structure memory area for storing variables and the results of arithmetic assignment statements. This pointer mechanism can substantially increase the overall execution efficiency of a numerical code. The C compiler will arrange for the values of the pointer variables to be held temporarily in registers in the CPU architecture of Figure 1.6 and then to be dereferenced by loading their values into the MAR for use as memory addresses.

A variable declaration statement in C such as

```
double x,y,z;
```

could be seen logically in terms of this image as allocating three boxes, each of which will contain a double-precision floating-point value to be used in calculations. Each box will be referred to by its own numerical storage address in the compiled processor code, like the numbered addresses on the front of the post office box, but referred to symbolically in the C code by the variable identifiers x, y, and z in the compiler's *symbol table*. Each symbolic box may actually require different numbers of bytes of physical memory to be allocated to it, according to the size of the numeric encoding, as discussed in Section 2.1. However, the contents of those memory locations are in an indeterminate state after just a declaration.

Then another declaration and assignment statement such as

```
double *p;
p = &x;
```

is logically allocating another box of storage symbolically referred to by the identifier p, but this box is allocated to store a pointer value, and is capable of holding one of the postal box numbers printed on the front of a box rather than a data number. In this case, the compiler makes a symbol table entry so that the location information stored in the pointer box will be a location address of a box that contains a double-precision floating-point number. Then the assignment statement fills the pointer box with the particular address of the box containing the x variable.

Pointer variables in C give the programmer more direct control over those registers in the central processing unit (CPU) block diagram in Figure 1.6, which contain the addresses in memory where variable data is stored than do most other higher level programming languages. That control is a two-edged sword. It can be used to write more efficient code for elaborate numerical algorithms, but it can also lead the novice programmer into program bugs that are especially difficult to identify and correct, even with the aid of a debugger tool. If programming with pointers is not yet second nature to the reader, the following illustrations should be studied carefully.

Figure 3.3 shows statements in C used to declare symbolic variable names of different types, and corresponding gray rectangles representing storage space in memory allocated by the compiler. The first four variables are among the fundamental arithmetic types defined in C, a character, an integer, a single-precision floating point, and a double-precision floating point. The last four variables are pointer variables to the four basic numerical types. The relative sizes of the rectangles in the figure represent the size of the variable storage required, in terms of the number of bytes typically needed. In the

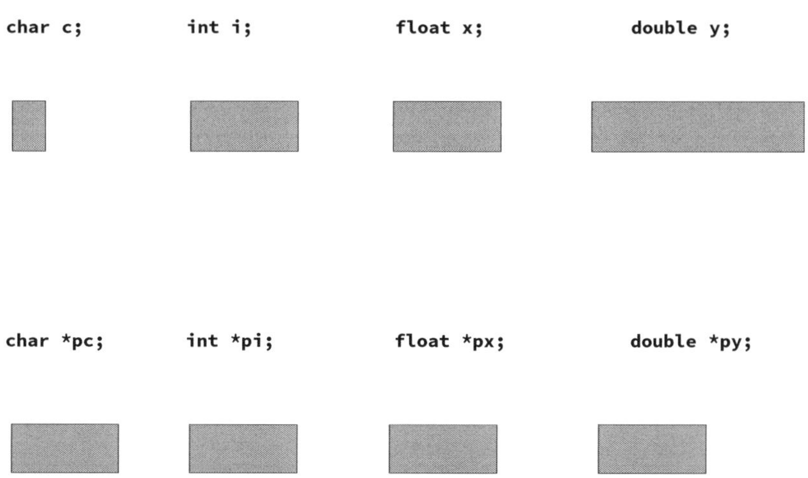

Figure 3.3 Allocated but not yet initialized memory for eight variable types.

uninitialized state, the contents of the pointer variables are undefined, and do not necessarily contain an address of usable memory.

The empty gray rectangles in Figure 3.3 emphasize that there is nothing of meaning yet stored at the corresponding memory locations, or in the corresponding postal boxes. If a subsequent C statement is encountered containing an arithmetic expression that makes a reference to one of the data variable values stored in the uninitialized storage, then the results of the arithmetic evaluation will be undefined and meaningless. Although this sort of bug would almost assuredly produce some meaningless numerical results, the program will likely continue executing. However, if instructions follow that attempt to dereference one of the uninitialized pointer variables, the memory access logic in the CPU will attempt to read or write from some memory location the operating system has not planned for. The operating system maintains a tight control over what memory segments belong to what processes that are currently running, and almost assuredly an undefined pointer dereference will trigger a memory segment access violation fault. The operating system will abort the execution of the program and report a "segmentation fault" to the user, which can be a rather obscure bug to track down.

It is almost universal that programmers who accept the challenge, and reap the powerful benefits, of efficient use of pointers in C and C++ code, will at some point encounter segmentation faults as their code is being developed. Sometimes the use of an uninitialized pointer can be found by careful inspection, but at other times a powerful debugging tool such as `gdb` can speed the debugging process greatly. Fortunately, the `gcc` and other C compilers will often detect and warn the programmer at compile time of potential access to all types of uninitialized variables, including pointers. You will also find examples throughout the code packaged for this textbook of another useful feature of the C library, the `assert()` macro [14, Chap. 7]. For example, suppose that at some point in our program structure, perhaps at the entry point of a function, we assume a given pointer variable *p*, perhaps a formal function argument, has been set to a proper memory location. In general, any time we dereference a pointer variable, it is critical to the program operation that the variable points to the proper, intended location in memory where data is stored. If a pointer variable has been declared but not yet been assigned a proper value, it will hold the value of NULL, as defined in the system header file `stdio.h`. We may place calls to the `assert()` macro with a test condition anywhere in our code; for example,

```
assert(p != NULL);
```

This call will test any condition supplied as an argument, in this example, if the value of the pointer variable is not NULL, at the current place in the code execution, and only allow the program to continue if the condition is true. If it is false, in this example, if the pointer variable *p* is undefined, the program will terminate and report a helpful debugging message to the user. Sprinkling calls to `assert()` throughout your code can provide a very powerful self-debugging capability during development.

Figure 3.4 shows arithmetic assignment statements, optionally with a mathematical expression to evaluate, used to assign a value to a variable. The binary values associated with each variable according to the encoding systems of Section 2.1 are stored into the memory location allocated for the variable. The decimal equivalents are seen in the figure now occupying the boxes of memory locations. Then the four-pointer variables are initialized to memory locations that point to the character and three numerical variables. These binary address values are symbolized by the black squares in the figure occupying the pointer variable rectangles. In order to generate an address to initialize the pointer variables, the & operator resolves to the allocated memory location associated with the variable identifier following the & character. At this point, dereferencing these pointers with the * operator will let the CPU hardware refer to usable memory locations that belong to the program's process, and no segmentation faults will occur.

A pointer variable need not point to only data variables, but can also be a pointer to another pointer variable. That is, one pointer may hold the box address or memory location of another box where another pointer is stored. Pointers may be nested in this way to any level, as in Figure 3.5. This figure shows a sequence where each pointer points to the next pointer, and eventually pointing to the location of the actual data item, in this case a double-precision floating-point variable. The variable *y* is said to be of type `double`, the variable py is said to be of type "pointer to double," or `double *`, the variable ppy is said to be of type "pointer to pointer to double," or `double **`, and so on. Each application of the * dereferencing operator in the variable declaration adds one additional level of pointer indirection, and each additional * operator in an expression adds one additional level of pointer dereferencing.

A test program that illustrates the process of navigating such a sequence of pointers is shown in Listing 3.1. In general, there isn't much practical use of pointers nested beyond two or perhaps three levels, but for the purpose of illustration, this program carries the process even beyond that. Evaluating the expression ****ppppy in the call to `printf()` starts at the pointer ppppy and follows a chain of four pointer dereferences before finally arriving at the memory location holding the floating-point value in the variable *y*.

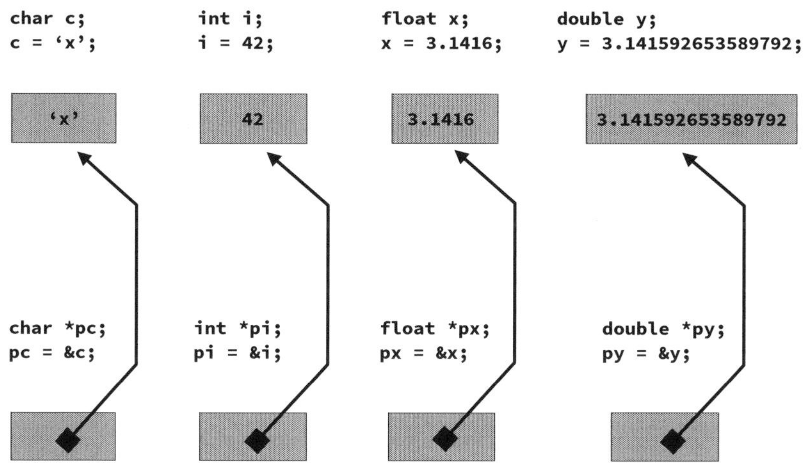

Figure 3.4 Initialized memory for eight variable types.

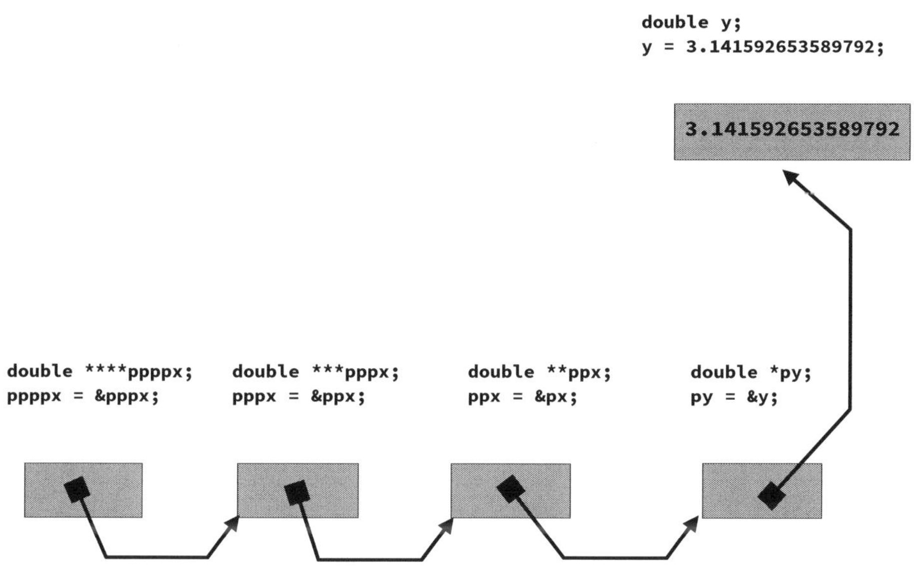

Figure 3.5 Successive sequence of pointers eventually ending at a double.

```c
// File: pointers1.c
#include <stdio.h>
#include <stdlib.h>

int main(int argc,char *argv[]) {
  double y,*py,**ppy,***pppy,****ppppy;

  y = 1.23;
  py = &y;
  ppy = &py;
  pppy = &ppy;
  ppppy = &pppy;
  printf("%.4g %.4g %.4g %.4g %.4g\n",y,*py,**ppy,***pppy,****ppppy);
  exit(0);
}
```

Listing 3.1 C program illustrating a successive sequence of pointers to a double

Now compiling and running this program at the terminal shell outputs the same floating-point value five times, after five different ways of arriving at its memory location:

```
$ gcc -Wall -Wshadow -O2 -o pointers1 pointers1.c -lm
$ ./pointers1
1.23 1.23 1.23 1.23 1.23
```

An array of numerical variables may be declared, and the memory allocation spans sufficient memory locations to consecutively store all the elements of the array of the specified type. In Figure 3.6 a five-element array of double-precision floating-point values is declared, and the compiler will allocate a space in memory of sufficient size to store the elements.

The symbolic identifier of an array declaration is then defined to be of a type of a pointer to the array element type, and the value of that pointer is the location of the first element of the array. That pointer may then be assigned to a pointer variable, as in Figure 3.7. Note that the array identifier *a* has a value of type "pointer to float," and can be assigned to the variable `py` of the same type.

Elements of an array are stored in consecutive memory locations, with each location being sized to hold a single array element, whatever its type. This is analogous to allocating a sequence of postal boxes in Figure 3.1 at successively higher box addresses, to hold each array element. If we had a pointer to the first box in a sequence, in order to arrive at any particular box in the array, we merely have to add the index of the element of interest to the address of the initial box. Naturally, the first element is then at an index of 0, and the last element is at an index of $N - 1$, where N is the number of elements in the array.

In the same way, we can use the pointer to the first element of an array in C to calculate a pointer to any arbitrary index array element. The identifier *a* in Figure 3.7 is a pointer to the first element, so we could dereference that pointer with the construct `*a` to refer to the value of the first element. Adding one to the pointer is analogous to adding one to the postal box address, so the construct `*(a + 1)` would be the value of the second array element. In general, then, the value of any element with index *i* would be `*(a + i)`. Since this syntax is somewhat clumsy, the C language defines an equivalent and more convenient syntax for array indexing with the `[]` characters. The first element is `a[0]`, the second `a[1]`, the *i*th is `a[i]`, and the last `a[N - 1]`, as in Figure 3.7. The pointer calculations are identical to the more explicit syntax with the `*` operator.

It should be understood that the pointer arithmetic in these examples always refers to the arithmetic needed to arrive at a particular element in an array of any data type, no matter how many physical bytes of memory each element requires. The C compiler removes from the programmer any burden of keeping track of physical memory bytes. We may simply conceptualize array elements as general storage containers, such as the postal boxes, that may hold values of various data types, such as integers or floating-point numbers. Moving from one box to the next in an array always involves simply incrementing a pointer by one.

Using an array indexing operator such as `a[i]` refers to the value of the *i*th array element. We may use this language construct to, for example, retrieve the value from the array to form an operand in an arithmetic expression, or to pass its value as an argument to a function, or to set its value on the left side of an assignment statement. But what if we want to form a function argument with the *pointer* to the *i*th element in the array, instead of its value? We may, for example, want

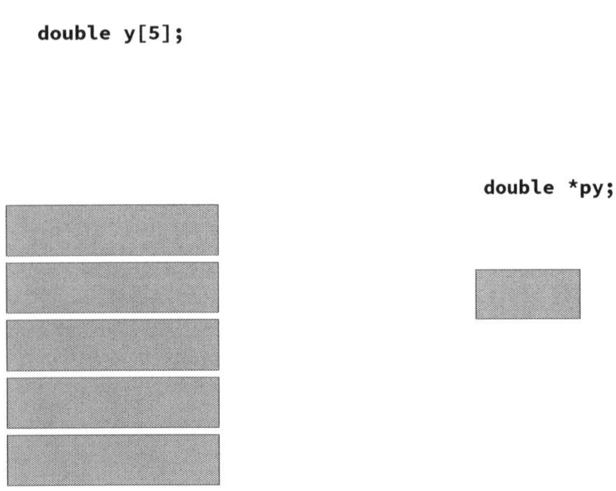

Figure 3.6 Allocated but not yet initialized memory for a double-precision floating-point array.

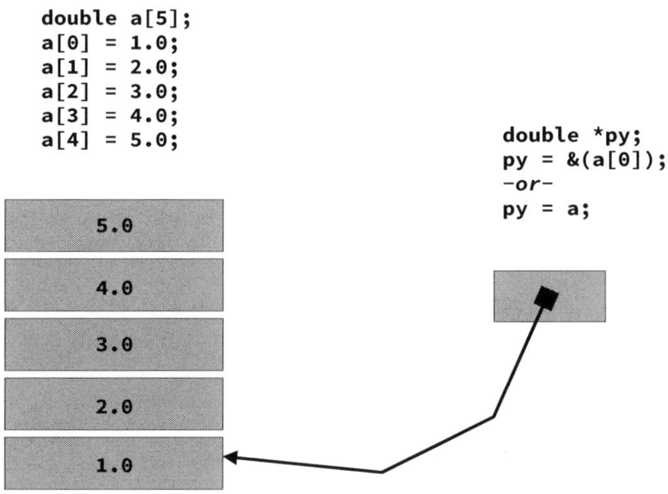

Figure 3.7 Initialized memory for a double-precision floating-point array.

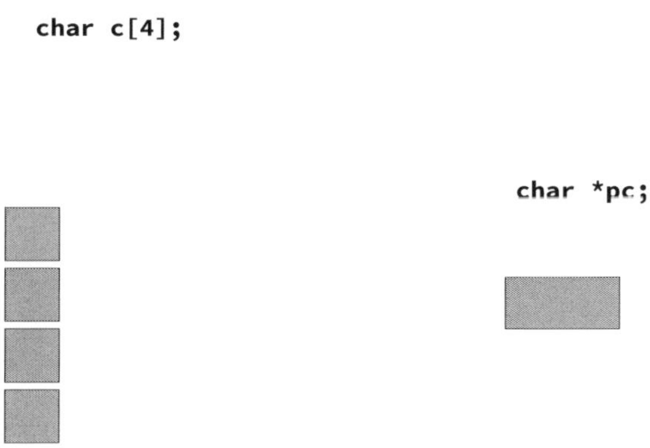

Figure 3.8 Allocated but not yet initialized memory for a character array.

to pass a pointer of the element to a function so that the function may change its value. In that case, we would stay with the pointer arithmetic form of `a + i` without any dereferencing or indexing. That is, adding the index number in the integer variable *i* to the array base location in the identifier *a* results in a pointer value to the specified element, rather than its value. So `a + i` can be passed as an argument to a function, assigned to a pointer variable within the function, and then be dereferenced within the function's code.

Arrays of character variables can be declared to store character strings, terminated with a null character. Characters are represented by small binary numerical values, according to some character encoding standards, such as ASCII or UTF-8. The smallest of these standards allows one of 256 characters to be encoded in a single memory byte. An example character string declaration is shown in Figure 3.8.

The call to the system library function `strncpy()` can be used to initialize a character array with a string of characters and its terminating null character. Characters in the string are filled into memory bytes starting from the lowest address up to the null terminating byte at the highest address, as shown in Figure 3.9. In this case, the array identifier `c` is of type "pointer to char," or `char *`, and can then be assigned to the variable `pc`, which is declared to be of that same type.

In an example of a very useful case of multiple levels of pointers, in Figure 3.10, we store multiple strings in any number of character arrays, and then store a pointer to each string in an array of pointer variables. The variable `pc` is declared to be an array of three elements, each element a "pointer to char." This is actually the structure that the operating system will prepare in memory for every program that is run from a command line, where the strings contain the sequence of characters in each of the command-line program arguments. In this case, the type associated with the symbol `pc` would be "pointer to pointer to char".

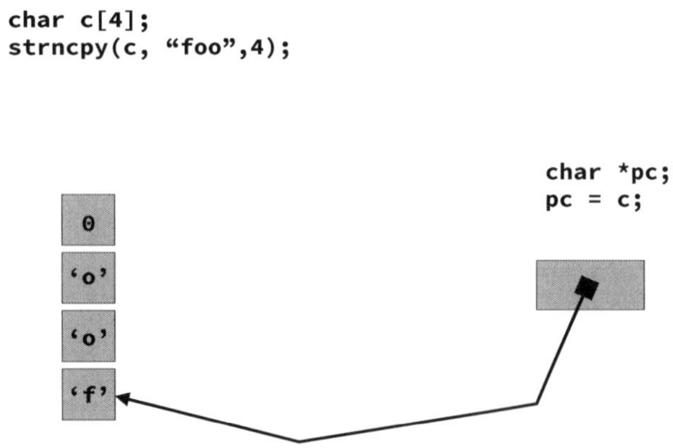

Figure 3.9 Initialized memory for a character array.

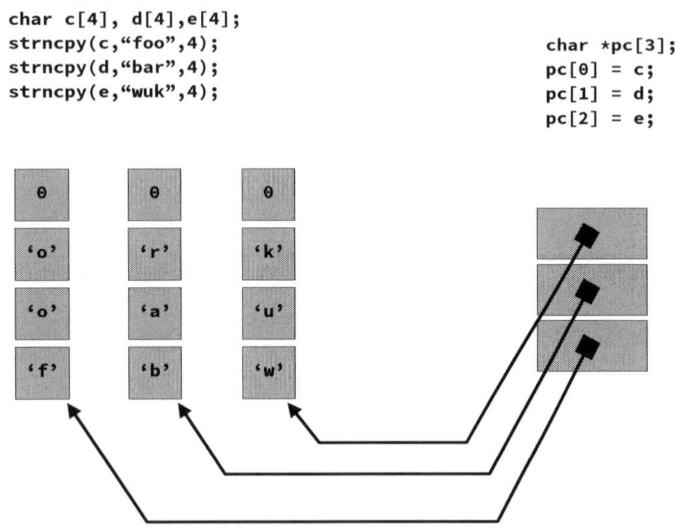

Figure 3.10 Initialized memory for an array of pointers, each pointing to a character array.

3.1.2 Custom Data Types

We are not confined to only the data types that the C language defines as its fundamental set. To allow very flexible data types for our solutions of physical equations, we may aggregate any number of variable types together in a structure. This lets us bundle a collection of variables into a package that represents some conceptual description and refer to them together as a custom data type [14, Chap. 11]. The first step is to define the variable types and names that form the members of the structure, and then declare memory space with this newly defined variable type. As before, when only declared with an identifier, the structure member variables are uninitialized. In Figure 3.11, an example structure with a tag name of `foo` is first defined, and then the structure is given the equivalent status of a new data type with a name of `FOO`. (The use of lower and upper case characters in these names follows a long-standing convention used in the C library routines.) In this example, the structure comprises three double-precision floating-point values and an integer value, but the member variables may be any arrangement that is needed for the physical computation being prepared. Then an instance of this structure can be declared and given the symbolic name `a`, using a declaration statement of a similar format to those for fundamental variable types. The new custom, or "derived," data type may be referred to either by the `struct` keyword followed by the tag name, or, more conveniently, by the use of the C language keyword `typedef` indicating a definition of a custom data type, simply by the new data type name. The compiler will allocate sufficient memory to hold all the values making up the structure.

Then to assign or retrieve values to and from the variables that are members of a structure, we make use of the `.` member operator together with the symbolic identifier of the structure. For example, in Figure 3.11 an instance of a structure was

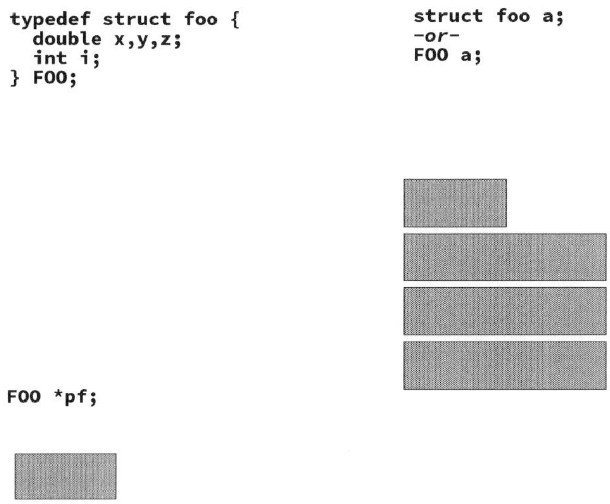

Figure 3.11 Allocated but not yet initialized memory for a structure.

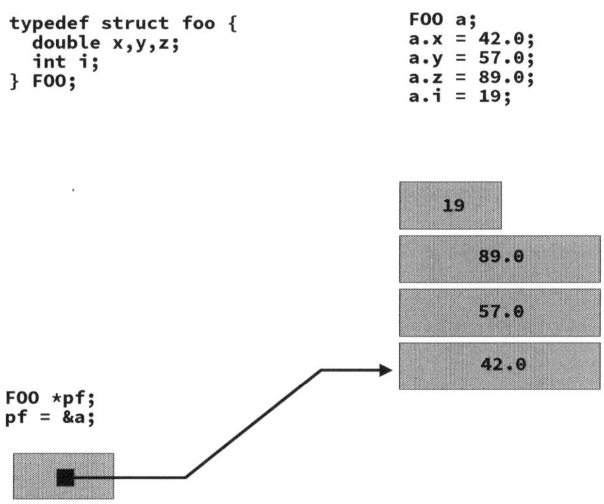

Figure 3.12 Initialized memory for a structure.

allocated space in memory and given the symbolic name *a*. We would refer to only the floating-point value for the *x* member of that instance with the construct `a.x`, and to the integer member `i` in that instance as `a.i`. An example is shown in Figure 3.12.

It is important to understand that the symbolic names and data types of the members of a structure refer to only a template pattern specifying how memory will be used for data values within the memory allocated for each independent instance of a structure. That means that we can allocate any number of structure instances we need with different identifiers, and the composite member variables in each instance hold completely independent values in different memory locations. For instance, we could declare other instances of our new data type `FOO` named, say, *b* and *c*, and the member variables `b.y` and `c.y` would hold values independent from `a.y`.

To continue the sense that these custom data types are similar in language constructs to fundamental variable types, the C language allows us to equate variables that are defined as instances of custom data structures. For example, if both *a* and *b* are defined to be instances of our data type `FOO`, then once we assign values to the members in *a*, we could simply write the C statement

```
b = a;
```

and code will be compiled to equate all the members in the *b* instance to their corresponding member values in *a*.

In the beginning of this section, one example of the utility of pointers was to pass aggregations of data as an argument to a function. This is formed as a pointer to a structure instance. Also shown in Figure 3.12 is a declaration of a variable named `pf` that is declared to be a pointer to our new data type `FOO`. The pointer variable `pf`, which of course has its own memory space allocated for it, is initialized to the address of the structure instance *a*. If we pass the value of the pointer `pf` to a function as one of its arguments, we are passing a way for that function to refer to any member variable in that structure instance. We are conceptually passing the entire instance to the function, without requiring the hardware to spend machine cycles to copy each and every instance member value from one memory location to another. The instance members may simply reside where they are in its allocated memory, and we pass to the function a road map to all those memory locations given in the pointer to the beginning of the memory area and the knowledge of the layout in memory within the structure.

Inside the function, the lines of code could refer to a member variable by combining the `*` dereferencing operator with the `.` member isolation operator. For example, given a pointer `pf` to an instance of the `FOO` type, we could refer to its *x* member as `(*pf).x`. (The parentheses are required because of the operator precedence rules of C.) This syntax is somewhat clumsy; however, so a new operator is defined in C for this purpose. The more readable and convenient syntax `pf->x` is available to the programmer, and this is introduced in the examples in Figure 3.13.

How may we combine the power and convenience of arrays and structures? The C language allows the declaration and use of arrays of structures, as well as structures that hold arrays as members.

We have made use of the capability C gives us to define a new data type from a structured aggregation of other data types. A newly defined data type may take the place of any fundamental data type in any syntactical element in the language. Since we can declare arrays of, for instance, characters, integers, and floating-point variables, we can use the same syntax to declare arrays of structures, as shown in Figure 3.14.

Figure 3.15 shows how the particular structure element in an array is located with an array index operator, just as with the arrays of simple variables in Figure 3.7. Then once a particular structure element is indexed, particular members within that structure can be specified by appending the `.` member operator.

Pointer arithmetic in C works for pointers to structures and defined data types as well as for fundamental data types. That is, we may form a pointer to a particular structure in this array by adding an integer to the array base location. So, for example, a pointer to the third structure would be either `b + 2` or `pf + 2`. Since we have declared the pointer variable `pf` to be of a type that will always point to our new data type `FOO`, whose structure is known to the compiler, the number of memory bytes occupied by that data type is also known, and the compiler will know how to perform the pointer arithmetic to locate each array element.

In another example of two levels of pointers, a structure may be defined with a data member being an array. In the example of Figure 3.16, the structure is declared with two members, one an array of `doubles` and the other an array of `ints`.

These may be initialized as shown in Figure 3.17. Note that we first apply the `.` member operator on the structure identifier to locate the member of interest, and then apply the array index operator to locate a particular array element. In this example, we also see using the `->` operator to arrive at a structure member from a pointer to that structure instead of the structure identifier.

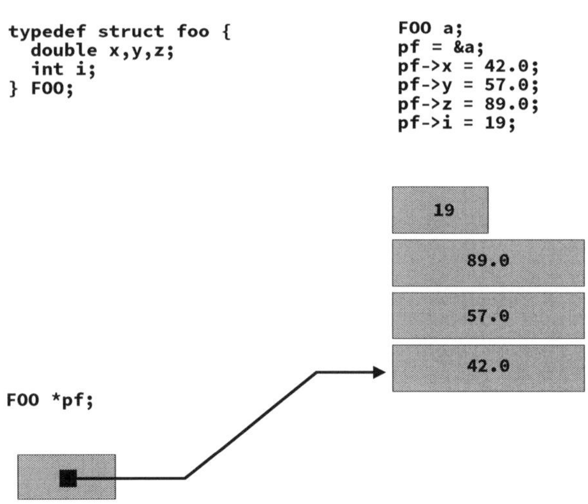

Figure 3.13 Initialized memory for a structure referred to by a pointer variable.

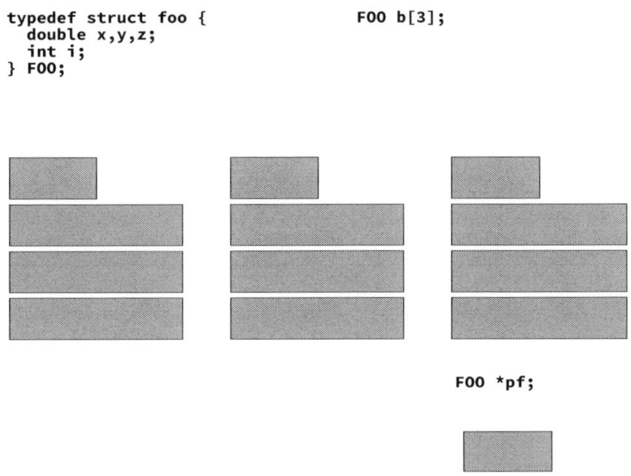

Figure 3.14 Allocated but not yet initialized memory for an array of structures.

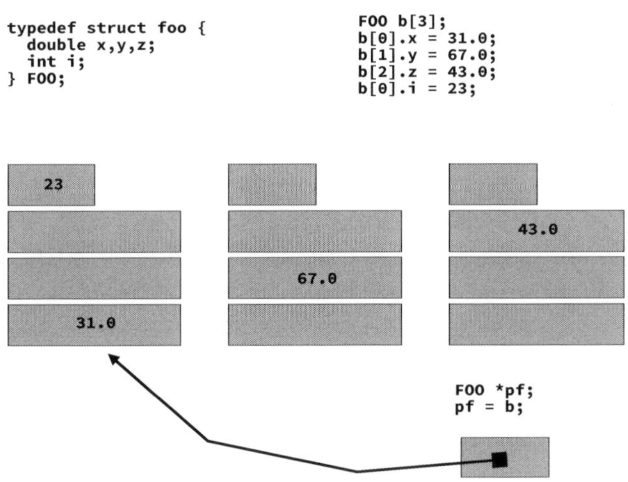

Figure 3.15 Initialized memory for an array of structures.

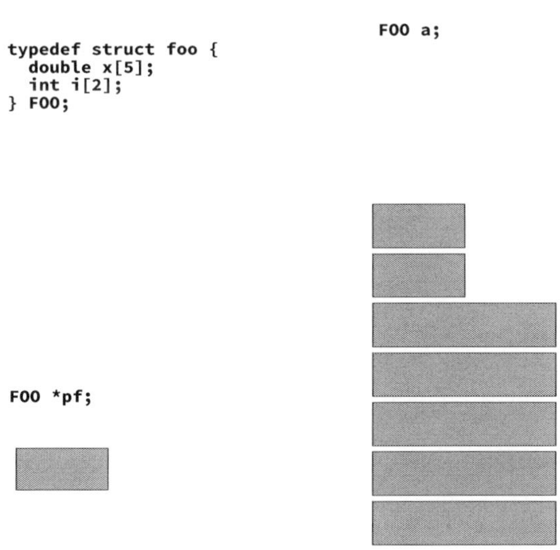

Figure 3.16 Allocated but not yet initialized memory for a structure containing two arrays.

```
typedef struct foo {
  double x[5];
  int i[2];
} FOO;
```

```
FOO a;
pf = &a;
a.x[0] = 42.0;
a.x[2] = 53.0;
a.x[4] = 79.0;
pf->i[0] = 19;
pf->i[1] = 23;
```

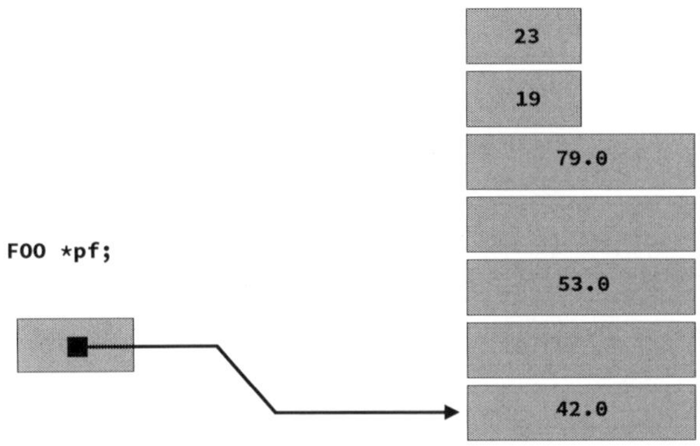

Figure 3.17 Initialized memory for a structure containing two arrays.

3.1.3 Dynamic Memory Allocation

The arrays in the previous subsection were all of some size, that is, with some number of elements that was known when the program was coded. Therefore, an appropriate amount of memory storage could be planned for by the C compiler at the time the program is compiled, and allocated to be used by the program when the operating system first loads it into memory to be executed. This is referred to as allocating memory "statically."

However, there are many numerical problems in the physical sciences where the count of elements in an array of variables is not known at the time the program is compiled. For example, we will be writing, compiling, and running many example programs in this text, and almost all of them are programmed to accept various numerical arguments on the command line that are typed into a shell prompt to run the program. There may be a command-line argument that specifies the number and location of point charges in an electrostatic simulation, or the number of bodies with mass and their initial locations and velocities in a gravitational system simulation, or some other description of an arbitrary number of objects. For the program to be of the most general applicability possible, it should be coded to accommodate any arbitrary number of participating objects that it finds described on its initiating command line when it starts to execute. Or perhaps when a program starts to execute, it first reads a data file that specifies some large and arbitrary number of data points describing a function generated by another program that must be all read in and stored in memory to be processed.

Therefore, operating systems provide an alternative way to allocate memory for storing variables other than coding an allocation of a fixed size in a line of source code. Memory may be allocated "dynamically," meaning at program run time rather than at compile time, from available memory that the operating system manages among all the processes running on a machine. The `malloc()` C library function will allocate a number of bytes of memory and return a pointer to that memory array. This is another example where the language construct of pointers allows powerful and convenient management.

The pointer returned from the `malloc()` function is not intrinsically declared to be a pointer to any particular data type. However, in order to use the constructs of pointer arithmetic, the C compiler must know what type of data a pointer variable points to, as it must know the memory size in bytes for each data item to compute memory addresses properly. So for each call to `malloc()` function, we need to assign the function return value to a pointer variable that has been declared to point to an explicit variable type. For example, when calling `malloc()` to allocate memory for an array of double-precision floating-point variables, the return value should be assigned to a pointer variable of type `double *`.

In small computing environments with limited memory it is possible that the operating system will not be able to fulfill a program's request for memory resources. In that case the `malloc()` function will return a constant pointer known as the `NULL` value. This outcome may be tested for after the function returns and appropriate failure steps taken.

We should be aware, however, that the processor time needed to allocate memory is not negligible. The request is made to the operating system, out of the control of the user's program, and considerable searching or shuffling of memory resources may be required for the operating system to satisfy the program's requests. So in practice, the number of memory requests should be minimized.

Consider the sequence described in Figures 3.18 through 3.20. The program is coded to allocate a pointer variable statically at compile time. When the program first starts executing, the data memory capable of storing only that pointer is known to the program, as shown in Figure 3.18.

Then at some point in the execution, the program comes to know how much storage space is required to store its computation variables, perhaps by examining its own command line to find the number of arguments, or processing a data file and reading the number of lines, or some other computed size. It then calls the `malloc()` library function to request the required memory from the operating system, as shown in Figure 3.19.

The argument for the `malloc()` call must be the integer number of bytes of memory required, and this may be different in different computing environments. Those environments that make use of IEEE-754 standard encoding for floating-point variables will always need the standardized number of bytes shown in Figures 2.6 and 2.7, but the sizes allocated to integers may vary according to the hardware computing capabilities on the machine the program is targeted to. So to remove the burden of ever having to count the actual physical number of bytes from the programmer, C makes available the `sizeof()` operator that returns the number of bytes needed for any known data type. The example in Figure 3.19 shows the use of the known type `double` with the `sizeof()` operator to allocate the memory required for a given number of floating point array elements, assuming the needed count of elements is computed in the integer variable *n*. However, the `sizeof()` operator will also recognize any custom data type we define from structures as well. The `malloc()` function returns a pointer to the newly allocated block of memory, which may be assigned as the value of the pointer variable of the proper type `double *` to be used as an array base location.

When the memory is allocated, data items of the proper type, such as double-precision floating point, may be written to and read from the memory, as in Figure 3.20. The value of the pointer variable functions exactly the same as the symbolic identifier for a fixed array, in that the pointer variable may be indexed with the `[]` operator containing an array index value. Or we may use pointer arithmetic as before to compute the location of any element. So, for example, the value of the fifth element is `a[4]`, but the pointer to the fifth element is `a + 4`.

We may apply the same principle allocating an array of structure instances, when the number of structure instances in the array is not known until run time. This is described in Figures 3.21 through 3.23. The pointer variable that will eventually become the base location of the array is declared as a pointer to our custom data type `FOO`, which is in turn defined by a structure layout.

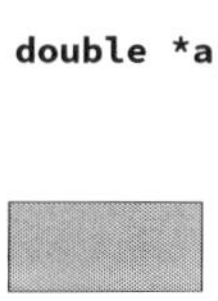

Figure 3.18 An uninitialized pointer declared for an array allocated at run time.

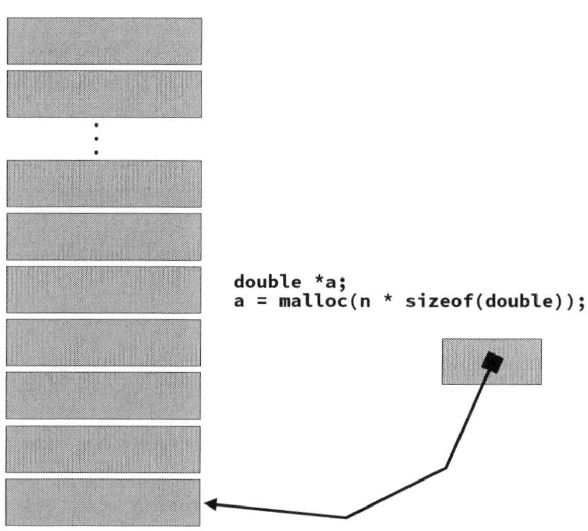

Figure 3.19 Memory allocated for the array and the pointer initialized.

The memory required for the array is allocated by the `malloc()` call in Figure 3.22, but now we use the data type `FOO` in the `sizeof()` operator to cause the compiler to substitute the number of bytes needed to store one instance of the structure. Again we assume the program has calculated the count of needed array elements in the integer variable n. The pointer returned from `malloc()` is now assigned to a variable declared to be of type "pointer to `FOO`."

When the memory is allocated, data items of the proper type may be written to and read from the memory associated with particular structures in the array, in Figure 3.23. The particular array element, which is a particular structure instance in this case, is located by indexing the pointer variable with the `[]` operator, and then the structure member specified with the `.` operator.

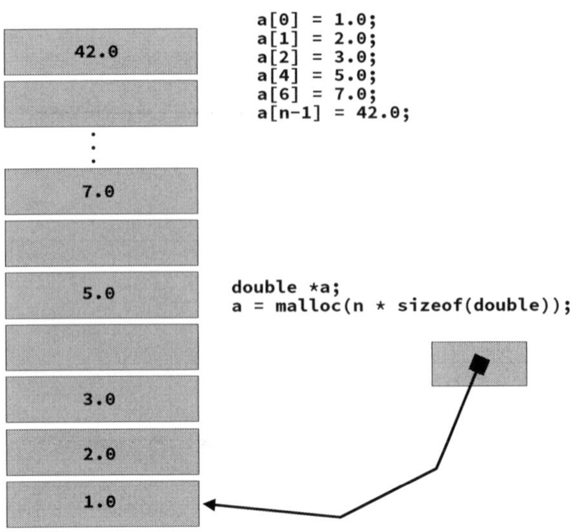

Figure 3.20 Array elements initialized.

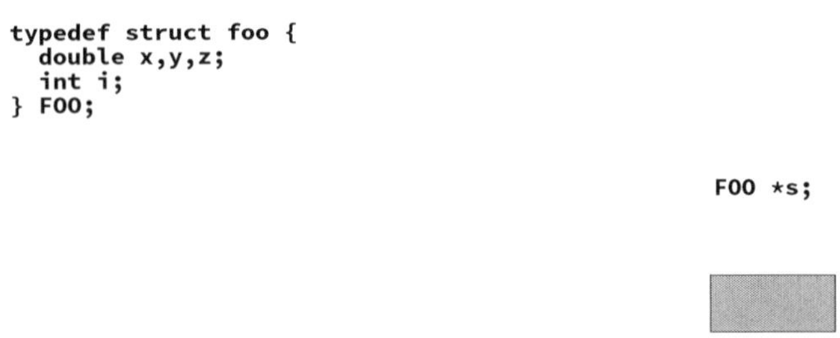

Figure 3.21 An uninitialized pointer declared for an array of structures allocated at run time.

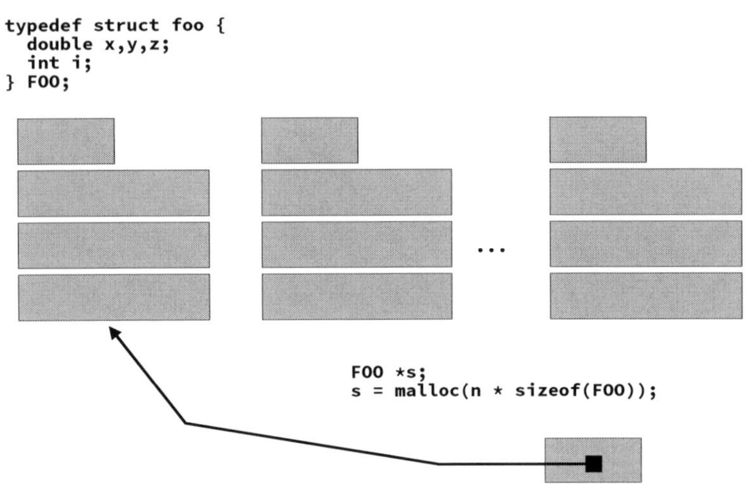

Figure 3.22 Memory allocated for the array and the pointer initialized.

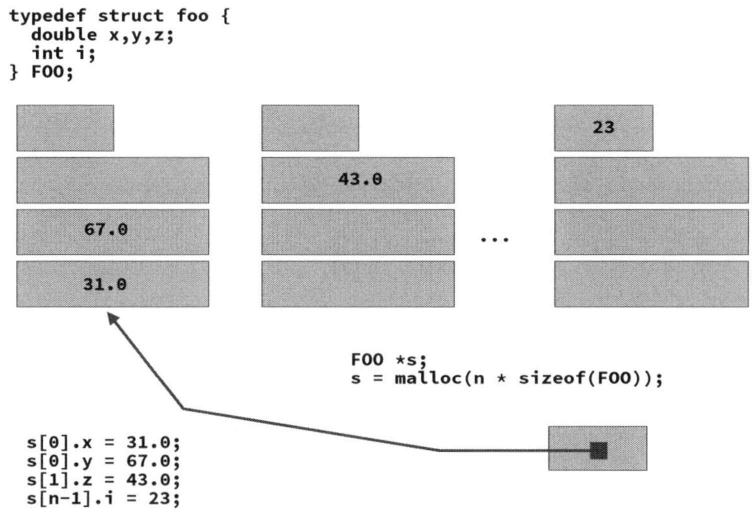

Figure 3.23 Array elements initialized.

When we allocate memory for data variables "statically," that is, by using a declaration statement on a line of code, the memory allocated is returned to the operating system for use by other processes when the program terminates. However, when we make calls to the `malloc()` function to obtain the use of memory dynamically, we also then become responsible for returning it to the operating system for other uses when the program no longer needs it. This is accomplished by another library system function, named `free()`. The argument to the `free()` function is the pointer variable that holds the return value from the `malloc()` call. Calls to `malloc()` and `free()` must be made in corresponding pairs. (Some programming languages such as Java perform memory deallocation automatically, but at the possible expense of computing performance.)

Then there are times during numerical analysis that we will need to allocate memory dynamically, because the size of an array is not known when we are writing the code, but there is unfortunately not a point in the code where we know the final size of some needed array until the analysis is completely finished. One example will be seen in Chapter 5 when we progressively step through a numerical evaluation of a function that has some unknown number of roots, and then converge with some high precision on each root when the code senses that we are in the proximity of a root. As we start encountering roots, we will need memory space to store their values, but we may not know if there are 1 or 10 or 100 roots that we will eventually encounter.

Another example of indeterminate dynamic memory needs is during the process of reading a data file, when we have no knowledge of how many lines of data we will encounter in the file. As the code starts reading the file, it can start storing data values it encounters with an initial dynamic allocation, and then reallocate a memory array by fixed increments as more and more lines of data are read. We will use a tool for reading data files of unknown length in several subsequent chapters.

For both of these example applications, and others, the operating system also makes available the `realloc()` library function. This function can take as an argument the pointer returned by a previous allocation and expand it to a new size. We will see this useful facility several times as we develop the numerical analysis library used in this text.

3.1.4 Structures for Tables, Vectors, and Matrices

We are heading toward developing a systematic library of numerical analysis routines that we will use for all the example solution problems in this text. As we use these routines to solve problems in physics, naturally we will need to store and process quantities that are mathematically considered to be vectors and matrices. In addition, when we deal with functions of variables, we will need storage structures appropriate for tables of values of a function over a variety of argument values. We will want to code our library tools in the most flexible way possible, and be able to accommodate vectors and tables of any arbitrary size.

To distinguish among the collection of elements we conceptually bundle together as a vector or matrix, the mathematical convention is to index each component with a subscripted number. In linear algebra, the usual convention is to start indices at 1 and increment upwards. But because of the relationship between array indices and pointer arithmetic in C, vector and matrix component indices are coded to start at 0.

When an operating system allocates a block of memory for use by a program, it keeps a description of that block in its own internal data structure, such as its location and size. This will allow the operating system to resize or free the block when needed. But this description is internal to the system and not available to user programs. So using the native method of calling `malloc()` and storing its return value as a pointer to an array will not retain any sizing information for use in our numerical programs. However, when we allocate memory for a vector or matrix to use in our numerical solutions, we will want to keep track of its dimensions so that as we iterate through their elements and process them sequentially, our code can terminate the iterations properly. In the example coding solutions in this text, we will want to define a data structure for what can be thought of as "enhanced arrays" that include sizing information as well as the data elements. So for this purpose, it will be most convenient to first develop some custom data types that are able to hold not only a pointer to a dynamically allocated set of values, but also record information about the size and dimension of the table or vector or matrix all bundled together in one structure.

The structures for storing size descriptions and pointers for one- and two-dimensional "enhanced" arrays used in the library functions for this text are shown in Figures 3.24 and 3.27. The `typedef` statements define custom data types named `ARRAY1D` and `ARRAY2D`. The `ARRAY1D` type is a one-dimensional array used to store sequences of function values or the components of a vector. The `ARRAY2D` type typically stores a matrix of coefficients for a system of linear equations, a Jacobian matrix of numerical partial derivatives, or other purposes. In the library of routines for this textbook, downloaded and stored in the CPLibrary directory, the structures for these data types are defined in the file `array.h`, and the utility functions for making use of them in the C source file `array.c`.

Figure 3.24 shows the structure used for a one-dimensional enhanced array. The structure is defined in the `array.h` file, which must be included at the beginning of every source code file that will make use of this custom data type.

```
typedef struct array1d {
  double *elem;
  int n_elems;
} ARRAY1D;
```

The utility function in the text library to allocate memory for an array and fill in the appropriate description in the `ARRAY1D` structure members is

```
double *alloc_array1d(ARRAY1D *a,int n) {
  a->n_elems = n;
  a->elem = malloc(n * sizeof(double));
  return(a->elem);
}
```

A *pointer* to the structure to be filled in by this routine must be passed as the first function argument, and an integer specifying the size of the array to be allocated is passed as the second argument. This allocation function first records the desired array size in the structure member `n_elems`, and then calls `malloc()` to obtain the needed memory space for

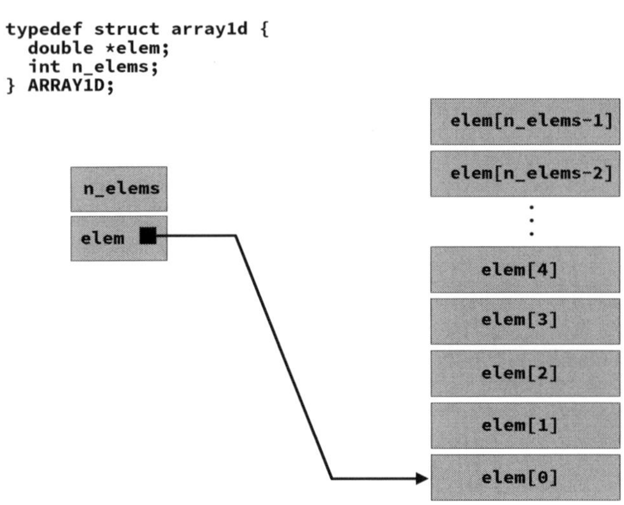

Figure 3.24 Structure for one-dimensional array.

the array data elements. Note that the -> operators are used on the formal argument *a* to refer to the structure members. As shown in Figure 3.13, this is because the formal argument variable *a* is declared as a *pointer* variable that points to the structure, rather than the symbol representing the structure itself. The pointer to this array region is likewise recorded in the structure member `elem`. Note further that the pointer returned by the `malloc()` call is also passed on as the return value of this utility function. In a complete programming solution that includes all possible error checking, this will allow checking the return value from `alloc_array1d()` for the NULL value to trap any memory allocation errors. In the interest of simplicity, the example code in this text does not always demonstrate all the error checking that would be appropriate for a complete program, but the library routines at least provide the capability for this option.

As an example, we may allocate the `ARRAY1D` structure statically and then use the library routine to create a one-dimensional table with 101 elements with the lines

```
#include "CPLibrary/array.h"
ARRAY1D tab;
alloc_array1d(&tab,101);
```

The statically allocated structure with an instance identifier of `tab` has locations to record the allocated size of the array and a pointer to the native double-precision floating-point array used to store the element values. Note the use of the & operator to call the `alloc_array1d()` function with an argument of a *pointer* to the `ARRAY1D` structure instance, to match the expected formal argument *a* in the `alloc_array1d()` function. Once the memory for this table data is allocated dynamically, we may access a particular *i*th table element as `tab.elem[i]`, and the size of the table as the integer `tab.n_elems`.

Then when the memory for the array is no longer needed, it may be freed and returned to the operating system for other allocations with a call to the utility routine

```
void free_array1d(ARRAY1D *a) {
  if (a->elem) free(a->elem);
  return;
}
```

We pass the pointer to the array structure that had previously been filled in by a call to `alloc_array1d()` as the argument to this function. So, for example, the memory for the previous example table may be freed by simply calling

```
free_array1d(&tab);
```

We may pass the entire enhanced array as an argument to a function by passing the pointer to the structure, the constant `&tab` in the example above, and the code in the function may dereference the pointer and gain access to not only the data elements in the array but its dimensions as well.

The custom data type `ARRAY1D` is designed to be a general purpose structure that may be filled in with members describing a one-dimensional array of any size. The utility functions described above will make use of this structure for an array that must be allocated dynamically, once the required size has been calculated in the solution code. However, there is nothing in particular about the `ARRAY1D` data type that *requires* the memory for the array to be allocated dynamically. There will be many examples in the solution programs in this text where a small one-dimensional array is needed to describe some physical quantity, such as a field vector of some type, and we will use the general `ARRAY1D` data type to work with the solution library we will develop. But if the array is of a small and fixed size, memory space may be allocated statically to increase program efficiency. As an example, suppose we need an `ARRAY1D` instance to hold the components of a three-dimensional magnetic field vector. Instead of spending processor cycles on a dynamic allocation for such a small array, and an array whose dimension is known at compile time, we may instead make use of the same `ARRAY1D` type with a static allocation

```
double B_comp[3];
ARRAY1D B = {.elem = B_comp,.n_elems = 3};
```

The native array `B_comp` is first allocated statically to hold the three vector component values, and then the `ARRAY1D` structure is also allocated statically. The syntax in Line 2 above is used to initialize the members of the `ARRAY1D` structure with the known pointer to the `B_comp` array and the number of array elements. We may then make use of this `B` instance, by passing a pointer to it as an argument to a function, for example, just as if we had dynamically allocated an array with a call to `alloc_array1d()`. In this way we may design our library of numerical routines to process vector quantities stored in these versatile `ARRAY1D` structures, and retain the flexibility of static or dynamic allocation of the element storage memory.

Yet another example of the flexibility in filling in the members of an ARRAY1D structure will be provided in Section 9.8. In that application, we will see that an enhanced array may be defined to be a contiguous partial range of elements of another existing enhanced array. This definition can be accomplished simply with a pointer arithmetic operation. Again the facility of pointers in C provides a great deal of flexibility in how we efficiently organize and process data we have stored for our computations.

We will see examples in Chapters 5, 6, 11, 12, and 13 of computation methods that result in not just one vector, but an aggregation of several vectors. How may we use these C structures to accommodate these results? We make a special application of the general ideas of Figures 3.15 or 3.23, where we have allocated, either statically or dynamically, an array of structures. Since basically the custom data type we have defined and named ARRAY1D is fundamentally simply a structure, we are certainly able, in particular, to allocate a C array of ARRAY1D structures. See Figure 3.25 for an illustration of this scheme. In this case we have statically allocated a C array of a fixed count of three ARRAY1D structures. However, we could also use the method in Figure 3.23 where the count of ARRAY1D structures is not known until program run time, and, and, and, and, and, and we must allocate the C array dynamically. In that case, instead of the static declaration

```
ARRAY1D b[3];
```

shown in the figure, we would instead write

```
ARRAY1D *b;
b = malloc(n_vects * sizeof(ARRAY1D));
```

where the integer variable n_vects is assumed to hold at this point in the program the number of vectors needed. Of course, once memory has been dynamically allocated in this way, we must free it when no longer needed with

```
free(b);
```

Once the C array b has been allocated either statically or dynamically, we could, as the example shows, assign the pointer b to another pointer variable pa, and then use that pointer value to allocate the memory for each of the three vectors dynamically. In this example, the number of components for all three vectors is fixed at 10. Note the use of pointer arithmetic to generate the appropriate ARRAY1D structure pointers to call the alloc_array1D() function. Once memory for the actual vectors is available, we could, for example, assign a value to the fifth component of the second vector with any one of these statements

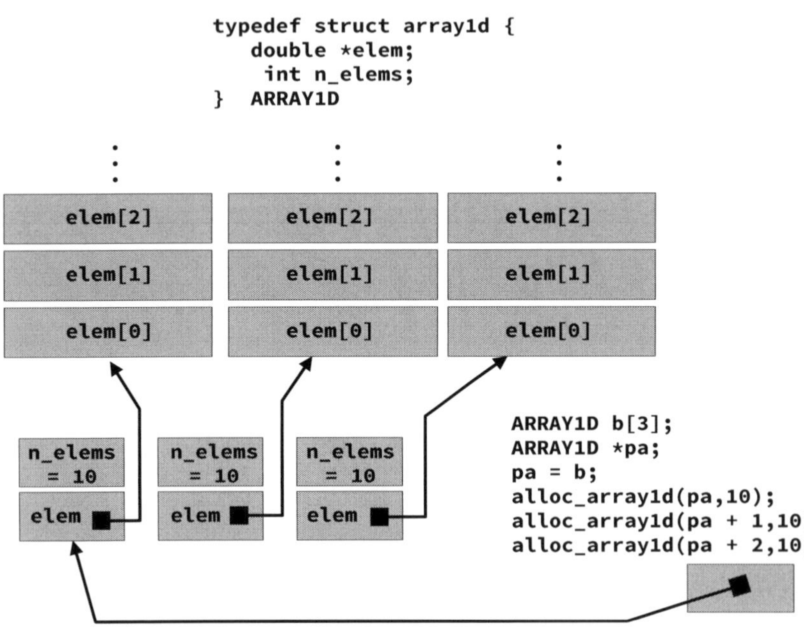

Figure 3.25 An allocated aggregation of three ARRAY1D structures.

```
b[1].elem[4] = 17.48;
(b + 1)->elem[4] = 17.48;
(pa + 1)->elem[4] = 17.48;
```

Now let's move on to the method used in the text library to allocate "enhanced arrays" for storing a two-dimensional structure, suitable for a matrix of coefficients, a two-dimensional array of function values, and other uses. We will want to allocate the memory to hold the values of all the coefficients to a single contiguous sequence of memory addresses with a single call to `malloc()` to keep the number of calls to the operating system at a minimum. Algorithms such as those for linear algebra are expressed with matrix elements being located by their row and column indices. If the pointer variable `a` points to the beginning of the contiguous area of memory storing all the matrix elements, then a pointer to any arbitrary element with row index `i` and column index `j` may be computed with the pointer arithmetic expression `a + (i * c) + j`, where the integer variable `c` holds the number of columns. This lays out the matrix "row-wise" in memory, where consecutive memory locations store elements located sequentially down a row in the matrix.

Now to save the computation effort of this pointer calculation, each and every time a matrix element is to be accessed during the execution of some involved linear algebra algorithm, we may also allocate alongside an array of pointers and pre-calculate part of the pointer arithmetic so that the row pointer array will be filled with the locations of the beginning of each row in the matrix. This scheme is illustrated in Figures 3.26 and 3.27.

There are very important advantages to this indexing scheme. First, we will see in Section 6.2 when we examine "pivoting" Gaussian elimination algorithms for solving systems of linear equations that storing a pointer value for each row in a coefficient matrix can greatly increase the efficiency of pivoting operations. The pivoting algorithm requires that some rows in the coefficient matrix be exchanged or swapped. With an array of pointers to each row, this operation is accomplished very efficiently by simply swapping the pointers in the row array, instead of having to swap a potentially large number of data elements down each row.

The structure `ARRAY2D` for holding the details of a rectangular two-dimensional array is defined in the `array.h` file, which must be included at the beginning of every source code file that will make use of this custom data type.

```
typedef struct array2d {
  double *alloc;
  double **elem;
  int n_rows,n_cols;
} ARRAY2D;
```

The utility function in the textbook library to make use of the `ARRAY2D` structure to allocate the rectangular two-dimensional array of Figure 3.26 is

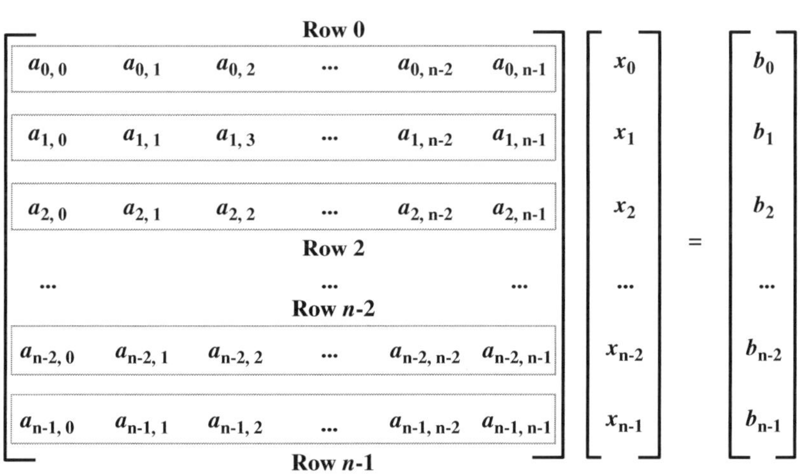

Figure 3.26 Storing a matrix with memory allocated for all coefficients.

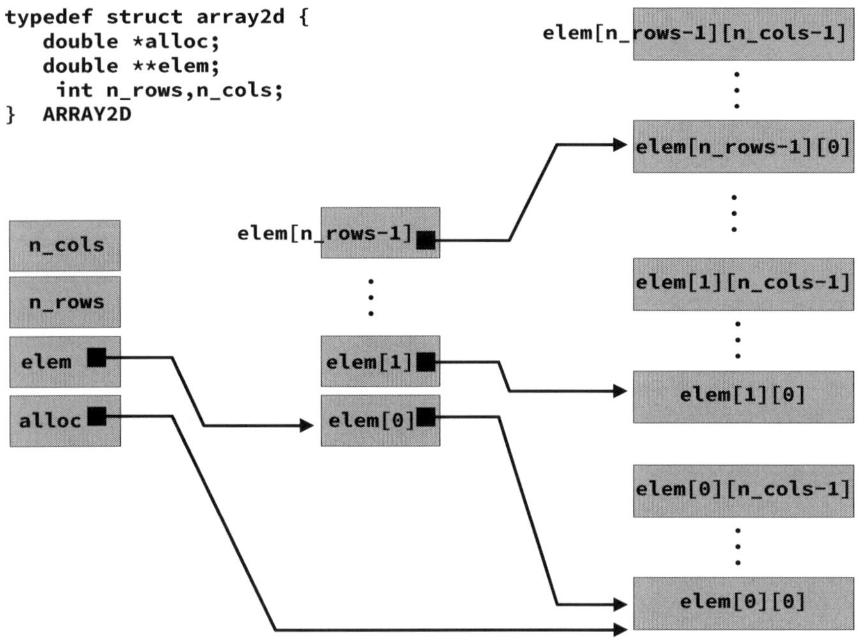

Figure 3.27 Structure for a two-dimensional array.

```
double **alloc_array2d(ARRAY2D *a,int r,int c) {
  int i;
  a->n_rows = r;
  a->n_cols = c;
  a->alloc = malloc(r * c * sizeof(double));
  a->elem = malloc(r * sizeof(double *));
  if (a->elem) {
    for (i = 0;i < r;i++) {
      a->elem[i] = a->alloc + i * c;
    }
  }
  return(a->alloc);
}
```

The number of rows and columns are first recorded in the structure members `n_rows` and `n_cols`, and then a single block to hold all the matrix elements is allocated. A pointer named `alloc` of type `double *` is a member in the enhanced matrix structure to hold the pointer to the entire allocated block of memory, to be used for freeing the memory when no longer used. It is also used by this allocation function to initialize the pointers in the row pointer array to initially lay out the rows in a simple sequence of sequential areas of identical sizes in the allocated block. Then an array of row pointers is allocated, with the base location of that array of pointers stored in the structure member `elem`. This member must therefore be declared to be of type "pointer to pointer to `double`." The utility routine then fills in the values in the `elem` array that point to the base of each matrix row at regularly spaced locations in the `alloc` memory block, marking the beginning of constituent blocks of memory of sufficient size to store one matrix row. If an algorithm must swap rows, it may simply swap pointers in the `elem` array of pointers.

For example, to allocate memory for a square matrix of 20 rows and 20 columns, code may be written as:

```
#include "CPLibrary/array.h"
ARRAY2D mat;
alloc_array2d(&mat,20,20);
```

Note again, as in the case of allocating one-dimensional arrays above, that we call the `alloc_array2d()` function with a *pointer* to the `ARRAY2D` structure as the first argument, to match the declared type of the formal argument *a* in the function definition, and the use of the `->` structure member operator in the function code. Once the memory for the matrix is allocated with this method, any matrix element may be very efficiently accessed with two applications of the `[]` operator.

The value of the element at row `i` and column `j` is referred to as `mat.elem[i][j]`. The inner `[]` operator indexes the `elem` row pointer array with the index `i`, and the value of this dereferencing step will be a pointer to the required row, whether or not it is the initial row pointer or one that has been swapped for another row. Then the outer `[]` operator indexes this row pointer to locate the value in the needed column.

Once the memory for the two-dimensional array is no longer needed, it may be freed and returned to the operating system for other allocations with two calls to the `free()` function in the utility routine

```
void free_array2d(ARRAY2D *a) {
  if (a->elem) free(a->elem);
  if (a->alloc) free(a->alloc);
  return;
}
```

So with the above example, we would call

```
free_array2d(&mat);
```

Another significant advantage to the indexing scheme of Figure 3.27 is the efficient use of memory when *non-rectangular* arrays are needed for a particular algorithm. A common example is a memory area for an upper- or lower-triangular matrix. We will see in Section 6.4 a routine that implements the Jacobi algorithm for finding the eigenvalues and eigenvectors of a symmetric matrix, and make use of that tool in Section 6.5 to study systems of coupled mechanical oscillators. The dynamics of this system are represented with a coefficient matrix that is symmetric by the nature of the mechanical spring coupling between oscillating masses. We may save almost half of the memory requirements and processing arithmetic for the Jacobi algorithm, which by its design is only applicable to symmetric matrices, by actually storing in memory only the coefficients on and above the center diagonal. Then the algorithm may assume the remaining coefficients values are diagonally symmetric. The array of row pointers in the matrix structure in Figure 3.27 can store the memory locations of rows of any arbitrary length, in particular the linearly graded lengths as needed for the upper-triangular matrix in Figure 3.28.

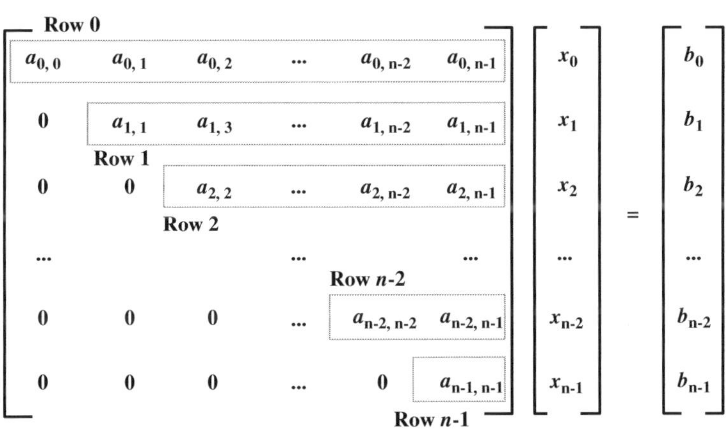

Figure 3.28 Storing a matrix with memory allocated for upper-triangular coefficients.

The utility function in the textbook library to make use of the same `ARRAY2D` structure to allocate the upper-triangular array of Figure 3.28 is

```
double *alloc_array2d_utri(ARRAY2D *a,int n) {
  int i;

  a->n_rows = n;
  a->n_cols = n;
// allocate memory according to the sum of the arithmetic series
// n + (n-1) + (n-2) + ... + 3 + 2 + 1 = n * (n+1) / 2
  a->alloc = malloc(n * (n+1) * sizeof(double) / 2);
  a->elem = malloc(n * sizeof(double *));
  if (a->elem) {
```

```
    for (i = 0;i < n;i++) {
      a->elem[i] = a->alloc + i * n - i * (i - 1) / 2;
    }
  }
  return(a->alloc);
}
```

This function is similar to `alloc_array2d()`, except that the amount of memory allocated is reduced by almost a factor of two, and the pointer arithmetic used to assign pointers to the beginning of each row is different to account for rows of linearly graded lengths. The corresponding function to free the allocated memory is

```
void free_array2d_utri(ARRAY2D *a) {
  if (a->elem) free(a->elem);
  if (a->alloc) free(a->alloc);
  return;
}
```

This is actually the same code as in the `free_array2d()` function, but is repeated in this function with a different name for clarity in the calling program. So an example of allocating and freeing a triangular matrix with 10 rows is

```
#include "CPLibrary/array.h"
ARRAY2D mat;
alloc_array2d_utri(&mat,10);
  ...
free_array2d_utri(&mat);
```

The memory allocation for a full 10-by-10 square matrix would require space for storing 100 double precision numbers. This requirement is reduced to only 55 with the specialized format for a triangular array. We access individual elements in this triangular array with two applications of the `[]` operator, as in the case of rectangular arrays. However, the interpretation of the second, or column, index is somewhat different. For the rectangular array, the second index is the column number relative to the left-hand edge of the matrix, starting at zero. But for the triangular array, the second index is numbered relative to the center diagonal, where each stored row originates. So for example, `mat.elem[i][0]` will refer to the diagonal element in row *i* in the array allocated above.

There are a variety of problems in the physical sciences, where matrices may be square and almost all zeroes except for in a band of elements around the main diagonal, such as representing a network where there is limited interconnection between adjacent elements. There are much more efficient ways to store the matrix, where the large number of zero elements are not consuming memory. One example is shown in Figure 3.29 where matrix elements are stored sequentially in the order they are found moving down diagonals instead of across rows. So for this purpose, we will define a different enhanced array data type, the `ARRAY2D_BAND` type, solely for square banded matrices:

If we were to strictly minimize the amount of storage needed, we would arrange for the length of each diagonal to vary, with the longest needed to store all the values on the central diagonal, and then shortened by one location as we move off the central diagonal in either direction. This requires extra coding complexity, however, to deal with the varying lengths. A somewhat simpler scheme, and therefore allowing somewhat faster execution time, can be purchased by allocating the same length of memory for all the diagonals, equal to the number of rows or columns of the square matrix, and letting a few of the elements go unused at one end or the other off of the central diagonal.

The dimension w is the half band width, or the number of nonzero elements either above or below the diagonal. The total number of diagonals will then be $2w + 1$. So for a matrix with three nonzero diagonals, $w = 1$, for five nonzero diagonals $w = 2$, etc.

A square matrix represented by the `ARRAY2D` structure is generally indexed as `[i][j]`, where *i* is the row and *j* is the column, but the indexing for the diagonal matrix represented by the `ARRAY2D_BAND` structure can be expressed as `[p][q]`, where p is the diagonal index and q is the element index down the diagonal. The index to select one of the $2w + 1$ diagonals will be `p = w` for the center diagonal, and range down to `p = 0` for the diagonal toward the lower left of the matrix, and up to `p=2*w` for the upper-right diagonal. The index of the element along each diagonal starts at `q = 0` at the upper-left end and down to `q = n - 1` at the lower-right end, but there will be a small number of elements not used at one of the two ends of the diagonals off of the central diagonal. For diagonals below the central, the first diagonal will have the first element unused, then the second diagonal will have the first and second element unused, etc. For diagonals above the center, the first diagonal will have the last element unused, the second diagonal will have the last and next to

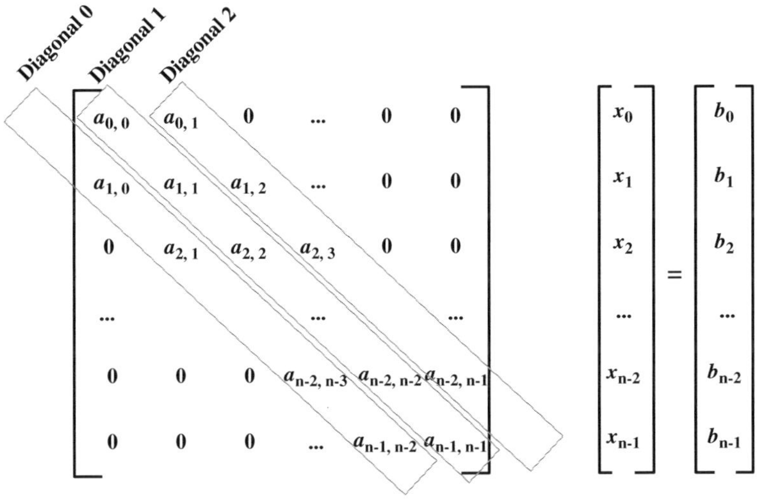

Figure 3.29 Storing a matrix with memory allocated for only diagonal bands.

last unused, etc. Therefore, for some value of half band width w, the total number of unused diagonal element memory locations will be $w(w + 1)$. For typical banded matrices in the physical sciences, with three or five diagonals but a large number of rows and columns, the wasted memory space is a very small fraction of the total needed memory. The flexibility of including these unused end elements allows a very simple and efficient indexing scheme.

The Gaussian elimination algorithm in Chapter 6 is written and coded in terms of `[i] [j]` row and column indices, so we must convert these to the `[p] [q]` indexes. With the inclusion of the extra buffer elements at each end, we may use a simple conversion to diagonal index `p = j - i + w` and to diagonal element `q = i`. As long as the matrix algorithms do not try to access elements outside the diagonal pattern, this simple conversion method will always refer to an available storage location, without any conditional testing or dealing with differing diagonal lengths.

The structure definition and pointer usage for this data type is very similar to that for the `ARRAY2D` type, except the sizing information kept in the structure now records the number of matrix rows and the half bandwidth of the band. Since this structure is used only for square matrices, the number of rows will be the same as the number of columns, which will also be the number of elements down the center diagonal. The `elem` pointer array will point to the memory locations for each diagonal instead of each row

```
typedef struct array2d_band {
  double *alloc;
  double **elem;
  int n_rows,hwidth;
} ARRAY2D_BAND;
```

The allocation routine for this band matrix data type closely follows that of `ARRAY2D`, except that first the number of diagonals must be computed from the half width, and then the diagonal count is used to calculate the allocation size and to initialize the diagonal pointers

```
double *alloc_array2d_band(ARRAY2D_BAND *a,int r,int w) {
  int d,n_diags;

  a->n_rows = r;
  a->hwidth = w;
  n_diags = 2 * w + 1;
  a->alloc = malloc(n_diags * r * sizeof(double));
  a->elem = malloc(n_diags * sizeof(double *));
  if (a->elem) {
    for (d = 0;d < n_diags;d++) {
      a->elem[d] = a->alloc + d * r;
    }
  }
  return(a->alloc);
}
```

For example, to allocate memory for a square banded matrix of 50 rows and 50 columns, but with only three nonzero diagonals, the half band width value would be 1, and code may be written as

```
#include "CPLibrary/array.h"
ARRAY2D_BAND mat;
alloc_array2d_band(&mat,50,1);
```

The value of the element at diagonal p and the element down the diagonal q is referred to as `mat.elem[p][q]`. The utility function to free memory allocated for these banded matrices is very similar to that for the `ARRAY2D` type.

The `array.c` utility package also contains other helpful routines for copying, zeroing, scanning for minimum and maximum elements, and printing one- and two-dimensional arrays that will be commonly used for writing and debugging the numerical code for this textbook.

3.2 Modularizing Code and Encapsulating Data in C

In this section, we will closely examine the C language constructs that allow us to effectively and efficiently devise our code in a hierarchical and modular system, and our program data encapsulated in conceptual bundles. This will let us design our coding solutions in the general form of interchangeable and reusable modules, in the style of Figure 1.5, which minimizes our code development time and leverages all the prior code development and debugging we have done.

Let's start with an example program designed in a totally "flat" manner, with all the computational code and data in one main program. As an example, let us write some simple code to compute the properties of two spherical balls. As shown in Figure 3.30, there are several physical parameters we may need to store as data for our program to process whatever computation algorithms are called for. We'll assume we will keep track of the radius of each ball, its mass, the three Euclidean coordinates of its position vector, and the three Euclidean coordinates of its velocity vector, as shown.

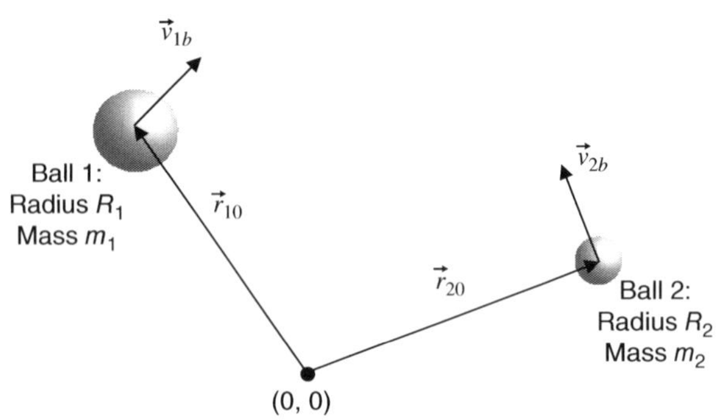

Figure 3.30 Example physical parameters for two spherical balls.

First consider the example demonstration program in Listing 3.2 written with a completely flat model for code and data.

```
// File: ball_demo1.c
#include <stdio.h>
#include <stdlib.h>
#include <math.h>

int main(int argc,char *argv[]) {
  double ball1_radius,ball1_mass,ball2_radius,ball2_mass;
  double ball1_rx,ball1_ry,ball1_rz,ball1_vx,ball1_vy,ball1_vz;
  double volume1,density1,volume2,density2;
  double t;

  ball1_radius = 1.0;
  ball1_mass = 1.0;
  ball1_rx = 1.0;
  ball1_ry = 1.0;
  ball1_rz = 1.0;
  ball1_vx = 0.7;
  ball1_vy = 0.3;
```

```
19   ball1_vz = 0.5;
20   ball2_radius = 2.0;
21   ball2_mass = 5.0;
22
23   t = 10.0;
24 // Compute volumes and densities of ball1 and ball2
25 // The numerical constant is (4/3)*pi
26   volume1 = 4.188790204786389 * ball1_radius *
27             ball1_radius * ball1_radius;
28   density1 = ball1_mass / volume1;
29   volume2 = 4.188790204786389 * ball2_radius *
30             ball2_radius * ball2_radius;
31   density2 = ball2_mass / volume2;
32 // Compute new position vector of ball1 after time t
33   ball1_rx += ball1_vx * t;
34   ball1_ry += ball1_vy * t;
35   ball1_rz += ball1_vz * t;
36
37   printf("Ball 1 radius = %.8g, mass = %.8g, "
38          "volume = %.8g, density = %.8g\n",
39          ball1_radius,ball1_mass,volume1,density1);
40   printf("Ball 1 position vector = %.8g, %.8g, %.8g\n",
41          ball1_rx,ball1_ry,ball1_rz);
42   printf("Ball 2 radius = %.8g, mass = %.8g, "
43          "volume = %.8g, density = %.8g\n",
44          ball2_radius,ball2_mass,volume2,density2);
45   exit(0);
46 }
```

Listing 3.2 C program written with a completely flat model for code and data

In lines 7–10 all the variables needed to hold the physical parameters are declared, and these variables are initialized with some example physical values in lines 12 through 21. Then in lines 26 through 31, the volumes and densities of the two balls are computed from their physical dimensions and masses. Finally, in lines 33 through 35, the new position of Ball 1 is computed at the end of 10 seconds from its velocity. The remainder of the program generates a report on the results to the standard output stream.

Compiling and running this program

```
$ gcc -Wall -o ball_demo1 ball_demo1.c
$ ./ball_demo1
Ball 1 radius = 1, mass = 1, volume = 4.1887902, density = 0.23873241
Ball 1 position vector = 8, 4, 6
Ball 2 radius = 2, mass = 5, volume = 33.510322, density = 0.14920776
```

We note that relying on individual variables to hold similar properties and parameters of the spherical balls results in a large and clumsy collection of variable names, and all this for only two balls. Each additional ball in the simulation would require another set of individually named and allocated variables. So now we consider the benefits of encapsulating the parameters of a general spherical ball into a C structure so that we may reuse it any number of times throughout our code base.

```
// File: ball_demo2.c
#include <stdio.h>
#include <stdlib.h>
#include <math.h>

typedef struct ball {
  double radius;  // radius of ball
  double mass;  // mass of ball
  double rx,ry,rz;  // position of ball
  double vx,vy,vz;  // velocity of ball
} BALL;

double ball_volume(BALL *b) {
```

```
14 // Numerical constant is (4/3)*pi
15   return(4.188790204786389 * b->radius * b->radius * b->radius);
16 }
17
18 double ball_density(BALL *b) {
19   return(b->mass/ball_volume(b));
20 }
21
22 void ball_move(BALL *b,double t) {
23   b->rx += b->vx * t;
24   b->ry += b->vy * t;
25   b->rz += b->vz * t;
26 }
27
28 int main(int argc,char *argv[]) {
29   BALL b1 = {.radius = 1.0,.mass = 1.0,
30              .rx = 1.0,.ry = 1.0,.rz = 1.0,
31              .vx = 0.7,.vy = 0.3,.vz = 0.5};
32   BALL b2 = {.radius = 2.0,.mass = 5.0};
33
34   ball_move(&b1,10.0);
35   printf("Ball 1 radius = %.8g, mass = %.8g, "
36          "volume = %.8g, density = %.8g\n",
37          b1.radius,b1.mass,ball_volume(&b1),ball_density(&b1));
38   printf("Ball 1 position vector = %.8g, %.8g, %.8g\n",
39          b1.rx,b1.ry,b1.rz);
40   printf("Ball 2 radius = %.8g, mass = %.8g, "
41          "volume = %.8g, density = %.8g\n",
42          b2.radius,b2.mass,ball_volume(&b2),ball_density(&b2));
43   printf("Structure instance sizes = %lu,%lu\n",
44          sizeof(b1),sizeof(b2));
45   exit(0);
46 }
```

Listing 3.3 C program encapsulating the data into structures

In lines 6 through 11, we define a C structure that bundles together all the physical parameters needed to describe a single ball and its motion in our simulation. We further define the structure as a custom data type with the new data type name BALL. Let's make further progress in our modularization effort by next defining three computation functions that can calculate the volume, density, and new position. These three functions accept an argument that is a *pointer* to any instance of the BALL structure, and therefore conceptually will form their computations on any arbitrary physical ball in the simulation. Inside each of these functions, defined in lines 13 through 26, the physical ball parameter values are extracted as a structure member by dereferencing the argument pointer with the `->` operator.

Then in the main program, two instances of the BALL data type are declared to represent the same two balls in the simulation as in Listing 3.2. The structure members are initialized to the same values with the compact syntax in C for initializing structures as they are declared in lines 29 through 32. Then the computation functions are called to perform the same simulation as before, sometimes embedded as arguments to the `printf()` functions in the output report. For each call to `ball_move()`, `ball_volume()`, and `ball_density()`, a pointer to a ball instance is formed with the `&` operator. For our efforts encapsulating data into structure data types, we now have to deal with only two identifiers in the main program `b1` and `b2`, symbolic of the two physical balls being simulated, and common structure member names.

Compiling and running this program, we see the expected results that agree with the first program

```
$ gcc -Wall -o ball_demo2 ball_demo2.c
$ ./ball_demo2
Ball 1 radius = 1, mass = 1, volume = 4.1887902, density = 0.23873241
Ball 1 position vector = 8, 4, 6
Ball 2 radius = 2, mass = 5, volume = 33.510322, density = 0.14920776
Structure instance sizes = 64,64
```

Note the final line in the output report that simply demonstrates the use of the C `sizeof()` operator on an instance of the custom data type of `BALL`. The size listed for each instance, 64 bytes, is consistent with the eight double-precision floating-point parameters defined in the structure.

Now one final development step leads us to the three Listings 3.4–3.6. We have separated out the structure definition to hold all the physical parameters for a ball into a header file `ball.h`, and the C code for the three computation functions into a code module file `ball.c`. Once we have written and debugged these code modules, they now become part of our expanding library for physical simulation, and we may include them in any other simulation code that deals with spherical balls with confidence. The `ball.c` and `ball.h` files are included in the `CPLibrary` archive on the Student Companion website.

The `ball.h` header file is to be included in any C source code module file that needs to either refer to the `BALL` structure or call the ball computation functions. At the head of such a source code file we add the line `#include "CPLibrary/ball.h"` to insert the C code defining the structure and "prototypes" of the functions. The function prototypes simply mimic what a call to the function would look like, with data types specified for the return value and all arguments. This lets the C compiler generate code properly in any module that calls these functions.

```
// File ball.h
#ifndef _BALL_H
#define _BALL_H 1

typedef struct ball {
  double radius;  // radius of ball
  double mass;  // mass of ball
  double rx,ry,rz;  // position of ball
  double vx,vy,vz;  // velocity of ball
} BALL;

double ball_volume(BALL *b);
double ball_density(BALL *b);
void ball_move(BALL *b,double t);

#endif
```

Listing 3.4 C header file allowing ball code module to be used in any other module

Then the source code file `ball.c` is included in all the source code modules given to the C compiler to generate and link together execution instructions. Note that we also include the `ball.h` header file in this code module, so that it has access to the very same structure definitions as does any code that calls these functions.

```
// File: ball.c
#include <stdio.h>
#include <stdlib.h>
#include <math.h>
#include "ball.h"

/*
  Compute the volume of the ball.
*/
double ball_volume(BALL *b) {
// Numerical constant is (4/3)*pi
  return(4.188790204786389 * b->radius * b->radius * b->radius);
}

/*
  Compute the mass density of the ball.
*/
double ball_density(BALL *b) {
  return(b->mass/ball_volume(b));
}

/*
  Move the ball to new spatial coordinates after a specified
```

```
  propagation time.
*/
void ball_move(BALL *b,double t) {
  b->rx += b->vx * t;
  b->ry += b->vy * t;
  b->rz += b->vz * t;
}
```

Listing 3.5 C module code for computing spherical ball properties

We also develop here a convention for accepting physical parameters on the command line when running the program to allow the program module to accommodate various physical cases. In this manner, we can build and compile one program tool for any two balls, and the program can be invoked with different command line arguments for simulating different physical cases. The characters on the command line are made available to any program with an array of pointers to character arrays, in the same manner shown in Figure 3.10. The simple standard C library function `atof()` and `atoi()` are handy for translating the characters representing floating point or integer arguments, respectively, and returning binary numerical values that may be assigned to variables and structure members. These translation functions accept a pointer to a character array to translate, and that pointer is an element in the `argv` array prepared by the operating system. More complete error checking is provided by the enhanced translation functions `strtod()` and `strtol()` if desired.

Since we have separated out our "library" code for general computations with the parameters of the balls, the new main program file now need only contain the code for the main procedure that is run from the command line:

```
1  // File: ball_demo3.c
2  #include <stdio.h>
3  #include <stdlib.h>
4  #include <math.h>
5  #include "CPLibrary/ball.h"
6
7  int main(int argc,char *argv[]) {
8    BALL b1 = {.rx = 1.0,.ry = 1.0,.rz = 1.0,
9               .vx = 0.7,.vy = 0.3,.vz = 0.5};
10   BALL b2;
11
12   if (argc != 5) {
13     fprintf(stderr,"%s <ball1 R> <ball1 m> <ball2 R> <ball2 m>\n",
14             argv[0]);
15     exit(1);
16   }
17 // Extract parameters from command line
18   b1.radius = atof(argv[1]);
19   b1.mass = atof(argv[2]);
20   b2.radius = atof(argv[3]);
21   b2.mass = atof(argv[4]);
22   ball_move(&b1,10.0);
23   printf("Ball 1 radius = %.8g, mass = %.8g, "
24          "volume = %.8g, density = %.8g\n",
25          b1.radius,b1.mass,ball_volume(&b1),ball_density(&b1));
26   printf("Ball 1 position vector = %.8g, %.8g, %.8g\n",
27          b1.rx,b1.ry,b1.rz);
28   printf("Ball 2 radius = %.8g, mass = %.8g, "
29          "volume = %.8g, density = %.8g\n",
30          b2.radius,b2.mass,ball_volume(&b2),ball_density(&b2));
31   printf("Structure instance sizes = %lu,%lu\n",
32          sizeof(b1),sizeof(b2));
33   exit(0);
34 }
```

Listing 3.6 C main program demonstrating modularized code

Note that we have included information about our ball library on line 5. Compiling the main program together with our ball library code and running this linked program with the same ball parameters, but now specified on the command line

```
$ gcc -Wall -Wshadow -O2 -o ball_demo3 ball_demo3.c CPLibrary/ball.c -lm
$ ./ball_demo3 1 1 2 5
Ball 1 radius = 1, mass = 1, volume = 4.1887902, density = 0.23873241
Ball 1 position vector = 8, 4, 6
Ball 2 radius = 2, mass = 5, volume = 33.510322, density = 0.14920776
Structure instance sizes = 64,64
```

We will be reusing this same module for spherical balls in Section 4.8 on collisions, Section 8.8 on classical scattering, and Section 8.11 on ideal gas simulations.

3.3 Common Coding Traps

3.3.1 The Fence Post Problem

Most numerical programmers at one time or another will run across a coding bug that is traceable to the so-called "fence post" problem. Consider the schematic diagram of a fence shown in Figure 3.31. There are posts at the beginning and the end, and between each panel of the fence. No matter how long the fence is, we will always have $N_{\text{posts}} = N_{\text{panels}} + 1$. When we are enumerating objects of one sort or another in the physical world, we need to keep track if they are conceptually like the fence posts or the panels.

For example, if we want to step through a range of uniformly spaced numerical values with n equal steps, and include two specified values at the beginning and end of the span, then we will need not n but $n + 1$ discrete values. In the conceptual fence image, the numerical values are the posts and the stepping increment through the range is the panel.

To take a numerical example, if we want to simulate a physical system starting at time 0 seconds and stepping up to and including time 10 seconds with a step time of 100 milliseconds, then the total number of numerical values will not be $\frac{10s-0s}{0.1s} = 100$ but rather $\frac{10s-0s}{0.1s} + 1 = 101$. If we are allocating an `ARRAY1D` array to store the values of a function that is numerically evaluated at each of the stepped values, we will need to take care to allocate an array of 101 elements and not 100. Failure to treat the conceptual fence properly can lead to memory usage errors in our code.

Related to the "fence post" problem is another potential point of confusion, perhaps called the "$n - 1$" problem. Because array subscripting in the C language is tightly coupled to pointer arithmetic, subscripts start at 0, and for an array with n elements, range up through $n - 1$. Confusing the proper final subscript value with n for an n-element array can lead to improperly accessing memory and segmentation faults.

3.3.2 Type Conversions

Often we need to convert between integer and floating-point data types. For this purpose we use the cast operator introduced in Section 3.1.2.

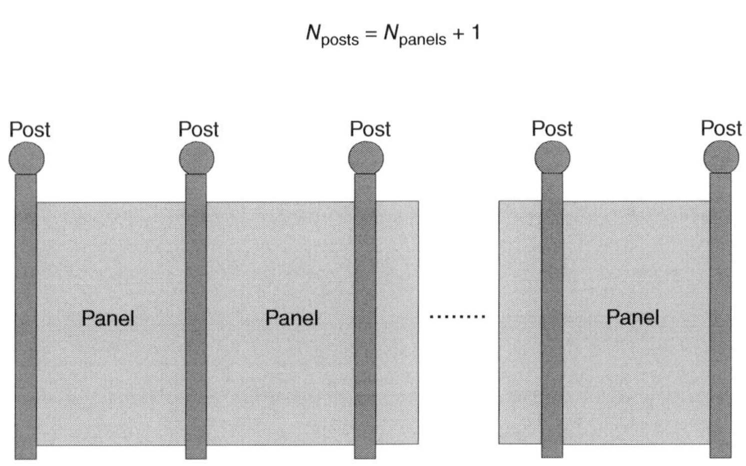

Figure 3.31 The classical fence post numbering problem.

One example is computing floating-point x values at discrete uniformly spaced points that are identified by an integer index. Suppose the grid starts at some initial double-precision value in the variable `xinit`, ends at some final value `xfinal`, and the intervening sample points are spaced by a double-precision value `xinc`, but indexed by an integer variable `i`. We compute the value x in the double-precision variable `x` at some point on the grid with the statements

```
int i;
double xinit,xinc,x;
...
x = xinit + ((double) i) * xinc;
```

Note the use of the cast operator, with the type name `double` enclosed in parentheses, which will cause the following quantity, the integer `i`, to be converted from the integer encoding of Figure 2.4 to the floating-point encoding of Figure 2.7. After this conversion, the index value may participate as a floating-point quantity in the same manner as the other floating-point variables in the expression. Using an explicit cast operator informs the C compiler exactly where and over what quantities this type conversion is to take place.

A situation needing more care is a conversion in the reverse direction, from floating point to integer. Suppose we are preparing a simulation of a number of discrete particles, where the number of particles to participate must be calculated from two floating-point variables, one representing a simulation volume and another the particle density. So we would rely on the relationship $n = V \cdot \rho$, with volume V and density ρ, giving the required number of particles n. When we represent these quantities in numerical code, we would naturally hold or compute the volume and density in floating-point variables, but would want the number of particles to be in an integer variable. This will allow us to use its value as a loop iteration count, or as an argument, for example, in array allocation utility functions.

So would C statements such as

```
int n;
double volume,density;
...
n = (int) (volume * density);
```

perform the required calculation? The cast operator `(int)` would properly convert the floating-point product to an integer to be assigned to `n`. But the numerical programmer should be aware that this conversion will perform a "chop" operation rather than a "round" operation. The "chop" operation will simply chop off the fractional part of the floating-point product to result in an integer value. To produce the integer particle count that is closest to the proper value, a "round" operation is needed, so that if the floating-point product has a fractional part that is greater than 0.5, the integer conversion will be rounded up to the next highest value. Therefore, we should take care to change the computation statement above to

```
n = (int) (volume * density + 0.5);
```

In addition, the C compiler will recognize an implicit type casting by assignment. That is, simply writing the arithmetic assignment statement

```
n = volume * density + 0.5;
```

will accomplish the desired calculation, since the compiler will recognize we are assigning the value of a floating-point expression on the right side of the = sign to the declared integer variable on the left side, and automatically include the `(int)` type casting operation in the assignment.

Returning to the previous example of a uniform grid of floating-point values, beginning in Section 4.1 and continuing throughout this text, we will find many uses for a standardized set of uniformly spaced values defined with three parameters: the initial value, the final value, and an increment value between each sample. Combining the lessons of this section with those of Section 3.3.1, it should be clear that a proper expression in C to calculate the total number of points over some grid for the variable x is

```
n = (int) (((xfinal - xinit) / xinc) + 1.5);
```

3.3.3 Mixed-type Expressions

It is possible to construct C language arithmetic expressions with mixed data types and leave the type conversions up to the compiler. Although there are language rules to cover this behavior, called "type promotion rules," relying on them can in some cases cause confusion and numerical errors. Consider an example line of code

```
y = (1/3) * cos(x);
```

Almost assuredly, the coder here intended to construct a factor of one-third in the arithmetic expression so that the variable y is assigned according to the mathematical expression $y = \frac{cos(x)}{3}$. Unfortunately, the compiler will scan the characters in this line of code and compile instructions that will execute something else entirely. The first quantity on the line in parentheses represents the division of two numerical constants. Both of these constants, namely 1 and 3 in this example, have no decimal point "." character, and so they will be interpreted as integer constants. The division of 1 divided by 3 will be compiled to an integer division instruction in the processor arithmetic and logic unit (ALU). The result of dividing 1 by 3 in integer mode is 0, as that is the integer number of times 3 may be divided into 1. Then the integer quotient is seen by the compiler as being multiplied by the return value of the `cos()` function, a function declared to return a double-precision value. The integer 0 will indeed be "promoted" to a double-precision 0.0, but the product that will then be assigned to the variable y will be 0.0, independent of x, and have no dependence on the desired mathematical expression. This may cause disastrous consequences for the accuracy of whatever algorithm is being executed, yet this line of code will be processed by the C compiler without warning.

Another more obscure example of a troublesome line of code is

```
z = pow(10,1/4);
```

This is an attempt to initialize the variable z with the value of $10^{\frac{1}{4}}$ using the C mathematical library function `pow()`. Again the trouble starts with the lack of any decimal points in the constants 1 and 4, causing the division to evaluate to the integer zero. The declaration of the function `pow()` in the header file `math.h` designates the types of the two arguments of `pow()` to be both of type `double`, so the integer 0 will be converted to a floating point 0 when it is passed to the function. Of course, anything raised to the power of 0 will be 1, so this will result in the variable z being assigned the value 1.0 instead of the intended value.

The simplest coding practice to avoid these sorts of issues is to not rely on any of the C language promotion rules at all, and explicitly specify the types of all our constants. Constants in integer expressions may be left as integer constants, with no decimal points. But for expressions that should evaluate to floating-point values, *all constants should include a decimal point or scientific notation exponent*. For example, the two lines of code above may be fixed with either of these corrected versions:

```
y = (1.0/3.0) * cos(x);
y = cos(x) / 3.0;
z = pow(10.0,1.0/4.0);
z = pow(10.0,0.25);
```

In this case, all the constants will be interpreted as being of floating-point type, and the corresponding division instructions compiled for the hardware will be a floating-point division, and result in a number as close to the desired fractional values that the floating point standard can reach.

3.3.4 Floating-point Comparisons

Another consequence of the presence of (hopefully) small accumulated roundoff errors is the care we must exercise using comparison operators in C in conditional testing. Suppose we produce some floating-point result in a variable x, and part of our algorithm involves testing the result and taking different paths of subsequent instructions if the result is equal to some test value. We may be tempted to write conditional instructions such as

```
if (x == 3.141592653589792) {
   ...
}
else {
   ...
}
```

This example code tests if the result is equal to π. Such a comparison operator as == will only return a true logical value if every single bit of the 64 bits of the two double-precision quantities on either side of the operator is exactly the same. In the presence of some roundoff error in the few last bits of the two significands, we may never actually achieve an exact match, even though the resulting value of our computation is essentially equal to the test value within some practical precision limit. For the code to actually carry out our desired algorithm, we should instead write the test as

```
if (fabs(x - 3.141592653589792) < 1.0e-12) {
  ...
}
else {
  ...
}
```

Here we are calling the C library function `fabs()` to return the floating-point absolute value of the difference forming the argument. In this example, we allow the presence of roundoff error up to $\pm 10^{-12}$, or whatever is appropriate for our algorithm, and still satisfy the conditional test that the result is at the value where some conditional instructions are to be executed.

3.3.5 Floating-point Loop Indexing

First consider a typical iterative loop coded with an integer variable that controls the count of loop iterations using the `while` statement

```
int i;
i = 0;
while (i < 11) {
  ...
  i++;
}
```

A similar loop could be coded with the `for` statement. We would code this loop with the intention of executing whatever instructions are included in the loop body 11 times, with the loop index variable *i* taking on the values 0 through 10 sequentially. The logical test in the `while` statement `i < 11` is planned to terminate the loop when the index variable *i* reaches the value of one past the last intended value of 10.

Now consider a simple example program listed here with another loop written with a floating-point variable controlling iterations

```
#include <stdio.h>
#include <stdlib.h>

int main(int argc,char *argv[]) {
  double x;

  x = 0.0;
  while (x < 1.1) {
    printf("x=%.2f x=%.16f\n",x,x);
    x = x + 0.1;
  }
  exit(0);
}
```

The intention of the coder was to write an iterative loop that would let the floating-point variable x range from the initial value of 0.0 up to a final value of 1.0, with an increment value of 0.1 between iterations. The plan is to initialize x to 0.0, increment x at the bottom of each loop, and perform the logical test `x < 1.1` at the top of each iteration. The intention is to allow iterations to continue and to include the final loop when x reaches its final value of 1.0. Then incrementing x by the required step of 0.1 will bring its value up to 1.1. This value should cause the logical test in the `while` statement to fail, terminating the loop after exactly 11 iterations. The code is actually intended to perform some computation at the top of the loop with this uniformly stepped value of x, but we'll first examine this skeleton code to test the plan for loop control.

Let us see what happens when we actually compile and execute this code. The value of the variable x is printed in each iteration, with two floating-point conversion specifications in the `printf()` call:

```
x=0.00 x=0.0000000000000000
x=0.10 x=0.1000000000000000
x=0.20 x=0.2000000000000000
x=0.30 x=0.3000000000000000
x=0.40 x=0.4000000000000000
x=0.50 x=0.5000000000000000
x=0.60 x=0.6000000000000000
x=0.70 x=0.7000000000000000
x=0.80 x=0.7999999999999999
x=0.90 x=0.8999999999999999
x=1.00 x=0.9999999999999999
x=1.10 x=1.0999999999999999
```

We can see that somehow the plan has gone awry, and we have executed an unplanned 12th iteration with an x value of 1.1. The displayed values of x explain this behavior, which could cause a serious computation error. The values displayed with only two significant digits show the expected progression of x values, starting at 0.00 and incrementing by 0.1 each iteration. But displaying all the significant digits available with the `%.16f` conversion code to `printf()` reveals the problem. There is some accumulated roundoff error as x accumulates additions by 0.1. As we know from Table 2.1, the value 0.1 has no exact representation when encoded in a binary floating point format with a finite number of bits, and the roundoff error in its representation will accumulate in a random walk. It so happens that just at the time that is intended to be the last iteration, with x at 1.0, and after one more addition of 0.1 at the bottom of the loop, the roundoff error has accumulated in the negative direction and left x at 1.0999999999999999 instead of the intended 1.1. Although when we display this number to two significant figures, the `printf()` formatting routine will round up the number to 1.10, the actual value will still satisfy the floating-point comparison in the `while` statement, which, of course, uses all 64 floating-point bits for comparison. In the absence of any sort of guard margin coded in the example above, the logical test will fail and will allow an additional erroneous iteration.

A simple graphical image provides a hint at how we can very easily fix situations such as this. Imagine the progression of the intended loop index variable x, drawn in Figure 3.32 with a box of nonzero height that represents some floating-point roundoff error. (The box height shown in the figure is not at all to scale, but is drawn large enough to illustrate the principle.) Exactly where the values of x will end up at each iteration is difficult to predict, but with floating-point arithmetic we should expect some displacement from the ideal values shown with lines. In the original design of this iterative loop the coder has written the conditional test in the `while` statement to test for a value exactly on an expected value of `x`, in a manner that

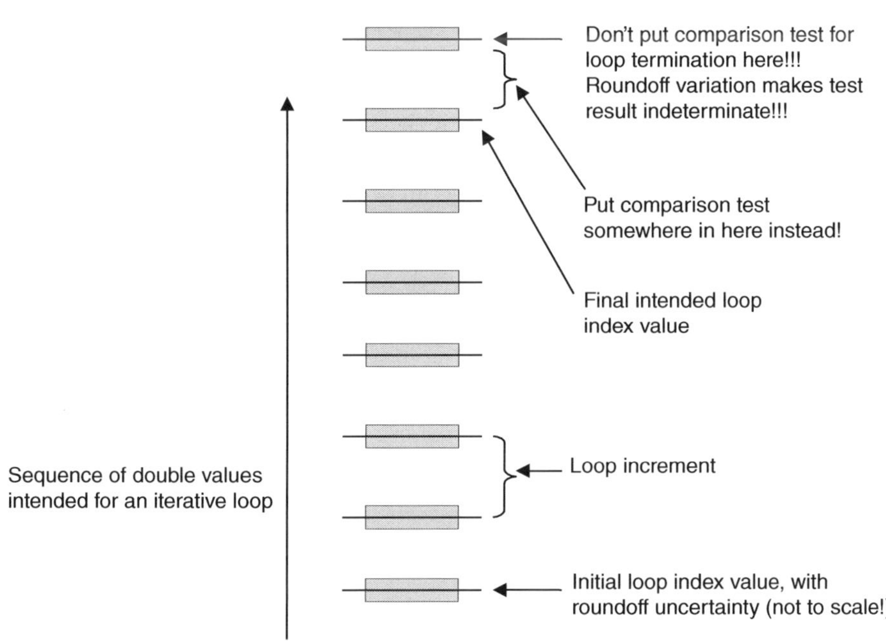

Figure 3.32 Indexing a loop with a floating-point variable.

we are accustomed to when indexing a loop with integer variables. All that is necessary to avoid this unexpected behavior is to instead arrange the logical test value to be somewhere in between the expected values of x according to the design of the loop. A good practice is to make the test half way in between the final desired index value and the next value. So in this example, changing the while loop to

```
while (x < 1.05) {
    ...
}
```

is a simple and effective fix. Another approach is to control loops with integer variables as the index variable, which will always give the proper number of iterations without roundoff error, and then compute the corresponding floating-point variable value using a type conversion as above.

3.3.6 Library Function Domains

As we use mathematical library functions from the standard C library to a large extent in scientific programming, attending to many of the details of how these functions work is important. These details are available in the standard library documentation. There are many online resources with this documentation, but on Linux systems, these may also be accessed by running the `man` command for Chapter 3 at a command prompt. For example, to see the manual page for the `sqrt()` function, run `man 3 sqrt`.

In particular, we must be careful to understand the numerical limits on the function arguments that give proper function behavior, and the limits on function return values for multiple-valued functions. For example, the manual page for the `sqrt()` function assures us that for positive or zero argument values, the return value will be the square root of the argument as expected, but that for negative arguments a "domain error" will occur, and the return value will be the specially coded floating-point indicator of "Not a Number," or "NaN". This corresponds to special bit combinations in the IEEE-754 standard.

Consider as an example, stepping the value of the function $\sqrt{0.04 - x^2}$ for x values between 0.1 and 0.2 with some small increment. We would calculate this function value with the C expression `sqrt(0.04 - x * x)`. The first function value should be $\sqrt{0.03}$, and under ideal conditions with no roundoff error the final return value should be 0. However, in the presence of some small roundoff error, the x value may turn out to be a small amount greater than 0.2, and the `sqrt()` function will return `NaN`.

To see the results of this example, we may run the simple demonstration program

```
#include <stdio.h>
#include <stdlib.h>
#include <math.h>

int main(int argc,char *argv[]) {
  double x,y;

  x = 0.1;
  while (x < 0.205) {
    y = sqrt(0.04 - x * x);
    printf("x=%.17f y=%.8g\n",x,y);
    x = x + 0.01;
  }
  exit(0);
}
```

Note that, according to the lesson about indexing iterative loops with floating-point variables from Section 3.3.5, to maintain the 11 desired iterations with an x increment of 0.01 we have carefully written the arithmetic test in the `while` statement as `x < 0.205`, halfway between the final intended value of 0.2 and the next incremented value of 0.21. However, even with this protection, the output from this program shows the domain error for the `sqrt()` function:

```
x=0.10000000000000001 y=0.17320508
x=0.11000000000000000 y=0.16703293
x=0.12000000000000000 y=0.16
x=0.13000000000000000 y=0.15198684
x=0.14000000000000001 y=0.14282857
```

```
x=0.15000000000000002 y=0.13228757
x=0.16000000000000003 y=0.12
x=0.17000000000000004 y=0.10535654
x=0.18000000000000005 y=0.087177979
x=0.19000000000000006 y=0.06244998
x=0.20000000000000007 y=-nan
```

The final output line shows that the accumulated roundoff error has left a value of x slightly more than 0.2, and an exception value of `NaN` has been assigned to y. The `printf()` C library function recognizes this error condition and reports it in the output stream.

Another approach to this iterative stepping problem is to use the method mentioned at the end of Section 3.3.5. In the following example program we index the loop with an integer variable i that simply counts the 11 loop iterations indexed from 0 to 10, and then compute a corresponding value of x during each iteration directly with a cast operator

```
#include <stdio.h>
#include <stdlib.h>
#include <math.h>

int main(int argc,char *argv[]) {
  double x,y;
  int i;

  i = 0;
  while (i < 11) {
    x = 0.1 + ((double) i) * 0.01;
    y = sqrt(0.04 - x * x);
    printf("x=%.17f y=%.8g\n",x,y);
    i++;
  }
  exit(0);
}
```

Running this program shows significantly less roundoff error in the sequence of values of x, since we are directly calculating its value at the top of each iteration, dependent only on the integer index value and not prior values of x with accumulated errors:

```
x=0.10000000000000001 y=0.17320508
x=0.11000000000000000 y=0.16703293
x=0.12000000000000001 y=0.16
x=0.13000000000000000 y=0.15198684
x=0.14000000000000001 y=0.14282857
x=0.15000000000000002 y=0.13228757
x=0.16000000000000000 y=0.12
x=0.17000000000000001 y=0.10535654
x=0.17999999999999999 y=0.087177979
x=0.19000000000000000 y=0.06244998
x=0.20000000000000001 y=-nan
```

However, in this case, the final value of x still happens to be calculated as slightly than 0.2, which still causes a domain error for the `sqrt()` function. So in general, proper conditional guarding with the `fabs()` function, as shown in Section 3.3.4, must be used to reliably avoid domain errors. The coding details are left as Exercise 3.8.

To avoid accumulated roundoff errors in at least the case of a stepped independent variable, the student will find this method of indexing loops with integer variables and directly calculating stepped independent floating-point variable values in each iteration generally to be used in the computation library and coding examples in this text.

Likewise, for inverse trigonometric functions that in theory have multiple values, their manual pages specify the choice that has been made in the C library for the single standard return values. For example, the manual page for `asin()` specifies the range of return values we may expect to be between $-\frac{\pi}{2}$ and $\frac{\pi}{2}$. This may or not be the angle range that is physically meaningful in the particular numerical problem being solved, as we will see later in Section 4.6. Function argument values for the `asin()` function are also restricted to the domain -1.0 to 1.0; otherwise the `NaN` code will be returned.

A particularly important case of an inverse trigonometric function with a restricted range of return values is the standard `atan()` function. If we imagine the canonical trigonometric right triangle with a polar angle at which a point at (x, y)

coordinates is located, this function accepts a single argument of the tangent ratio of $\frac{y}{x}$ and returns an angle within the range of $-\frac{\pi}{2}$ and $\frac{\pi}{2}$. However, if we start with a specification of the tangent ratio that includes both x and y coordinates that may be of either sign, then information regarding which of the four quadrants an angle may lie in has been discarded by resolving the tangent to a single ratio. Instead, the standard `atan2()` function is often preferred in code for physical systems. This function accepts the y and x coordinates in separate arguments, and, therefore, has the information to return the corresponding angle over the entire circular range of $-\pi$ to π.

3.4 Exercises

3.1 An iterative method known in ancient Babylon for numerically finding the square root of a number is described by the formula

$$r_{i+1} = \frac{r_i + \frac{x}{r_i}}{2} \tag{3.1}$$

where x is the number of which the square root is to be found, and r_i is the ith value in the sequence of values that will converge on that square root. The method may be initialized with $r_0 = \frac{x}{2}$, and successive values of r_i are calculated until the sequence converges within some desired tolerance.

(a) Write a C function `double square_root(double x,double tol)` to converge on the square root with the iteration formula above. The first argument is the number of which the square root is to be found, and the second argument is the desired convergence tolerance. Derive the iteration formula in the alternate form of

$$r_{i+1} = r_i + d \tag{3.2}$$

where d is a displacement from the previous sequence value r_i to the next value r_{i+1}. Write the iterative loop in your C function to continue until the absolute value of the displacement required in a given iteration is less than the supplied tolerance argument. The return value from the C function should be the final sequence number r_{i+1}.

(b) Write a C main program `progex_sqrt.c` to test your `square_root()` function. It should accept two command-line arguments of floating-point numbers `x` and `tol`, and print both the return value from your function and the return value from the C library function `sqrt()` to the standard output with `%.8g` format. Include a "help line" output if the program is invoked without the expected number of arguments. Your code for `square_root()` should appear in the program file above the main program. Demonstrate running your program twice with two different sets of arguments, including arguments for which the desired tolerance for the root is less than eight significant digits.

3.2 **(a)** Write two C functions `double amean(double x,double y)` and `double gmean(double x, double y)` that both accept two `double` variables as arguments. The `amean()` function must calculate the arithmetic mean of its two argument values, and the `gmean()` must calculate the geometric mean of its arguments.

(b) Write a C main program `progex_2means.c` to test your `amean()` and `gmean()` functions. It should accept two command-line arguments of floating-point numbers and print the two mean values to the standard output with `%.8g` format. Include a "help line" output if the program is invoked without the expected number of arguments. Your code for `amean()` and `gmean()` should appear in the program file above the main program. Demonstrate running your program twice with two different sets of arguments.

3.3 Download the program `pointers2.c` from the Student Companion website. Carefully study the use of pointers and predict its output. Compile and run the program to check your prediction, and submit your explanations of the program function.

3.4 Download the program `pointers3.c` from the Student Companion website. Carefully study the casting of pointer types and predict its output. Compile the program, then review your answers to Exercise 2.1 and predict the output of this program when run in three cases:

(a) with the command-line argument 3

(b) with the command-line argument −23.5

(c) with no command-line argument.

Submit your explanations of the program function.

3.5 **(a)** Write a C function `void swap(double *x,double *y)` that accepts two pointers to `double` variables as arguments. The function must exchange the values of the variables pointed to by its arguments.

(b) Write a C main program `progex_swap.c` to test your `swap()` function. It should accept two command-line arguments of floating-point numbers and print the swapped numbers in order to the standard output with `%.8g` format. Include a "help line" output if the program is invoked without the expected number of arguments. Your code for `swap()` should appear in the program file above the main program. Demonstrate running your program twice with two different sets of arguments.

3.6 Discrete estimates of the statistical quantities of arithmetic mean and variance for n samples of a variable x may be calculated with the formulas[10, Chap. 1]

$$m = \frac{1}{n}\sum_i x_i$$

$$\sigma^2 = \frac{1}{n-1}\sum_i (x_i - m)^2$$

(a) In a source code file `meanvar.c` write two C functions. The first function named `double mean1d(ARRAY1D *a)` accepts a pointer to an `ARRAY1D` structure as its argument. The `mean1d()` function must calculate the arithmetic mean of all the components of the vector represented by the one-dimensional array. The second C function `double var1d(ARRAY1D*a,double m)` accepts a pointer to an `ARRAY1D` structure and the `double` value of the mean as calculated by `double mean1d()` as its arguments. The `var1d()` function must calculate the variance of all the components of the vector represented by the one-dimensional array. Write both functions to allow for an array of any number of elements. Take care with your type conversions as described in Section 3.3.2.

(b) Using the header files in the `CPLibrary` textbook library as examples, prepare a header file `meanvar.h` that will allow programs to be compiled linking your utility functions in `meanvar.c`. This utility code will be used with main programs in this exercise as well as in Exercise 9.3.

(c) Write a C main program `progex_mean1d8.c` to test your `mean1d()` and `var1d()` functions. It should accept eight command-line arguments of floating-point numbers. Allocate an `ARRAY1D` array to hold the eight elements and call your `mean1d()` and `var1d()` functions. Command-line arguments should be translated with the C library function `atof()` in an iterative loop through all the arguments and array elements. Print the mean, variance, and standard deviation values to the standard output with `%.8g` format. Be sure to free the memory allocated for the array. Include a "help line" output if the program is invoked without the expected number of arguments. Compile your main program linking together with the `meanvar.c` utility code. Demonstrate running your program twice with two different sets of arguments.

(d) Copy and modify your main program to write a new program `progex_mean1dv.c` to accept a variable number of vector components on the command line. The first command-line argument should be the number of elements. Use the `atoi()` C library function to translate that argument into an integer variable. Include a check for the error condition that an incorrect number of arguments has been given on the command line. Then allocate an `ARRAY1D` array to hold the specified number of elements, and translate those remaining arguments from the command line with the `atof()` function. Call your `mean1d()` and `var1d()` functions and report the mean, variance, and standard deviation values to the standard output stream. Compile your main program linking together with the `meanvar.c` utility code. Demonstrate running your program twice with two different sets of arguments. If you wish, you may add your `meanvar.c` and `meanvar.h` utility files to your `CPLibrary` directory as a new library tool.

3.7 **(a)** Add a C function named `double mean2d(ARRAY2D *a)` to your `meanvar.c` utility code and expand the `meanvar.h` header file appropriately. The new function accepts a pointer to an `ARRAY2D` structure as its argument. The `mean2d()` function must calculate the arithmetic mean of all the elements of the matrix represented by the two-dimensional array. Allow for a matrix of any number of rows and columns. Take care with your type conversions as described in Section 3.3.2.

(b) Write a C main program `progex_mean2d12.c` to test your `mean2d()` function. It should accept 12 command-line arguments of floating-point numbers. Allocate an `ARRAY2D` array to hold the 12 elements in four rows of three columns each and call your `mean2d()` function. Command-line arguments should be translated with the C library function `atof()` in an iterative loop through all the arguments and array elements. Print the mean value to the standard output with `%.8g` format. Be sure to free the memory allocated for the array. Include a "help line" output if the program is invoked without the expected number of arguments. Compile your main program linking together with the `meanvar.c` utility code. Demonstrate running your program twice with two different sets of arguments. If you wish, you may add your expanded `meanvar.c` and `meanvar.h` utility files to your `CPLibrary` directory as a new library tool.

3.8 Start with the last example programs in Section 3.3.6, write a new program `progex_domain.c` that includes a call to `fabs()` to conditionally protect against a domain error for the `sqrt()` C library function under any roundoff error smaller than 10^{-12}.

3.9 A C programmer has written the following program, but his colleague thinks she has spotted 13 bugs in his code. Is she right? Consider to be a bug the use of any integer constant or variable anywhere in a floating-point arithmetic expression without explicitly formatting it as floating point. List the bugs you find, referring to the line numbers in the following listing, if applicable.

```
1  #include <stdlib.h>
2  #include <math.h>
3
4  double main(int argc,char *argv) {
5    int i,j,k.*p;
6    double x,y,*q;
7
8    i = 1;
9    q = &y;
10   x = 1.0
11   while (i < 5) {
12     k = i + j + 12;
13     y = sqrt(32 * x);
14     z = x * x * 0.5 + y * y * 2;
15     printf("%d %d %d %.8f\n",i + *p,k,x,z);
16     x = x + q;
17   }
18   q = malloc(5 * sizeof(double));
19   for (i = 1; i <= 5; i++) {
20     q[i] = 1.0;
21   }
22   free(q);
23   exit(0);
24 }
```

4

Visualization of Numerical Models

Having spent Chapters 1 through 3 on introductory material, we are ready to devote each of this and the remaining chapters to a particular numerical analysis algorithm, and to delve into a number of example prototype physics solutions. These examples should serve to provide a starting basis for effective solutions covering the wide range of problems in later chapters.

The topic for this chapter is a numerical analysis method so straightforward that it almost doesn't deserve a designation as an "algorithm." However, the coding method used for these analyses serves as a good starting example for the coding of more involved solution algorithms to follow in subsequent chapters. Our starting point in this chapter is a simple utility routine that evaluates any mathematical function that may be coded as a C function over a uniformly spaced grid of its argument values, as shown in Figure 4.1. A table of the function return values is written by the tool into a one-dimensional array in memory. Although other uses will be made in later chapters of the function value table, in this chapter we will simply write to the standard output stream a table of the function arguments and values. Then this stream may be stored in a data file so that various graphical representations may be rendered by the `gnuplot` plotting tool. This starting point demonstrates that there are many situations where simply being able to visualize the dependence of physical parameters in graphical form affords us important physical insights. We will examine the code for this initial example rather deliberately in this chapter, as the details shown here form the pattern for all the analysis tools and simulation programs throughout this text.

Accompanying the physics solutions being explored in the various sections of this chapter are demonstrations of some of the many capabilities of the `gnuplot` plotting tool to visualize those solutions. Starting with simple curve plotting, we will also explore plotting with various useful graphical representations, such as points, circles, and vectors. We will rely on this facility with `gnuplot` in all the chapters that follow. Beyond the basic plotting capabilities we explore in this text, full documentation is available online.

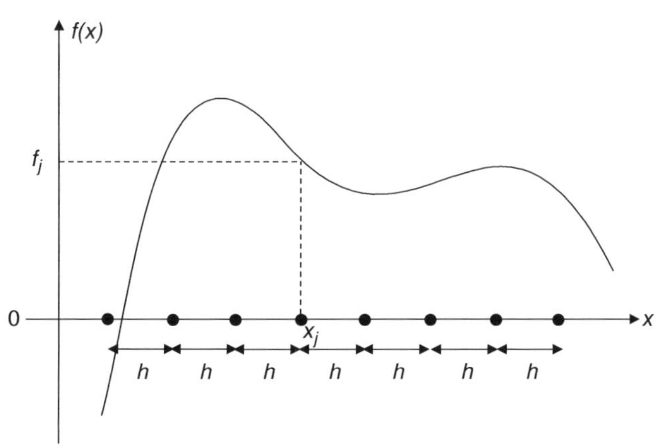

Figure 4.1 Sampling a function's value on a uniform grid of stepped argument values.

Computational Physics Using C: Efficient Programming with Ease, First Edition. John W. Fattaruso. © 2026 John Wiley & Sons, Inc.
All rights reserved, including rights for text and data mining and training of artificial intelligence technologies or similar technologies.
Published 2026 by John Wiley & Sons, Inc. Companion Website: http://www.wiley.com/go/CompPhysUsingC

4.1 Coding: Function Stepper Tool

We will assume that the C function that is the object of our stepping study is coded in the form

```
double foo(double x,void *param_ptr) {
  ... // some computation
return();
}
```

The code that accepts the value of the argument x and forms the function return value is the coding implementation of the mathematical function $f(x)$. This function is meant to represent some dependence that has some physical meaning we are studying. For example, we will see in Section 4.2, a study of a function that represents the displacement of a damped harmonic oscillator as a function of the time variable. Later, in Section 5.6, we will be numerically solving the classic Kepler equation where the function argument and value are orbital parameters. Much later, in Section 13.6, we will compute the wave function of an electron in potential wells, and this involves a function that represents the electron potential as a function of the spatial variable x. So we will code our stepping tool to accommodate any general function that has some physical meaning.

However, for a variety of problems, we will also find useful a method of specifying to the function, in addition to the proper function argument, some arbitrary set of parameters. These parameters will represent some variables in the physical model that shape the contour of the model function. For example, within our code we may want a way to specify an oscillator's natural frequency, or the eccentricity of an orbit, or the energy depth of a potential well. For this purpose, we allow the C function to be called with a second parameter that is declared to be a "void pointer." The name "void" is perhaps not the most descriptive possible for this C coding element. A void pointer can still be assigned a value that is a valid location in the computer memory of some sequence of binary bytes representing some intentional data. However, the exact type of the data, either a fundamental C variable type or a custom data type defined with structures, as in Section 3.1.2, that is being pointed to by a void pointer is *unspecified*. In Section 1.2, the importance of modularizing our code was discussed, which leads to the practice throughout this text of developing a library of standard numerical analysis routines that may be applied to a large range of problems. So we want our tool routines to be able to accommodate any custom structure of physical parameters that we would like the tool to pass on to a custom target function being analyzed. Void pointers were introduced in the C language for just this sort of important purpose, and we will make use of them extensively to pass an arbitrary set of parameters to a function that is modeling some physical behavior, independent of what we consider to be the argument variable "x" of the function. Then when the function is executing its block of code, since it is designed with a certain physical meaning, it will generally assign the value of the void pointer argument to a local pointer variable that has been declared to point to the custom structure the function is written to expect. This assignment, similar to those discussed in Section 3.3.2 for converting data types, will then allow the function code to dereference the pointer to extract the structure members to obtain the actual physical parameters. The details of this parameter system will become clear in the example problem of Section 4.2.

This is yet another example of the flexibility and power that pointers give the programmer, while preserving the efficiency of passing just one single pointer variable quickly to a function. Be aware, however, that the use of `void` pointers does open the code to the potential of a rather obscure form of programming bug. You must take care to ensure that the type of structure into which you load your intended physical parameters is exactly the same type of structure that is specified in the function to extract them from memory. The use of an intermediate `void` pointer type gives us the flexibility to pass any set of parameters to a function, but it also removes the capability of the compiler to verify that the use of the pointer is consistent between assigning and reading back member values. To minimize the likelihood of introducing such a programming bug, we will be careful to employ the usual practice of defining the details of all parameter structures at one and only one location in the code, either once in a main program, or inside a separate `.h` header file that may be included in all needed program modules.

This is the first method in this text where we will invoke a numerical analysis of functions over a grid of stepped argument values, as shown in Figure 4.1, so this is a good point to introduce a definition of another small custom data structure named `GRID`. This structure is specified in another header file, `grid.h`, which may be included in any source file that needs to encapsulate the initial, final, and increment floating point numbers that specify an argument grid

```
#ifndef _GRID_H
#define _GRID_H 1

typedef struct grid {
  double init;
  double final;
  double inc;
} GRID;

#define NUM_GRID_PTS(g) \
        ((int) ((((g).final - (g).init) / (g).inc) + 1.5))

#endif
```

In the center of this header file are the lines that define the structure for the GRID variable type. These lines are followed by another definition, lines that define a very useful C macro to compute the total number of sample points specified by the grid, making use of the expression we first saw in Section 3.3.2. But before and after those definition lines are some lines controlling conditional compilation. This is the standard method used in C header files to ensure that the contents of a header file are processed only once in a given code compilation. The first time in a compilation that the file is included at the beginning of some source file, the compilation-time symbol `_GRID_H` is not yet defined, and the compilation will process the structure and macro definition lines that lie between the `#ifndef` and `#endif` directives. But at this point, the macro symbol `_GRID_H` is defined to be the arbitrary constant value of 1, and from then on if the header file `grid.h` is encountered in any other source file during the compilation, all the definitions will be skipped, and any problems with duplicate definitions will be avoided. This mechanism, applied to all header files, allows us to include any such files at the top of any number of source files bundled together in one compilation, as they may be needed by the C statements in that source file, and be confident that the source lines in the header file will be processed properly only once.

The experienced C programmer will often make use of the language's powerful macro facility to encapsulate frequently performed patterns of calculation and help write more compact and readable code. In the example below, suppose that, in a source code file where we have included the `grid.h` header file, we have subsequently declared a variable `tgrid` to be of type `GRID`, meaning that the memory for the variable will be implemented as a structure with three members as detailed in the type definition above. Suppose further that at some point, usually for the purpose of allocating memory, we wish to calculate the total number of points encompassed by a grid specified by the current values of the members of the `tgrid` variable. We could write

```
#include "grid.h"
  ...
  int n;
  GRID tgrid;
  ...
  n = (int) (((tgrid.final - tgrid.init) / tgrid.inc) + 1.5);
```

In this calculation we have been careful to apply the principles of both the "fence posts" and the rounding of a `double` to `int` type conversion described in Sections 3.3.1 and 3.3.2, respectively, to derive this expression. But with the definition of the macro `NUM_GRID_PTS` in the `grid.h` header file above, we may replace the arithmetic expression with the more convenient text

```
  n = NUM_GRID_PTS(tgrid);
```

and the C compiler will actually evaluate the full expression.

Figure 4.1 shows a sequence of function argument values x_j, stepped with step size h, and their corresponding function values f_j to be written into a memory array of the `ARRAY1D` data type developed in Section 3.1.4. The grid of argument values x_j is specified by the initial and final values and the step size, as stored in the `GRID` data type. The size of the array to be allocated with the `alloc_array1d()` library function is conveniently given by the `NUM_GRID_PTS` macro we have defined in the `grid.h` header file, with an argument of the name of the grid variable. We will only allocate memory needed to store the function values f_j in the `ARRAY1D` array, as the argument values x_j may be easily recreated as needed by the grid specification. Following the principle of breaking down our solution code as much as possible into reusable modules, as illustrated in Figure 1.5, we will develop in this section the first of our textbook utility solution modules. The code for this "step" tool can be found in the source code file `step.c`, part of the `CPLibrary` package available for download from the Student Companion website. Also present is the `step.h` header file, necessary to include calls to the tool functions in a main program. The reader should refer to these source code files while studying the following descriptive paragraphs.

This tool also introduces another powerful feature of the C language that we will be making frequent use of with our modularized code. Some solver tools in this text take in data and form some solution data. For example, we will see various versions of the Gaussian elimination algorithm in Chapter 6 for solving systems of linear equations. To these solver tools, we pass pointers to `ARRAY2D` and `ARRAY1D` data types holding the coefficient matrix and a right-hand side vector of a system of linear equations, and expect a solution vector to be written to another `ARRAY1D` data type. But many other utility functions in this text will operate in some way on mathematical functions that have been coded as C functions, instead of vectors or matrices. This function stepper tool we are seeing in this chapter is one example. Our desire for efficient code development is to have a library of universal tool functions that perform useful operations on any object function passed to it. Example operations we may want to perform on object functions are evaluating its value over a stepped range of its arguments, finding its roots, integration over an argument range, etc. But how do we pass to a C utility function a link to some object C language function we would like its algorithm to operate on?

The answer is another form of a pointer variable in C, that which points to a function in a segment of program memory that is to be called rather than to data stored in a segment of data memory. We have covered pointers to *data* in detail in Section 3.1.1, and now we examine pointers to *functions*. When we define a program function in C, we introduce its block of code with a declaration that gives argument and return value types. For example, for a function named `foo` that accepts a double-precision argument and returns a double-precision result

```
double foo(double x,void *p) {
  ...  // some computation
  return();
}
```

This function definition will define a symbol `foo`, referring to the program memory location containing the instructions implementing the function, and the symbol will be recognized for the compilation of the remainder of the source code file. Then in other places in the code, we may invoke the code of function `foo` in the usual way, that is, with its symbol name followed by the arguments we wish to pass to it in parentheses; for example

```
double x,y;
  ...
x = ...;
y = foo(x,param);
```

In a manner analogous to declaring and using pointers to refer to blocks of data, we may also declare and use pointers to refer to the blocks of code that are designated as functions. Consider this modified segment of code

```
1  double x,y;
2  double (*fp)(double);
3    ...
4  fp = foo;
5  x = ...;
6  y = (*fp)(x,param);
```

The somewhat obscure syntax of line 2 above shows another type of pointer declaration, one that declares a pointer variable named `fp` that is of type "pointer to function.". That function must accept a double argument and returns a double result. In line 4 above, we assign to this function pointer variable, a constant pointer to the function `foo`. In a manner similar to the declared name of a native array having the value of a pointer to its first element, the declared name of a function, when referred to in an assignment statement, evaluates to the value of a pointer to the block of code in a program memory segment for that function. Therefore, it may be assigned to a pointer variable, such as `fp` in the example above, that has been declared as a pointer to a function matching the same type. Pointers to function in C give the programmer direct control over those registers in the central processing unit (CPU) block diagram (Figure 1.6) that contain the addresses in memory where program instructions are stored – addresses that will be loaded into the Program Counter (PC) register to call a function. Then to invoke a call to that function, we dereference the function pointer with the "*" operator, just like referring to a data item to which we have a pointer, as in the last line above.

Now with this mechanism, we have a method of passing to any utility function a pointer to another function we wish to have the utility code treat as an object function to operate on. In the `step.c` code, three utility functions are included: named `step()`, `step_vect()`, and `step2d()`. The first is a one-dimensional stepping tool for a single-valued function, the second is a one-dimensional stepping tool for a function of one variable that returns a table or vector with any number of component values, and the third is for stepping a single-valued function of two arguments in two dimensions. We will examine the second and third tools in later sections, but at this stage we will consider some examples of using step() tool for single-valued functions of a single argument, as above.

The function definition of step() takes the form

```
void step(double (*func)(double,void *),GRID *xgrid_ptr,void *param,
          ARRAY1D *ytab_ptr) {
  ...
}
```

In this definition, the first argument to the tool is declared as a pointer to a function that takes a double argument and returns a double value. We can then call this tool and pass as a first argument the coding symbol name of any function we have written that computes some mathematical function we wish to study. Inside the step() code the formal argument named func holds the pointer to the object function, whatever the code passed to it on any particular invocation, and it may call that function over the stepped value of its argument and accumulate a table of values with the dereferencing expression (*func)(). Thus conceptually, we may treat an entire mathematical function as a "data" item with this powerful pointer mechanism. In this chapter we use this mechanism with the utility tool to generate a table of stepped function values, but in later chapters we will use the same pointer mechanism with other utility tools to locate roots of functions, integrate functions, solve differential equations described by functions, and other applications.

The second argument to the step() utility is a pointer to an instance of the GRID data type that has been filled in with the initial, final, and increment values that specify the grid over which we want the object function evaluated. The GRID argument is passed to a formal parameter named xgrid in the stepping tool code, and the initial, final, and increment parameters are extracted with pointer dereferencing as xgrid->init, xgrid->final, and xgrid->inc.

The third argument is a pointer to any structure or custom data type that holds the physical parameters that we will want to pass on to our object function. The stepper tool will simply pass on this pointer argument directly to the object function as its second argument without modification for each and every call to the function as it steps through the requested grid points of the function argument variable. In order to create a general purpose function stepping tool in our library applicable to any C function of one independent variable modeling any physical process, the structure pointer must be cast to a void pointer type inside our stepping tool code so that a pointer to any parameter structure of any format will be able to be passed through our library tools.

The fourth and final argument is a pointer to an ARRAY1D data type where the step() tool will record all the sampled function values at the grid argument values. After this utility routine returns, the elem array in the ARRAY1D structure will hold the sampled values $f(x_j)$ of the object function, and the n_elems member will hold the size of this array. The sample index j corresponds to the independent variable value $x = x_{\text{initial}} + jx_{\text{increment}}$. It is important to note that the allocation (and subsequent freeing) of the memory pointed to by the ARRAY1D structure that is needed to hold all the sampled values *is the responsibility of whatever code module calls the* step() *tool.* Calculation of the size of this allocation is made convenient by the definition of the NUM_GRID_PTS macro above, referring to the same grid structure that is passed to the step() tool, as this ensures that the ARRAY1D will be allocated with sufficient size to accommodate $f(x_j)$ values up to and including x_{final}, where x_{initial}, x_{final}, and $x_{\text{increment}}$ are specified in the GRID structure.

The iterative loop over which all the function values are evaluated is written in step() as

```
for (j = 0; j < ytab_ptr->n_elems; j++) {
  x = xgrid_ptr->init + ((double) j) * xgrid_ptr->inc;
  ytab_ptr->elem[j] = (*func)(x,param);
}
```

Within each iteration, the corresponding function argument value is computed in the variable x, with a type conversion covered in Section 3.3.2. The function is invoked in the next line by dereferencing the function pointer passed to this tool, and the return value stored in the corresponding array element. Referring to the n_elems member of the ARRAY1D sample values structure as an iteration limit ensures that the only memory written to by this tool will be that memory allocated by the calling code.

We intend to call this function stepping tool from either a simple main program, as shown in the example of Section 4.2, or from another program function. In order to compile any module of code that calls this tool as the C function step(), we must make available the step.h header file that declares the variable types involved in the call. The process of allocating memory, invoking the step() tool, using the sampled results, and freeing memory will typically follow a sequence like

```
ARRAY1D ytable;
GRID xgrid;
...
alloc_array1d(&ytable,NUM_GRID_PTS(xgrid));
step(f,&xgrid,&param,&ytable);
...
free_array1d(&ytable);
```

4.2 Application: Damped Harmonic Oscillator

Let us now gather all the concepts detailed in the previous pages and assemble one complete example of the simulation of physical systems. Consider a simple example of a model of a damped harmonic oscillator. This is the behavior of any of a multitude of physical systems where a displacement x from a nominal condition is described by the differential equation (4.1).

$$\frac{d^2x(t)}{dt^2} + 2\zeta\omega_0\frac{dx(t)}{dt} + \omega_0^2 x(t) = 0 \tag{4.1}$$

The second term in this equation is due to the damping forces in the oscillator system from any mechanism that is able to dissipate the oscillation energy into other forms. For a mechanical oscillator, damping may be due to frictional forces, or for an electrical oscillator, damping may be due to resistance in the circuit. We will study the displacement of the oscillator as a function of time, so the time variable is designated as the function argument. However, we will also want to have the ability to set three physical parameters of the oscillator, the natural frequency ω_0, the damping factor ζ, and the initial displacement x_0, to some arbitrary values in the program. The dimensionless damping factor ζ must be less than 1 for an oscillatory response. In that case, the damped oscillation frequency $\omega_d = \omega_0\sqrt{1-\zeta^2}$. For an initial displacement of x_0 and an initial velocity of 0 at time $t = 0$, the oscillation displacement over subsequent time values can be found with a straightforward solution of the differential equation (4.1), and will follow Equation (4.2)

$$x(t) = x_0 e^{-\zeta\omega_0 t}\cos(\omega_d t) \tag{4.2}$$

To visualize the damped oscillatory behavior at various values of the damping coefficient, we can assemble a program application with four components: a definition of a structure holding physical parameters, a C function that evaluates the time dependence of displacement Equation (4.2), another C function to generate a stream of output text to report the results of the simulation, and main program code, to interpret parameters from the command line and call the function stepper and output routines. This program, `damped_oscillator.c`, available for download from the Student Companion website, follows a general form used in almost all the example solutions in this text, and represents the minimum new code we need to assemble to model the oscillator. Refer to the downloaded source code and the excerpts here in the following discussion. The example downloadable source code file includes copius comment lines that will describe to the student the basic pattern of the source code for virtually all the coding examples and exercises in this textbook.

After including the required header files at the top of the source code, a structure is defined to hold the physical oscillator parameters and is given the data type name `OSC`

```
typedef struct osc_param {
  double x0;  // initial displacement
  double w0;  // natural frequency
  double wd;  // damped frequency
  double zeta;  // damping factor
} OSC;
```

We will use these structure members to store the oscillator initial displacement x_0, the natural frequency ω_0, the damped frequency ω_d, and the damping factor ζ. This structure is defined at the top of the source code file so that in all the code that follows it, the C compiler will understand the internal layout of the structure. This will let us use a pointer variable that has been declared to be of the type `OSC *`, or "pointer to `OSC`," in any following code to locate any member of the structure by dereferencing the pointer.

After this structure definition comes the C function named `osc()` that models the displacement solution in Equation (4.2). It is to this function that a `void` pointer to the parameters will be passed. This pointer allows the function to extract and make use of the parameters.

```
double osc(double t,void *param_ptr) {
  double disp;
  OSC *p;

  p = param_ptr;
  disp = p->x0 * exp(-p->w0 * p->zeta * t) * cos(p->wd * t);
  return(disp);
}
```

For the C compiler to know the layout of the binary bytes in the parameter structure, we assign the `void *` pointer function argument `param_ptr` to the local pointer variable p, which has been declared to point to the `OSC` structure holding the oscillator physical parameters that this computational model expects. Then we are able to include dereference operators such as `p->w0` in this function code in arithmetic expressions to evaluate to the oscillator natural frequency ω_0. We use the "`->`" operator here, as shown in Figure 3.13, rather than the "`.`" operator. This syntax is chosen because the variable p is a *pointer* to the structure, rather than a symbol representing the structure itself. Other parameters such as ω_d, ζ, and x_0 are extracted in the same manner.

Next in the source code comes a C function named `gen_output()` to generate text lines in the standard output stream representing the simulation results. This assumes two arguments are passed to `gen_output()`, the first a pointer to an `ARRAY1D` structure holding all the simulated displacements at the time sample points, and a pointer to a `GRID` structure describing the time sampling grid. Using an iterative loop, the simulation function time argument and displacement values are formatted as a two-column text stream. The format string in the call to `printf()` will translate the computed time and the stored displacement at each grid point into a line with two text fields with up to eight significant digits, and with a space character separating them into two columns.

```
void gen_output(ARRAY1D *ytab_ptr,GRID *tgrid_ptr) {
  int i;
  double t;

  for (i = 0; i < ytab_ptr->n_elems; i++) {
    t = tgrid_ptr->init + ((double) i) * tgrid_ptr->inc;
    printf("%.8g %.8g\n",t,ytab_ptr->elem[i]);
  }
}
```

Finally comes the main program, the code block that the operating system will initially pass execution to when the compiled program is run. In keeping with good coding practices, a typical main program is written to describe an overall strategy for the simulation task, and more elementary functions are called to realize each component step. The first task is to declare some variables to hold numerical values for the simulation. Among these are instances of the `GRID` structure to hold the time variable grid specification, the `OSC` structure to hold the physical parameters of the oscillator, and the `ARRAY1D` structure that the `step()` tool will fill with function values.

In this text, we will be assembling many main programs of this sort that solve prototypical physics problems, and when we sit down at a command prompt to run these programs, we have the flexibility to supply various physical parameters on the command line. It is often very difficult to remember exactly what the sequence of parameters on the command line each program is expecting. For that reason, each of the main programs here will be coded with a simple "help" line that is sent to the standard error stream in the case that the program is invoked with an incorrect number of command line arguments, or with no arguments at all. This is sensed by the value assigned to the integer variable `argc`, which is set by the operating system to be the number of tokens on the command line, being other than that expected. The lines

```
if (argc != 6) {
  fprintf(stderr,"%s <f0> <zeta> <x0> <tfinal> <tinc>\n",argv[0]);
  exit(1);
}
```

perform this function. Note that in the case of this test for a proper argument count failing, the "help" line text is sent to the standard error stream, which by default will be displayed to the user in the terminal window regardless of any redirection of the standard output stream into a data file.

However, if the test of the proper number of command line tokens is passed and execution continues, then the token strings are translated into variables that hold the parameter the user wishes to set with calls to the C library function `atof()`. (In other programs in this text, some parameters will naturally have integer values, and we will see these translated by the `atoi()` library call to integer variables.) Some values are stored in individual local variables; some are stored in the `GRID` structure members; and others are stored in the `OSC` structure members.

After assigning values to the needed variables, we may proceed with the core computational part of the code. The line

```
alloc_array1d(&ytable,NUM_GRID_PTS(tgrid));
```

will properly allocate the required memory for the `ARRAY1D` elements. Next the textbook library function `step()`, with a pointer to our `osc()` function as its first argument, is called to compute and store the sampled displacement values

```
step(osc,&tgrid,&param,&ytable);
```

supplying four pointers as arguments. The first is the pointer to the `osc` function to be stepped; the second is the pointer to `GRID` structure instance describing the time grid to step the oscillator simulation; the third is the pointer to the `OSC` structure instance to pass on the oscillator parameters; and the fourth is the pointer to the `ytable` array that has been allocated to use for a table of function values. Then a pointer to the `ARRAY1D` structure that is filled in by the `step()` function is passed on to the `gen_output()` function as the simulation result in the line

```
gen_output(&ytable,&tgrid);
```

The output stream of the program is generated in the `gen_output()` function, and in this first example simulation tool, with the intention of simply plotting the time dependence of oscillator displacement, the output stream is a two-column tabular listing of the time and displacement values formatted for `gnuplot`. After the output stream is generated, the memory allocated by the `alloc_array1d()` function is freed with the line

```
free_array1d(&ytable);
```

The following shows an example dialog at a Linux or MSYS2 command prompt terminal compiling and running this main program

```
$ gcc -Wall -Wshadow -O2 -o damped_oscillator damped_oscillator.c CPLibrary/step.c
    ↪ CPLibrary/array.c -lm
$ ./damped_oscillator
./damped_oscillator <f0> <zeta> <x0> <tfinal> <tinc>
$ ./damped_oscillator 1 0.1 1 10 0.1 >| damped_oscillator_z0p1.dat
```

In line 1, we build the executable program with the `gcc` compiler, gathering together the main program and the other source code modules for the function stepper tool and the array utility code. The compiler will generate its output into the executable file `damped_oscillator` in the current working directory. In line 2, we invoke the executable compiled program in the current directory with no command-line arguments at all. That requests the program to respond with a simple reminder line of what parameters are expected. For this program, the parameters are the natural frequency, the damping factor, the initial displacement, the final time and the time step, all expressed in some consistent system of units for time and displacement. Finally in line 4, we invoke the program with some desired parameters, such as a natural frequency of 1 inverse time units, a damping factor of 0.1, an initial displacement of 1 unit, and a time grid of 0 to 10 stepped every 0.1 time units. The redirection characters `>|` instruct the command shell to save the standard output data stream into a data file. Taking a look at that output file, we see the first and last several lines in the format

```
0 1
0.1 0.76148368
0.2 0.27780341
0.3 -0.2484756
0.4 -0.62341821
0.5 -0.73031213
...
9.5 -0.0024431406
```

```
9.6 -0.0022745938
9.7 -0.0013094986
9.8 1.1666538e-05
9.9 0.0011726294
10 0.0017755875
```

The format of this file is as prepared by the call to `printf()`

```
printf("%.8g %.8g\n",t,ytab_ptr->elem[i]);
```

in the `gen_output()` function. Each line represents a sample on the grid of time values, with the two columns being the time and the displacement values. The first line is, of course, the initial displacement of 1 unit at a time value of 0 units.

It should be mentioned here that the most efficient use of a computer system for the process of compiling C source files is to compile the various code modules independently and save the resulting intermediate binary files, called "object files," from each module. These object files may then be linked together to form the final binary executable file. In the process of program development, if edits or changes are made to only one or some of the source code modules, processor time is not wasted on recompiling a module that has not changed. However, for the purpose of this text, we will be working with relatively simple source code files that compile quickly (yet in many cases execute slowly), and mention of the rich set of development tools that are available for tracking multiple code modules in a larger project, such as the `make` tool or various integrated development environments, is omitted here for simplicity. A single invocation of the `gcc` command is all that is required to compile and link all the code modules.

4.3 Coding: The `gnuplot` Plotting Tool

First let us start with an introduction to the `gnuplot` plotting tool we will be using for rendering our graphical data plots. Typing the command `gnuplot` at a command prompt with no additional command line arguments will start an interactive `gnuplot` session. While in interactive mode, the tool will display a prompt of `gnuplot>` in the terminal window and wait for user commands. A very useful first command to learn is

```
gnuplot> test
```

In response, `gnuplot` will display a graphics window with a demonstration array of the default line colors, point types, line dash patterns, and line widths that are available for plotting numerical data in curves or points. Each graphic type is labeled with a numerical code that will be recognized in plotting commands.

There are plentiful resources online where full documentation on this powerful graphical tool may be found, but for our purposes we may learn the basics with several examples. An example interactive session formatting a useful plot of the oscillator displacement versus time consists of the commands

```
gnuplot> set title 'Damped Harmonic Oscillator, damping factor = 0.1'
gnuplot> unset key
gnuplot> set xlabel 'Time (seconds)'
gnuplot> set ylabel 'Displacement (centimeters)'
gnuplot> set linetype 1 linecolor -1 linewidth 2 dashtype 1
gnuplot> plot 'damped_oscillator_z0p1.dat' using 1:2 with points pointtype 7
```

In line 1, we set the overall plot title. The plotting examples for the damped harmonic oscillator in this section include overall plot titles to illustrate this capability of `gnuplot`, and this feature is useful when generating a standalone image file that may be attached to an email message or used as a presentation slide. After this section, however, the overall titles will be omitted in the plots included in this text, as the figure captioning mechanism in a text document may also be used to provide an overall title. Our first graphical plot will consist of only one time-dependent curve, so there is no need for the `key` feature where curve titles are shown. This is turned off in line 2. Lines 3 and 4 set the labels for the x and y axes. Line 5 describes the type of line that will be graphically rendered to describe the displacement dependence versus time. In this example, a line color code of -1 is specified for a standard black line, with a thickness code of 2. These are the parameters that have been used for the plots in the pages of this text. Finally the curve is plotted with the `plot` command in line 6. This command takes a number of arguments on its command line, the first of which is the name of the file containing the columns of data. The `using` keyword specifies how the data columns in the file are mapped to plot axes. In this case, we map column 1 to the x axis and column 2 to the y axis, but any arbitrary mapping may be specified. The final keywords

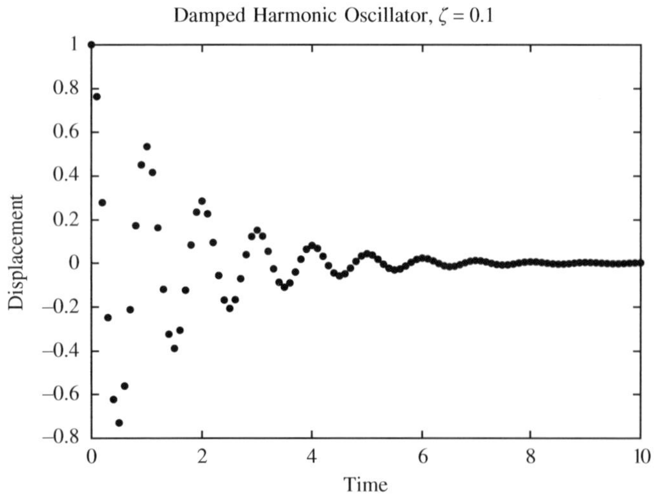

Figure 4.2 Example point plot of the damped harmonic oscillator simulation, $\zeta = 0.1$.

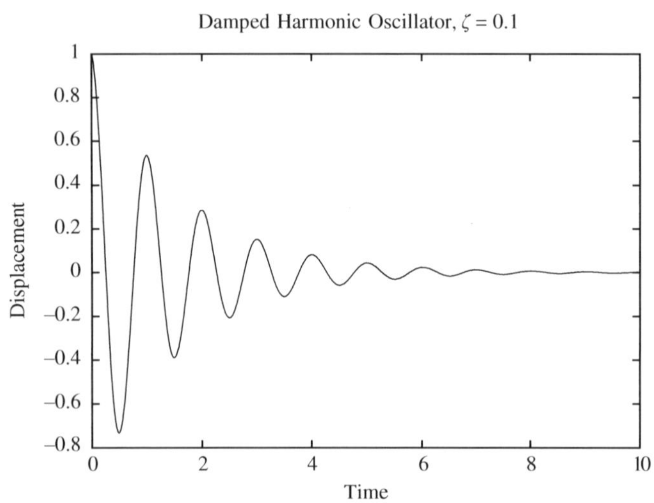

Figure 4.3 Example line plot of the damped harmonic oscillator simulation, $\zeta = 0.1$.

request a plot where the data points on each line in the data file are plotted individually by points. (The default plot point type is a small cross symbol, and this is overridden with a point type of a small dot.) Note that character strings such as file names or labels are enclosed in single or double quotes. The result is the plot seen in Figure 4.2.

Of course the real behavior of an oscillator governed by Equation (4.1) is a displacement that follows a continuous function of time. To visualize a better approximation to this continuous behavior within the resolution of the pixels in plotting images, we may rerun the program with a decreased time step specified on the command line

```
$ ./damped_oscillator 1 0.1 1 10 0.01 >| damped_oscillator_z0p1.dat
```

and replot with `line` mode instead of `point` mode, so that the data points on each line in the data file are connected by (very short) lines.

```
gnuplot> plot 'damped_oscillator_z0p1.dat' using 1:2 with lines
```

The resulting image is shown in Figure 4.3.

Let's appreciate more features of the `gnuplot` tool with the intention of visualizing the effect of various values of the damping factor ζ. First we will generate two other two-column data files with different damping factors, as specified on the command line, by rerunning our simulation tool

```
./damped_oscillator 1 0.05 1 10 0.01 >| damped_oscillator_z0p05.dat
./damped_oscillator 1 0.025 1 10 0.01 >| damped_oscillator_z0p025.dat
```

Then we may overlay three curves representing the three damping factors with the set of `gnuplot` commands

```
gnuplot> set title 'Damped Harmonic Oscillator, damping coefficient = 0.1, 0.5, 0.025'
gnuplot> set key default
gnuplot> set xlabel 'Time'
gnuplot> set ylabel 'Displacement'
gnuplot> set linetype 1 linecolor -1 linewidth 2 dashtype 1
gnuplot> set linetype 2 linecolor -1 linewidth 2 dashtype 2
gnuplot> set linetype 3 linecolor -1 linewidth 2 dashtype 4
gnuplot> plot 'damped_oscillator_z0p1.dat' using 1:2 with lines title '$\zeta = 0.1$',\
'damped_oscillator_z0p05.dat' using 1:2 with lines title '$\zeta = 0.05$',\
'damped_oscillator_z0p025.dat' using 1:2 with lines title '$\zeta = 0.025$'
```

We are making use of some additional plotting features in this session. We intend to generate a plot overlaying three curves representing the three simulated damping factors, and including a legend in the corner of the plot that labels each line with the corresponding value of ζ. So after a new overall title line is defined in line 1, in line 2 we reset the position of the plot "key" to its default position in the upper-right corner. After the axis labels are specified as before, we describe the desired plot appearance with a more elaborate version of the `plot` command. This example illustrates the use of the backslash character \ as a line continuation character when we have several curves to specify. The curve specifications are separated by the comma character. Each curve description consists of the same sequence as before, with a file name, a column mapping, and a plot mode. Here we add one more clause to the description, and that is introduced by the keyword `title`. The character string that follows is the title we want to label this line in the key legend. So we may see the plot that this session produces in Figure 4.4.

When multiple curves are plotted, by default `gnuplot` will change the line type sequentially to help us distinguish them. In lines 5 through 7, we have defined the desired sequence of line types to be rendered, using different dash type codes, from solid, to dashed, to dash-dot. Again, this is shown in this example `gnuplot` dialog to mimic what was used for the print pages of this text, which are limited to grayscale images only. On the other hand, most users will find rendering plots with lines that are all solid but of different colors to be more clear when viewing simulation results on a display screen or for other applications. For that case, some alternative line type definitions are as follows

```
set linetype 1 linecolor 'red'
set linetype 2 linecolor 'green'
set linetype 3 linecolor 'blue'
set linetype 4 linecolor 'cyan'
set linetype 5 linecolor 'violet'
set linetype 6 linecolor 'brown'
set linetype 7 linecolor 'red'
set linetype 8 linecolor 'green'
set linetype 9 linecolor 'blue'
set linetype 10 linecolor 'cyan'
set linetype 11 linecolor 'violet'
set linetype 12 linecolor 'brown'
```

The `gnuplot` tool includes a mechanism to perform these sorts of initializations in a more convenient way than having to type them each time an interactive session is started. You may simply include any `gnuplot` commands, such as those above for setting line types, in a text file named `.gnuplot` in the user's home directory, and `gnuplot` will automatically read the file and process the contained commands as if they were typed into an interactive session. If the initializations above were specified in this manner, then there would be no need for any interactive `set linetype` commands, and the plot in Figure 4.4 would be rendered on a display screen with a red line for the first curve, green for the second, and blue for the third. You may customize the color sequence to your needs with this configuration file.

To save typing effort in an interactive session, `gnuplot` also allows us to abbreviate many of the keywords, such as *u* for `using`, *w* for `with`, *p* for `points`, *l* for `lines`, `lt` for `linetype`, `pt` for `pointtype`, and *t* for `title`. These abbreviations are illustrated in the examples in the following sections. Additional commonly used features of `gnuplot` will be covered in Section 4.4.

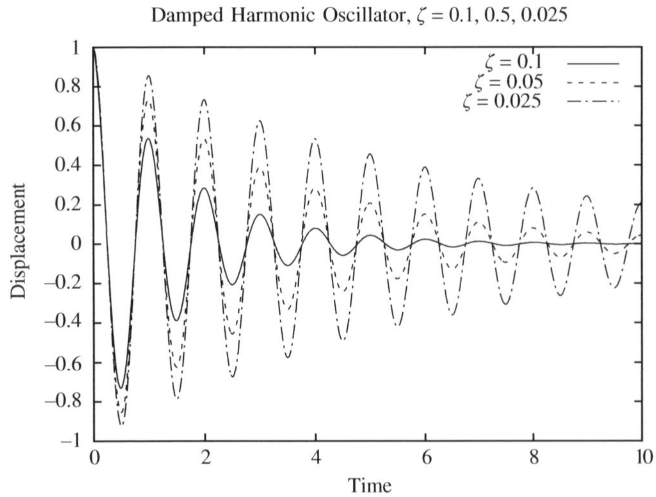

Figure 4.4 Harmonic oscillator simulation with three damping factors.

Often we make use of the `gnuplot` tool initially in interactive sessions such as those described above, making trial adjustments to various parameters, labels, and titles, until a plot that serves our needs is rendered. However, the `gnuplot` program may also be run in a non-interactive, or "batch" mode, taking its commands from either an input script file or its standard input stream. For example, to save our final version of the plot description as a programmed script for future use, we may copy the text lines

```
set title 'Damped Harmonic Oscillator, $\zeta = 0.1, 0.5, 0.025$'
set key default
set xlabel 'Time'
set ylabel 'Displacement'
plot 'damped_oscillator_z0p1.dat' using 1:2 with lines title '$\zeta =
    ↪ 0.1$','damped_oscillator_z0p05.dat' using 1:2 with lines title '$\zeta =
    ↪ 0.05$','damped_oscillator_z0p025.dat' using 1:2 with lines title '$\zeta = 0.025$'
```

from our terminal window into a file we may name `damped_oscillator_z3.gp` with any text editor tool. This script assumes we have set our desired line types in a `.gnuplot` file in our home directory. Then simply running the one command

```
gnuplot -p damped_oscillator_z3.gp
```

at a Linux or MSYS2 command prompt with the script file name as a command line argument. With the command line option `-p`, an abbreviation for "persistent," the plot image will be displayed in a separate window on the screen, and remain persistently displayed until closed. In addition, other commands are available to render the graphics output into image files in formats such as PNG, EMF, EPS, and others. This allows us to include our plot images in web pages, documents, projection slides, email messages, etc. For this purpose, we set two other `gnuplot` variables named `terminal` and `output`. The `terminal` variable designates the destination mode for the rendered graphics, such as a PNG file. (The plot images for this textbook were rendered with the `epslatex` terminal type to be included in a text assembled as LaTeX code.) The `output` variable specifies the name of the file to be written with the graphical rendering. So as an example, if the script file above were edited to include two additional `set` commands for these variables before the `plot` command

```
set title 'Damped Harmonic Oscillator, $\zeta = 0.1, 0.5, 0.025$'
set key default
set xlabel 'Time'
set ylabel 'Displacement'
set terminal png size 640,480
set output 'damped_oscillator_z3.png'
plot 'damped_oscillator_z0p1.dat' using 1:2 with lines title '$\zeta =
    ↪ 0.1$','damped_oscillator_z0p05.dat' using 1:2 with lines title '$\zeta =
    ↪ 0.05$','damped_oscillator_z0p025.dat' using 1:2 with lines title '$\zeta = 0.025$'
unset output
```

then running the Linux or MSYS2 command

```
gnuplot damped_oscillator_z3.gp
```

this time, without any `-p` option needed, will generate the PNG graphics file named `damped_oscillator_z3.png`, with a resolution of 640 by 480 pixels, in the current directory for further use. We should also include an `unset output` command after the `plot` command to close the written graphics file properly.

This scripting capability of the `gnuplot` tool opens up very flexible possibilities of programming our C simulation code to write out a plotting script file itself, including its own simulation data. This will be demonstrated in Section 4.13, and lead us to the ability of generating animated graphics as well as simple static images.

We will revisit the damped harmonic oscillator at the end of Section 5.1, Exercise 5.8, Section 11.11, and Exercise 11.2.

4.4 Application: The Helmholtz Coil

For experimental work in electricity and magnetism, a spatial region containing a uniform magnetic field is sometimes required. Historically, it has been known that a simple arrangement to establish a good approximation to a uniform field over a limited region is the Helmholtz coil[6, Chap. 3; 7, Chap. 4; 8, Chap. 8]. Two circular coils of wire carrying an electric current through multiple turns are arranged with geometry as shown in Figure 4.5.

In this geometry, the planes of the two coils lie parallel to the x-y plane in the three-dimensional coordinate system, and are situated at a distance D both above and below the x-y plane. Each circular coil has radius R, and carries a current I_c through n turns. We'll make the approximation that the n turns occupy essentially the same space, so this is equivalent to a single turn in each coil carrying a current $I = nI_c$. The magnetic field in the space near the center of the coordinate system will be oriented almost entirely vertically, and to a good approximation will be of uniform strength. In the center of the region, the field is a superposition of the fields from the two coils. How may we use the visualization tools developed in the previous sections to graphically appreciate the approximation of a uniform field?

We will start by restricting our calculation of the magnetic field to that only along the central z axis, where by symmetry, the field will be directed exclusively in the z direction. Along this line we may apply some basic magnetic field theory and form an integral of the field due to an assembly of infinitesimal current elements around each circular loop using the Biot–Savart law[6, Chap. 2; 7, Chap. 4; 8, Chap. 8; 15, Chap. 5]. Along the axis of the coils, the integral is analytically tractable, and we can derive the field at any z coordinate as the sum of the fields from the two coils. (Later on, in Chapter 10, we will return to the Helmholtz coil and extend our analysis with a numerical integration over the entire region.) So we first derive the field along the z axis for a single coil located at an arbitrary z coordinate, as shown in Figure 4.6.

It is possible to derive the geometric condition for the derivative with respect to z of the total field due to two such coils located at $z = \pm D$ evaluated at $z = 0$ is zero. That condition is that $D = \frac{R}{2}$, half the radius of each coil. So let us use our numeric tools to visualize how the optimization of field uniformity appears through the region along the axis. This will also allow us to visualize the disturbance from a uniform field if the distances D are not quite optimum.

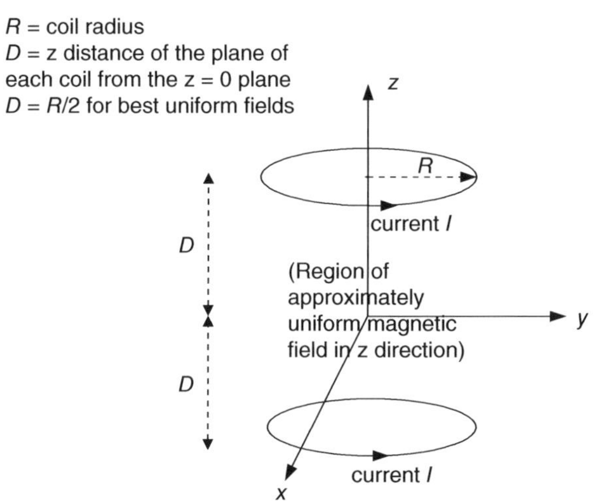

Figure 4.5 Geometry of a Helmholtz coil.

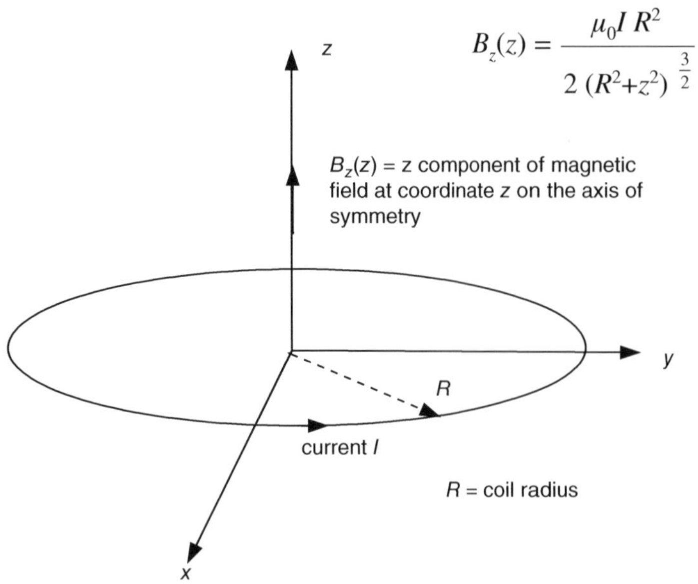

Figure 4.6 Magnetic field along axis of a circular coil.

For the simple example of a damped harmonic oscillator in Section 4.2 we looked at a main program coded to represent physical quantities, but in arbitrary units. For this example, magnetic field problem, and for many others in this text, we will be using SI physical units. An optional header file we may make use of for many programs is one that makes symbolic equivalents to the physical constants we will need in our computations. The following is an example header file that makes use of constant macros for this purpose, named `constants.h`, which is included in the `CPLibrary` archive on the Student Companion website and used in many example programs here. The reader may, of course, add other constants to it as the need arises

```
#define BOLTZMANN 1.380648813e-23
#define STEFAN_BOLTZMANN 5.670374419e-8
#define PLANCK 6.6260695729e-34
#define PLANCK_REDUCED 1.05457172647e-34
#define GRAVITY 6.67428e-11
#define GRAVITATIONAL_ACCEL 9.80665
#define CHARGE 1.602e-19
#define PERMITTIVITY 8.854e-12
#define PERMEABILITY 1.2566370614e-6
#define SPEED_OF_LIGHT 299792458.0
#define PI 3.14159265358979323846264338327950288
#define TWO_PI 6.283185307179586
#define FOUR_PI 12.56637061435917
#define HALF_PI 1.570796326794897
#define QUARTER_PI 0.7853981633974483
#define RADTODEG 57.29577951308234
#define DEGTORAD 0.01745329251994329
```

Near the top of any file of program code we may simply add the line

```
#include "CPLibrary/constants.h"
```

and we will be able to refer to these physical constants symbolically. For instance, we can just use the constant `GRAVITY` in an arithmetic expression for Newton's gravitational constant, rather than having to look up all the significant figures for each program.

The code for `helmholtz_axis_step.c`, available for download from the Student Companion website, uses the same overall structure as that analyzing the damped oscillator in `damped_oscillator.c` from Section 4.2, except now we are stepping various functions modeling a magnetic field over a z spatial coordinate instead of time in the case of the oscillator.

The structure data type named HELM is defined at the beginning of the program to hold the physical parameters describing the coils, namely the radius of each coil, the distance each coil is situated from the *x*-*y* plane, and the effective electric current passing through both coils

```
typedef struct helm_param {
  double radius;  // radius of circular coils
  double distance;  // distance from xy plane to each coil
  double current;  // electric current flowing in coils
} HELM;
```

The universal function stepping tool in `step.c` is used again, except it is called three times in this program. It is called first passing a pointer to an object function to step that only models the contribution of the magnetic field from the upper coil, located at $z = +D$, and then a second time passing a similar object function for the field from the lower coil at $z = -D$. Then a final call is made passing a pointer to an object function that sums the two field contributions.Our intention is to plot the three functions together with overlaid curves so we may visualize the superposition of the fields from the two coils forming the total magnetic field. At the optimum coil distances, this superposition results in a fortunate field dependence with respect to the z coordinate that graphically appears quite constant. An example of the object function found in the source code is the following function, computing the field only from the upper coil

```
double helmholtz_upper(double z,void *param_ptr) {
  double b,d;
  HELM *p;

  p = param_ptr;
  d = z - p->distance;
  b = PERMEABILITY * 0.5 * p->current *
      p->radius * p->radius /
      pow((p->radius * p->radius + d * d),1.5);
  return(b);
}
```

In each of the three object model functions, including in the example above, the `void` parameter pointer `param_ptr` passed to it is assigned to a local pointer variable *p* declared to be of type `HELM *`.

```
HELM *p;
p = param_ptr;
```

This assignment allows the physical parameter members to be evaluated in the arithmetic expressions within each stepping object function. For example, the radius of the current loop is extracted from the HELM parameter structure with the dereferencing operator `p->radius`. Note that after each call to the `step()` utility, a table of z coordinates and the magnetic field is sent to the standard output stream with a call to the `gen_output()` function. This is an example of the benefits of writing code with small elemental tasks, such as generating an output stream from a table of values, into separate functions. After that elemental function is defined once, our code is concisely written calling it multiple times. After each table is output, the character string "`\n\n''` is then output. This generates two consecutive blank lines in the output stream, and we will see shortly that this separates the three tables into three "datasets" as interpreted by `gnuplot`. A single ARRAY1D instance `htable` is allocated for use by all three calls to the `step()` tool, and then freed after its last use by `gen_output()`.

We will simulate the case of each coil having a radius R of 10 cm. To see the optimum solution, first we plot the field dependence versus z for the coil separation somewhat closer than optimum, with D of 4 cm, in Figure 4.7. Then we try the field dependence for coils separated wider than optimum, with D of 6 cm, in Figure 4.8. Finally we can see the optimum configuration giving a uniform field for a good portion of the center region in Figure 4.9, for D of 5 cm.

Consider the following interactive session at a Linux or MSYS2 command shell to compile and run this program. The execution command line arguments are for the nominal case of 5-cm coil distance from the origin. Of course, the physical parameters on the command line are assumed to be in SI units, meters and amperes.

```
$ gcc -Wall -Wshadow -O2 -o helmholtz_axis_step helmholtz_axis_step.c CPLibrary/step.c
    ↪ CPLibrary/array.c -lm
$ ./helmholtz_axis_step
./helmholtz_axis_step <radius> <distance from center> <current> <zinit> <zfinal> <zinc>
$ ./helmholtz_axis_step 0.1 0.05 1 -0.1 0.1 0.001 >| helmholtz_axis_step.dat
```

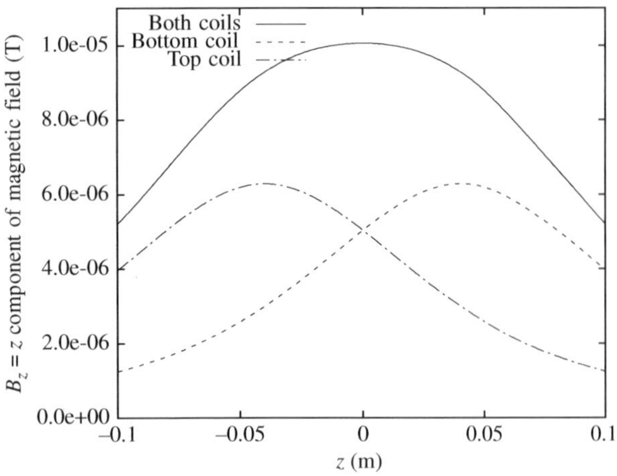

Figure 4.7 Helmholtz coil field along axis with close separation: radius = 10 cm, distance = 4 cm.

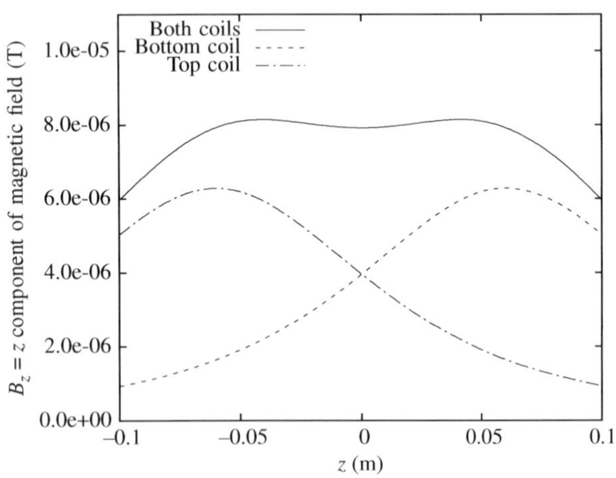

Figure 4.8 Helmholtz coil field along axis with far separation: radius = 10 cm, distance = 6 cm.

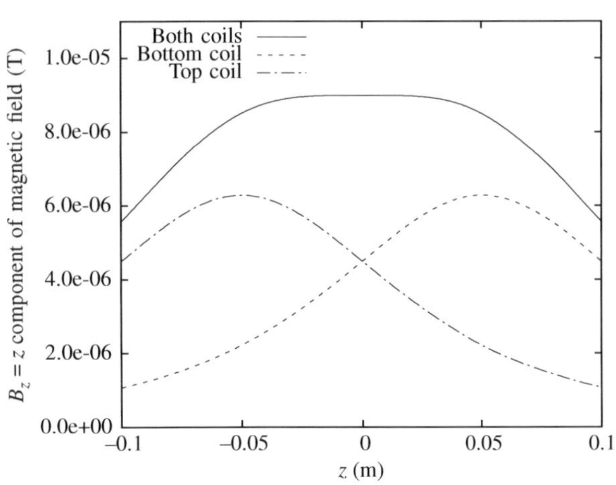

Figure 4.9 Helmholtz coil field along axis with optimum separation: radius = 10 cm, distance = 5 cm.

A good practice when writing numerical programs is to find some sample in its output that corresponds to an analytically tractable condition and verify at least one numerical value. This practice can alert us to the case where an unfortunate programming bug has crept into even our best coding efforts. Scanning the output data file from the program run above with the Linux `less` tool, we can find in the third dataset a calculation of the z component of the magnetic field in Teslas at a z coordinate of 0 on the center line of

```
-0.002 8.9917612e-06
-0.001 8.9917628e-06
0 8.9917629e-06
0.001 8.9917628e-06
0.002 8.9917612e-06
```

By summing two instances of the expression for the magnetic field along the axis of a circular loop of current given in Figure 4.6 for the two coils of an optimally spaced Helmholtz arrangement, we can find that the magnetic field at the center point of the structure at $z = 0$ must be

$$B = \frac{\mu_0 I}{\left(\frac{5}{4}\right)^{\frac{3}{2}} R} \tag{4.3}$$

Evaluating Equation (4.3) with a radius R of 10 cm, we verify that the program correctly calculates the center field value, lending us confidence in the plotted results of Figures 4.7 through 4.9.

The following is an interactive session with `gnuplot` used to render the plot in Figure 4.9. This dialog illustrates a few more features of `gnuplot`, such as setting a specific location for the curve legend key, specific ranges for the x and y axes, and a specific display format for numerical marks along an axis. In addition, note some new features illustrated in this `plot` command. First there is a new keyword included, abbreviated with the single "*i*" character, but really standing for the keyword "index." Data files processed by `gnuplot` may consist of multiple "datasets," which are collections of lines in the file separated by two consecutive blank lines. These are numbered with a dataset index starting from 0 at the beginning of the file. We refer to a particular dataset in the plot command with an "index" or "*i*" keyword followed by the set index. Since we are plotting sets from the same file, for convenience we only need specify the data file name once at the beginning of the `plot` command. Subsequent file names can simply be left as a blank string with two consecutive single or double quote characters if the corresponding curve is to be drawn from the same file as previously named. This feature allows the user to save many keystrokes when typing interactive plot commands.

```
gnuplot> set key left top
gnuplot> set xlabel 'Z-coordinate (m)'
gnuplot> set xrange [-0.1:0.1]
gnuplot> set ylabel 'Z-component of magnetic field (T)'
gnuplot> set format y "%0.1e"
gnuplot> set yrange [0:1.1e-5]
gnuplot> plot 'helmholtz_axis_step.dat' i 0 w l t 'Upper coil',\
>'' i 1 w l t 'Lower coil','' i 2 w l t 'Both coils'
```

4.5 Application: The Maxwell–Boltzmann Distribution

We will treat different aspects of the classical Maxwell–Boltzmann distribution at several points in this text, specifically in Sections 8.11 and 11.9, as well as in Exercise 10.3. This distribution gives the probability of a particle or molecule of mass m in an ideal gas at a given temperature T having a particular speed, per unit speed v.

$$\frac{n(v)dv}{N} = 4\pi\left(\frac{m}{2\pi kT}\right)^{\frac{3}{2}} v^2 \exp\left(\frac{-mv^2}{2kT}\right) \tag{4.4}$$

At this point, it will be helpful to visualize the shape of the distribution function at various temperatures, as plotted in Figure 4.10.

The coding of this visualization task is left as Exercise 4.1. As detailed in the exercise, this can be most easily accomplished by copying and editing the code for the examples of `damped_oscillator.c` and `helmholtz_axis_step.c` above.

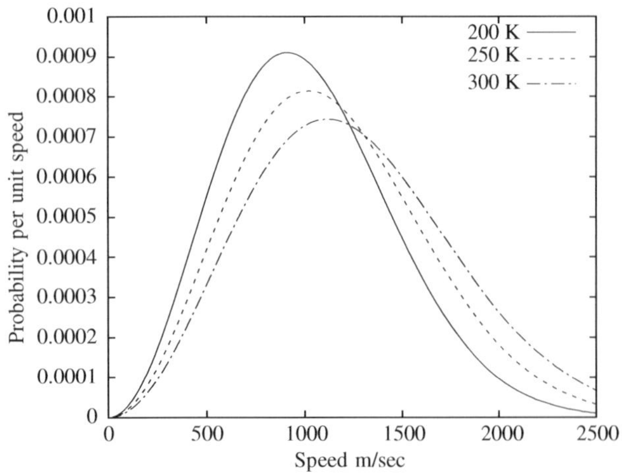

Figure 4.10 Classical Maxwell–Boltzmann distribution at three temperatures.

Figure 4.11 Rainbow over the Grand Canyon, Arizona.

4.6 Application: Rainbows

Sightings of rainbows, such as the one pictured in Figure 4.11 arching over the Grand Canyon in Arizona, or the one in the mist of Niagara Falls in Figure 4.12, are beautiful and captivating. What are the optical laws that govern their creation? This question was solved some centuries ago by René Descartes [16], by a straightforward geometrical analysis that we can revisit with simple numerical code.

We start with the assumption that surface tension holds each falling raindrop in a region of rainfall in an approximately spherical shape, as shown in Figure 4.13. Then we introduce rays of sunlight laterally that are uniformly distributed through all the y coordinates that will impinge on the raindrop. Since we are interested in rays of light that will reflect as shown at elevations below the raindrop to an observer on the ground, we will consider light rays incident with y coordinates in the range of 0 to r, where r is the raindrop radius. For simplicity, we treat only two dimensions here, but with the understanding that when the ray tracing exercise that follows is extended to three dimensions, we will observe the circular pattern shown in Figures 4.11 and 4.12.

So let us follow the horizontal sunlight rays shown in Figure 4.13 as they pass into the raindrop, are reflected from the water–air interface on the back side, and pass back out of the drop towards the observer [17, Chap. 34; 18, Chap. 23; 19, Chap. 33; 20, Chap. 35]. We assume that the index of refraction of the water at the particular wavelength of the light ray is

Figure 4.12 Rainbow in the mist of Niagara Falls, Ontario.

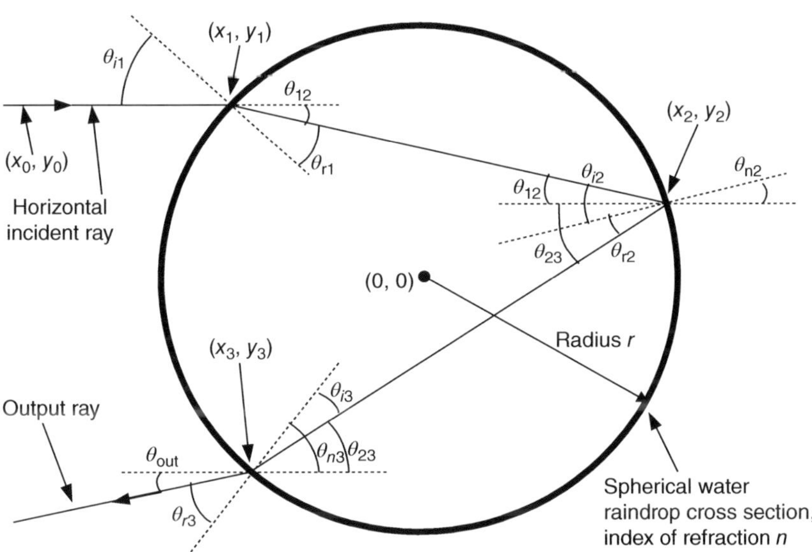

Figure 4.13 Light ray trajectory angles for a rainbow.

n, and that in air is 1. If we let the y coordinate of the incident ray be y_0, then the coordinates (x_1, y_1) of the point where the ray passes through the first air–water interface and its incident angle are

$$x_1 = -\sqrt{r^2 - y_0^2}$$

$$y_1 = y_0$$

$$\theta_{i1} = \arctan\left(\left|\frac{y_1}{x_1}\right|\right)$$

Then applying Snell's Law for the refraction through the first interface, we find the refracted angle through the first interface as

$$\theta_{r1} = \arcsin\left(\frac{1}{n}\sin(\theta_{i1})\right)$$

Next we must solve for the slope m_{12} of the ray line to the point (x_2, y_2) where the ray intersects the second water–air interface, on the back side of the raindrop.

$$\theta_{12} = \theta_{i1} - \theta_{r1}$$

$$m_{12} = \tan(-\theta_{12})$$

$$x_2 = \frac{x_1(m_{12}^2 - 1) - 2m_{12}y_1}{m_{12}^2 + 1}$$

$$y_2 = y_1 + m_{12}(x_2 - x_1)$$

The reflection off of the backside point will be symmetrical about the normal angle

$$\theta_{n2} = \arctan\left(\frac{y_2}{x_2}\right)$$

Then we figure the next point (x_3, y_3) where the ray will intersect the third water–air interface

$$\theta_{i2} = \theta_{n2} + \theta_{12}$$

$$\theta_{23} = \theta_{n2} + \theta_{i2} = 2\theta_{n2} + \theta_{12}$$

$$m_{23} = \tan(\theta_{23})$$

$$x_3 = \frac{x_2(m_{23}^2 - 1) - 2m_{23}y_2}{m_{23}^2 + 1}$$

$$y_3 = y_2 + m_{23}(x_3 - x_2)$$

Finally the ray will refract again through the third water–air interface, and we may calculate the output angle formed between the refracted ray and horizontal by again applying Snell's Law

$$\theta_{n3} = \arctan\left(\frac{y_3}{x_3}\right)$$

$$\theta_{i3} = \theta_{n3} - \theta_{23}$$

$$\theta_{r3} = \arcsin\left(n \sin(\theta_{i3})\right)$$

$$\theta_{\text{out}} = \theta_{n3} - \theta_{r3}$$

To visualize the resulting ray path, we can assemble a simple program to evaluate this sequence of expressions inside an iterative loop that steps the incident y coordinate through the top half of the raindrop. At each step that calculates the (x, y) coordinates of a point where the light ray changes direction, the program can send the coordinates to the standard output stream. This will allow a data file to be written that `gnuplot` may connect together with lines that trace the light ray path. For visualization purposes, we will simply set the raindrop radius r to 1 and use arbitrary spatial units. The stepping grid of the program iterates over the incident ray y coordinate, ranging from 0.05 to 0.95 in arbitrary spatial units.

Since we are modeling physical behavior that is dependent on the wavelength of the light rays, we will need to model in the program code the slight dependence of the index of refraction of water on wavelength. This effect is called chromatic dispersion. Let us study the paths of the rays at three different light colors, with their wavelengths and indices of refraction given in Table 4.1.

Table 4.1 Index of refraction of water n at three light colors.

Color	Wavelength (nm)	n
Violet	400	1.339
Green	525	1.334
Red	700	1.331

As we can see, the variation of the index of refraction is small, but the geometry of the refractions and reflections of the light rays in a spherical raindrop tends to magnify this small dependence into an observable phenomenon. Figures 4.14 through 4.16 show the graphics rendered by `gnuplot` describing the resulting ray paths.

The difference between these plots is quite subtle, and can mostly be seen at the bottom where the trajectories exit the plot window. From these ray plots we can visualize one feature of the appearance of the rainbow. In this particular geometry, the refractions and reflections work to cluster together several of the incident sunlight rays into tight bundles at the bottom of the plot. This accounts for the patterns of somewhat brighter light when we observe the rainbow. Therefore with a few dozen lines of C code, we have recreated what Descartes had recorded about his discovery: "I took my pen and made an accurate calculation of the paths of the rays which fall on the different points of a globe of water to determine at which angles, after two refractions and one or two reflections they will come to the eye, and I then found that after one reflection and two refractions there are many more rays which can be seen at an angle of from forty-one to forty-two degrees than at any smaller angle; and that there are none which can be seen at a larger angle."

But how do we explain the physical origin of the most striking feature of the rainbow, the prismatic color separation? A careful examination of the representation of Figures 4.14 through 4.16, will show a small migration in the location where the bundle of rays intercepts the bottom x axis. But beyond this general observation, it is difficult to distinguish any quantitative difference in the optical behavior dependent on wavelength. Exercise 4.3 for the student makes use of the `step()` utility to generate function plots that should bring clarity to the question at hand. The three datasets generated by the solution program in this exercise are plotted with `gnuplot` in Figure 4.17.

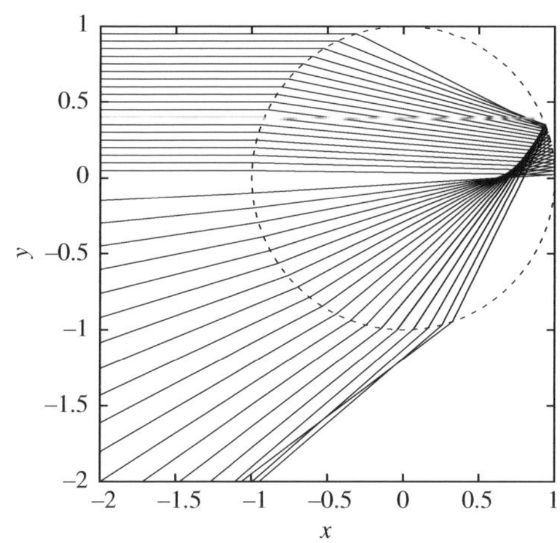

Figure 4.14 Refraction and reflection rays at violet light, wavelength = 400 nm.

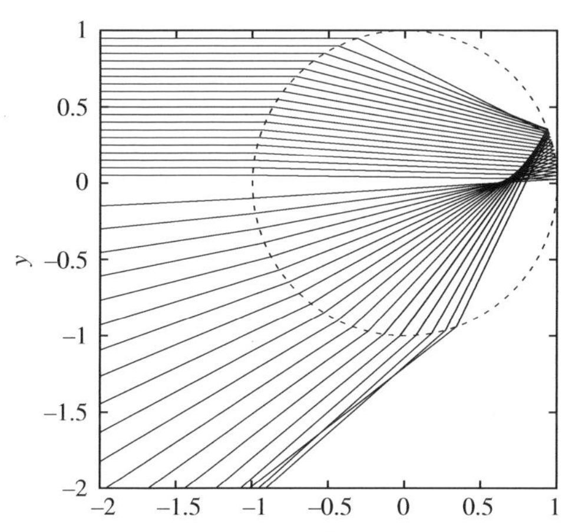

Figure 4.15 Refraction and reflection rays at green light, wavelength = 525 nm.

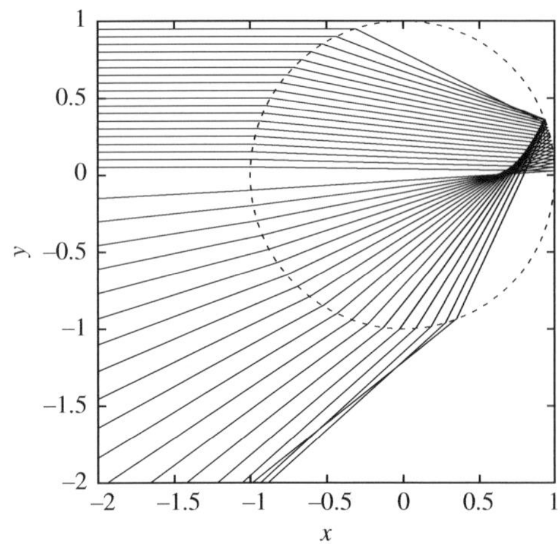

Figure 4.16 Refraction and reflection rays at red light, wavelength = 700 nm.

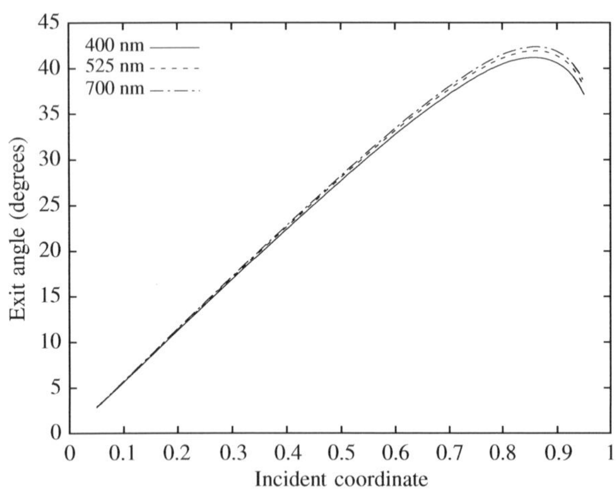

Figure 4.17 Raindrop exit angles creating a rainbow at three light wavelengths – 400, 525, and 700 nm.

The same effect of several of the rays bundling together in Figures 4.14 through 4.16 is now apparent in Figure 4.17 as the curling of the tops of the three curves. But with this numerical analysis, the dependence of the peak angle where the rays bundle on the light wavelength is manifest.

The wavelength dependence now clear in Figure 4.17 is what brings about the appearance of the rainbow. If we imagine the collective pattern of light from three raindrops at different altitudes with just the required angles as shown in Figure 4.18, we see the angular dependence on wavelength that gives the prismatic light pattern we observe. That only a portion of the ray distribution is bundled together at their peak accounts for the somewhat pastel coloration of the rainbow.

4.7 Application: Diffraction Patterns

In this section, we will examine problems in diffraction[17, Chap. 37; 18, Chap. 22; 19, Chap. 33; 20, Chap. 38] of light rays through a two-dimensional circular aperture, and introduce the two-dimensional function stepping tool `step2d()` in the `step.c` utility code. Instead of the object function taking a single argument value that we have been stepping up to this stage, now we introduce another function type, one that may be a function of any number of arguments. Or we can consider it to realize a function $f(\vec{x})$ that accepts a vector of any dimension as its argument and returns a function value of a scalar. In our C code we assume the function is in the form

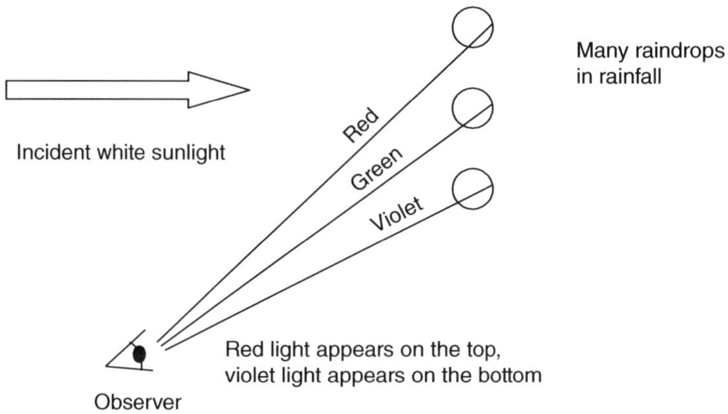

Figure 4.18 Cumulative effect of light rays into an observed rainbow.

```
double foo(ARRAY1D *x,void *p) {
   ...  // some computation
return();
}
```

The first argument `x` is a pointer to an `ARRAY1D` structure that holds the components of the function argument vector, together with its size. In the same manner as the functions stepped above, the second argument `p` is a pointer to an arbitrary structure holding any physical parameters the function needs.

The `step2d()` utility function will accept a pointer to any object function of this form as a first argument, and then followed by two arguments that are pointers to two `GRID` structures. The first of these is the grid over which the function will step the first component of the object function argument vector, and the second grid specifies how the second vector component is stepped. The object function argument vector may have any number of components, but only the first two may be stepped by the two-dimensional nested loop algorithm in `step2d()`. This stepping algorithm produces a two-dimensional table of function values, rather than the one-dimensional table from `step()`. This output table is stored in the array pointed to by the instance of the `ARRAY2D` structure that forms the final argument to `step2d()`. As with the `step()` tool, the code that calls the `step2d()` tool is responsible for allocating the memory space for the function values, filling in the pointers in the `ARRAY2D` structure with a call to `alloc_array2d()`.

We can use this two-dimensional stepping tool to study the optical properties of a small aperture that is comparable in size to the wavelength of light passing through it. This arrangement can be easily treated analytically for the ideal case where the distances between the aperture, the light source, and the surface where the light intensity is measured are very long compared with the wavelength. However, there are many instances where the dimensions of the structure are comparable with the wavelength, such as imaging systems on microelectronic chips. In this case, the complexity of the summation of all the light wavelets at their different phases requires the analysis to be done numerically.

So consider the geometry shown in Figure 4.19. A circular aperture is present in an otherwise opaque surface in the x–y plane, located at z coordinate of 0. The aperture is centered at point (0, 0, 0) and has radius R. A point light source is located at some point below the aperture at a point $(x_0, y_0, -D_0)$. The "observation screen," where the light intensity is to be measured, is on a plane with z coordinate of $+D_1$. We will be numerically tracing wavelets of a given wavelength from the point source to some point at polar coordinates (r_1, θ_1) in the aperture and then to the observation screen at (x, y, D_1). As the light wave propagates through distances d_0 and d_1, the wave phase will progress. We may follow Huygens' Principle to form the sum of all the wavelets over points in the aperture, and the waves will then interfere in a way to form a light pattern on the screen. This is illustrated in Figure 4.20 for two example point wave sources inside the aperture. In this figure, we let $\vec{E}$ be the electric field vector at the observation point, and we know that $\vec{E} = \sum \vec{E}_j$ over all point sources j. We'll assume uniform polarization for all the $\vec{E}_j$ contributions, so for instance, assuming polarization in the $\vec{x}$ direction, $\vec{E}_j = E_{jx}\vec{x}$. Following Huygens' Principle, $E_{jx} = E_{j0}e^{i(\omega t-\phi)} = E_{j0}e^{i\phi}e^{i\omega t}$ where $E_{j0}e^{i\phi}$ is the "phasor" from the source, and ϕ is the phase shift due to path length differences from the source to the observation point. We may simulate that pattern to discern the optical resolution of this geometry. We'll use arbitrary spatial units in the simulation code for simplicity.

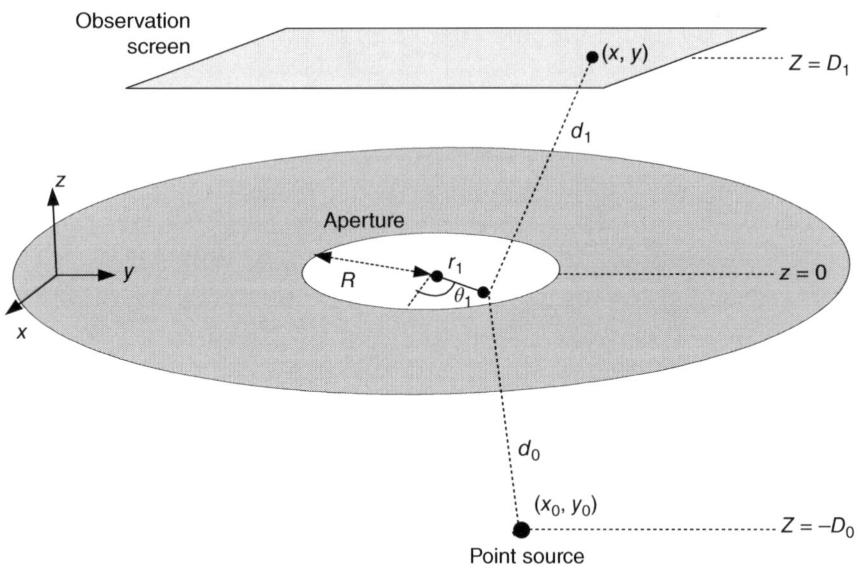

Figure 4.19 Geometry for visualizing diffraction patterns through aperture.

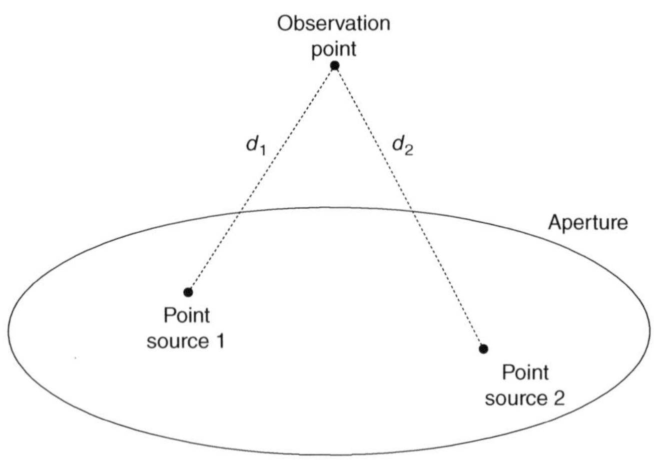

Figure 4.20 Predicting diffraction patterns with light ray interference.

See the program code in `diffraction2D_step_1src.c`, available for download from the Student Companion website. Several physical parameters are needed to describe the geometry of the diffraction system using the structure `DIFF1` in the lines

```
typedef struct diff_1src_param {
  double xsrc;  // x coordinate of light source
  double ysrc;  // y coordinate of light source
  double D0;  // distance from aperture plane to source plane
  double D1;  // distance from aperture plane to observation plane
  double R;  // radius of circular aperture
  double lambda;  // wavelength
} DIFF1;
```

The main program performs the diffraction simulation with a call to the `step2d()` tool, passing a pointer to the function `diffraction2D_1src()`. This function is written to compute the sum of light rays over the aperture. That sum is computed over two-dimensional nested loops that will step over all the wavelet points in the aperture. We assume a uniform polarization of the electromagnetic wave here.

```
double diffraction2D_1src(ARRAY1D *x,void *param_ptr) {
  double d0,d1,d,phi,r1,theta1,sumEi,sumEq,max;
  DIFF1 *p;

  p = param_ptr;
  sumEi = 0.0;
  sumEq = 0.0;
  max = 0.0;
  r1 = 0.05 * p->R;
  while (r1 < (1.001 * p->R)) {
    theta1 = 0.0;
    while (theta1 < (TWO_PI + 0.01)) {
      d1 = sqrt(((x->elem[0] - r1 * cos(theta1)) *
                (x->elem[0] - r1 * cos(theta1))) +
                ((x->elem[1] - r1 * sin(theta1)) *
                (x->elem[1] - r1 * sin(theta1))) +
                (p->D1 * p->D1));
      d0 = sqrt(((r1 * cos(theta1) - p->xsrc) *
                 (r1 * cos(theta1) - p->xsrc)) +
                 ((r1 * sin(theta1) - p->ysrc) *
                 (r1 * sin(theta1) - p->ysrc)) +
                 (p->D0 * p->D0));
      d = d0 + d1;
      phi = TWO_PI * (d - p->D1 - p->D0) / p->lambda;
      sumEi += cos(phi);
      sumEq += sin(phi);
      max += 1.0;
      theta1 += 0.1;
    }
// Increment radius to keep a constant area of each small patch of
// aperture
    r1 += p->R * p->R / (400.0 * r1);
  }
  return(((sumEi * sumEi) + (sumEq * sumEq)) / (max * max));
}
```

Each wavelet point should represent a small patch within the aperture, and each patch should cover the same amount of area to keep the electromagnetic wave power uniform over all wavelets. The increment value for the θ_1 coordinate is uniform through the loop. Therefore the increment value for the r_1 polar radial component is variable according to the current r_1 coordinate in the loop with the line

```
r1 += p->R * p->R / (400.0 * r1);
```

Once all the wavelet endpoint coordinates are calculated, the distances d_0 and d_1 are then known, and the relative phase difference for each wavelet can be calculated for a given wavelength. Note the careful displacement of the floating point loop limits away from expected looping values, as covered in Section 3.3.5.

With the `diffraction2D_1src()` object function defined above, the main program can invoke the `step2d()` tool with the lines

```
alloc_array2d(&ztable,NUM_GRID_PTS(ygrid),NUM_GRID_PTS(xgrid));
step2d(diffraction2D_1src,&xgrid,&ygrid,&param,&ztable);
gen_output(&ztable,&xgrid,&ygrid);
free_array2d(&ztable);
```

Note that, for the conceptual stepping of any function of two components of an argument vector $f(x_0, x_1)$, the `step2d()` function expects that the calling program has allocated the two-dimensional result array with the number of columns set to the number of grid points for the first "x_0" argument and the number of rows set to the number of grid points for the second "x_1" argument. As in the one-dimensional examples in the previous sections, a `gen_output()` function is also defined to generate tabulated data in the output stream, although in this case structured in three columns.

A shell dialog to compile and run this code is

```
$ gcc -Wall -Wshadow -O2 -o diffraction2D_step_1src diffraction2D_step_1src.c CPLibrary/step.c
    ↪ CPLibrary/array.c -lm
$ ./diffraction2D_step_1src
./diffraction2D_step_1src <aperture radius> <wavelength> <screen distance> <source distance>
    ↪ <source x> <source y> <xinit> <xfinal> <xinc> <yinit> <yfinal> <yinc>
$ ./diffraction2D_step_1src 1 0.2 20 20 0 0 -6 6 0.2 -6 6 0.2 >| diffraction2D_1src_0_0.dat
```

We first look at the situation where the single coherent light source illuminates the aperture from a location on its axis, or directly underneath the center, at coordinates $(0, 0, -D_0)$, with $D = 20$ units. The results of the summation performed by the code are shown in Figure 4.21, and represent a single visible spot of light with the fringing we expect from diffraction. Since the light source is located directly on the axis of the aperture, the spot of light on the screen is, as expected, also centered on the axis of the screen directly above the source.

With this exercise we take the opportunity to explore the capability of `gnuplot` to produce projections of a three-dimensional plot.

```
gnuplot> unset key
gnuplot> set hidden3d
gnuplot> set xlabel 'X-coordinate'
gnuplot> set ylabel 'Y-coordinate'
gnuplot> set zlabel 'Relative intensity' rotate by 90
gnuplot> set ticslevel 0.25
gnuplot> splot 'diffraction2D_1src_0_0.dat' u 1:2:3 w l
```

The `plot` command is replaced by `splot` to render a projection of a surface plot, where the surface of a function of the x and y coordinates is plotted in the xyz system. Note that there are now three column numbers specified in the `using` instruction, mapping those data file column numbers to x, y, and z coordinates, respectively. The surface is rendered with lines connecting adjacent points. In an interactive plotting session with `gnuplot`, the user may rotate the viewing angle around in three dimensions with the mouse, providing a good view of regions of interest. Looking at the code in `diffraction2D_step_1src.c`, note that there is a nested structure of two loops at the end of the main program that writes the x and y coordinates and the function value to the output stream in three columns. At each termination of the inner loop an extra new line character is sent to the output, and this is the indicator to `gnuplot` as to where to separate each contiguous line on the plot surface.

Next we performed the numerical experiment of moving the light source to two other locations along the x axis. Figures 4.22 and 4.23 show the resulting spot of light on the screen when the source is located at $(3, 0, -D_0)$ and $(-3, 0, -D_0)$, respectively. Note that when the light source is located on one side of the x axis, the spot of light moves on the screen in the other direction along the axis. This is as expected, as the light rays that pass through the aperture are now directed partly in a lateral direction, and that the light rays must follow the shortest path to the screen. Therefore, with this simple simulation code and these figures, we have numerically verified that Fermat's Principle follows from Huygens' Principle.

Next we add a second light source to study the limits imposed by diffraction on the resolution with which we can distinguish two closely separated sources. See the code in `diffraction2D_step_2src.c`, available for download from the Student Companion website.

The structure to hold the physical system parameters has been extended to `DIFF2` as

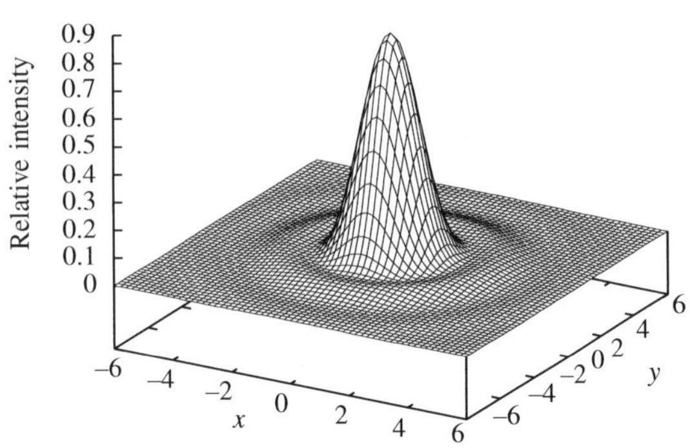

Figure 4.21 Diffraction through circular aperture with light source at (0, 0, −20).

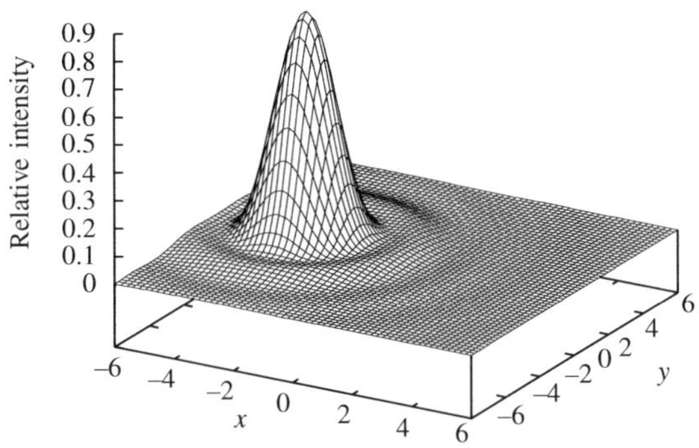

Figure 4.22 Diffraction through circular aperture with light source at (3, 0, −20).

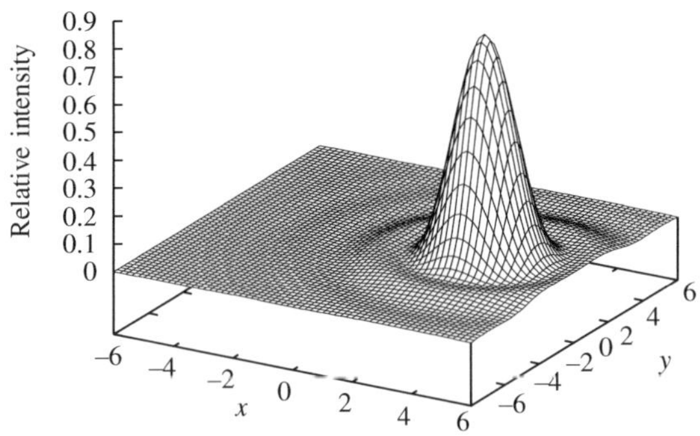

Figure 4.23 Diffraction through circular aperture with light source at (−3, 0, −20).

```
typedef struct diff_2src_param {
  double xsrc1;  // x coordinate of light source 1
  double ysrc1;  // y coordinate of light source 1
  double xsrc2;  // x coordinate of light source 2
  double ysrc2;  // y coordinate of light source 2
  double D0;  // distance from aperture plane to source plane
  double D1;  // distance from aperture plane to observation plane
  double R;  // radius of circular aperture
  double lambda;  // wavelength
} DIFF2;
```

Now the function `diffraction2D_2src()` is passed to the `step2d()` utility, with the modifications to add in the squared intensity from a coherent second source. With the sources initially widely separated, the two light sources are quite distinguishable on the observation screen of Figure 4.24. For this simulation, the two sources are located at $(3, 0, -D_0)$ and $(-3, 0, -D_0)$.

However, if the two light sources are moved closer together, at some point they will become indistinguishable, as in Figure 4.25. These two sources are at $(1.5, 0, -D_0)$ and $(-1.5, 0, -D_0)$, and are closer than what may be effectively resolved by this optical system.

So clearly the limit on resolution for this optical system is somewhere between the separations of Figures 4.24 and 4.25. Some trial runs of the simulation at various separation distances can be used to visualize and measure the limit of resolution for this optical system in Figure 4.26.

Another variation is to invert the circular aperture into a circular shadowing disc, as shown in Figure 4.27. The diffraction pattern from a single coherent source centered on its axis is a simulation of the historical Poisson spot. The experimental observation of this spot of light in 1818 was a verification of the wave theory of light. This numerical problem now involves forming the diffraction sum over radii extending from the radius of the disc outward to a large distance, and still adjusting the radial increment for each iteration to keep the patch area uniform. Writing the code for this simulation is Exercise 4.4 for the student, and can be accomplished most straightforwardly by modifying the code in `diffraction2D_step_1src.c`. This computation involves a very large number of iterations, and generating the well-defined simulation plot in Figure 4.28

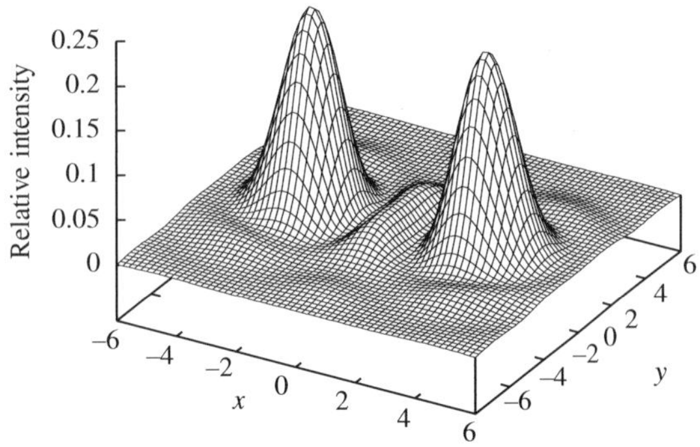

Figure 4.24 Diffraction with two light sources at (−3, 0, −20) and (3, 0, −20).

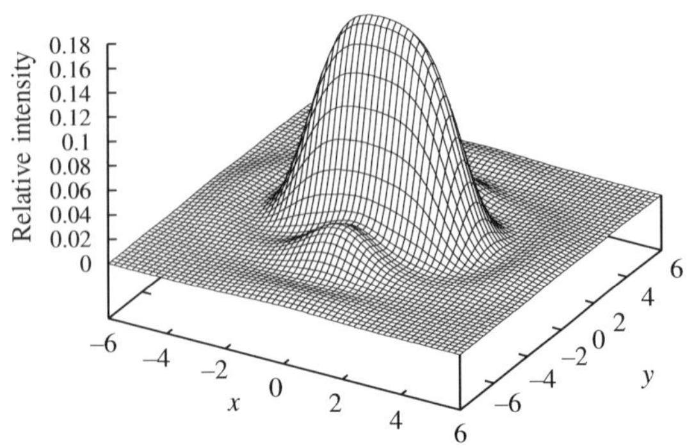

Figure 4.25 Diffraction with two light sources at (−1.5, 0, −20) and (1.5, 0, −20).

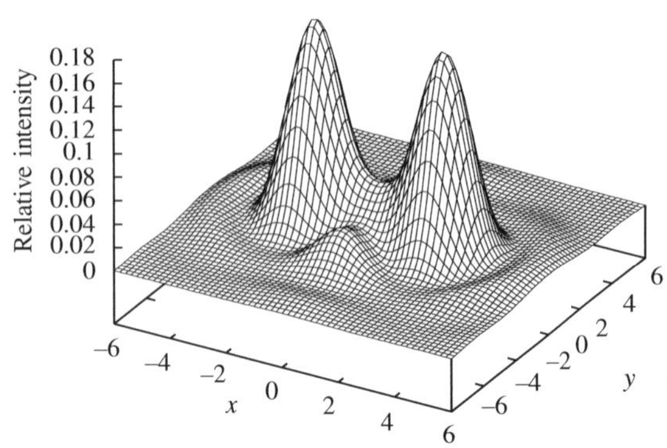

Figure 4.26 Diffraction with two light sources at (−2, 0, −20) and (2, 0, −20).

takes a large amount of processing time. Without the careful attention to computing efficiency from Chapters 2 and 3, we would be required to wait an impractical length of time for this simulation task to complete.

4.8 Application: Collisions

There are many interesting physical consequences of the collision of two or more objects. For instance, planets are created after many smaller objects collide and coalesce; the pressure of confined gasses is due to an enormous number of molecular collisions; and subatomic structures are probed with collisions. Let us now use our numerical tools to visualize some properties and consequences of colliding objects.

Consider the problem of the collision of two spherical balls that are able to move freely in three dimensions. Momentum will be preserved before and after the collision, and we'll assume a perfectly elastic collision where energy is conserved as well. The two balls, numbered 1 and 2, are characterized by radii R_1 and R_2 and masses m_1 and m_2. As illustrated in

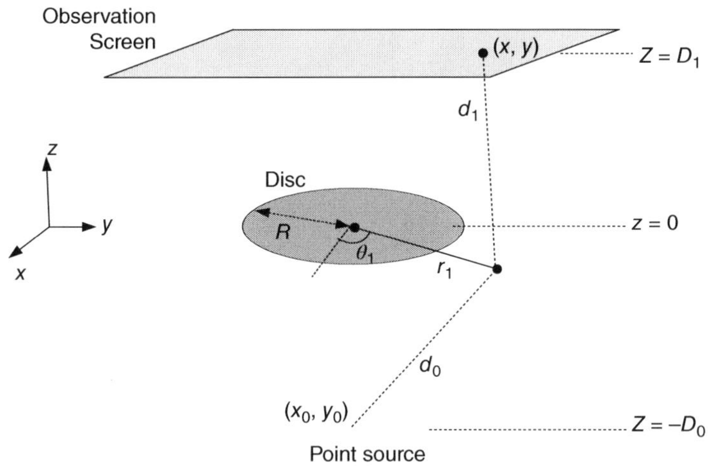

Figure 4.27 Geometry for visualizing diffraction pattern around a shadowing disc.

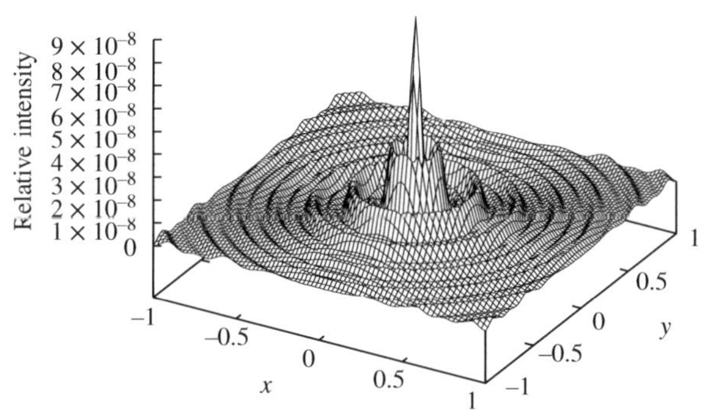

Figure 4.28 Diffraction around a circular disc showing the Poisson spot.

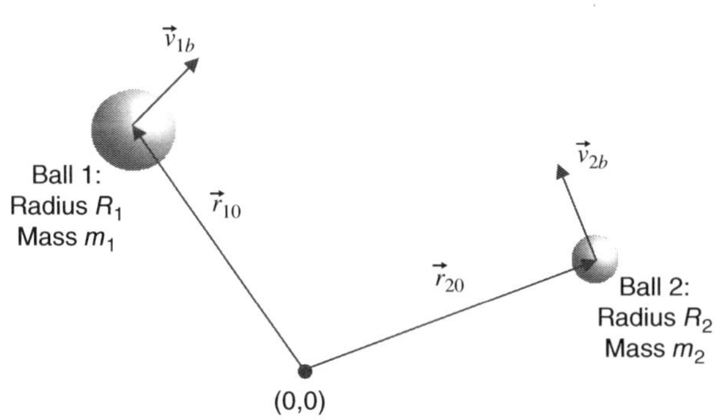

Figure 4.29 Initial configuration of two balls at time $t = 0$.

Figure 4.29, initially, for time $t = 0$, the centers of the balls are located at position vectors $\vec{r}_{10}$ and $\vec{r}_{20}$, and are moving with velocity vectors $\vec{v}_{10}$ and $\vec{v}_{20}$.

At any time *t before* any collision occurs, the balls will be located at $\vec{r}_{10} + \vec{v}_{1b}t$ and $\vec{r}_{20} + \vec{v}_{2b}t$. If this initial condition does not result in a collision, these expressions will predict the location of the balls for all time $t > 0$. However, if a collision does ensue, then at the instant of the first contact the geometry is illustrated in Figure 4.30.

If this collision occurs, we can solve for the time of the collision t_c when the centers of the two balls are separated by exactly the sum of the two radii, $R_1 + R_2$. For the geometry in Figure 4.30 to be satisfied, there must be a time when the distance between the centers of the two balls $|\vec{r}_{10} + \vec{v}_{1b}t_c - \vec{r}_{20} - \vec{v}_{2b}t_c|$ is just equal to the sum of the two radii $R_1 + R_2$.

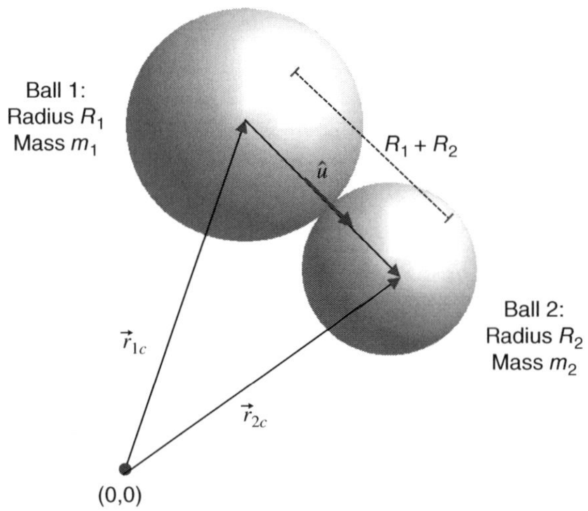

Figure 4.30 The geometry of the collision between two balls.

This equation also must hold for the squares of the distances as well. So we may write in the three components of the position vectors

$$(r_{10x} + v_{1bx}t_c - r_{20x} - v_{2bx}t_c)^2 + (r_{10y} + v_{1by}t_c - r_{20y} - v_{2by}t_c)^2 + (r_{10z} + v_{1bz}t_c - r_{20z} - v_{2bz}t_c)^2 = (R_1 + R_2)^2 \tag{4.5}$$

By expanding the squares and collecting terms that are either constant, linear in t_c, or quadratic in t_c, we can express Equation (4.5) in a standard quadratic form

$$At_c^2 + Bt_c + C = 0 \tag{4.6}$$

where the coefficients are

$$\begin{aligned} A &= (v_{1bx} - v_{2bx})^2 + (v_{1by} - v_{2by})^2 + (v_{1bz} - v_{2bz})^2 \\ B &= 2(r_{10x} - r_{20x})(v_{1bx} - v_{2bx}) + 2(r_{10y} - r_{20y})(v_{1by} - v_{2by}) + 2(r_{10z} - r_{20z})(v_{1bz} - v_{2bz}) \\ C &= (r_{10x} - r_{20x})^2 + (r_{10y} - r_{20y})^2 + (r_{10z} - r_{20z})^2 - (R_1 + R_2)^2 \end{aligned} \tag{4.7}$$

The standard quadratic formula can then be used to find the collision time t_c, under the condition that a collision actually occurs. Here we use the form of the quadratic formula that allows us to cancel out the factors of two in the expression for B in Equation (4.7) and save a bit of computation.

$$t_c = \frac{-\left(\frac{B}{2}\right) \pm \sqrt{\left(\frac{B}{2}\right)^2 - AC}}{A} \tag{4.8}$$

The relevant solution will be that with the negative sign in front of the square root, as this will represent the first time the geometric conditions are satisfied. (The later time the geometry is satisfied with the positive sign corresponds to the balls mysteriously passing through each other and just emerging from an overlap condition, which is not a particularly meaningful situation.) We can test for the condition that the geometry of Figure 4.30 is consistent with the initial conditions by testing the sign of the discriminant $D = \left(\frac{B}{2}\right)^2 - AC$. If D is positive, then a collision indeed occurs, and we may solve for the collision time t_c with Equation (4.8). Then at the collision time, we denote the ball positions as

$$\begin{aligned} \vec{r}_{1c} &= \vec{r}_{10} + \vec{v}_{1b}t_c \\ \vec{r}_{2c} &= \vec{r}_{20} + \vec{v}_{2b}t_c \end{aligned} \tag{4.9}$$

During a collision event that occurs between times t_1 and t_2 the momentum of the balls will be changed by an impulse vector that acts on the two balls

$$\vec{J} = \int_{t1}^{t2} \vec{F}(t)dt \tag{4.10}$$

In the most general collision case, there are more variables than equations, and a unique solution cannot be found. We will revisit this collision process later in Section 13.3, where we solve the differential equations describing a gradual impulse with "springy" surfaces of the balls. But at this stage, we can make an ideal assumption about the direction of an instantaneous impulse of the collision, and a unique solution with straightforward expression evaluation may be found.

We assume that the surface of the balls is perfectly smooth, like that of an ideal ball for the game of pool. Although any force directed normal to the surfaces may be supported, motion tangential to the surfaces will assumed to slip without friction. We can also assume that the duration of the impulse $t_2 - t_1$ is negligible compared to the travel time of the balls, so we can simply consider the action of an instantaneous vector impulse on the two balls in a direction normal to the surfaces, and at a single instantaneous point of contact, at the instant illustrated in Figure 4.30. In Section 13.3, we'll see numerically that this ideal assumption is the limit of the differential equation solutions as the stiffness of "springy" ball surfaces increases. But for now, we'll assume that the impulse vector can be expressed as $\vec{J} = J\hat{u}$, where J is a scalar impulse magnitude that has yet to be solved for, and $\hat{u}$ is the unit vector directed on a line between the centers. Having solved for the position vectors $\vec{r}_{1c}$ and $\vec{r}_{2c}$ with Equation (4.9) at the point of the collision, we can easily calculate this unit vector as

$$\hat{u} = \frac{\vec{r}_{2c} - \vec{r}_{1c}}{R_1 + R_2} \tag{4.11}$$

With the ideal impulse assumption, the conservation of momentum expressions are

$$\begin{aligned} \vec{p}_{1a} &= \vec{p}_{1b} - J\hat{u} \\ \vec{p}_{2a} &= \vec{p}_{2b} + J\hat{u} \end{aligned} \tag{4.12}$$

or, broken down into x, y, and z components

$$\begin{aligned} p_{1ax} &= p_{1bx} - Ju_x \\ p_{1ay} &= p_{1by} - Ju_y \\ p_{1az} &= p_{1bz} - Ju_z \\ p_{2ay} &= p_{2by} + Ju_y \\ p_{2ay} &= p_{2by} + Ju_y \\ p_{2az} &= p_{2bz} + Ju_z \end{aligned} \tag{4.13}$$

In order to solve for the scalar impulse magnitude J, and assuming a perfectly elastic collision, we substitute each of these momentum components into the conservation of energy relationship

$$\frac{p_{1bx}^2}{m_1} + \frac{p_{1by}^2}{m_1} + \frac{p_{1bz}^2}{m_1} + \frac{p_{2bx}^2}{m_2} + \frac{p_{2by}^2}{m_2} + \frac{p_{2bz}^2}{m_2} = \frac{p_{1ax}^2}{m_1} + \frac{p_{1ay}^2}{m_1} + \frac{p_{1az}^2}{m_1} + \frac{p_{2ax}^2}{m_2} + \frac{p_{2ay}^2}{m_2} + \frac{p_{2az}^2}{m_2} \tag{4.14}$$

By collecting terms we arrive at the result

$$J = \frac{2m_2}{m_1 + m_2}(\vec{p}_{1b} \cdot \hat{u}) - \frac{2m_1}{m_1 + m_2}(\vec{p}_{2b} \cdot \hat{u}) \tag{4.15}$$

where the "·" operator denotes a vector dot product. Then once J is computed from Equation (4.15), the two momentum vectors after the instantaneous collision may be calculated from the conservation of momentum equations (4.12) or (4.13). After the collision, as shown in Figure 4.31, the balls follow their trajectories with velocities $\vec{v}_{1a}$ and $\vec{v}_{2a}$, so at any time t *after* the collision, the balls will be located at $\vec{r}_{1c} + \vec{v}_{1a}(t - t_c)$ and $\vec{r}_{2c} + \vec{v}_{2a}(t - t_c)$.

Again with the intention of developing our base of simulation code in small modules, and enabling as much as possible the reuse of already written and debugged code, we will now collect the code that models this collision process.

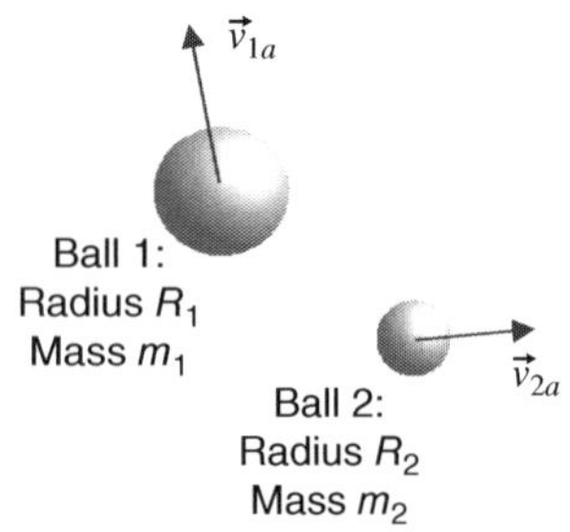

Figure 4.31 The configuration after a collision between two balls.

First we revisit small modules initially developed in Section 3.2 for defining and tracking the motion of spherical balls, in Listings 3.4 and 3.5. Then another module, listed below, follows in the code `collide_balls.c` in the CPLibrary package, available for download from the Student Companion website, as a utility function that simulates the collision of two balls, following Equations (4.7), (4.8), (4.11), (4.15), and (4.12). This utility's corresponding header file is in `collide_balls.h`.

```
int collide_balls(BALL *b1,BALL *b2,double *tc) {
  double A,Bover2,C,D,ux,uy,uz,J;

  A = (b1->vx - b2->vx) * (b1->vx - b2->vx) +
      (b1->vy - b2->vy) * (b1->vy - b2->vy) +
      (b1->vz - b2->vz) * (b1->vz - b2->vz);
  Bover2 = (b1->rx - b2->rx) * (b1->vx - b2->vx) +
           (b1->ry - b2->ry) * (b1->vy - b2->vy) +
           (b1->rz - b2->rz) * (b1->vz - b2->vz);
  C = (b1->rx - b2->rx) * (b1->rx - b2->rx) +
      (b1->ry - b2->ry) * (b1->ry - b2->ry) +
      (b1->rz - b2->rz) * (b1->rz - b2->rz) -
      (b1->radius + b2->radius) * (b1->radius + b2->radius);
      D = Bover2 * Bover2 - A * C;
  if (D > 0.0) {
    *tc = (-Bover2 - sqrt(D)) / A;
    ball_move(b1,*tc);
    ball_move(b2,*tc);
    ux = (b2->rx - b1->rx) / (b1->radius + b2->radius);
    uy = (b2->ry - b1->ry) / (b1->radius + b2->radius);
    uz = (b2->rz - b1->rz) / (b1->radius + b2->radius);
    J = ((2.0 * b1->mass * b2->mass) / (b1->mass + b2->mass)) *
        ((b1->vx * ux + b1->vy * uy + b1->vz * uz) -
        (b2->vx * ux + b2->vy * uy + b2->vz * uz));
    b1->vx -= J * ux / b1->mass;
    b1->vy -= J * uy / b1->mass;
    b1->vz -= J * uz / b1->mass;
    b2->vx += J * ux / b2->mass;
    b2->vy += J * uy / b2->mass;
    b2->vz += J * uz / b2->mass;
    return(BALLS_COLLIDE);
  }
  else {
    return(BALLS_DO_NOT_COLLIDE);
  }
}
```

The utility function `collide_balls()` illustrates a typical coding tactic for modeling a process that may involve a number of possible outcomes. In this case, there are two possible scenarios, where the initial conditions of the two balls may or may not result in a collision. So we make use of the return value from this utility function, declared as an `int` type, as simply an integer code that indicates the result. It is always more convenient while writing our programs to refer to integer codes as a symbolic name, rather than always having to remember some integer value with no mnemonic value. So we use two lines in `collide_balls.h` to define two constant macros

```
#define BALLS_COLLIDE 0
#define BALLS_DO_NOT_COLLIDE 1
```

These constants are used as return values from `collide_balls()` according to the conditional test on the discriminant $D = \left(\frac{B}{2}\right)^2 - AC$ above. Since any part of the code that makes use of the `collide_balls()` utility function will also include the `collide_balls.h` header file to process the prototype for the function, those same symbolic constants may be used by the calling code to test the function return value and execute its code conditionally on the two possible outcomes. Then we also pass a pointer to a double precision variable to the function, and if the simulations indicate a collision does occur, then the function will make use of the pointer as a location to store the collision time t_c. It will then make use of the `ball_move()` utility function in the `ball.c` package to move the ball to the position of the instantaneous contact in Figure 4.30. This is necessary to update the position vectors of the balls, and from these positions calculate the impulse unit vector $\hat{u}$. With that unit vector and the impulse computed, the velocity vectors of the two balls are then updated to their values after the collision. However, if a collision does not occur, nothing is stored at the collision time pointer location, and no modifications are done to the two ball structures. Two different integer return values are used according to the action taken in the function. We will make extensive use of this tactic for integer return values in Chapter 5, as root-finding algorithms may or may not converge properly on roots of a nonlinear function.

Let us exercise this collision simulation utility with a simple program that models a collision of two balls that are rolling on a flat and level plane of negligible friction. This physical scenario is otherwise known as a shot in the game of pool. Consider the collision between a cue ball and a target ball as shown in Figure 4.32. This only exercises the model in two of the three dimensions, and is a simulation of a pool player making a delicate side pocket shot. It is well known that such shots at wide deflection angles must be made with a careful and gentle impulse from the cue ball, as this will let the collision more closely follow the ideal of a complete slip in the circumferential direction, and allow the collision impulse to act only in the radial direction, albeit slowly. (Alternatively, we could also consider these geometries to be what we would expect for collisions of circular pucks on a flat frictionless air table.)

The main program code is in the program `two_pool_balls.c`, available for download from the Student Companion website. Notable in this main program is a slightly different method to extract numerical values from the text arguments listed on the program's command line. Up to this point, we have simulated systems that can be described by several numerical parameters, and assigning each parameter to an isolated command line word, and then translating that word to a numerical value using the `atof()`, function is practical. This collision system, however, requires many parameters to specify the relevant initial conditions, and having each parameter occupy a separate command line word can grow to be inconvenient. So this main program illustrates making use of the C library `sscanf()` function to scan and translate the text in a command line word comprising several parameters, separated by a comma character. The expected command line format with three argument words specifying the initial conditions for the motion of the balls is in the form

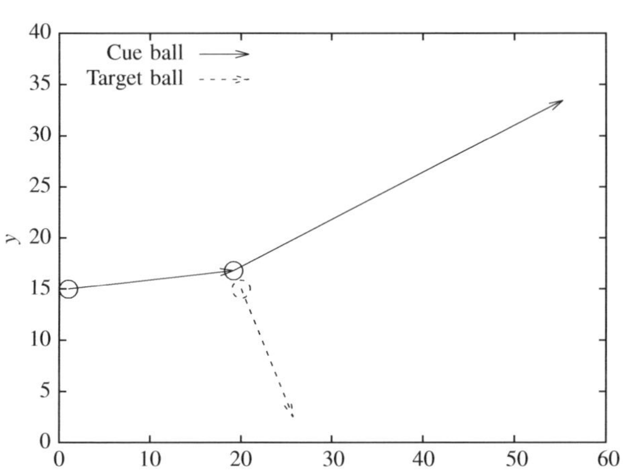

Figure 4.32 Example collision path simulating a side pocket pool shot.

```
$ ./two_pool_balls <r10x,r10y,r10z,v1bx,v1by,v1bz,R1,m1> <r20x,r20y,r20z,v2bx,v2by,v2bz,R2,m2>
    ↪ <tfinal>
```

and these parameters are translated and stored into their corresponding double-precision variables with the lines

```
sscanf(argv[1],"%lf,%lf,%lf,%lf,%lf,%lf,%lf,%lf",
       &ball1.rx,&ball1.ry,&ball1.rz,&ball1.vx,&ball1.vy,&ball1.vz,
       &ball1.radius,&ball1.mass);
sscanf(argv[2],"%lf,%lf,%lf,%lf,%lf,%lf,%lf,%lf",
       &ball2.rx,&ball2.ry,&ball2.rz,&ball2.vx,&ball2.vy,&ball2.vz,
       &ball2.radius,&ball2.mass);
tfinal = atof(argv[3]);
```

The `sscanf()` function expects as a first argument a pointer to a text string to scan, in this case a pointer to a particular command line word. Then the next argument is a text string that specifies the format of fields to guide the scanning process, in a syntax similar to the conversion strings in `printf()` function. This is followed by a sequence of *pointer* arguments giving the locations where the scanned numeric values are to be stored. Unlike with `printf()`, the variables must be specified with pointers because the `sscanf()` function must store values there, rather than simply invoke a value already stored.

So we build and run the program, together with the code for both utilities of `ball.c` and `collide_balls.c`, with a shell dialog of

```
$ gcc -Wall -Wshadow -O2 -o two_pool_balls two_pool_balls.c CPLibrary/collide_balls.c
    ↪ CPLibrary/ball.c -lm
$ ./two_pool_balls
./two_pool_balls <r10x,r10y,r10z,v1bx,v1by,v1bz,R1,m1> <r20x,r20y,r20z,v2bx,v2by,v2bz,R2,m2>
    ↪ <tfinal>
$ ./two_pool_balls 1,15,0,1,0.1,0,1,1 20,15,0,0,0,0,1,1 60 >| side_pocket.dat
```

Then a dialog with `gnuplot` generates a visualization of the side pocket shot

```
gnuplot> set key left top
gnuplot> set xrange [0:60]
gnuplot> set yrange [0:40]
gnuplot> set xlabel 'X-coordinate'
gnuplot> set ylabel 'Y-coordinate'
gnuplot> plot 'side_pocket.dat' u 1:2:(1) w circle lc 'red' t 'Cue ball','' u 1:2:4:5 w vect lc
    ↪ 'red' notitle,'' u 7:8:(1) w circle lc 'green' t 'Target ball','' u 7:8:10:11 w vect lc
    ↪ 'green' notitle
```

This dialog introduces two different plotting modes for the `plot` command. Instead of connecting adjacent plot coordinates with lines with the `line` mode, we first draw a circle at each coordinate with the `circle` mode. In conjunction with the program code that outputs the position of the balls at the critical turning points, this makes the plot more graphically appealing as something like a "stroboscopic" view of the process. The `circle` mode requires three data elements to plot, as specified by the first `using` or *u* phrase. The first two are the *x* and *y* coordinates of the center of the circle, and the third is the radius. In the line above, we map columns 1 and 2 in the data file to the center coordinates, and then a constant expression evaluating to 1 for the radius, matching the radius specified on the command line to run the simulation. Then we use the `vector` plot mode as well to add directed arrows into the plot to emphasize the motion of the balls. This mode requires four data elements, the first two being the *x* and *y* coordinates of the start of the vector, and the last two being the *x* and *y* components of the vector length. These are mapped from columns 1, 2, 4, and 5 in the data file, as prepared by the simulation code. The color of each plotting symbol is explicitly specified by the `linecolor` clauses, here conveniently abbreviated with `lc`. Although we can render color plots suitable for a graphical display with the `gnuplot` dialog above, Figure 4.32 shows an equivalent plot with different line colors and types suitable for a black-and-white printed page.

As an exercise, the student may rerun the program with a slightly different initial velocity vector for the first ball, representing a narrowly missed shot

```
$ ./two_pool_balls 1,15,0,1,0.11,0,1,1 20,15,0,0,0,0,1,1 60 >| side_pocket.dat
```

Replotting the output vectors will show the trajectory of the cue ball without a collision.

In Section 8.8, we will again use this ideal collision model in all three dimensions in combination with Monte Carlo simulation to visualize classical scattering. Then in Section 8.11, we will use this model in a three-dimensional Monte Carlo simulation to visualize the velocity distribution of an ideal gas coming to an equilibrium with the statistics predicted by the Maxwell–Boltzmann law.

4.9 Application: Quantum Wave Packets

The pioneers of quantum physics opened our understanding of the duality of matter and waves, and the mechanics of particles at energy levels on atomic and nuclear scales. We may solve the Schrödinger differential equation to find wave solutions giving the probability density of the existence of a particle possibly under the influence of a potential field. For simplicity, let us start with a numerical study of solutions representing movement in a single dimension, where the particle with energy E is moving in a potential with one-dimensional dependence given by the function $V(x)$.

In later sections, we will be finding numerical solutions of the time-independent Schrödinger equation in one dimension

$$\frac{d^2\psi(x)}{dx^2} = -\frac{2m}{\hbar^2}\left[E - V(x)\right]\psi(x) \tag{4.16}$$

for various interesting potential profiles. In this section, however, we will be examining only the regions where particles are free to move and not subject to forces, or where $V(x) = 0$ and $E > 0$. In these regions, solutions of Equation (4.16) are in the form of the complex exponentials

$$\psi(x) = Ae^{+ikx} + Be^{-ikx} \tag{4.17}$$

where A and B are constants, possibly complex-valued, evaluated from the spatial boundary conditions. This makes the complete solution of the wave equation, including the time dependence, in the form

$$\Psi(x,t) = Ae^{+i(kx-\omega t)} + Be^{-i(kx+\omega t)} \tag{4.18}$$

This single wave solution with some particular wave number k and time frequency ω is associated with a continuous flux of particles in the positive or negative x direction, with momentum p and energy E, where $k = \frac{p}{\hbar}$ and $\omega = \frac{E}{\hbar}$. This means that for wave solutions of the Schrödinger equation, the important dispersion relationship between ω and k is

$$\omega = \frac{\hbar}{2m}k^2 \tag{4.19}$$

The phase velocity v_p of a *single wave* with particular values of k and ω is the velocity of the point in space that maintains a constant sinusoidal phase

$$v_p = \frac{\omega}{k} = \frac{\hbar}{2m}k \tag{4.20}$$

Generally, however, instead of single wave solutions representing a flux of particles, we are interested in solutions that are associated with individual particles in motion [1, Chap. 3; 2, Chap. 4; 3, Chap. 2; 4, Chap. 1] that are localized in space within an allowed certainty limit. This scenario may be treated by forming solutions that are continuous linear combinations of multiple waves in the form of packets where

$$\Psi(x,t) = \int A(k)\Psi_k(x,t)dk \tag{4.21}$$

To visualize how summing a number of constituent waves at different wave numbers can lead to a localization in space, consider the plots in Figure 4.33. The five constituent cosine functions in the top plot, none of which show any particular localization in space, are summed to produce the packet shown in the bottom plot. Note that at x coordinate values around 0, the phases of the constituent waves all align to produce the highest localized peak in the summed packet. At x coordinates farther from 0, the phase correlations become more scattered, and the oscillatory peaks in the summed packet fall off in magnitude. We are summing only five waves in this simple demonstration, and all the constituent waves contribute to the sum with equal weighting, so in this simple example there is still a considerable oscillatory nature in the packet. But this principle may be extended to a large number of constituents with a prescribed weighting function to produce a smoother and more localized packet.

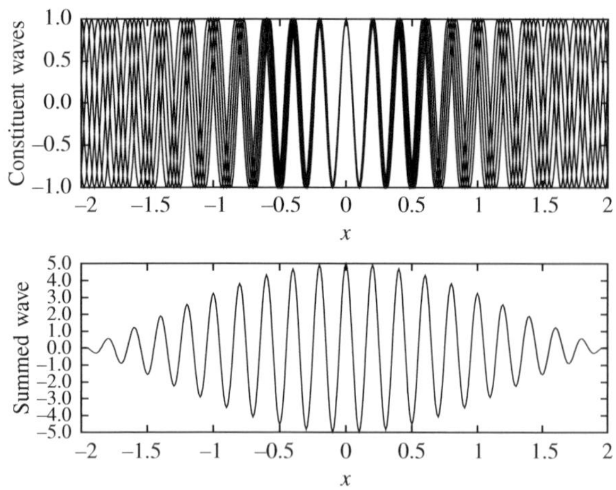

Figure 4.33 A wave packet consisting of the sum of five equally weighted sinusoids at different spatial frequencies.

It is mathematically convenient to consider Gaussian wave packets where we introduce a parameter a to specify the width of a Gaussian distribution of wave numbers about a central wave number k_0. With this formulation of the wave packet coefficients, as the parameter a increases, the spread of wave numbers constituting the packet decreases:

$$A(k) = e^{-a^2(k-k_0)^2} \tag{4.22}$$

The spatial localization of the existence probability mathematically comes from the constructive interference of the phases of all the constituent waves in the packet in the vicinity of one particular location, and destructive interference elsewhere. The group velocity v_g of a wave packet is the velocity of that peak of spatial distribution, and it is a consequence of the dispersion relationship of Equation (4.19) that

$$v_g = \left.\frac{d\omega}{dk}\right|_{(k=k_0)} = \frac{2\hbar}{2m}k_0 = 2v_{p0} \tag{4.23}$$

where v_{p0} is the phase velocity of the single wave at the center of the k distribution with wave number k_0. Note that

$$v_g = \frac{2\hbar}{2m}k_0 = \frac{\hbar}{m}\frac{p_0}{\hbar} = \frac{p_0}{m} \tag{4.24}$$

So it is the wave packet group velocity that corresponds to the velocity $\frac{p}{m}$ in the classical limit.

For numerical work, the continuous wave packet may be approximated by a finite sum

$$\Psi(x,t) = \sum_k A_k \Psi_k(x,t) \tag{4.25}$$

where the discrete weighting coefficients A_k forming the packet follow the Gaussian distribution

$$A_k = e^{-a^2(k-k_0)^2} \tag{4.26}$$

What is an appropriate value of the width parameter a in Equation (4.26)? For numerical work, we will construct a wave packet with the discrete sum of Equation (4.25), with a distinct minimum and maximum value of the wave number k contributing to the finite sum. As an example, let's consider including wave numbers k less than and greater than the central wave number k_0 by a symmetrical half-width, and displaced from k_0 sufficiently to cause the coefficient A_k to fall from 1 at the center of the distribution down to $\frac{1}{100}$. Evaluating the weighting coefficient A_k according to Equation (4.26) at the extremes of $k = k_0 \pm \frac{\Delta k}{2}$, we see that

$$\frac{1}{100} = e^{-a^2\left(\frac{\Delta k}{2}\right)^2} \tag{4.27}$$

from which we can solve for a required value of a as

$$a = \frac{2\sqrt{-\ln(0.01)}}{\Delta k} \approx \frac{4.2919321}{\Delta k} \tag{4.28}$$

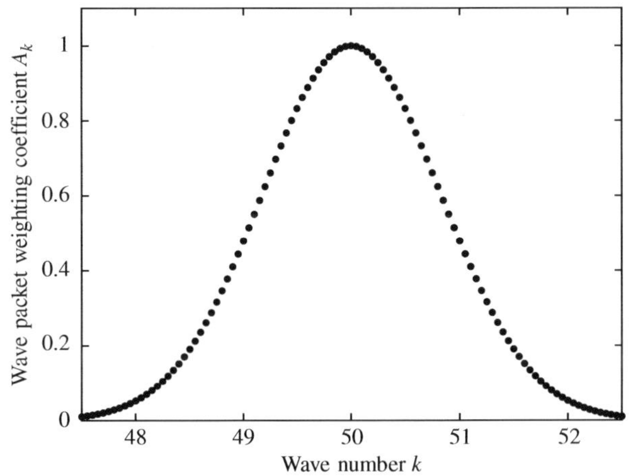

Figure 4.34 Weighting coefficients A_k of an example discrete wave packet with $k_0 = 50$ and $\Delta k = 5$.

For the example values of a central wave number $k_0 = 50$ in some system of units and a total wave number spread of $\Delta k = 5$, the wave numbers k range from 47.5 to 52.5. If we form a discrete wave packet from the summation in Equation (4.25) with a total of 101 values of the wave number k, then the weighting coefficients A_k, ranging from 0.01 up to 1 and back down to 0.01, will be those plotted as in Figure 4.34.

We must keep in mind that only a wave packet formed from an infinite number of constituent waves with a continuous distribution of wave numbers mathematically will be able to represent a particle that is localized to the neighborhood of one limited region in space. However, for numerical work, we are limited to packets constituted from a finite number of waves with discretely spaced wave numbers, and we will explore the appropriate number of waves to assemble in our numerical packet for a given simulation task in Exercise 4.6.

Let's use the structure of a function stepping utility tool we have developed in previous sections to evaluate a discrete wave packet in the form of Equation (4.25) and visualize its propagation through space as time progresses. However, these general solutions of the Schrödinger equation are complex valued, and the software routines we have developed to this point are all real valued, represented by double-precision floating point quantities. So the `step_complex.c` and `step_complex.h` source code files in the CPLibrary download package show a straightforward modification of our stepping tools, with the assumed `double` data types for the function value replaced by the standard C data type for complex values, named `double complex`. For this data type, both real and imaginary components are represented by double-precision quantities, and the arithmetic operations on these variables are implemented. In addition, versions of the C library transcendental functions are available for complex values as well. The stepping utility also makes use of the source code file `array_complex.c`, containing complex versions of the enhanced array utilities described in Section 3.1.4.

Then the main program in `wave_packet.c`, available for download from the Student Companion website, establishes the parameter structure data type `WAVEPACK` and a complex-valued object function to be stepped named `wave_packet()` that computes the discrete sum of wave function solutions of Equation (4.25). The student may inspect that source file example for more detailed comments, but the essence of the object function code is as follows.

```
typedef struct wp_param {
  double a;  // Gaussian spread parameter for wave number k
  double vp0;  // Phase velocity of center wave number k0
  double t;  // Time
  GRID kgrid;  // Sampling grid for wave number k
} WAVEPACK;

double complex wave_packet(double x,void *param_ptr) {
  double hbarover2m,k,k0,kp,omega,A;
  double complex psi;
  int i,n;
  WAVEPACK *p;

  p = param_ptr;
  k0 = (p->kgrid.init + p->kgrid.final) * 0.5;
```

```
  hbarover2m = p->vp0 / k0;
  psi = 0.0;
  n = NUM_GRID_PTS(p->kgrid);
  for (i = 0; i < n; i++) {
    k = p->kgrid.init + ((double) i) * p->kgrid.inc;
    kp = k - k0;
    omega = hbarover2m * k * k;
    A = exp(-p->a * p->a * kp * kp);
    psi += A * cexp(I * (k * x - omega * p->t));
  }
  return(psi);
}
```

We can think of our program as a numerical laboratory with which we can explore the general properties of quantum wave packets. For this purpose, arbitrary units are assumed, and no normalization is performed. An iterative loop in `wave_packet()` steps through each wave number in the sum of Equation (4.25). For each wave number, the corresponding value of ω is computed by the dispersion relationship of Equation (4.19), and each Gaussian weighting coefficient is computed by Equation (4.26) using the parameters stored in the structure data type `WAVEPACK`. Note the inclusion of a `GRID` structure as one of the members of the `WAVEPACK` structure to store the grid parameters for the wave number stepping. The complex exponential function is computed by the complex-valued `cexp()` library function. The time parameter in Equation (4.18) is accepted as a third grid specification in command line arguments of the main program. The wave packet stepping is repeated for each time value in its grid. The time parameter for each simulation biases all the phases of the complex exponential terms in Equation (4.18) so that the constructive interface occurs at the elapsed time t. This phase interference in the propagating wave group corresponds to free motion in the classical limit.

By calling the `gen_output()` function, the main program writes all of the values in the tables prepared by `step_complex()` to the output stream as usual, but the table entry is complex valued. So there are actually four output columns, the first being the x coordinate, the second and the third being the real and imaginary parts of the complex wave function value, and the fourth being the square of the complex magnitude. This gives us a choice of how to subsequently plot and graphically present these values. We can select the time value corresponding to a given plot by selecting the dataset index in the `plot` command.

How many waves must be practically assembled in a finite wave packet to approximate the continuous wave packet that represents a localized particle in motion? Let us compile this simulation program and use it to prepare a data file representing the state of the wave packet at time $t = 0$. The command line arguments specify a simulation with a wave number spread of $\Delta k = 5$, with 101 waves in the discrete packet with the central wave number $k_0 = 50$, ranging from $k = 47.5$ to $k = 52.5$ with a wave number step value of 0.05, and a phase velocity at the center of the packet $v_{p0} = 1$, all in some arbitrary units

```
$ gcc -Wall -Wshadow -O2 -o wave_packet wave_packet.c CPLibrary/step_complex.c
    ↪ CPLibrary/array_complex.c -lm
$ ./wave_packet
./wave_packet <vp0> <kinit> <kfinal> <kinc> <xinit> <xfinal> <xinc> <tinit> <tfinal> <tinc>
$ ./wave_packet 1 47.5 52.5 0.05 -8 8 0.01 0 0 1 >| wave_packet_dk5.dat
```

The example discrete wave packet is plotted in Figures 4.35 and 4.36, as real and imaginary components first, and then the complex magnitude squared, representing the probability distribution of the particle.

Heisenberg's uncertainty principle describes the tradeoff in our limits of knowledge of the particle's position and momentum. How do we visualize this uncertainty? Figure 4.37 shows the tradeoff of spread of the particle's probable position at time $t = 0$ with the width of the Gaussian distribution of wave numbers. This will be explored quantitatively with a propagating wave packet in Exercise 4.7.

We continue our visualization of the propagation of the wave packet by plotting the magnitude squared of the complex-valued wave at various snapshots in time. By the principles of quantum mechanics, the probability density of the particle's existence is given by the squared magnitude of the complex-valued wave function. This quantity is, of course, output in the fourth column of the data file. An example program execution and `gnuplot` dialog to plot the wave state at the four times $t = 0$, $t = 5$, $t = 10$, and $t = 15$ is

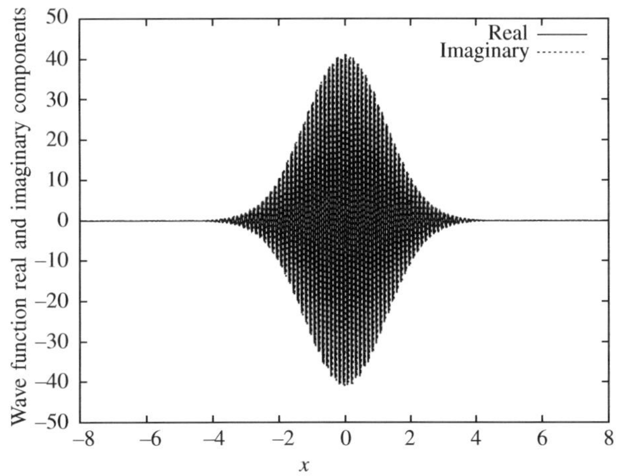

Figure 4.35 Real and imaginary components of complex quantum wave packet.

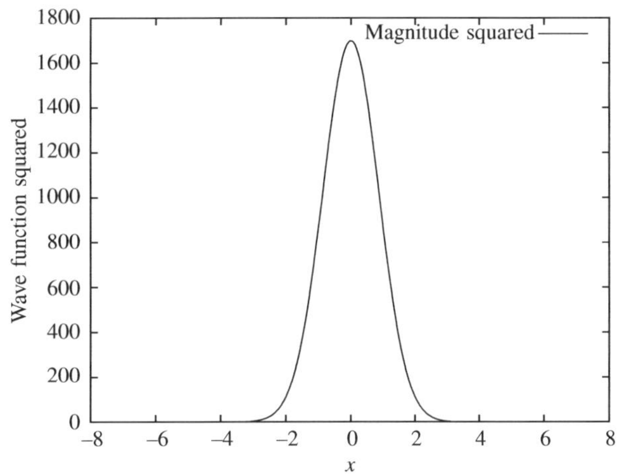

Figure 4.36 Magnitude squared of complex quantum wave packet.

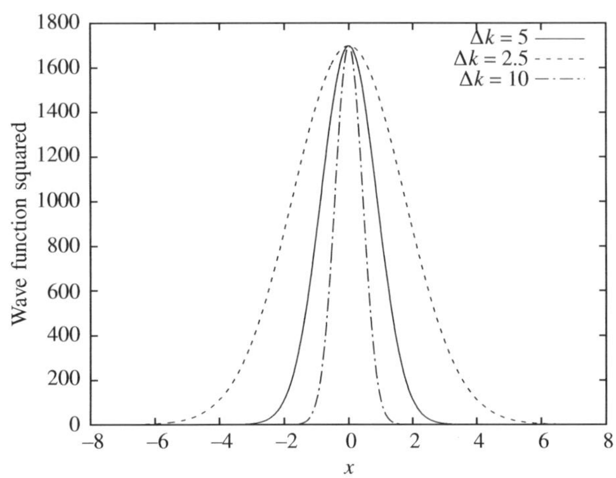

Figure 4.37 Quantum wave packet spread for three different values of $\Delta k = 5$, 2.5, and 10.

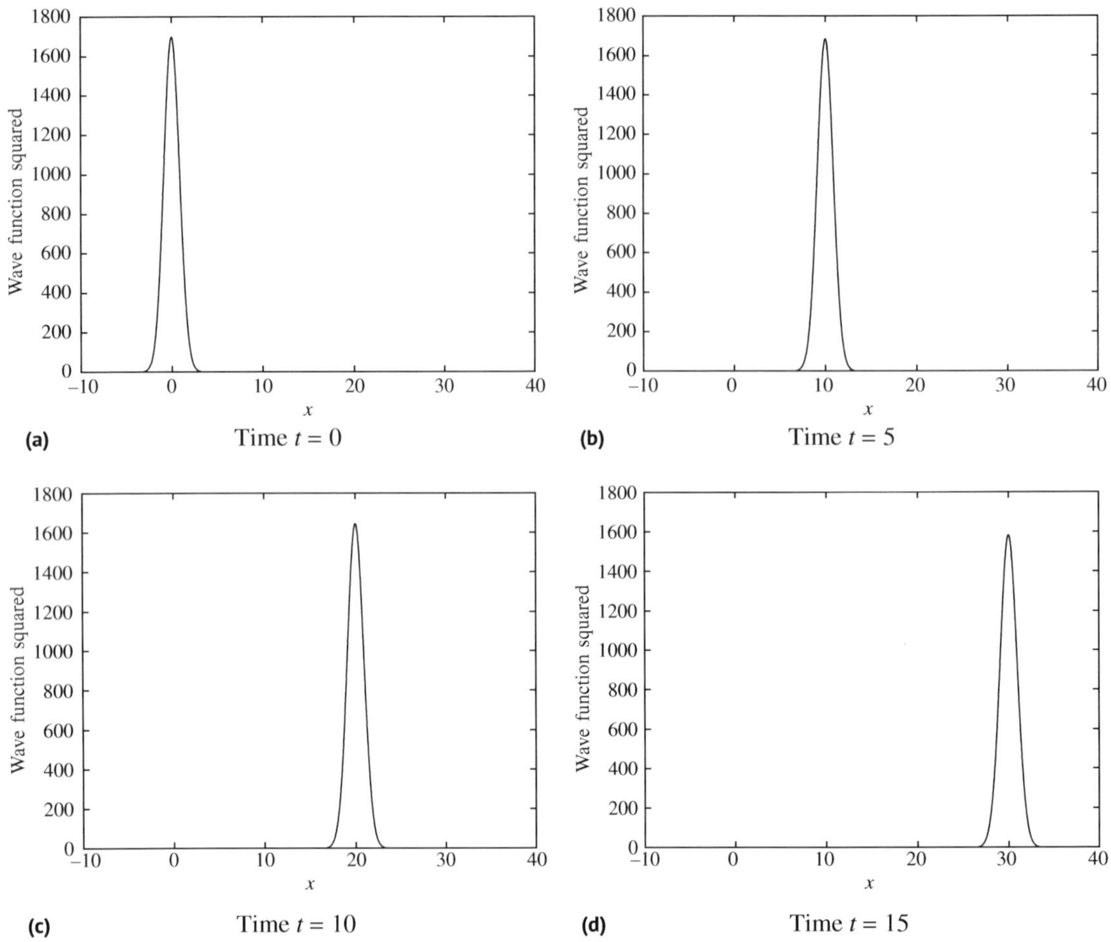

Figure 4.38 Quantum wave packet magnitude squared propagation over time.

```
$ ./wave_packet 1 47.5 52.5 0.05 -10 40 0.01 0 15 5 >| wave_packet_dk5.dat
```

```
gnuplot> set xlabel "X-coordinate"
gnuplot> set ylabel "Wave function magnitude squared"
gnuplot> plot 'wave_packet_dk5.dat' i 0 u 1:4 w l
gnuplot> plot 'wave_packet_dk5.dat' i 1 u 1:4 w l
gnuplot> plot 'wave_packet_dk5.dat' i 2 u 1:4 w l
gnuplot> plot 'wave_packet_dk5.dat' i 3 u 1:4 w l
```

These plots are all shown together in Figure 4.38. At a given time, the wave phases combine constructively centered around a central location, and we can clearly visualize the migration of that location in the positive x direction as time progresses. Note from the command line arguments that the wave phase velocity was set to a nominal value of 1 in arbitrary units. But we see in the plots of Figure 4.38 that for every five units of time elapsed, 10 units of the spatial coordinate are covered by the packet's migration. This numerically verifies the relationship of Equation (4.23).

We may also make use of the powerful internal iteration capability of `gnuplot` to see the progression of these wave packets plotted together in the same graph with

```
gnuplot> set xlabel "X-coordinate"
gnuplot> set ylabel "Wave function magnitude squared"
gnuplot> plot for [i=0:3] 'wave_packet_dk5.dat' i (i) u 1:4 w l
```

Another consequence of the dispersion relationship of Equation (4.19) is the property of quantum wave packets to disperse, or spread out in space, as the particle continues in free motion. Heisenberg's uncertainty principle establishes only

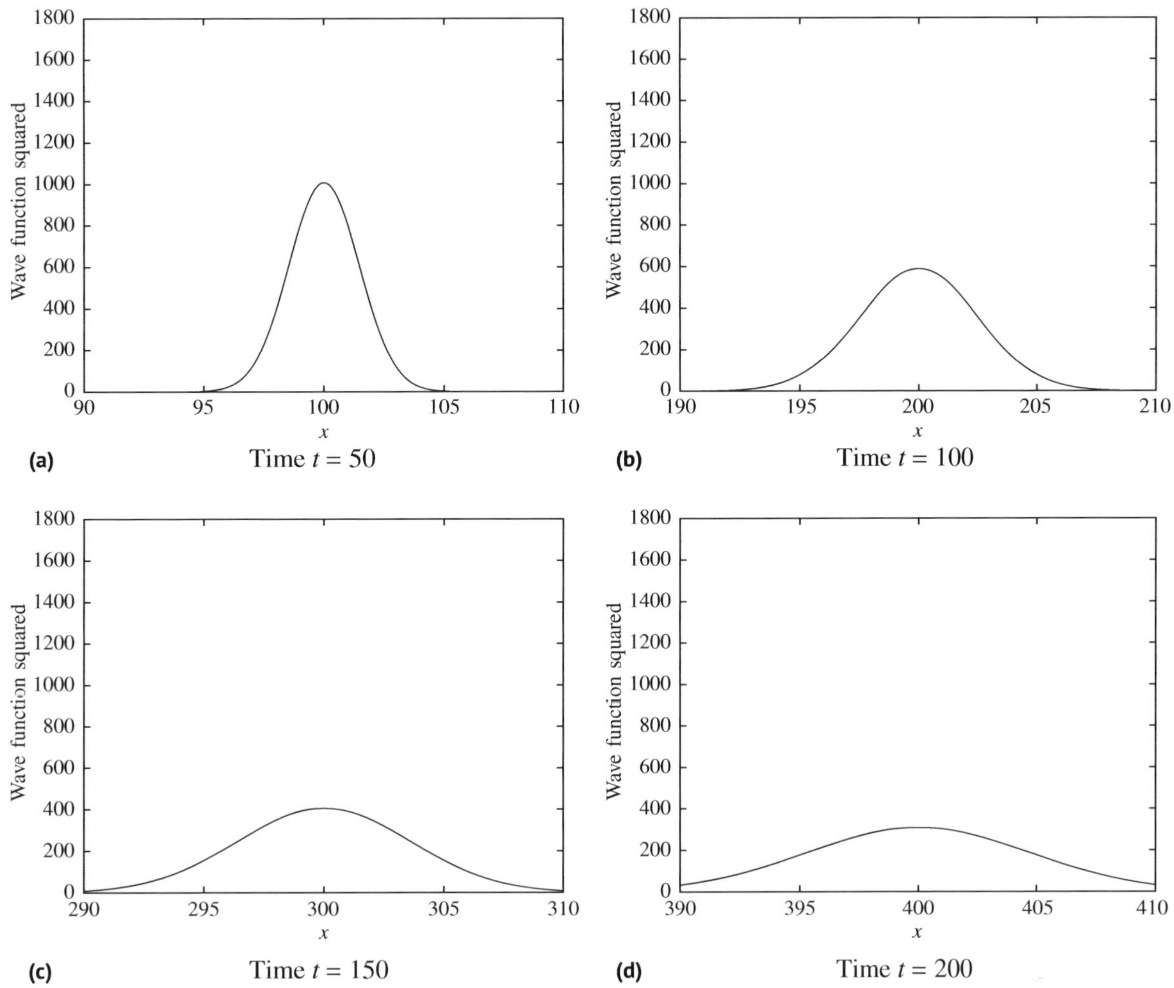

Figure 4.39 Spreading of quantum wave packet over time.

the lower bound for the position uncertainty of a particle, and this uncertainty will grow with time. We can see this mathematically in the sequence of four plots in Figure 4.39, where the simulated time values are stepped over longer intervals.

Further opportunity to quantitatively explore wave packet spreading, and to visualize the diminishment of spreading in the classical limit, is provided in Exercise 4.6. We will continue to make use of discrete Gaussian quantum wave packets in Section 4.10, and then revisit this topic in Chapter 13 with numerical solutions of the Schrödinger equation for interesting potential profiles.

4.10 Application: Quantum Scattering

A view of quantum physics that generally benefits from visualization tools is the correspondence between limits of the mathematical solutions of Schrödinger's equation and the behavior of classical objects we are familiar with. In Section 4.9, we assembled solutions of the time-independent Equation (4.16) into one-dimensional wave packets representing a particle localized in space and were able to visualize their propagation with time. What is the behavior of such a particle in a scenario similar to the collisions of classical objects analyzed in Section 4.8? The solution of Schrödinger's equation with a wave packet propagating toward the abrupt step of higher potential diagrammed in Figure 4.40 is straightforward, and visualizing the behavior of this quantum scattering problem is the object of this section. As shown in the figure, we will assume the particle energy E is less than the height of the potential step V_0, and that the particle is incident from the left and propagates to the right.

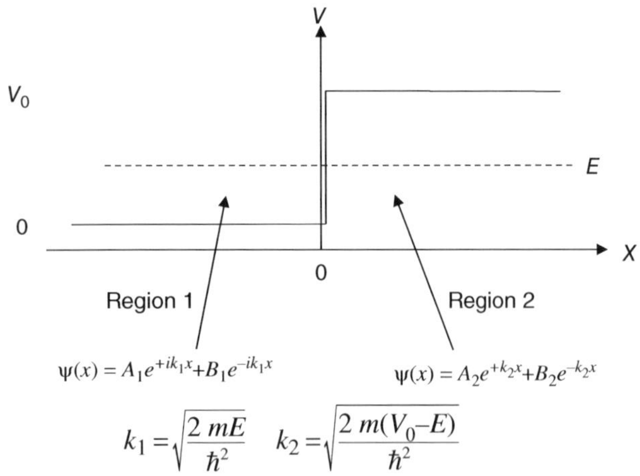

Figure 4.40 Potential of an abrupt step.

Since the potential in Region 1 is constant, the time-independent Schrödinger equation in one dimension, Equation (4.16), is a simple differential equation with constant coefficients. Therefore, we know in Region 1 its solution must be of the complex oscillatory form

$$\psi(x) = A_1 e^{+ik_1 x} + B_1 e^{-ik_1 x} \tag{4.29}$$

where $k_1 = \frac{\sqrt{2mE}}{\hbar}$. The coefficient A_1 determines the flux of particles incident from the left, and the coefficient B_1 scales the flux of particles scattered by the collision with this potential barrier and reflected back to the left. Then in Region 2, since we are considering only wave functions that are normalizable, we set the coefficient $A_2 = 0$, and the solution must follow the decaying exponential form

$$\psi(x) = B_2 e^{-k_2 x} \tag{4.30}$$

where $k_2 = \frac{\sqrt{2m(V_0 - E)}}{\hbar}$. Enforcing the continuity constraints at $x = 0$ of the complex wave function $\psi(x)$ and its first derivative $\psi'(x)$ allows us to solve for the ratios $\frac{B_1}{A_1}$ and $\frac{B_2}{A_1}$. With straightforward algebra [1, Chap. 6; 2, Chap. 8; 3, Chap. 4; 4, Chap. 1] we obtain

$$\begin{aligned} \frac{B_1}{A_1} &= \frac{k_1 - ik_2}{k_1 + ik_2} \\ \frac{B_2}{A_1} &= \frac{2k_1}{k_1 + ik_2} \end{aligned} \tag{4.31}$$

Examining the complex magnitude of the first ratio above, we see that

$$\left| \frac{k_1 - ik_2}{k_1 + ik_2} \right| = 1 \tag{4.32}$$

for any values of k_1 and k_2. Since the magnitude of this reflection coefficient is 1, we know that the flux of particles reflected off of the barrier will equal that incident, and all particles will (eventually) be reflected. Thus, as the incident and reflected particle waves superpose, we should expect a total solution in Region 1 that is a standing wave, with an oscillatory stationary dependence on the x spatial coordinate. We can see this by assigning the ratio $\frac{B_1}{A_1}$ to the expression

$$\frac{B_1}{A_1} = e^{-2i\theta} \tag{4.33}$$

for some phase shift angle θ, since its complex magnitude is 1. This makes the stationary wave solution in Region 1

$$\psi(x) = A_1 e^{-i\theta}(e^{+i\theta} e^{ik_1 x} + e^{-i\theta} e^{-ik_1 x}) \tag{4.34}$$

or, since $e^{i\alpha} + e^{-i\alpha} = 2\cos\alpha$ for any α, in Region 1

$$\psi(x) = 2A_1 e^{-i\theta}\cos(k_1 x + \theta) \tag{4.35}$$

with the expected oscillatory dependence. By equating Equations (4.31) and (4.33), we may solve for the required phase shift angle θ satisfying

$$\tan(2\theta) = \frac{2k_1k_2}{k_1^2 - k_2^2} \tag{4.36}$$

Given some arbitrary convenient choice of A_1, the square of the magnitude of the stationary wave function solutions as a function of x in both regions can be directly computed from the expressions above for any particle energy E. Three example solutions of probability are plotted in Figure 4.41 for the case of an electron approaching an abrupt potential step of 10-eV height. Three scattering events are simulated, with example electron energies of 2, 5, and 8 eV.

The phase angles required by the continuity constraints for these three cases are 63.43, 45, and 26.57°, respectively. This is consistent with the plots of the numerical solution, as the height of the decaying exponential that penetrates into Region 2 grows higher at higher electron energy levels. We interpret this dependence as visualizing that the probability of the electron being found penetrating into the classically forbidden Region 2 is higher with higher impact energies. The closer the phase shift angle approaches zero, the closer the peak of the cosine function dependence approaches the $x = 0$ transition coordinate, and the longer the statistical time the electron may spend in the forbidden region before being reflected.

As the stationary wave functions pictured in Figure 4.41 are all solutions of the time-independent Schrödinger equation, we are free to form linear combinations of these solutions and those of any other energy level so that the statistical behavior of an individual particle may be visualized, just as in Section 4.9 for the free particle. Figure 4.42 plots 50 of the 101 stationary wave function solutions for an abrupt potential step. These solutions will be combined into a wave packet that represents an approaching particle in the program `quantum_abruptstep_packet.c` from the Student Companion download website.

We may construct a wave packet centered at some nominal energy level E_0 that is less than the height of the potential step V_0, find the wave number $k_0 = \frac{\sqrt{2mE_0}}{\hbar}$ that corresponds to this single wave solution, and then construct a weighted sum of wave solutions. The wave numbers k in the packet are uniformly and symmetrically surrounding k_0 with the coefficient following Equation (4.26). Then each constituent wave is calculated with Equation (4.35) at the corresponding energy level $E = \frac{\hbar^2k^2}{2m}$. When the wave packet is constructed with the Gaussian weighting coefficients, the A_k in Equation (4.25), and the time-dependent factor included, we can visualize the time sequence of the particle approaching and rebounding from the potential step in Figure 4.43. Compiling and running the example program

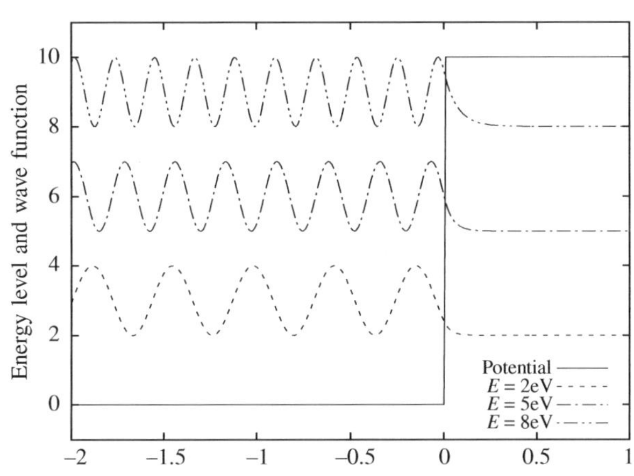

Figure 4.41 Wave function squared for three energies of an electron scattering from an abrupt 10-eV potential step.

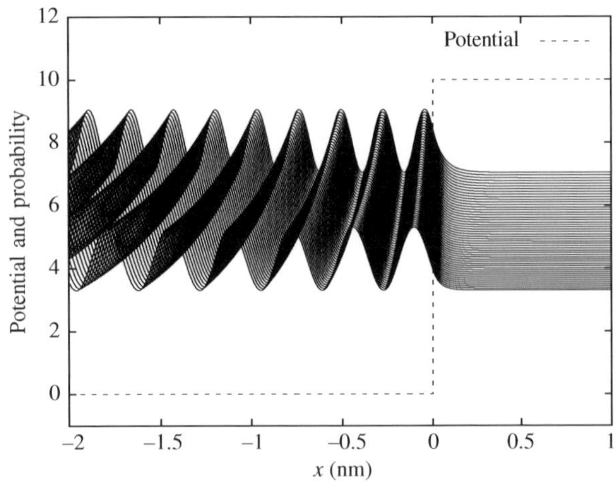

Figure 4.42 The family of stationary wave solutions constituting a wave packet for an abrupt potential step.

```
$ gcc -Wall -Wshadow -O2 -o quantum_abruptstep_packet quantum_abruptstep_packet.c
    ↪ CPLibrary/step_complex.c CPLibrary/step.c CPLibrary/array_complex.c CPLibrary/array.c -lm
$ ./quantum_abruptstep_packet
./quantum_abruptstep_packet <height in eV> <mass * c^2 in Mev> <E0> <a in nm> <xinit in nm>
    ↪ <xfinal in nm> <xinc in nm>
$ ./quantum_abruptstep_packet 10 0.5110 5 1 -10 1 0.0025 >| quantum_abruptstep_packet.dat
1 dataset output with potential
Packet center wavelength=0.54853423 nm
Alias distance=146.39527 nm
Wave function 1: Energy=3.3020246 eV, Reflection phase shift=54.9261 degrees Weight=0.0100
Wave function 51: Energy=5 eV, Reflection phase shift=45.0000 degrees Weight=1.0000
Wave function 101: Energy=7.0489643 eV, Reflection phase shift=32.9041 degrees Weight=0.0100
101 datasets output with time-independent wave function solutions
Velocity = 1326203.8 m/s  Distance = 1e-08 m  Simulation time from -7.5403195e-15 to 7.9173355e-15
    ↪ sec
21 datasets output with time-dependent wave packet solutions
```

The first dataset prepared for `gnuplot` in the output stream is for plotting the stepped potential as a helpful background reference on the graphs. Datasets 1 through 101 are the 101 curves describing the stationary wave solutions to be assembled in the wave packet, like those plotted in Figure 4.42. In order not to clutter the plot too densely, every odd-numbered dataset may be plotted with the iterative capability built into the `plot` command of `gnuplot` with the following example dialog

```
gnuplot> set xrange [-4:1]
gnuplot> set xlabel 'X (nm)'
gnuplot> set yrange [-1:12]
gnuplot> set ylabel 'Potential and probability'
gnuplot> plot 'quantum_abruptstep_packet.dat' i 0 u 1:2 w l lt 2 t 'Potential',for [i=1:101:2] ''
    ↪ i (i) u 1:4 w l lt 1 notitle
```

Finally, datasets 102 through 122 are the 21 curves from the wave packet simulations as time is stepped. At a given simulated time t, the packet is summed from the constituent stationary waves with Gaussian weighting, and also multiplied by the time factor $e^{i\omega t}$, which was originally separated out of the full time-dependent Schrödinger equation with the separation of variables technique. The angular frequency ω corresponding to each constituent wave must be calculated from each wave's energy level as $\omega = \frac{E}{\hbar}$. Since the stationary waves plotted in Figure 4.42 are constructed from Equation (4.35) so that the zero phases of all the constituent waves line up at $x = 0$, to start the packet in motion at the left end of the simulated space we must initialize t to a negative time at the beginning of the simulation, and then let time proceed forward. The time increment between time steps is calculated from the wave packet group velocity so that 10 time steps would be required to cover the distance from the left side of the simulated space to the potential step at $x = 0$, and another 10 steps back if

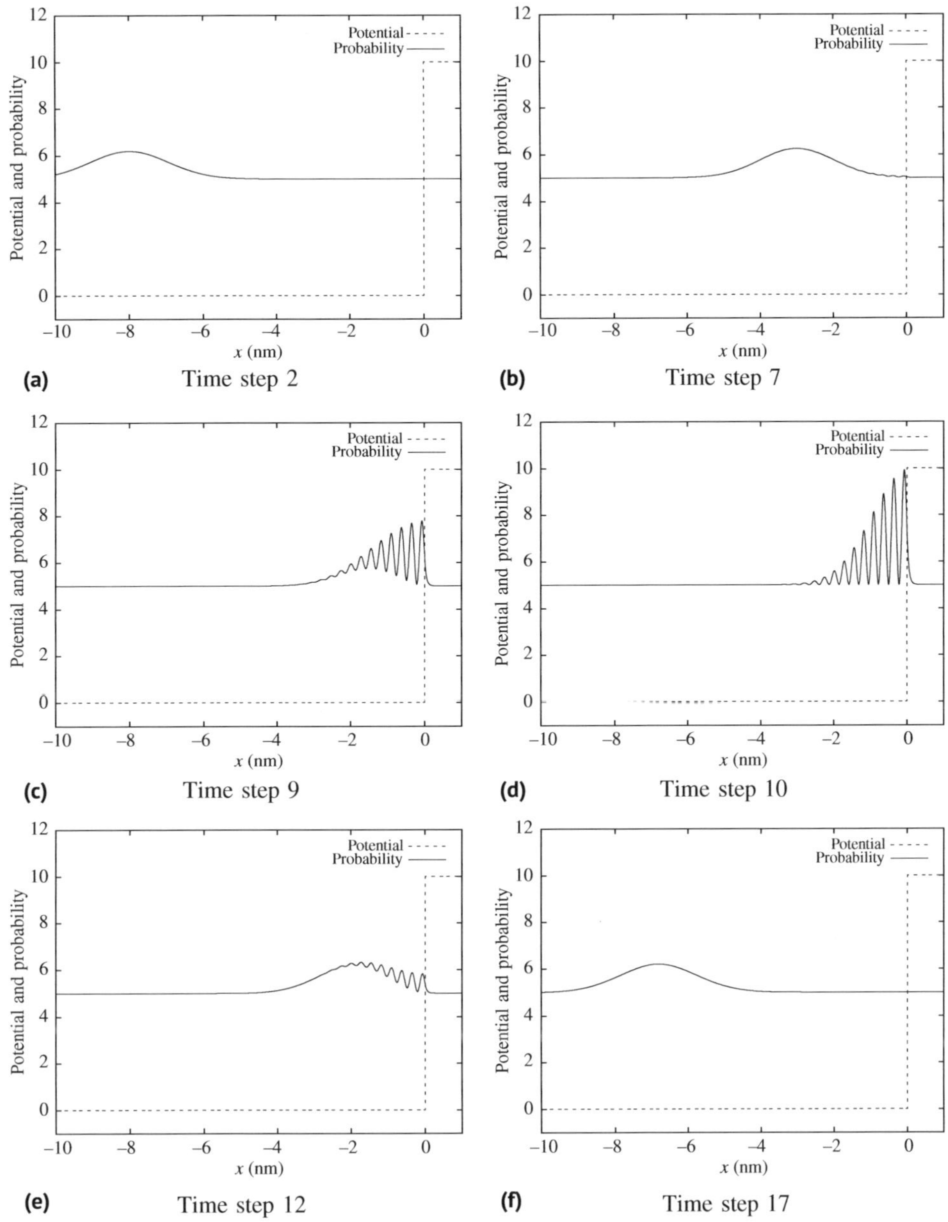

Figure 4.43 Time solution of an electron rebounding from an abrupt quantum potential step.

the wave reflection were to take place instantaneously. This progression of wave packets may be plotted with a `gnuplot` dialog starting with:

```
gnuplot> set xrange [-4:1]
gnuplot> set xlabel 'X (nm)'
gnuplot> set yrange [-1:12]
gnuplot> set ylabel 'Potential and probability'
gnuplot> plot 'quantum_abruptstep_packet.dat' i 0 u 1:2 w l lt 2 t 'Potential','' i 104 u 1:4 w l
    ↪ lt 1 t 'Probability'
```

for time step 2, and then repeating the `plot` command with successively larger dataset indices for later time steps.

This simulation provides us a statistical description of the likely position of the electron as it approaches and rebounds from the collision. It also allows us to imagine behavior that would approach our expectations of a classical particle with larger momenta with much smaller de Broglie wavelengths colliding with a barrier. Another significant feature we can visualize with these simulations is the nonzero probability that the electron exists beyond the potential barrier for $x > 0$, in the classically forbidden region. The time delay this penetration involves before the packet reflects is determined by the phase shift angle θ. With higher particle momenta and shorter wavelengths associated with classical particles, the oscillatory and penetrating behavior at the scattering event would become insignificant as the particle behaves according to the classical laws of motion. In Exercise 4.9 we will rerun the simulation for the case of an incident proton and begin to see the classical behavior approached.

We should keep in mind that these potential profiles with idealized abrupt transitions are good starting points for analysis, but are not particularly realistic, as abrupt potential transitions would require an unbounded level of force being applied to the particle over an infinitesimal distance. So we will return to more complete simulations of this quantum mechanical scattering scenario in Section 13.5. With the development of numerical methods to directly solve the time-independent Schrödinger equation, we will be able to examine the behavior of wave packets scattering against more realistic gradual potentials and visualize a greater reflection time delay due to the particle momentarily "rolling up" the potential hill.

4.11 Application: Field Vectors

In this section we use our coding and plotting tools to visualize classical electrostatic fields. This exercise will also illustrate some additional features of `gnuplot` for expressing information coded as line colors.

We will follow the general plan that was used for solving mechanical problems with spherical balls; namely, we will prepare a module of general purpose tool code that computes an electric field vector due to any number of point charges at known locations. This tool module is written and debugged once, but following the practice of building our solution code in reusable modules, it can then be linked to any main program. Additionally, we will reuse this point charge module in Section 8.12, numerically exploring Gauss' Law, and in Section 13.4, numerically drawing classical field lines.

Consider the geometry shown in Figure 4.44. The electric field is shown being calculated at an observation location by the vector $\vec{r}$ for a single point charge located at a position vector $\vec{p}$.

We know from basic electrostatic theory that the electric field vector $\vec{E}$ is given by [6, Chap. 2; 7, Chap. 1; 8, Chap. 4; 15, Chap. 1]

$$\vec{E} = \frac{1}{4\pi\epsilon_0}\frac{q}{|\vec{u}|^2}\hat{u} \tag{4.37}$$

where $\vec{u} = \vec{r} - \vec{p}$ is the vector directed from the point charge to the observation point, and $\hat{u}$ is the unit vector along the $\vec{u}$ direction. So we would compute $\hat{u}$ from the expression

$$\hat{u} = \frac{\vec{u}}{|\vec{u}|} = \frac{\vec{u}}{\sqrt{|\vec{u}|^2}} \tag{4.38}$$

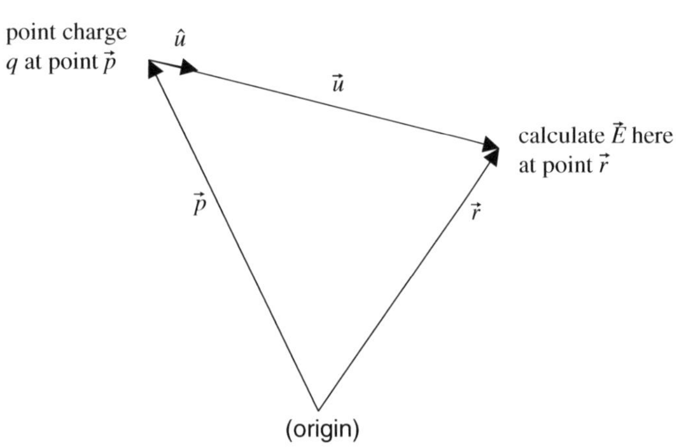

Figure 4.44 Geometry for the electric field due to a single point charge.

In Equation (4.38), we have expressed the vector magnitude in the denominator in the way it would actually be computed. That is, we would first compute $|\vec{u}|^2$ from the sum of the squares of its components, and then take its square root. This makes the expression for the electric field $\vec{E}$, in a manner suitable for the most efficient computation

$$\vec{E} = \frac{1}{4\pi\epsilon_0}\frac{q\vec{u}}{|\vec{u}|^2 \cdot \sqrt{|\vec{u}|^2}} \tag{4.39}$$

First we would compute $|\vec{u}|^2$ and save its value, and then take its square root, and finally multiply the two quantities together to form the denominator.

The electric field computation function named `efield_charges()` is part of the CPLibrary package from the Student Companion website in the file `point_charges.c`, together with its header file `point_charges.h`. This header file defines a data type structure named `PNTCHG` appropriate for defining a single point charge located in three-dimensional space. The `double` member variable `q` holds the amount of charge, possibly negative or positive, and the `x`, `y`, and `z` variables hold the position coordinates. A second structure named `CHGS` is then defined, similar to the `ARRAY1D` structure in `array.c`. The `CHGS` structure holds a pointer to an array of `PNTCHG` structures and a count of charges. So the parameters associated with any arbitrary arrangement of point charges can be stored by allocating an array of `PNTCHG` structures, with an array element for each charge, and filling in the position and charge properties in each structure instance. Then the number of charges and a pointer to the array is stored in the `CHGS` structure. A pointer to an instance of this overall structure can be passed as a parameter set to any function coded to compute an electric field vector at some point in space.

The function `efield_charges()` computes $\vec{E}(\vec{r})$, the vector electric field as a function of the three-dimensional position vector $\vec{r}$. So the function is declared with a return value type of `void`, meaning there is no return value at all, and three arguments. This is somewhat different from other functions we have examined up to this point in the text. For example, in the case of the damped oscillator of Section 4.2, the function passed to the stepper utility modeled the oscillator displacement versus time, with a `double` argument for time and returning a `double` type for displacement. In Section 4.9, the quantum wave packet function again accepted a `double` argument, in this case for the x coordinate, and returned a value of type `double complex`. However, in this case of the electric field, the modeling function accepts a pointer to an `ARRAY1D` argument `r` whose elements are the components of the position vector $\vec{r}$. Technically, C would compile code for a function that accepts structures as an argument and returns a structure. However, relying on this feature is usually a waste of the many processor cycles required to move all the members of structures back and forth in memory in and out of functions. So efficiency is much improved by passing only the pointer to an `ARRAY1D` position vector, and then passing a pointer to another `ARRAY1D` vector into which the function is to write the components of the computed electric field vector, essentially the function return value. After the `CHGS` parameter structure pointer `p`, the third argument is the pointer to `ARRAY1D` structure `E`, encapsulating an array allocated by the calling code, and the function uses it as memory space for the three field components. Since this is actually the computed result of the function, the formal C return value mechanism is not needed, and the function return type may be declared as `void`.

The main program for studying the electric field due to any number of point charges is in `efieldvectors.c`, available for download from the Student Companion website. This program accepts the specifications of any number of charges with their locations on the command line, allocates an appropriate number of `PNTCHG` structures, and fills in member values from the command line text. Then the `point_charges()` function is called to compute the total electric field over a two-dimensional grid of observation points. The coordinates are in the y-z plane of a three-dimensional system, at an x coordinate value of 0. The grid specifications in the y and z directions are also expected on the program command line. An example compilation and execution session for a simple electric dipole is

```
$ gcc -Wall -Wshadow -O2 -o efieldvectors efieldvectors.c CPLibrary/point_charges.c
    ↪ CPLibrary/array.c -lm
$ ./efieldvectors
./efieldvectors <num charges> <yinit> <yfinal> <yinc> <zinit> <zfinal> <zinc> <y1,z1,q1>
    ↪ <y2,z2,q2> ...
$ ./efieldvectors 2 -2 2 0.1 -2 2 0.1 0,0.95,1 0,-0.95,-1 >| efieldvectors_n2.dat
```

Visualizing a vector quantity such as the electric field in a two-dimensional plotting image requires some special techniques, more involved than straightforward curve plotting. Both the direction and magnitude of the vector at each spatial coordinate must somehow be graphically expressed. We will examine two possible approaches that are provided by `gnuplot`. We can use the `vector` plot mode to draw small arrows to represent the graphed direction, and the two plotting examples here represent the vector magnitude as either the length of the graphic in one color or its color with a constant length. Using vector lengths to represent magnitude means that an image may be rendered only using a black line color, and is suitable for the pages of this textbook. However, coding the vector magnitude in line color leads to an image that is only downloadable from the Student Companion website.

Careful examination of the main program in the program `efieldvectors.c` from the Student Companion website will show that there will be four datasets included in a data file captured from the output stream. The first and second sets are simple point coordinates of the point charges that were specified on the command line, so that they may be plotted in point mode by `gnuplot` to indicate in the image where the charges that are the source of electric flux are located. The points where positive charges are located are included in the first dataset, and the negative charges in the second set. For versions of the plot using color representation, these separate datasets allow `gnuplot` to use two different point colors for the two charge polarities. The third dataset lists the vector specifications of starting points and lengths for the case of the field vector magnitude represented by the vector length. The four data plotting coordinates are mapped from columns 1, 2, 3, and 4, and in the image printed here in this textbook, all black lines are used. In the fourth dataset the vectors are specified by fixed length vectors, but also include a fifth column. In that column, the program computes a numeric code for a color that `gnuplot` recognizes. The color number expresses the red-green-blue components of the line color used to draw each plotting vector. The colors are coded to suggest large field magnitudes with reddish colors and small field magnitudes with bluish colors. Mapping this fifth column to the line color is realized in `gnuplot` with the additional keywords `rgb variable` on the plot command line.

```
gnuplot> unset key
gnuplot> set xlabel 'Y-coordinate'
gnuplot> set xrange [-2:2]
gnuplot> set ylabel 'Z-coordinate'
gnuplot> set yrange [-2:2]
gnuplot> set linetype 1 linecolor -1 linewidth 2 dashtype 1
gnuplot> plot 'efieldvectors_n2.dat' i 0 w p pt 7 lt 1,\
> '' i 1 w p pt 7 lt 1,\
> '' i 2 u 1:2:3:4 w vectors lt 1'
gnuplot> plot 'efieldvectors_n2.dat' i 0 w p pt 7 lc 'red',\
> '' i 1 w p pt 7 lc 'blue',\
> '' i 3 u 1:2:3:4:5 w vectors lc rgb variable
```

The resulting example plots are shown in Figure 4.45 for the black-and-white case and in the downloadable file `efieldvectors_n2_color.png` for the color case. It should be emphasized that these plots are not the classical field lines where the density of the flux lines indicates the strength of the field. A numerical algorithm for rendering such plot images will be discussed in Section 13.4. With the plotting method in this section, the area density with which small vectors are drawn is constant, as the vectors are drawn on a uniform two-dimensional grid. The magnitude and direction of the field at each sampled grid point on the plane are represented by the graphical characteristics of each vector, either length or color, rather than line density [21, Chap. 6].

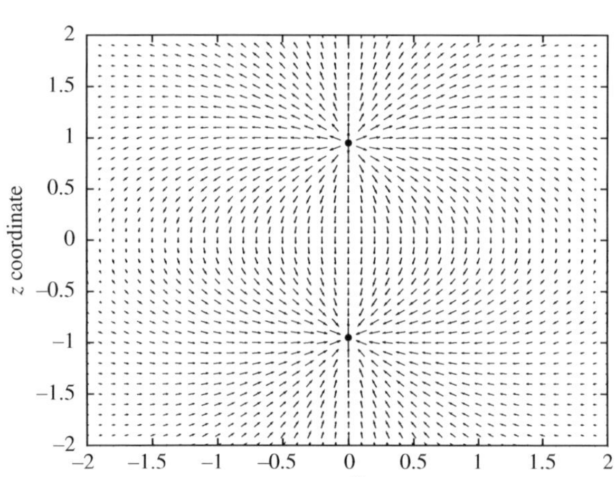

Figure 4.45 Electric field vectors around a dipole with coded lengths.

We will revisit this graphical technique in Sections 6.6 and 10.7. In Section 6.6, we will also plot the direction of a uniformly sampled $\vec{E}$ field, but a field that has been derived from an electrostatic potential that we will find with a solution of the Laplace equation. In Section 10.7, we will plot vectors representing the magnetic field from loops of electric current.

4.12 Application: The Thomson Problem

We now consider a classical electrostatics problem initially proposed by J. J. Thomson. We seek to find the distribution of locations of identical electrical charges on the surface of a sphere that will represent a minimum of electrostatic potential energy. While this problem was originally proposed in relation to the "plum-pudding" model of the atom, even though that model has been superseded by the development of quantum mechanics, it remains an interesting problem for other situations of the equilibrium distribution of macroscopic charges. Mathematically, the general Thomson optimization problem for an arbitrary number of charges remains unsolved, and even for small collections of a particular number of charges is very challenging. We may use the tools of this chapter to visualize the problem space for a limited case.

For any arbitrary collection of charges, the potential energy of the configuration may be calculated as

$$E = \sum_{i \neq j} \frac{q_i q_j}{4\pi\epsilon_0 r_{ij}} \tag{4.40}$$

where r_{ij} is the separation distance between charges q_i and q_j. For the purpose of the minimization of energy for identical charges, we may scale the charge magnitudes arbitrarily and define simply

$$E = \sum_{i \neq j} \frac{1}{r_{ij}} \tag{4.41}$$

It has been proven for the case of five charges on the surface of a unit sphere that the minimum energy corresponds to two of the charges located at the top and bottom of the sphere and three others distributed in an equilateral triangle about the equator, as shown in Figure 4.46, using a mixture of Cartesian and spherical coordinates.

To appreciate the mathematical difficulty of this problem, we may consider the simplified problem assuming that charges 0 through 3 have already been placed in what we know will be their optimum locations, and we seek the optimum location on the unit sphere of only charge 4. The coding and plotting of this simplified problem is available as Exercise 4.5 for the student.

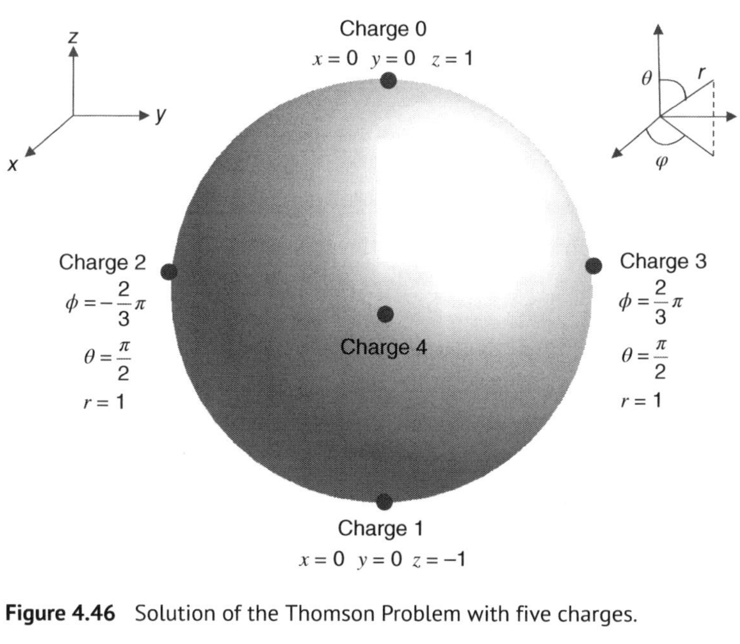

Figure 4.46 Solution of the Thomson Problem with five charges.

4.13 Coding: Generating Animated Graphics

We now explore some of the sophisticated capabilities of the gnuplot plotting tool that will expand our ability to render effective graphical results from our simulation code. The first is the ability to embed a long list of numeric plotting coordinates in the sequence of plot description commands. When we invoke a plot command such as the following

```
gnuplot> plot 'damped_oscillator_z0p1.dat' using 1:2 with points pointtype 7
```

which we used in Section 4.3, we are specifying a plot command that will process multi-column lines from the external file damped_oscillator_z0p1.dat, and interpret the numbers in column 1 as the *x* coordinate, and those in column 2 as the *y* coordinate of the points to be plotted. However, gnuplot also supports the capability of defining "named data blocks" along with plotting commands to hold the numerical coordinate data in a named variable, internal to gnuplot. Then a subsequent plot command can refer to the internally stored multi-column lines in a named data block instead of pointing to a separate file for reading. This is illustrated with the following example sequence in a gnuplot command script

```
$data << EOD
1.0 14.2 24.5
1.5 89.3 90.6
2.0 53.7 14.8
2.5 19.6 48.2
EOD
plot '$data' u 1:2 with points pointtype 7,'$data' u 1:3 with points pointtype 4
```

The first line alerts gnuplot that the following lines are to be processed as a sequence of data lines, to be stored in an internal variable named $data. The $ character is required at the beginning of a data block name. The data line sequence will be terminated by the first line starting with the key string EOD, which stands for "end of data." In this example, we have loaded four lines of data text, with three columns on each line. Then the subsequent plot command will invoke reading lines from this internal named data block for its coordinate data, instead of an external file. In this example, two sets of graphical points are plotted, the first using *x* and *y* coordinates in columns 1 and 2, and the second using *x* and *y* coordinates in columns 1 and 3, respectively.

Now let's see how we can make use of this capability as we revisit the simulation of the damped harmonic oscillator in Section 4.2. In Section 4.2, we wrote a program with a gen_output() function simply to send a text description of raw simulation result data to the output stream. This is useful if, as illustrated, we need to save a file copy of the data for other uses, such as importing into a spreadsheet, inputting into another analysis program, etc. However, if the only application of our simulation is to produce a graphical depiction of the motion, we can write another version of our program to directly generate plotting commands for gnuplot. For example, we may make a copy of the original source code file damped_oscillator.c and name it damped_oscillator_gp.c, meaning a program version written to directly generate gnuplot plotting commands instead of raw data. In this new version we replace the gen_output() function with the following function gen_output_gp(), and call this new function in the main program module to generate the output stream.

```
void gen_output_gp(ARRAY1D *ytab_ptr,GRID *tgrid_ptr) {
  int i;
  double t;

  printf("$data << EOD\n");
  for (i = 0; i < ytab_ptr->n_elems; i++) {
    t = tgrid_ptr->init + ((double) i) * tgrid_ptr->inc;
    printf("%.8g %.8g\n",t,ytab_ptr->elem[i]);
  }
  printf("EOD\n");
  printf("set linetype 1 linecolor rgb 'red' linewidth 2\n");
  printf("set xlabel 'Time\n");
  printf("set ylabel 'Displacement'\n");
  printf("unset key\n");
  printf("plot '$data' u 1:2 w l\n");
}
```

The entire new example program code in damped_oscillator_gp.c can be downloaded from the Student Companion website. Again, the example downloadable source code file includes frequent comment lines to guide the student in the manner used for coding virtually all the examples and exercises in this textbook. In keeping with the principles of modular code structures discussed in Section 3.2, the rest of the code can remain the same (except for now calling `gen_output_gp()` in the main program), and only the module that generates a different type of output stream replaced. The reader should work through this new `gen_output_gp()` listed above and understand how it now generates in the output stream a sequence of lines below to be interpreted by `gnuplot`. Note that for brevity, many data lines have been omitted in the definition of the named data block.

```
$data << EOD
0 1
0.01 0.99490117
0.02 0.98592017
0.03 0.97311664
0.04 0.9565649
0.05 0.93635354
...
9.96 0.041404819
9.97 0.042079156
9.98 0.042583724
9.99 0.042917599
10 0.043080537
EOD
set linetype 1 linecolor rgb 'red' linewidth 2
set xlabel 'Time
set ylabel 'Displacement'
unset key
plot '$data' u 1:2 w l
```

In compiling and running this program, we could redirect this standard output stream from our simulation program into a file, named, for example, `damped_oscillator_gp.gp`, and then supply that file as an input script to the `gnuplot` tool.

```
$ gcc -Wall -Wshadow -O2 -o damped_oscillator_gp damped_oscillator_gp.c CPLibrary/step.c
    ↪ CPLibrary/array.c -lm
$ ./damped_oscillator_gp 1 0.05 1 10 0.01 >| damped_oscillator_gp.gp
$ gnuplot -p damped_oscillator_gp.gp
```

However, another approach is simply to use a Linux pipe to directly connect the simulator code standard output into the `gnuplot` standard input, without the need to spend processor and file resources on writing and reading a separate file:

```
$ gcc -Wall -Wshadow -O2 -o damped_oscillator_gp damped_oscillator_gp.c CPLibrary/step.c
    ↪ CPLibrary/array.c -lm
$ ./damped_oscillator_gp 1 0.05 1 10 0.01 | gnuplot -p
```

What we generate above is another static graphical plot describing the oscillator displacement versus time. This physical system is simple enough so that we can easily imagine the motion by inspection of a plot. However, as we develop more complex numerical solutions, we will soon see other systems in this textbook with motions complex enough so that a direct animated graphical rendering is the most effective way to visualize or communicate behavior. The multiple oscillation modes of coupled spring-mass systems analyzed in Section 6.5, for systems with more than two or three masses, are an example. So we can now explore how the detailed programming approach used in this text, in conjunction with the `gnuplot` tool, can also be used to generate animated graphical simulation results. A video stream with animated graphics is composed of a large number of frame images displayed in a sequence that moves from frame to frame at a rate fast enough to appear to be continuous motion to the human eye. We will iteratively use `gnuplot` to create the graphics for each individual frame and store each frame in many individual image files. However, we will arrange for the data plotted in each frame to be only some portion of the overall simulation results that represents a momentary snapshot of the motion.

To make this clear with an example, we again make a copy of the original source code file `damped_oscillator.c` and name it `damped_oscillator_gpanim.c`. Again, this example downloadable source code file includes copious comment lines intended as a guide for the student. Suppose we would like a video animation where the state of the oscillator is represented by a circular dot on a plot with a y coordinate at the oscillator's displacement, but with an x coordinate that sweeps along as time progresses. We'll also demonstrate the capability of `gnuplot` to insert a text label at some desired location in the plotting area to display the simulation time in each frame. To achieve this animated view, in this new version of the code we replace the `gen_output()` function with the following function `gen_output_gpanim()`, and call this new function in the main program module to generate the output stream. The output function starts by writing lines of gnuplot commands to configure the appearance of the plot that will remain consistent in each frame, such as title and axis specifications. Then an iterative loop is set up to step through the simulated data and include some portion of it needed for each frame. In each iteration of the loop gnuplot, commands are written so that a new named data block will be created with the data for only each frame, and the coordinates will be plotted. In each iteration, a unique sequential output file name is defined as a destination for the PNG image graphics. The sequence of file names will be frame00001.png, frame00002.png, frame00003.png, ... up to the last frame image needed for all the simulation data.

```
void gen_output_gpanim(ARRAY1D *ytab_ptr,GRID *tgrid_ptr) {
  int i,f;
  double t;

  printf("set linetype 1 linecolor rgb 'red' linewidth 2\n");
  printf("unset key\n");
  printf("set xlabel 'Time'\n");
  printf("set xrange [%.3f:%.3f]\n",tgrid_ptr->init,tgrid_ptr->final);
  printf("set ylabel 'Displacement'\n");
  printf("set yrange [-1:1]\n");
  printf("set terminal png size 640,480\n");
  f = 1;
  for (i = 0; i < ytab_ptr->n_elems; i++) {
    t = tgrid_ptr->init + ((double) i) * tgrid_ptr->inc;
    printf("$data << EOD\n");
    printf("%.8g %.8g\n",t,ytab_ptr->elem[i]);
    printf("EOD\n");
    printf("set label 1 't=%8.4f seconds' at graph 0.5,0.95 center\n",
           t);
    printf("set output 'frame%05d.png'\n",f);
    printf("plot '$data' u 1:2 w p lt 1 pt 7\n");
    printf("unset output\n");
    printf("undefine $data\n");
    printf("unset label 1\n");
    f++;
  }
}
```

Compiling and running this animated simulator

```
$ gcc -Wall -Wshadow -O2 -o damped_oscillator_gpanim damped_oscillator_gpanim.c CPLibrary/step.c
    ↪ CPLibrary/array.c -lm
$ ./damped_oscillator_gpanim 1 0.05 1 10 0.01 | gnuplot
```

we can again use the method of piping the program's standard output directly in the standard input of gnuplot. The command line for damped_oscillator_gpanim specifies a 10-second simulation with a 0.01-second time step, so with this command we will generate 1001 frame image files named frame00001.png through frame01001.png in the current directory. The reader should again trace through the gen_output_gpanim() function above and understand how it will generate an output stream shown here

```
set linetype 1 linecolor rgb 'red' linewidth 2
unset key
set xlabel 'Time'
set xrange [0.000:10.000]
set ylabel 'Displacement'
set yrange [-1:1]
set terminal png size 640,480
$data << EOD
0 1
EOD
set label 1 't=  0.0000 seconds' at graph 0.5,0.95 center
set output 'frame00001.png'
plot '$data' u 1:2 w p lt 1 pt 7
unset output
undefine $data
```

```
unset label 1
$data << EOD
0.01 0.99490117
EOD
set label 1 't=  0.0100 seconds' at graph 0.5,0.95 center
set output 'frame00002.png'
plot '$data' u 1:2 w p lt 1 pt 7
unset output
undefine $data
unset label 1
  (...skipping to last frame...)
$data << EOD
10 0.043080537
EOD
set label 1 't= 10.0000 seconds' at graph 0.5,0.95 center
set output 'frame01001.png'
plot '$data' u 1:2 w p lt 1 pt 7
unset output
undefine $data
unset label 1
```

When gnuplot processes this stream piped into its standard input as commands, it will generate all 1001 PNG graphics files in the current working directory sequentially for all the video frames in the animation. Then the powerful and universally available tool ffmpeg may be applied to render a video stream composed of a continuous sequence of all the frame images. The ffmpeg command line may include a matching pattern to specify a sequence of any number of constituent frame image file names. The pattern in the example below is frame%05d.png, and resembles the printf() function format code that was used to generate the file names programmatically in the gen_output_gpanim() function. An example run of ffmpeg is shown below to render the video stream into an animated GIF file. These files are generally compact and may be viewed in any contemporary web browser, either from a local file or served on a web page, as well as in most presentation tools. The frame rate for the output video file is set to 30 frames per second with the -r option on the command line, but this parameter may be adjusted as needed. In order to keep the resulting animated GIF file compact, ffmpeg must build a common color palette for all the video frames it generates. This video tool is a fairly complex product of a long-term open-source software project, and over the past years its command line format has changed from time to time, but as of this writing the command line shown below will invoke its options for this color optimization. The ffmpeg tool is also capable of rendering the animation into a large variety of other video file formats to be viewed in media players, but navigating the complex space of video encoding is beyond the scope of this text.

```
$ ffmpeg -f image2 -i frame%05d.png -vf 'split[v],palettegen,[v]paletteuse' -r 30 -y
    ↪ damped_oscillator_gpanim.gif
$ rm -f frame*.png
```

The resulting damped_oscillator_gpanim.gif animated GIF file is available for download from the Student Companion website. After the ffmpeg command above, we take the important step to delete all the temporary frame graphics files after the animated video output file has been rendered, as the frame images potentially can represent a very large number of files occupying file system space.

Now suppose a more elaborate animation is desired, one that not only will depict the oscillator's displacement swept along in time, but that will also trace the history of displacement on the plotting axes. Then for each frame, not only do we need to use gnuplot to plot a large dot for the current displacement, but also to draw a plotting curve through all the previous points and the current point. So consider this new version of the gen_output_gpanim() function in the new source file, damped_oscillator_gpanim2.c

```
void gen_output_gpanim(ARRAY1D *ytab_ptr,GRID *tgrid_ptr) {
  int i,j,f;
  double t;

  printf("set linetype 1 linecolor rgb 'red' linewidth 2\n");
  printf("unset key\n");
  printf("set xlabel 'Time'\n");
  printf("set xrange [%.3f:%.3f]\n",tgrid_ptr->init,tgrid_ptr->final);
  printf("set ylabel 'Displacement'\n");
```

```
  printf("set yrange [-1:1]\n");
  printf("set terminal png size 640,480\n");
  f = 1;
  for (i = 1; i < ytab_ptr->n_elems; i++) {
    printf("$data << EOD\n");
    for (j = 0; j <= i; j++) {
      t = tgrid_ptr->init + ((double) j) * tgrid_ptr->inc;
      printf("%.8g %.8g\n",t,ytab_ptr->elem[j]);
    }
    printf("\n\n");
    printf("%.8g %.8g\n",t,ytab_ptr->elem[i]);
    printf("EOD\n");
    printf("set label 1 't=%8.4f seconds' at graph 0.5,0.95 center\n",
           t);
    printf("set output 'frame%05d.png'\n",f);
    printf("plot '$data' i 0 u 1:2 w l lt 1,"
           "'$data' i 1 u 1:2 w p lt 1 pt 7\n");
    printf("unset output\n");
    printf("undefine $data\n");
    printf("unset label 1\n");
    f++;
  }
}
```

Compiling and running this animated simulator with the same command sequence above would send a stream that begins with the lines below to the standard output, and into the pipe to the `gnuplot` input. Note that the named data block now includes two datasets, separated by the required two blank lines. The first set is the accumulated displacement path up to each frame's time point, to be plotted with an index code of 0 in line mode. The second dataset is simply the single current point coordinates, to be plotted as a large dot as before. The reader should verify their understanding of this result from the program code above.

```
set linetype 1 linecolor rgb 'red' linewidth 2
unset key
set xlabel 'Time'
set xrange [0.000:10.000]
set ylabel 'Displacement'
set yrange [-1:1]
set terminal png size 640,480
$data << EOD
0 1
0.01 0.99490117


0.01 0.99490117
EOD
set label 1 't=   0.0100 seconds' at graph 0.5,0.95 center
set output 'frame00001.png'
plot '$data' i 0 u 1:2 w l lt 1,'$data' i 1 u 1:2 w p lt 1 pt 7
unset output
undefine $data
unset label 1
$data << EOD
0 1
0.01 0.99490117
0.02 0.98592017


0.02 0.98592017
EOD
set label 1 't=   0.0200 seconds' at graph 0.5,0.95 center
set output 'frame00002.png'
plot '$data' i 0 u 1:2 w l lt 1,'$data' i 1 u 1:2 w p lt 1 pt 7
unset output
undefine $data
```

```
unset label 1
...
```

The resulting sequence of frame images may again be rendered into a video file with the `ffmpeg` tool as before. The resulting animated GIF file, `damped_oscillator_gpanim2.gif`, is available from the Student Companion website.

A final example of the capability of animated graphics to enhance the visualization of simulation data is shown in the downloadable animated file `diffraction2D_step_2src_gpanim.gif`. These videos are animated versions of the static plot seen in Figure 4.26 for the case of two sources of light being diffracted through a circular aperture. There will be a number of applications in this text of the three-dimensional surface plot capability of `gnuplot`, but the topology of such plots may obscure the view of interesting features around the plotting space. To help provide differing plot viewpoints, another coding technique we may apply in an output generation routine is to write `gnuplot` commands to render a video that rotates the apparent projection viewpoint around a circular path. This viewpoint is controlled by the `set view` command for `gnuplot`.

Consider the function that is included in the source code file `diffraction2D_step_2src_gpanim.c`. This code commands `gnuplot` to initialize the plot parameters and then load the entire three-dimensional surface data coordinates into a named data block. The interaction that writes all the frame image files repeatedly plots the data out of the same internal data block, but after rotating the plot viewpoint for each frame.

```
void gen_output_gpanim(ARRAY2D *ztab_ptr,GRID *xgrid_ptr,
                       GRID *ygrid_ptr) {
  int r,c,f;
  double x,y,angle;

  printf("set linetype 1 linecolor rgb 'red' linewidth 1\n");
  printf("set linetype 2 linecolor rgb 'green' linewidth 1\n");
  printf("unset key\n");
  printf("set hidden3d\n");
  printf("set xlabel 'X-coordinate'\n");
  printf("set ylabel 'Y-coordinate'\n");
  printf("set zlabel 'Relative intensity' rotate by 90\n");
  printf("set ticslevel 0.25\n");
  printf("set terminal png size 640,480\n");
  printf("$data << EOD\n");
  for (r = 0; r < ztab_ptr->n_rows; r++) {
    y = ygrid_ptr->init + ((double) r) * ygrid_ptr->inc;
    for (c = 0; c < ztab_ptr->n_cols; c++) {
      x = xgrid_ptr->init + ((double) c) * xgrid_ptr->inc;
      printf("%.8g %.8g %.8g\n",x,y,ztab_ptr->elem[r][c]);
    }
    printf("\n");
  }
  printf("EOD\n");
  f = 1;
  angle = 0.0;
  while (angle < 360.01) {
    printf("set view 50,%.2f,1.0,1.25\n",angle);
    printf("set output 'frame%05d.png'\n",f);
    printf("splot '$data' u 1:2:3 w l\n");
    printf("unset output\n");
    angle += 0.5;
    f++;
  }
  printf("undefine $data\n");
}
```

An example compilation and execution to render the video animation on the download website is shown below.

```
$ gcc -Wall -Wshadow -O2 -o diffraction2D_step_2src_gpanim diffraction2D_step_2src_gpanim.c
    ↪ CPLibrary/step.c CPLibrary/array.c -lm
$ ./diffraction2D_step_2src_gpanim 1 0.2 20 20 -2 0 2 0 -6 6 0.2 -6 6 0.2 | gnuplot
$ ffmpeg -f image2 -i frame%05d.png -vf 'split[v],palettegen,[v]paletteuse' -r 30 -y
    ↪ diffraction2D_step_2src_gpanim.gif
$ rm -f frame*.png
```

An excerpt from the resulting output stream that will be piped into `gnuplot` is shown here

```
set linetype 1 linecolor rgb 'red' linewidth 1
set linetype 2 linecolor rgb 'green' linewidth 1
unset key
set hidden3d
set xlabel 'X-coordinate'
set ylabel 'Y-coordinate'
set zlabel 'Relative intensity' rotate by 90
set ticslevel 0.25
set terminal png size 640,480
$data << EOD
-6 -6 0.00010405027
-5.8 -6 0.00015207976
-5.6 -6 0.00019615525
-5.4 -6 0.00023007282
-5.2 -6 0.00025003645
...
5.4 6 0.00020666305
5.6 6 0.00017618393
5.8 6 0.00013539295
6 6 9.0141044e-05

EOD
set view 50,0.00,1.0,1.25
set output 'frame00001.png'
splot '$data' u 1:2:3 w l
unset output
set view 50,0.50,1.0,1.25
set output 'frame00002.png'
splot '$data' u 1:2:3 w l
unset output
...
set view 50,359.50,1.0,1.25
set output 'frame00720.png'
splot '$data' u 1:2:3 w l
unset output
set view 50,360.00,1.0,1.25
set output 'frame00721.png'
splot '$data' u 1:2:3 w l
unset output
undefine $data
```

4.14 Exercises

4.1 Write a program `maxwell_boltzmann_step.c` to step the Maxwell–Boltzmann speed distribution. Use helium atoms with $m = 6.64647 \times 10^{-27}$ kg at T =200, 250, and 300 K. Step from 0 to 2500 m/sec with an increment of 20. Use the `step.c` tool to step a C function computing the probability over its argument, the particle speed. While the independent variable of the distribution function is the particle speed, we must be able to pass the particle mass and the gas temperature to the function being stepped with a physical parameter structure such as

```
typedef struct mb_param {
  double m;  // mass
  double T;  // temperature
} MB;
```

Plot with gnuplot with the three distribution curves overlaid with appropriate curve titles, as in Figure 4.10.

4.2 The files `prem_density.c` and `prem_density.h` in the CPLibrary from the Student Companion website provide a function `density()` that computes the mass density per unit volume of the Earth in units of kg per m^3 as a function of radial distance from the center according to the Preliminary Reference Earth Model (PREM) [22]. Note that the polynomial model segments are evaluated efficiently with the nested expressions of Horner's method as discussed in Section 1.3. Also provided is a second function `density_approx()` that computes a first-order linear approximation to the density. Write a main program `prem_step.c` to generate a plot of these two density models as a function of radius using the `step.c` tool. The various spherical layers of Earth's composition can be visualized clearly. We will revisit this density model to find the rotational inertia of the Earth in Exercise 10.4.

4.3 Write a program `raindrop_angle.c` to step the output angles of the reflected rays from the raindrop as plotted in Figure 4.17. Use the function stepping tool `step()` to sample the output angle as a function of input y coordinate. The object function being stepped calculates the series of angles as described in Section 4.6. This function accepts as an argument the y coordinate of an incident ray, and returns a computed value of the exit angle that the ray will be mapped to. The index of refraction should be passed to the function as a member of a parameter structure. Call the `step()` tool three times, each time after setting the index of refraction to the three values in Table 4.1. Call a `gen_output()` function three times to send the sampled angle values to the standard output stream. Separate each table of values with two newline characters so that `gnuplot` may plot the three dependencies as separate datasets. Generate the plot image as shown in Figure 4.17 with appropriate plot labels and key titles.

Hint: An important detail to understand is the adjustment of the return value range of the `atan()` function in the assignment of the `tn3` variable, which corresponds to θ_{n3} in the derivation above. This potential coding trap was detailed in Section 3.3.6. The standard return value range for the `atan()` function is $-\frac{\pi}{2}$ to $\frac{\pi}{2}$, but the geometry of Figure 4.13 requires that θ_{n3} range from 0 to π. If the C variable `tn3` holds the value of θ_{n3}, this correction can be adjusted with the conditional add operation with the source code lines

```
tn3 = atan(y_3 / x_3);
if (tn3 < 0.0) tn3 += PI;
```

4.4 Modify the code in `diffraction2D_step_1src.c` from Section 4.7 to write the new program `diffraction2D_disc_step_1src.c` to simulate the optical pattern showing the Poisson spot, using the geometry illustrated in Figure 4.27. Rendering the program output with `gnuplot` should generate the plot shown in Figure 4.28. Hint: To obtain a plot with a sharp definition of the central spot, it will be necessary to include radii out to the neighborhood of 50 times the radius of the disc in your phase summation. You may relax the iteration requirements by using somewhat larger patch areas than in the examples provided in Section 4.7. Expect this program to require a large amount of execution time.

4.5 Write a program `thomson_step.c` that calculates the potential energy of a collection of five charges on the surface of the unit sphere given by Equation (4.41). Assume that the first four charges have already been placed at their optimum positions as shown in Figure 4.46, but that charge 4 may be placed at any arbitrary location on the surface with spherical coordinates azimuthal angle ϕ, inclination angle θ, and radius $r = 1$. Use the two-dimensional step tool `step2d()` to step ϕ from $-\pi$ to π and step θ only from $\frac{\pi}{50}$ to $\frac{49\pi}{50}$ to avoid the charges at the polar locations, both with an increment of $\frac{\pi}{50}$. Use a `CHGS` structure pointing to an allocated array of `PNTCHG` structures to hold the Cartesian coordinates of each charge, and ignore the q member. Plot the potential as a two-dimensional surface plot with `gnuplot`. By restricting the range of the energy plotting axis, visualize the dependence of energy on position that allows you to verify the optimum location of the fifth charge as shown in Figure 4.46, even in the presence of another local minimum elsewhere.

4.6 Run the program `wave_packet.c` developed in Section 4.9, but with different sets of command line arguments to explore properties of the quantum wave packet, as follows.

(a) In Section 4.9, the discrete summation forming the wave packet was plotted in Figure 4.38 for four time values. The four time values of 0, 5, 10, and 15 units were chosen to demonstrate the propagation of the wave packet group at a velocity of twice the phase velocity of 1. The four plots in that figure were restricted to cover an x window of 50 units in width, and with a window that migrated to follow the motion. For that demonstration, we used a range of k values in the packet covering the range of 47.5 to 52.5 units with a step of 0.05. This resulted in packets with 101 constituent waves at discretely spaced wave numbers. Is this an appropriate number of waves to mathematically approximate a continuous distribution of wave numbers in a packet? Under what circumstances may we need more constituent waves? To answer these questions, plot the same wave packets as in Figure 4.38 over a wider range of x values. To cover a wide x range, allow the packet to propagate for a longer time with its group velocity of 2, and sample at 0, 50, 100, 150, and 200 time units. If the plot x range extends from 0 up to a large enough coordinate to capture the entire particle motion, you should discover multiple copies of the wave packet repeated in x with a spatial period of $\frac{2\pi}{\Delta k}$, where Δk is the step size in the wave number k. This signal processing property is called *aliasing*, and is due to the existence of periodic locations where the phases of the constituent waves in the packet again come into phase with each other. Increase the number of waves constituting the packet, but still maintaining the same range of wave number k, to find the number of discrete constituents required to approximate a continuous wave packet for this numerical experiment.

(b) You should also observe with a longer propagation time the gradual dispersion of a particle's probability location over time, as in Figure 4.39. This is a fundamental property of solutions of the Schrödinger equation due to the dispersion relationship of Equation (4.19). Change the command line arguments to `wave_packet.c` to increase the central value of the distribution of momentum of the wave packet, but still maintaining a width of 5 units of wave number across the packet, to demonstrate a particle approaching the classical limit. Observe the effect on particle dispersion.

4.7 The `wave_packet.c` wave packet program, as written, prints to the standard output stream the wave packet probability distribution as a function of x. Using this code as a starting point, write a new program `wave_packet_uncert.c` to send instead to the standard output stream statistical measurements of the wave packet probabilities *as a function of time*. The four output measurements should be: $\overline{(\Delta k)^2}$, the mean value of the squared variation in the wave number k; $\overline{x}$, the mean value of the x coordinate; $\overline{(\Delta x)^2}$, the mean value of the squared variation in the x coordinate of the wave function probability; and $\overline{(\Delta k)^2} \cdot \overline{(\Delta x)^2}$, the product of the two squared variations. Including the time coordinate, this will make a total of five output columns. You may calculate the statistical measurements across the wave packet as follows

$$\overline{(\Delta k)^2} = \frac{\sum_k A_k^2 (k - k_0)^2}{\sum_k A_k^2} \tag{4.42}$$

$$\overline{x} = \frac{\sum_x |\Psi(x)|^2 x}{\sum_x |\Psi(x)|^2} \tag{4.43}$$

$$\overline{(\Delta x)^2} = \frac{\sum_x |\Psi(x)|^2 (x - \overline{x})^2}{\sum_x |\Psi(x)|^2} \tag{4.44}$$

Run the program with command line arguments as developed for Exercise 4.6 so that the wave packets propagate over 200 time units with no aliasing. First plot the mean x coordinate column to verify that your program models a wave packet propagating to increasing x values at the expected group velocity. Then plot the last output column as a function of time samples to show the graphical relationship of the product of the two squared deviations with the lower limit established by the Heisenberg uncertainty relationship as the wave packet spreads.

$$\overline{(\Delta k)^2} \cdot \overline{(\Delta x)^2} \geq \frac{1}{4} \tag{4.45}$$

4.8 The provided source code `wave_packet.c` assembles discrete wave packets including only constituent waves propagating in the positive x direction. These packets are represented by solutions of the form of Equation (4.18) with the A coefficients nonzero and the B coefficients zero. Expand the `wave_packet.c` wave packet program into another program `wave_packet_bidir.c` to sum identical bidirectional constituent waves propagating in both the positive and negative x directions. By studying solutions that include bidirectional wave propagation, we are preparing for the general solutions of quantum scattering treated in Section 13.5.

4.9 In Section 4.10, the program `quantum_abruptstep_packet.c` was demonstrated for modeling quantum scattering of an electron from an abrupt potential step. Rerun that program for the case of an incident proton, still at a 5-eV energy level approaching a potential step of 10 eV. Plot the simulation results in a manner resembling Figure 4.43, and demonstrate behavior that is approaching the classical limit. That is, the proton should be seen rebounding directly from the potential step, with quantum mechanical effects seen only over very short distances. Hint: Since the proton has a mass of over 1800 times that of the electron, at the same energy level, the de Broglie wavelengths for the proton will be almost 43 times as short as that of the electron. Shrink the spatial parameters of the Gaussian a coefficient and the step size for the x coordinate by a factor of 40 relative to that shown in Section 4.10. To avoid the aliasing effect discovered in Exercise 4.6, run the simulation over an x range of -1 to 0.1.

4.10 Expand the program `quantum_abruptstep_packet.c` for modeling quantum scattering from an abrupt potential step presented in Section 4.10 to the new program `quantum_abruptstep_packet_gpanim.c` to render an animated graphics file showing the scattering motion of the wave packet. Expand the number of time steps from 21 to 401 and change the text in the output stream to form a script for `gnuplot` to plot the wave packet probability distribution over time in each of the 401 animation frames.

5

Roots of Nonlinear Functions

5.1 Algorithms: Root Finding

With the groundwork laid in the preceding chapters, we are ready to begin the study of more substantial numerical analysis algorithms and associated physics problems. The algorithms discussed in this chapter are for searching for and converging on a root of a nonlinear mathematical function. This object function will be coded as a C function, and a pointer to this function will be passed to a root-finding utility. Initializing the search requires one or two initial values for the function argument, and then the algorithm will generate a subsequent series of argument values that will hopefully converge to the root within some acceptable level of accuracy. Starting from the initial guess of the root of x_0 for Newton's method, or two initial guesses of x_0 and x_{-1} for the Secant method, or two initial "bracketing" points x_0 and x_{-1} that must surround the root for the Regula Falsi and Bisection methods, the sequence of values of the function argument x that hopefully will converge to the root of $f(x)$ is $x_1, x_2, \dots, x_n$, for n iterations. The algorithm must also monitor the progress of this sequence and prepare for the possibility that it fails to converge properly to a root. Each of the sections that follow describe the four algorithms we will study, each with different properties and applicabilities to different classes of problems.

5.1.1 Newton–Raphson Method

See Figure 5.1 to see how the Newton–Raphson method can converge on a root of a function given a sufficiently close initial guess. Equation (5.1) shows a Taylor series truncated to the first term to give an approximation for the function whose root is to be found. We assume that a known function argument x_i represents some argument in the search sequence, and the next argument x_{i+1} in the sequence is as yet unknown.

$$f(x_{i+1}) = f(x_i) + f'(x_i) \cdot (x_{i+1} - x_i) \tag{5.1}$$

If we set $f(x_{i+1}) = 0$ then x_{i+1} will be an approximation to the root, within the truncation error of the one-term Taylor series. Then solving Equation (5.1) for x_{i+1} gives Equation (5.2) as the iteration formula for Newton's method.

$$x_{i+1} = x_i - \frac{f(x_i)}{f'(x_i)} \tag{5.2}$$

Starting the sequence with some initial guess for the root as x_0, applying Equation (5.2) iteratively will give the sequence $x_1, x_2, \dots, x_n$ for n iterations. This sequence is illustrated in Figure 5.1, where x_0 is indicated with the index 0 on the x axis, x_1 is indicated with the index 1, etc. Starting from the function evaluated at x_0, the slope of the tangent line shown is given by the evaluation of the derivative at x_0. The iteration formula Equation (5.2) computes the next sequence argument value x_1 that geometrically can be seen as a corner of a triangle. If the algorithm converges, the iteration of this formula is graphically represented by the successively smaller triangular geometries in Figure 5.1. If the initial guess x_0 was sufficiently close to the actual root, this sequence will converge very quickly to the actual root, and we may increase the iteration count n until the displacement from x_{n-1} to x_n is sufficiently small to indicate we have reached our desired numerical precision.

Computational Physics Using C: Efficient Programming with Ease, First Edition. John W. Fattaruso. © 2026 John Wiley & Sons, Inc.
All rights reserved, including rights for text and data mining and training of artificial intelligence technologies or similar technologies.
Published 2026 by John Wiley & Sons, Inc. Companion Website: http://www.wiley.com/go/CompPhysUsingC

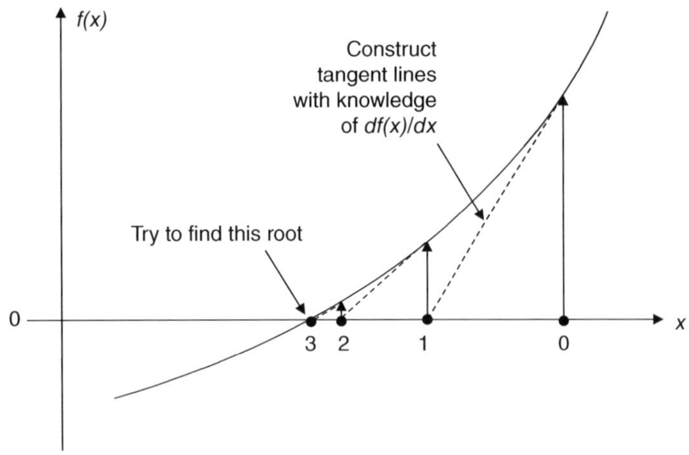

Figure 5.1 Newton's method for finding a root of a function.

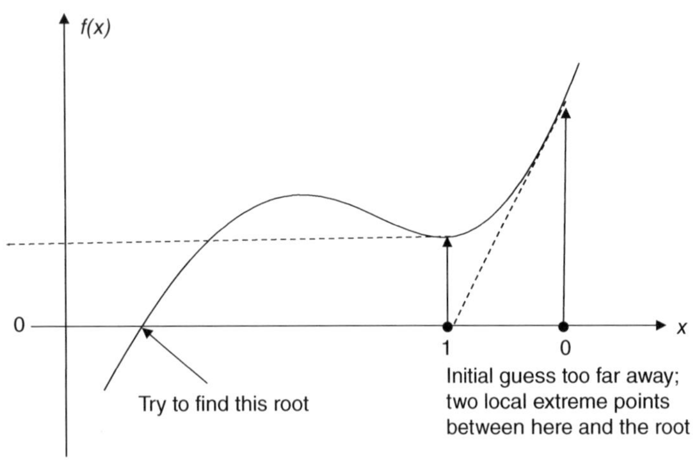

Figure 5.2 Potential divergence of Newton's method.

However, starting the algorithm with an initial guess too far away from the root along an unfortunate function contour can cause the algorithm to diverge or oscillate, as shown in Figures 5.2 and 5.3. Unstable, non-convergent behavior is generally a possibility when the sequence determined by Equation (5.2) strays into a region where the function's derivative has a small absolute value, and therefore the displacement to the next sequence point is very large, and far from the region of the desired root. Therefore, a practical implementation of this solving algorithm should include the capability by the user to set the maximum possible number of iterations before the solver surrenders and returns with an error code. The user must test for this condition and, if possible, alter the initial guess.

5.1.2 Secant Method

The Newton–Raphson algorithm is a powerful solver tool that can in most cases converge very rapidly on a desired root. However, it has the drawback of requiring that the user not only implement a C function that computes the object function, but also analytically perform a differentiation of the object function and implement a second C function that computes that derivative. For simple object functions, this presents only a small inconvenience to its use, but for many physics problems involving a complicated function, the requirement of an analytic differentiation may be troublesome. The Secant method provides an alternative where only an evaluation of the object function is required, and an approximation to that function's derivative is essentially provided by combining information from two consecutive evaluations. However, *two* initializing function argument values are required instead of the single value for the Newton–Raphson method.

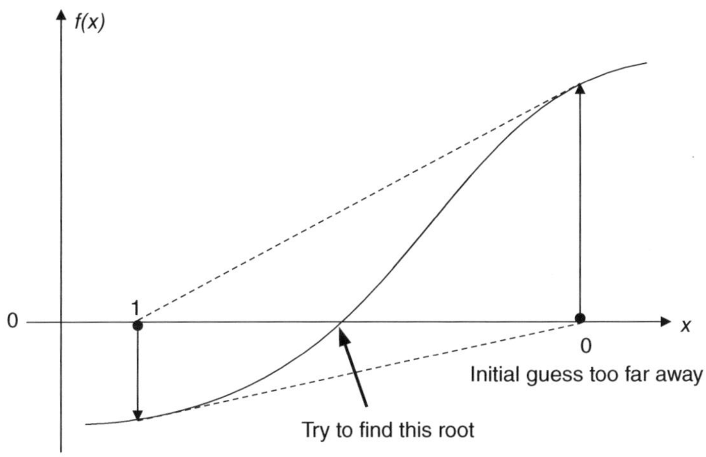

Figure 5.3 Potential oscillation of Newton's method.

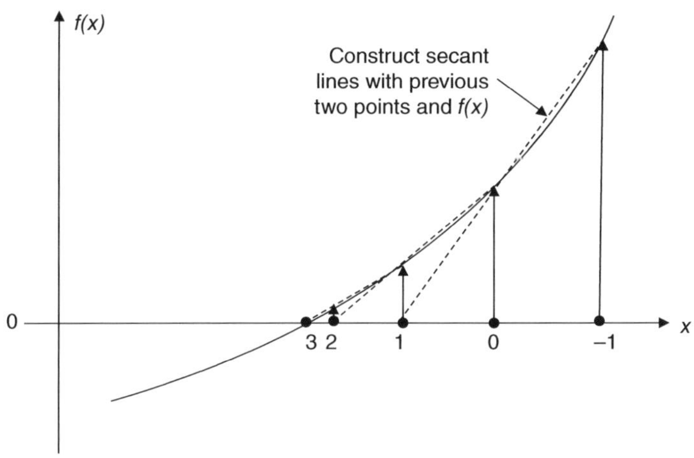

Figure 5.4 Secant method for finding a root of a function.

See Figure 5.4 for a geometric description of how the Secant method can converge on a root of a function given two initial values x_{-1} and x_0. An approximation to the tangent line of the Newton–Raphson algorithm is constructed as a *secant* line between the two points on the function. A similar triangular estimate of the root as in Newton–Raphson is constructed using this secant line, generating the next estimate of the root, x_1. Then, as shown in the figure, the secant line between the points x_0 and x_1 is used to generate the estimate x_2, and so on. As the sequence proceeds, the previous two function arguments and values are always kept in the algorithm to calculate what are hopefully better and better approximations to the root in the following iterations. Equation (5.3) describes the relationship between the sides of the triangle, which gives Equation (5.4) as the iteration formula for the Secant method.

$$\frac{x_{i-1} - x_i}{x_{i-1} - x_{i+1}} = \frac{f(x_{i-1}) - f(x_i)}{f(x_{i-1})} \tag{5.3}$$

$$x_{i+1} = x_i - \frac{x_i - x_{i-1}}{f(x_i) - f(x_{i-1})} \cdot f(x_i) \tag{5.4}$$

This iteration formula can be seen as essentially equivalent to Equation (5.2) for the Newton–Raphson method, except for the discrete approximation to the derivative of the object function $\frac{f(x_i) - f(x_{i-1})}{x_i - x_{i-1}}$.

As with the Newton–Raphson method, the iteration of the Secant method is terminated when the displacement from the previous step is within some precision specified by the user. The Secant method can be subject to the same sort of non-convergent behavior as the Newton–Raphson method, again in the vicinity where the object function's derivative is very small. Therefore, a maximum number of iterations expected by the user must again be enforced with the Secant method.

5.1.3 Regula Falsi Method

With a small modification to the method by which the previous two function points in the Secant algorithm are transformed into the next two points, we can greatly improve its convergence behavior, although in general at the expense of convergence rate. The term "Regula Falsi" means "False Point" in Latin, referring to an interpolated point on the x axis. For a root-finding algorithm, we arrange the projected, or "false," point to be a first-order approximation to where a function would cross the axis. This point is generated geometrically from two known points on the function with a triangular construction similar to that in the Secant algorithm. But for the Regula Falsi method, we make the additional constraint that the algorithm must be initialized with two known function points that are on the opposing sides of the root. This arrangement where two previous points surround, or "bracket," the root to be found must be maintained as the algorithm progresses. With this arrangement, the new interpolated point in the argument sequence will always lie somewhere between the two previous known points. Then the two points that are retained in the algorithm to determine one side of the next triangle are not necessarily the previous two points in the sequence as in the Secant method, but are rather the two previous points that bracket the root. This can be determined by always maintaining two points whose function values are of opposite sign. Since the method always starts and maintains two points that bracket the root, and then works inward from these two points to better approximations of the root, the algorithm will converge, in general, for most continuous functions that are meaningful in physical solutions.

See Figure 5.5 for an illustration of how the Regula Falsi method can iteratively converge on a root of a function given two initial values x_{-1} and x_0. Since the similar triangles are the same as in Figure 5.4, the iteration formula is the same as Equation (5.4). The two initial points at x_{-1} and x_0 determine the "false point" x_1, and then point x_0 is discarded and point x_{-1} is retained with x_1 for the next iteration, because these two points give opposite function signs. Then points x_{-1} and x_1 determine the "false point" x_2. For the next step, point x_{-1} is discarded and x_1 is retained, because x_2 and x_1 give opposite function signs. These determine point x_3, which is already quite close to the root, and so on. In this example, the two initial points are not particularly close estimates of the root, but they surround the desired root that is the object of the search. Between these two initial points there are two intervening regions with a zero function derivative that could easily cause divergence with the Newton–Raphson or Secant slope-based methods, but the Regula Falsi method successfully converges on the root.

In the example of Figure 5.6, the contour of the object function will cause the point at x_0 to always be retained in the algorithm as one of the root containment points, and the convergence to the root will always approach the root from the left side of the x axis. Although the convergence properties of the Regula Falsi method are in general very reliable, this example shows that the convergence rate on the root can be somewhat slower than with the Secant method for certain function contours. Moreover, an additional constraint for the user is the requirement to specify two initial points that bracket the desired root. Again, iteration is terminated when the root is found within some specified precision.

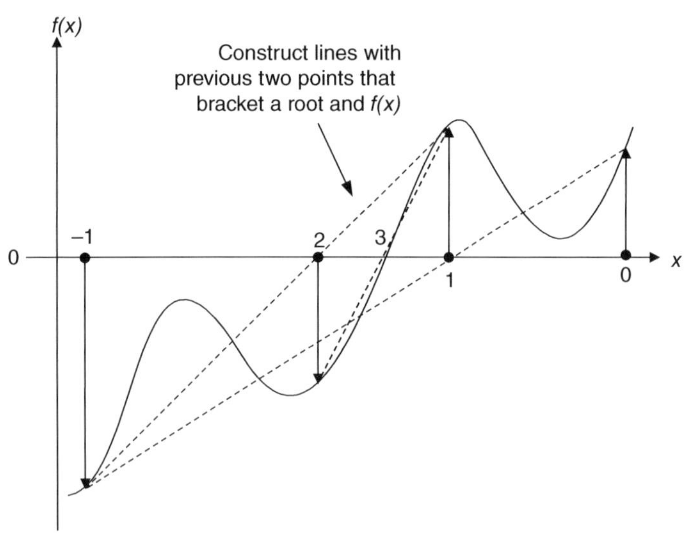

Figure 5.5 Regula Falsi method for finding a root of a function.

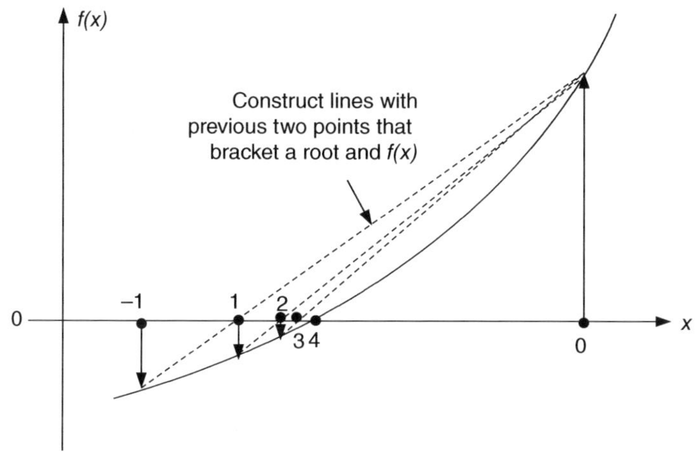

Figure 5.6 Regula Falsi method for finding a root of a function.

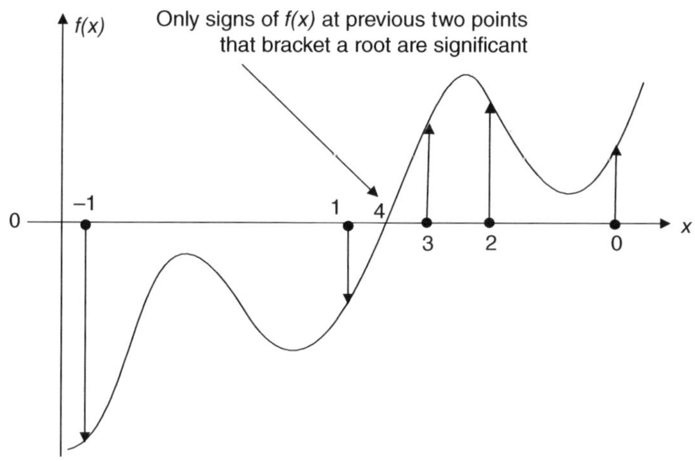

Figure 5.7 Bisection method for finding a root of a function.

5.1.4 Bisection Method

For converging on a root of a function with even more troublesome contours, with perhaps many local minima and maxima, we can consider the Bisection method. In this method, the values of the object function evaluated at the various argument points in the convergence sequence are discarded, and only their signs are significant. As with Regula Falsi method, the algorithm must be initialized with two argument values that are on opposite sides of the root to be found, and this bracketing is maintained until convergence. The new point in the sequence is simply computed as the bisection of the argument region defined by the two previous points that give opposite function signs, irrespective of the corresponding function values. Therefore, the iteration formula for the bisection method is simply computing the mean of the two previous points that bracket the root, as shown in Equation (5.5).

$$x_{i+1} = \frac{x_i + x_{i-1}}{2} \tag{5.5}$$

See Figure 5.7 to see how the bisection method can converge on a root of a function given two initial values. Since the algorithm deliberately ignores the function values, it cannot be confused by any local minima or maxima in the function. However, by discarding function information, helpful guidance that would speed convergence is also ignored, so convergence is rather slow, steadily plodding along in a binary search pattern.

It should be understood, however, that the bisection method can also converge on a singularity of a function rather than a root, as shown in Figure 5.8.

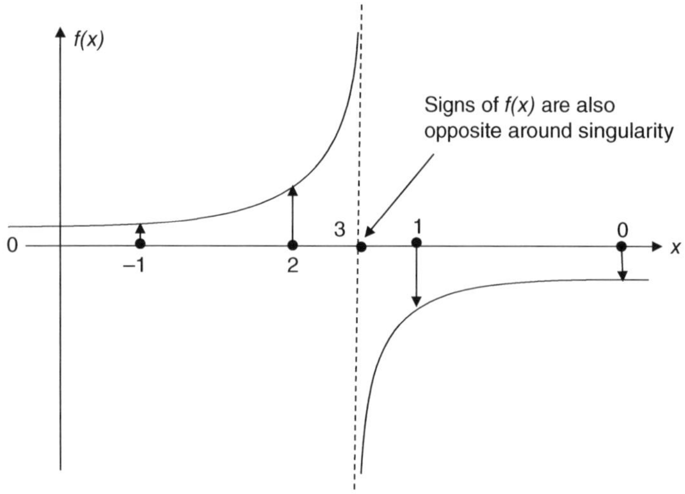

Figure 5.8 Bisection method converging on a singularity of a function.

In this section, we have examined the details of four different algorithms designed to converge on a root of a function. What conditions would guide us to choose one particular algorithm over the others? For functions that are generally smooth within some local neighborhood where a starting guess for the root is known, either the Newton–Raphson or Secant algorithms is appropriate. Both these algorithms usually converge quite rapidly to the desired root. The Secant method is a good choice when the object function's derivative is unknown or difficult to derive. On the other hand, if the convergence from the only known starting guesses to the root may traverse undulations with local minima or maxima of the object function, some convergence rate must be sacrificed with the slower Regula Falsi or Bisection methods in order to avoid divergence of the root search. The bisection method can be relied on to converge for the most troublesome functions.

To visualize the convergence process for two of these algorithms, we revisit the damped harmonic oscillator system treated in Section 4.2. In Section 4.2, the displacement of a general system governed by the differential equation of a standard second order oscillator with damping factor ζ was shown to be

$$x(t) = x_0 e^{-\zeta\omega_0 t}\cos(\omega_d t) \tag{5.6}$$

with the damped frequency $\omega_d = \omega_0\sqrt{1-\zeta^2}$. We can associate a natural period with the natural oscillation frequency as $T_0 = \frac{2\pi}{\omega_0}$. Suppose that, after the oscillator is initialized with some static displacement, we are interested in knowing the subsequent time t_z when the displacement first crosses zero. From the form of Equation (5.6), we know that this will be at a time when $\omega_d t = \frac{\pi}{2}$, or

$$t_z = \frac{T_0}{4\sqrt{1-\zeta^2}} \tag{5.7}$$

In the case of very light damping, the time dependence described by Equation (5.6) approaches a continuous cosine function, and the first zero crossing will occur at a time t_z very close to one quarter of the natural period. However, in the case of large damping, for example, in the design of a vehicle suspension system, the time to the first zero crossing will be lengthened. Refer to Figures 5.9 and 5.10 for an illustration of numerically converging on the zero displacement time, which is the root of the function of time in Equation (5.6). Figure 5.9 is a plot of the displacement versus time for the example case of an initial displacement of 1 unit, a natural period $T_0 = 1$ time unit, and a damping factor $\zeta = 0.75$. The two point plots in Figure 5.10 show the convergence process with both the Newton–Raphson and Bisection methods. The current values of the t argument at each iteration of the respective algorithm are plotted along the horizontal axis as the iteration index progresses up along the vertical axis. This function being well-behaved and easily differentiable, the Newton–Raphson algorithm converges rapidly. Even starting at a rather distant initial guess of $t = 0.05$, the trial t values quickly converge to the root at $t \approx 0.378$ time units with a tolerance of 10^{-3} only after four iterations. The two bracketing points on the t axis for the Bisection method are initialized at $t = 0.05$ and $t = 0.45$, and we can see the successive halving of the bracketing range as it contracts around the same root. It is clear that this more deliberate process of the Bisection method requires

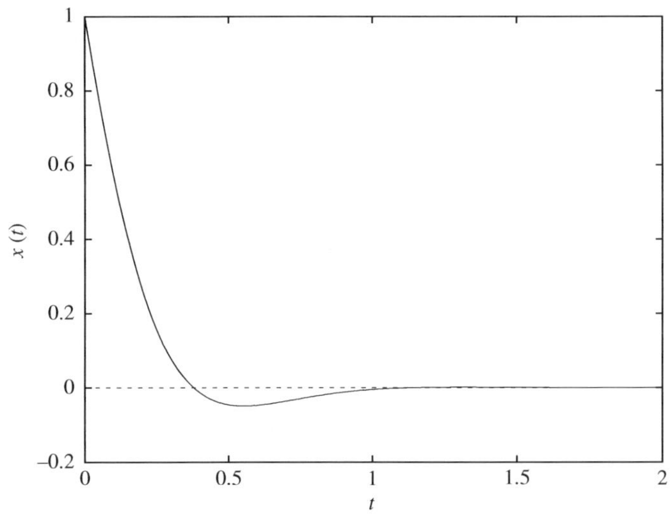

Figure 5.9 Test function for converging on the root of a nonlinear function.

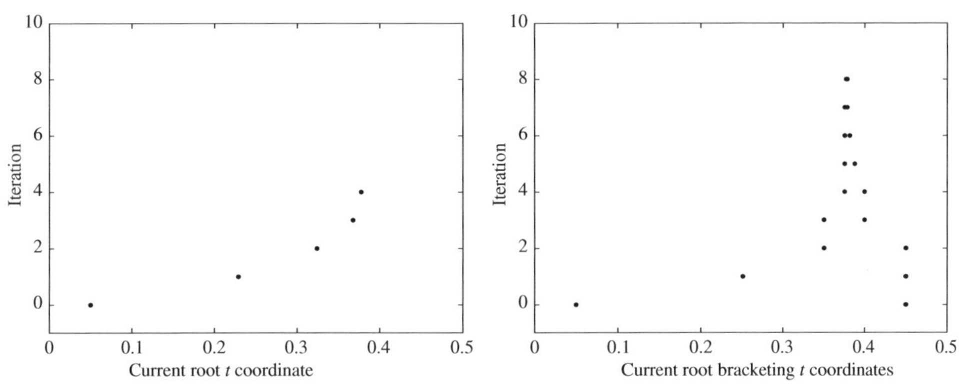

(a) Newton–Raphson algorithm **(b)** Bisection algorithm

Figure 5.10 Converging on the root with two algorithms.

many more iterations than for the Newton–Raphson algorithm. As a check, we can evaluate Equation (5.7) for the example values of $T_0 = 1$ and $\zeta = 0.75$, which predicts $t_z \approx 0.378$ as expected. Exercise 5.8 explores another metric of the oscillator time response that relies on numerical methods.

5.2 Coding: The Root Solver Tool

The code for the `root.c` tool containing four utility functions for converging on roots of functions, using the four different algorithms described in Section 5.1, is available for download from the Student Companion website in the CPLibrary package. The corresponding header file for including this tool in a full program, `root.h`, is also present. The function definition lines for the four routines are

```
int newton_raphson(double (*func)(double x,void *p),
                   double (*deriv)(double x,void *p),
                   double x,void *param,int nmax,double tol,
                   double min,double *root_ptr) {
  ...
}
int secant(double (*func)(double x,void *p),
           double a,double b,void *param,int nmax,double tol,
           double *root_ptr) {
```

```
  ...
}
int regula_falsi(double (*func)(double x,void *p),
                 double a,double b,void *param,int nmax,double tol,
                 double *root_ptr) {
  ...
}
int bisection(double (*func)(double x,void *p),
              double a,double b,void *param,int nmax,double tol,
              double *root_ptr) {
  ...
}
```

Each of the four utility functions accepts as its first argument a pointer to a C function, which must be coded to compute the object function for which a root is to be searched. The mechanism of the pointer to a function is described in Section 4.1. Note that a second function pointer is expected as the second argument for the `newton_raphson()` solver, which is the C function that evaluates the derivative of the object function.

Following the function pointer arguments, the utility tools accept one or two double-precision arguments that are the initial guess values the user wants the iterative root search to begin with. There is one initial value for the Newton–Raphson tool, and there are two initial values for the other tools.

For these root-finding tools, we will make the same assumption as in Section 4.1 for the function stepping tool, namely that the C functions that are the objects of our root searches are coded in the form

```
double foo(double x,void *p) {
  ... // some computation
  return();
}
```

The code in the function body accepts the value of the argument `x` and forms the return value of the object function whose root is to be found, or in the case of the Newton–Raphson algorithm, also the function's derivative. As in the case of the function stepping tool in Section 4.1, each object function also accepts a `void` pointer `p` as a second argument that may point to a structure storing any parameters needed by the physical model of the function. The next argument of each of the four root-finding routines in `root.c` is the pointer `param`, which is the parameter pointer value that is always passed on to the object functions at each iterative search step.

Then the following arguments to the utility tools specify the maximum number of iterations that each search algorithm will be allowed to make until returning to the calling code with a result code indicating a failure for lack of convergence. In addition, the Newton–Raphson solver tool also accepts an argument that specifies the minimum value of the object function's derivative for a search step to be made. Any derivative values encountered in the search below this minimum will also cause the search tool to terminate and return to the caller with an error code. The final argument of all four tools is a pointer to `double` that specifies where the solver function is to store the root if the algorithm successfully converges.

The header file `root.h` defines prototypes for the four root-finding utility functions, and also four symbolic integer constant macros that are used to report the results of the root iteration process back to the calling program.

```
#define ROOT_SUCCESS 0
#define ROOT_MAX_ITERATIONS 1
#define ROOT_MIN_SLOPE 2
#define ROOT_NOT_BRACKETED 3
```

A successful convergence on a root is reported with the constant `ROOT_SUCCESS` being returned as the `int` return type of the functions. However, if convergence fails, the return value reports the trouble encountered with other integer constants.

5.3 Application: The Catenary

Consider an ideal filament that is anchored at each end to two points at the same elevation, and is allowed to dangle between the anchor points. An ideal filament with some nonzero mass per unit length has no thickness and can bend freely transverse to its axis, but can support any amount of tension longitudinal to its axis without stretching. An approximation to this arrangement is a thin rope or cable dangling between two anchor points. Figure 5.11 shows a suspended line between two anchors. The anchor points are assumed to be located at x coordinates, $\pm x_1$, and y coordinate, y_1, and the bottom of the

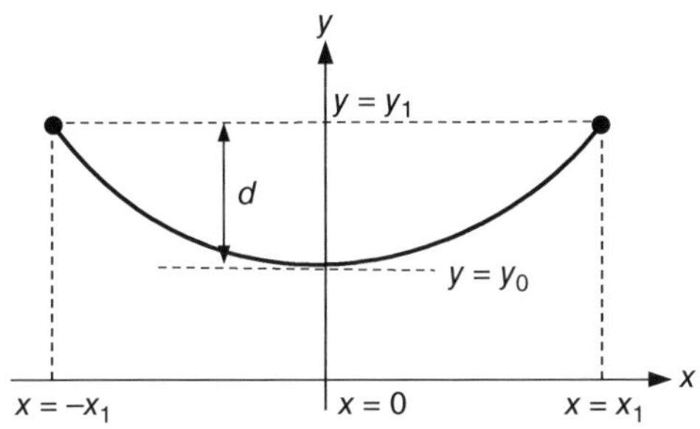

Figure 5.11 Geometry of the Catenary profile.

line droops down to the y coordinate, y_0, a distance d below elevation y_1. What profile does this dangling line describe? A spatial differential equation describing the static force relationships can be solved to define the solution, given the name of the Catenary curve. The elevation of this curve as a function of the x coordinate is given by Equation (5.8), where the parameter λ is a parameter that describes the degree of "droop" of the line. If $d << x_1$ then the line is stretched relatively tightly between the anchor points, and the parameter λ is large. If d is comparable to x_1, the line sags significantly, and the parameter λ is small.

$$y(x) = \lambda \cosh\left(\frac{x}{\lambda}\right) - \lambda + y_0 \tag{5.8}$$

From the differential equation solution also comes the total length of the line, given in Equation (5.9).

$$L = 2\lambda \sinh\left(\frac{x_1}{\lambda}\right) \tag{5.9}$$

Suppose the profile of a line is known, and the total length of a line must be calculated. We will need to solve Equation (5.8) for the parameter λ given x_1 and d, and then use this value of λ in Equation (5.9). Again, the transcendental nature of Equation (5.8) requires us to use numerical methods to find λ. Transforming the problem into a form of a root-finding algorithm, and noting that $y(x_1) = y_0 + d$, we need to find the root of the function $f(\lambda) = \lambda \cosh\left(\frac{x_1}{\lambda}\right) - \lambda - d$ treated as a function of λ. If we were to use the Newton–Raphson method to solve for this root, we would also need the derivative of $f(\lambda)$, which is easily derived to be $\frac{df}{d\lambda} = \cosh\left(\frac{x_1}{\lambda}\right) - \lambda \sinh\left(\frac{x_1}{\lambda}\right) \cdot \left(\frac{x_1}{\lambda^2}\right) - 1$. On the Student Companion website is the source code `catenary.c`, a C main program to perform this computation. To provide the most useful demonstration possible, this code makes use of all four root-finding algorithms developed in Section 5.1 to solve for λ. A convergence tolerance of 10^{-8} is specified for λ for all four methods. Arbitrary units are assumed for the distance values. The C parameter structure and the function whose root is to be found are coded in this example as follows

```
typedef struct catenary_param {
  double x1;  // Anchor x coordinate
  double droop;  // Vertical droop of line
} CAT;

double func(double lambda,void *param_ptr) {
  double f;
  CAT *p;

  p = param_ptr;
  f = lambda * cosh(p->x1/lambda) - lambda - p->droop;
  return(f);
}

double deriv(double lambda,void *param_ptr) {
  double f;
  CAT *p;

  p = param_ptr;
  f = cosh(p->x1/lambda) - lambda * sinh(p->x1/lambda) *
         (p->x1 / (lambda * lambda)) - 1.0;
  return(f);
}
```

The listing below shows the compilation of the program together with the root-solving tools at a Linux or MSYS2 shell prompt window. Running initially with no command line arguments reminds us that the code is written to accept the geometric parameters x_1 and d, and two guess values that bracket the root for the parameter λ on the command line, although only the first guess value is used for initializing the Newton–Raphson algorithm. Then two additional runs are shown here for two example cases of a large and small droop

```
$ gcc -Wall -Wshadow -O2 -o catenary catenary.c CPLibrary/root.c -lm
$ ./catenary
./catenary <x1> <droop> <a> <b>
$ ./catenary 30 20 10 30
lambda root from Newton-Raphson = 25.269947   total length = 75.121179
lambda root from Secant = 25.269947   total length = 75.121179
lambda root from Regula Falsi = 25.269947   total length = 75.121179
lambda root from Bisection = 25.269947   total length = 75.121179
$ ./catenary 30 5 50 150
lambda root from Newton-Raphson = 90.821312   total length = 61.097074
lambda root from Secant = 90.821312   total length = 61.097074
lambda root from Regula Falsi = 90.821312   total length = 61.097074
lambda root from Bisection = 90.821312   total length = 61.097074
```

These test cases are for an x_1 coordinate of 30 arbitrary units, and droop distances of both 20 and 5 units, respectively. Some trial and error may be needed to arrive at the two initial root brackets. Our physical reasoning would lead us to expect that for a tightly stretched line, the total length should approach $2x_1$, but that the length should be considerably longer for a more relaxed line with a larger droop. These expectations are demonstrated in the two solutions above. With this fairly simple and well-behaved function, all four algorithms converge to the same root.

5.4 Application: Kirchoff's Voltage Law

Kirchoff's Voltage Law (KVL) for lumped electrical circuits is illustrated in Figure 5.12. The circuit elements pictured are electrically connected in a single loop, and all the elements are carrying the same electric current I circulating around the loop. In the lumped circuit approximation, we assume that the electric potential at physical locations in the circuit does not change over the ideal electric conductors connecting the elements, but does differ on each side of the elements.

If all the reference polarities of the potential drops V_j across all circuit elements are in the same loop direction, then conservation of energy requires that the sum of all potential drops around the loop must equal 0. Assuming that the function relationships between I and the $V_j(I)$ for all the circuit elements are known, the current I can be found by solving for the root of Equation (5.10).

$$\sum V_j(I) = 0 \tag{5.10}$$

As an example of solving for the operating point of a nonlinear circuit with KVL, see Figure 5.13. Electric power from a 5-V source causes a current to flow through the light-emitting diode (LED), and generates light output. The current level is set by the linear resistance in the circuit loop. KVL requires that

$$V_{PS} + V_R + V_{\text{LED}} = 0 \tag{5.11}$$

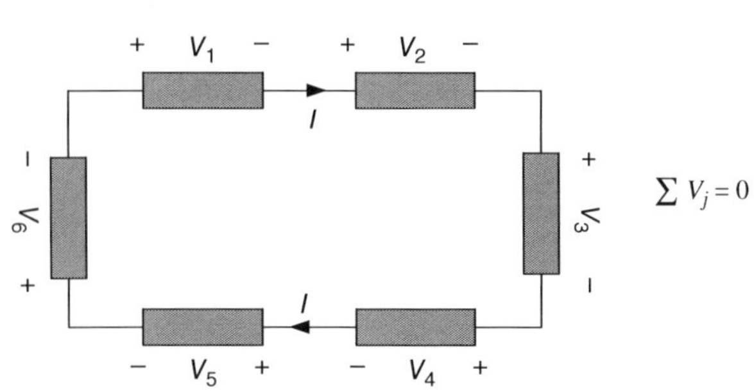

Figure 5.12 Kirchoff's Voltage Law.

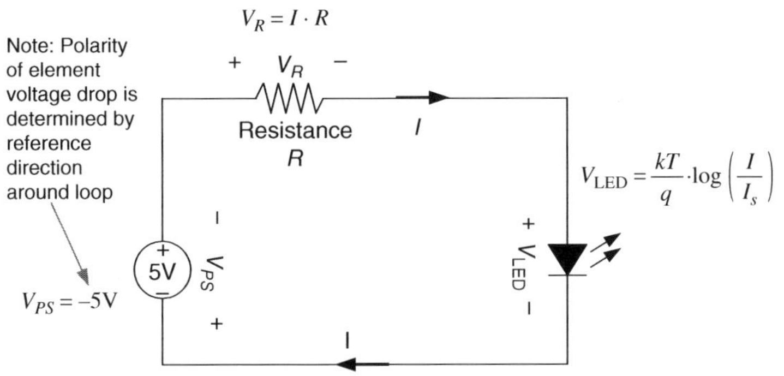

Figure 5.13 Series circuit with nonlinear light-emitting diode (LED).

The power in this circuit is supplied by some sort of voltage source, perhaps a physical battery or source subcircuit. Since a voltage source, by definition, raises the electric potential from one end to the other, the potential drop across a source is the negative of the source's voltage. Then using the linear relationship of the voltage drop across the resistor element and the nonlinear characteristics of the LED element with respect to the current flowing through them, the KVL relationship becomes

$$-5V + IR + \frac{kT}{q}\log\left(\frac{I}{I_S}\right) = 0 \tag{5.12}$$

where R is the resistance, T is the absolute temperature, q is the electron charge, and I_s is the saturation current of the LED junction. This is a transcendental equation in the current I, and the root of the function on the left side is therefore the physical current flowing in the loop.

Intending to use our numerical tools for a solution, how can we arrive at reasonable starting guesses for the unknown current? We may use some physical reasoning regarding the electrical behavior around this circuit loop. If the resistor has a resistance of 200 ohms, and if there is a negligible voltage drop across the LED, then the entire 5 V of the source would be dropped across the resistor and the loop current would be limited to $\frac{5V}{200\Omega} = 0.025$ A by Ohm's law. This current level can be used as an upper bracketing level. With a saturation current for the LED of $I_S = 6.18 \times 10^{-26}$ A and at a temperature of $T = 300$ K, the logarithmic characteristic of the diode drop voltage with respect to current would limit the voltage drop to the neighborhood of 1–2 V for any feasible current flow. So we may set the lower bracketing level at the current corresponding to a diode drop of 2 V, for example, or 0.015 A. The solution of this electrical problem, for completeness, using both the Newton–Raphson and the Secant algorithms, is left as Exercise 5.1.

5.5 Application: Mechanics Problems

This example problem illustrates how a geometrical mechanics problem that appears trivial on its surface can easily land us in the realm of nonlinear equations for which numerical solutions are the most practical approach. Let's say the Hollywood director of a film wants to stage a stunt with two colliding cars on a movie set. The geometry is illustrated in Figure 5.14. The first car, traveling at constant velocity $v_1 = 20$ m/sec, will be sent on a northeast trajectory at some unknown angle θ with respect to north, and passes a reference point at time $t = 0$. The second car, traveling at constant velocity $v_2 = 25$ m/sec, will be sent on a northwest trajectory at some unknown angle ϕ with respect to north and passes another reference point at time $t = 0$. The east-west distance between the two reference points is $D_1 = 200$ m and the north-south distances are $D_2 = 50$ m and $D_3 = 100$ m. At some unknown time $t_c > 0$ the collision will occur. In order to stage the scene, the director's assistant needs to solve for the unknown angles θ and ϕ and the unknown time t_c.

With the simplifying assumption that the sizes of the cars are negligible for the purpose of timing the collision, we may formulate a nonlinear equation to be solved for the collision time t_c with a simple geometric observation that the distances traveled by the two cars form the hypotenuse of two right triangles. That is, the distance $v_1 t_c$ forms the hypotenuse of a right triangle on the left, and similarly for the distance $v_2 t_c$ with a right triangle on the right. Using the knowledge of the lengths of the vertical triangle sides and the sum of the lengths of the horizontal sides, we can formulate a nonlinear function

$$f(t_c) = \sqrt{v_1^2 t_c^2 - (D_3 - D_2)^2} + \sqrt{v_2^2 t_c^2 - D_3^2} - D_1 = 0 \tag{5.13}$$

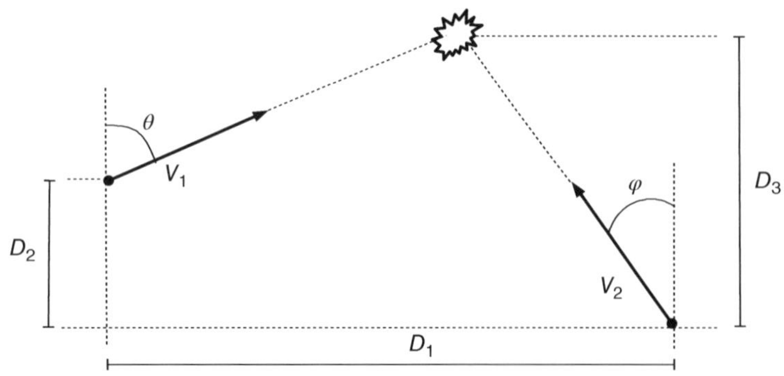

Figure 5.14 A simple nonlinear mechanics problem for staging a car stunt.

An attempt to solve this equation analytically will soon be bogged down with a cumbersome fourth-order polynomial, and it is more practical to find the root of this function with a numerical solving tool. Once the collision time t_c is found, the two unknown angles θ and ϕ may be found from the simple trigonometric relations

$$\begin{aligned} v_1 t_c \cos\theta &= D_3 - D_2 \\ v_2 t_c \cos\phi &= D_3 \end{aligned} \tag{5.14}$$

The solution of this mechanical problem is left as Exercise 5.2.

5.6 Application: Kepler's Equation

Figure 5.15 shows the geometry of Kepler's classic equation describing elliptical orbits of an object around a massive body located at one of the two foci of the ellipse. The orbital path is drawn with the thicker line in the figure, and the orbital period P is the time required to traverse the complete elliptical path once.

The lengths of the major and minor axes of the ellipse are the parameters a and b, respectively. The eccentricity of the ellipse is the parameter ϵ as shown in the figure, along with other useful metrics. In our own solar system we can observe orbits with eccentricities over a large range, with $\epsilon < 0.02$ for the Earth's orbit, and $\epsilon > 0.995$ for the Great Comet of 1811, with its orbital period estimated at almost 3000 years. For any orbital eccentricity, Kepler's laws give the orbital period P as

$$P = 2\pi\sqrt{\frac{a^3}{Gm_p}} \tag{5.15}$$

where a is the semi-major axis, m_p is the mass of the primary body, and G is the universal gravitation constant.

With somewhat anachronistic terminology, the *mean anomaly* M measures the uniform progression of time through an orbital period. We define a mean anomaly of 0 to correspond to the time $t = T$ that the orbiting object passes through the *periapsis*, or the point nearest to the primary object being orbited. As defined by Equation (5.16), a mean anomaly of π (or $-\pi$) corresponds to the orbiting object passing through the *apoapsis*, or the point farthest from the primary object. When $M = 2\pi$ the object again returns to the periapsis.

$$M = 2\pi\left(\frac{t - T}{P}\right) \tag{5.16}$$

With the geometry of Figure 5.15, the orbiting object's angular position can be determined by the *eccentric anomaly* E, an angle drawn from the center of the ellipse to the auxiliary circle. If the primary body being orbited is much more massive than the orbiting body, such as the case of the Earth around the Sun, we may consider the primary body to be essentially stationary.

For the case of a dominantly massive primary body, it is more meaningful to visualize the points along the orbit as being at a polar radius r relative to the center of the primary body and a polar angle ν, called the *true anomaly*, relative to the major axis of the ellipse. For example, it is more meaningful to consider the origin of the polar coordinate system of our own solar system to be at the focus of the elliptical orbits of planets and asteroids, essentially located at the center of the Sun,

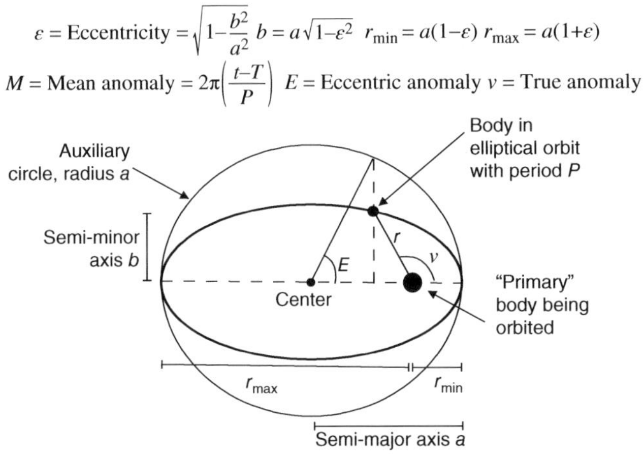

Figure 5.15 Geometry of Kepler's Equation.

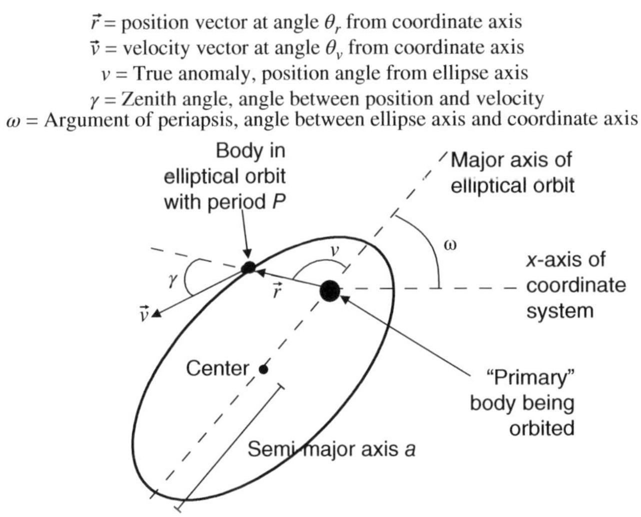

Figure 5.16 Position and velocity vectors on a Keplerian elliptical orbit.

rather than at the center of the elliptical orbits. For elliptical orbits, both r and ν will be functions of time. The conversion relationships to these coordinates are

$$\nu = 2\tan^{-1}\left[\frac{\sqrt{1+\epsilon}\sin\left(\frac{E}{2}\right)}{\sqrt{1-\epsilon}\cos\left(\frac{E}{2}\right)}\right] \tag{5.17}$$

$$r = \frac{a(1-\epsilon^2)}{1+\epsilon\cos(\nu)} \tag{5.18}$$

In Section 13.2, we will approach orbital systems from a different point of view and find the motion in n-body gravitational system by numerically solving the differential equations of motion. For that approach, we will need the correspondence between the Keplerian elements for an orbit and the mechanical position and velocity vectors of an orbiting body in Cartesian coordinates centered on the primary body. For simplicity, we will treat this problem in only two dimensions and assume that the plane of the orbit is the x–y plane of a coordinate system, but we wish to accommodate an orbit with any angle ω, called the argument of periapsis, between the axis of the ellipse and the x axis of the coordinate system. The coordinate system may then be aligned to some convenient reference direction in space, such as the direction defined by a planetary equinox. If the position vector $\vec{r}$ makes an angle θ_r with respect to the coordinate system, we see from Figure 5.16 that

$$\theta_r = \nu + \omega \tag{5.19}$$

and the position vector components from the center of the primary can be found with the expressions $r_x = r\cos(\theta_r)$ and $r_y = r\sin(\theta_r)$. For the velocity vector, further consequences of Kepler's laws are that once we know the radial coordinate r, we can find the magnitude of the orbital velocity around the elliptical path, also a function of time, as

$$v = \sqrt{Gm_p\left(\frac{2}{r} - \frac{1}{a}\right)} \tag{5.20}$$

The zenith angle, or the angle between the position and velocity vectors, as shown in Figure 5.16, is given by

$$\gamma = \tan^{-1}\left[\frac{1+\epsilon\cos\nu}{\epsilon\sin\nu}\right] \tag{5.21}$$

Therefore, while the magnitude of the velocity vector is given by Equation (5.20), we can calculate the angle of the velocity vector with respect to the x axis θ_v by

$$\theta_v = \nu + \gamma + \omega \tag{5.22}$$

Some care must be taken for a computational evaluation of Equations (5.17) and (5.21) to ensure that the proper domains of the trigonometric functions are accounted for and that the signs of the angles are preserved. These equations should be calculated using the `atan2()` C library function, as detailed in Section 3.3.6. For example, for Equation (5.17) we would use the statement

```
nu = 2.0 * atan2(sqrt(1.0 + p->eccentricity) * sin(0.5 * E),
     sqrt(1.0 - p->eccentricity) * cos(0.5 * E));
```

A prerequisite for the practical use of all this development in characterizing, for example, the motion around a simple orbit of a planet, asteroid, or comet under the gravitational influence of a single primary body such as a star, as a function of time, is the need to associate a given value of the mean anomaly M that linearly measures the progression of time with the corresponding value of the eccentric anomaly E so that all the above expressions may be evaluated. Kepler's iconic equation, derived in the early seventeenth century, describes the relationship between the eccentric anomaly E and the mean anomaly M as

$$M = E - \epsilon\sin(E) \tag{5.23}$$

However, the problem of finding a value of E for a given value of M cannot be solved analytically due to the transcendental nature of this equation. To visualize the behavior of Kepler's equation, the function visualization methods of Chapter 4 may be applied. The plot of Figure 5.17 is the result of stepping the right side of Kepler's equation as a function of E ranging from 0 to 2π for three different values of eccentricity. But note that the sampled function values are plotted with `gnuplot` so that the argument E lies on the *vertical* axis and the function value M lies on the *horizontal* axis. Then "solving" Kepler's equation, that is, finding the value of E that corresponds to a given target value of M, is graphically equivalent to looking up a value on this plot in the conventional manner, with the independent variable now on the horizontal axis and the dependent variable now on the vertical axis. The contour of this plot is qualitatively consistent with our sense of an object in an elliptical orbit, which speeds up through the periapsis and slows down through the apoapsis. As the passage of time moves the mean anomaly M uniformly along the horizontal plot axis, in the case of a highly elliptical orbit, we can see that the eccentric anomaly E changes rapidly through the periapsis at both ends of the plot. However, through the apoapsis at the center of the plot, the eccentric anomaly changes slowly. How do we use the numerical root-finding tools of this chapter to start with a value of M and arrive at a corresponding value of E to some arbitrary level of precision? We may transform the problem into one simply of finding the root of the function $f(E) = E - \epsilon\sin(E) - M_{\text{target}}$ treated as a function of E.

The main program to find this solution using the fast convergence of the Newton–Raphson algorithm is in the file `kepler_root.c`, available for download from the Student Companion website. The eccentricity of the orbit is accepted on the command line, as well as the target value of M for which an orbital E value is needed. These parameters are stored in a structure and passed to the root-finding tool, which in turn passes them on to the object function in each iteration. Two C functions are provided for the `newton_raphson()` solver tool, one coding the object function $f(E) = E - \epsilon\sin(E) - M_{\text{target}}$ and the other its derivative with respect to the function argument $f'(E) = 1 - \epsilon\cos(E)$.

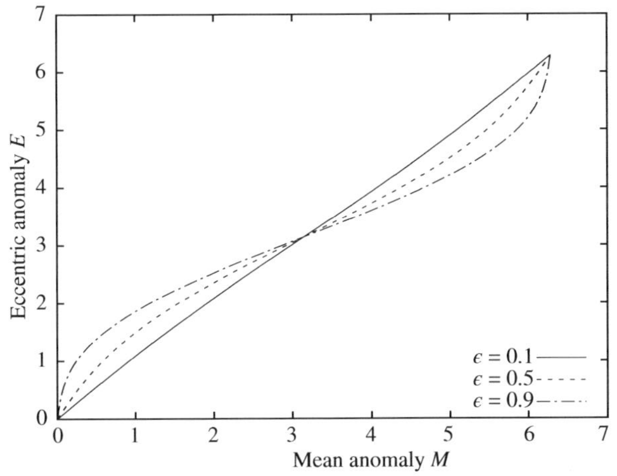

Figure 5.17 Dependency of Kepler's equation.

```
typedef struct kep_param {
  double eccentricity;  // Eccentricity of orbit
  double a;  // Semi-major axis of orbit
  double omega;  // Argument of periapsis of orbit
  double m_primary;  // Mass of the primary;
  double P;  // Period of orbit
// Target value of mean anomaly for solution of Kepler's equation
  double target_M;
} KEP;

double kepler(double E,void *param_ptr) {
  double M;
  KEP *p;

  p = param_ptr;
  M = E - p->eccentricity * sin(E) -
      p->target_M;
  return(M);
}

double keplerderiv(double E,void *param_ptr) {
  double dMdE;
  KEP *p;

  p = param_ptr;
  dMdE = 1.0 - p->eccentricity * cos(E);
  return(dMdE);
}
```

We must also provide a reasonable starting guess to the Newton–Raphson algorithm for successful convergence over the possible range of parameter values. Consider the first-order estimate function $\hat{E} = M + \sin(M)$, as plotted in Figure 5.18. Although for highly eccentric orbits, the estimate $\hat{E}$ is not particularly close to the true solution of E, we can see that the general contour of this estimate function follows the actual dependence in Figure 5.17, and in practice this estimate provides a reliable starting guess for the Newton–Raphson iterations.

Compiling this program together with the root-finding tool routines and running at a Linux shell prompt window with some example parameter values shows reliable convergence

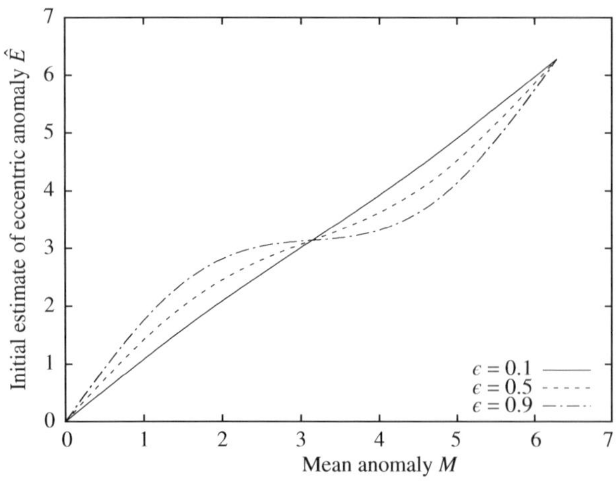

Figure 5.18 First-order estimate of the dependency of Kepler's equation.

```
$ gcc -Wall -Wshadow -O2 -o kepler_root kepler_root.c CPLibrary/root.c -lm
$ ./kepler_root
./kepler <eccentricity> <target M>
$ ./kepler_root 0.1 2
E = 2.086972
$ ./kepler_root 0.5 3
E = 3.047151
$ ./kepler_root 0.99 5
E = 4.158184
```

The `kepler_root.c` demonstration program finds a single value of the eccentric anomaly E that corresponds to a single value of the mean anomaly M. For a more complete visualization of the orbital trajectory that Kepler's laws predict, we may embed this numerical root solution code into a function `kepler_orbit()`. This function is written to compute first the eccentric anomaly from an argument of the mean anomaly. The argument M can be stepped from $-\pi$ to $+\pi$ to cover the time range of one orbit from apoapsis through periapsis and returning to apoapsis. Then with the value of eccentric anomaly, the function continues to compute the true anomaly ν, the orbital radial coordinate r, the velocity magnitude v, and the zenith angle γ for each eccentric anomaly using the conversion formulas in Equations (5.17), (5.18), (5.20), and (5.21). From these quantities the four polar coordinates of the position and velocity vectors (r, θ_r) and (v, θ_v) are calculated. Unlike most of the C functions with single return values that are the subject of sections in Chapter 4, this function has multiple return values mapped from a single argument value of mean anomaly M. These four return values are stored in the elements pointed to by an `ARRAY1D` structure that is essentially a function return table. A pointer to this return structure is accepted as the last function argument, and the formal return value of the function can be `void`. The computed orbital value of position radius r is returned in the array element with index 0, the position angle θ_r in element 1, the velocity v in element 2, and the velocity angle θ_v in element 3. Although this form of object function that returns multiple values is most often thought of as returning related components of some vector value, such as an electric or magnetic field, in this case we may use the same coding form to return a set of four scalar quantities in a table as well.

```
void kepler_orbit(double M,void *param_ptr,ARRAY1D *ytab_ptr) {
  double E,nu,gamma;
  KEP *p;

  p = param_ptr;
  p->target_M = M;
// Solve Kepler's equation with the Newton-Raphson algorithm to find
// eccentric anomaly
  if (newton_raphson(kepler,keplerderiv,M + p->eccentricity * sin(M),
                     p,100,1.0e-6,1.0e-12,&E) == ROOT_SUCCESS) {
      nu = 2.0 * atan2(sqrt(1.0 + p->eccentricity) * sin(0.5 * E),
            sqrt(1.0 - p->eccentricity) * cos(0.5 * E));
      ytab_ptr->elem[0] =
```

```
            p->a * (1.0 - p->eccentricity * p->eccentricity) /
            (1.0 + p->eccentricity * cos(nu));
        ytab_ptr->elem[1] = nu + p->omega;
        ytab_ptr->elem[2] = sqrt(GRAVITY * p->m_primary *
                            (2.0 / ytab_ptr->elem[0] - 1.0 / p->a));
        gamma = atan2(1.0 + p->eccentricity * cos(nu),
                      p->eccentricity * sin(nu));
        ytab_ptr->elem[3] = nu + p->omega + gamma;
    }
    else fprintf(stderr,"Kepler equation root failure\n");
}
```

The program `orbit_step.c`, available for download from the Student Companion website, provides an example of stepping the C function `kepler_orbit()` and generating a multi-column output stream suitable for plotting. In the main code module of the `orbit_step.c` program we make use of the CPLibrary tool `step_vect()` in the `step.c` source file to step the mean anomaly and store the four return values of the `kepler_orbit()` function into successive columns of an `ARRAY2D` structure. The two-dimensional array where the sampled values are stored must be allocated by the calling code to `step_vect()`. The number of rows must be the number of components of the vector-valued function passed to the library routine, and the number of columns must be the number of grid samples, similar to the number of elements in the single-dimensional result arrays passed to the `step()` library tool discussed in Chapter 4. For this purpose, the `NUM_GRID_PTS` macro is a convenient aid. Then this array of function return values is sent to a `gen_output()` function to be printed to the standard output stream. The code lines

```
alloc_array2d(&otable,4,NUM_GRID_PTS(Mgrid));
step_vect(kepler_orbit,&Mgrid,&param,&otable);
gen_output(&otable,&Mgrid);
free_array2d(&otable);
```

concisely accomplish these tasks.

An example of plotting 101 time samples of the position of the orbiting body around an origin at the primary body with various eccentricities and arguments of periapsis is shown in Figure 5.19. The semi-major axis parameter a for all of these example orbits is the approximate radius of the Earth's orbit, 1.5×10^{11} m, and the location of the primary body at the origin is plotted as a gray circle. Since the mean anomaly is stepped uniformly, the positions in these plots represent uniformly spaced orbit instances in time, and the slower and faster portions of the orbits around the apoapsis and periapsis, respectively, may be visualized. This example program will also be used in Section 11.13 and Exercise 11.8 to establish an initial guess for fitting a set of orbital parameters from astronomical observations, and then later in Exercise 13.1 to establish initial conditions for the solution of differential equations of motion of gravitation.

5.7 Application: Gravitational Lagrange Points

Simple orbital problems of a planet circling a star are solved with Newton's formulation of gravitational forces and Kepler's laws. An interesting mathematical consequence of the inverse square dependence of these forces is the existence of the so-called Lagrange points in the planetary orbit geometry. These are points off of the planetary path where a satellite will orbit the star with the same period as the planet. The James Webb Space Telescope (JWST) was launched into the Earth–Sun orbital system at Lagrange point 2 (L2). This positioning exploits the properties of the L2 point that the telescope satellite will remain stationary with respect to the Earth and shadowed from the Sun. However, solving for the exact location of the Lagrange point with respect to the Earth involves finding a root of a fifth-order polynomial. This can be easily accomplished using the numerical tools of this chapter.

We begin by deriving the properties of the simple Sun–Earth orbital system, shown in Figure 5.20. Assuming a circular orbit of period P_e and using simple Newtonian gravitational laws to equate the gravitational force on the Earth with the centripetal force on the Earth

$$F_e = \frac{Gm_s m_e}{R^2} = \frac{m_e v_e^2}{R} = \frac{4\pi^2 R m_e}{P_e^2} \tag{5.24}$$

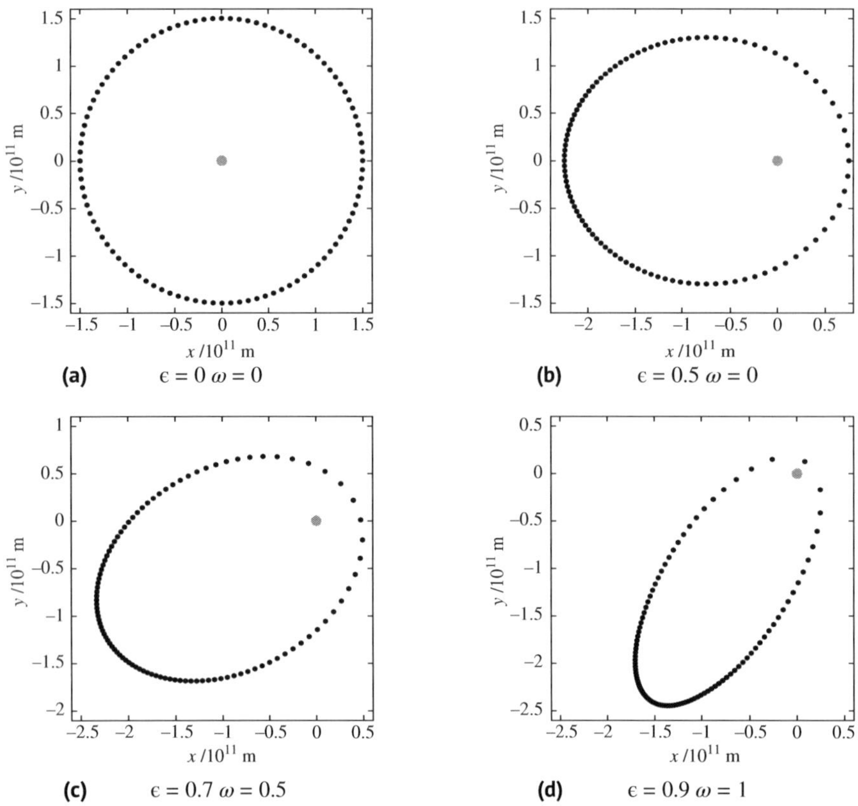

Figure 5.19 Orbital paths predicted by Kepler's laws for semi-major axis $a = 1.5 \times 10^{11}$ m and different values of the eccentricity ϵ and argument of periapsis ω, sampled over uniform time intervals.

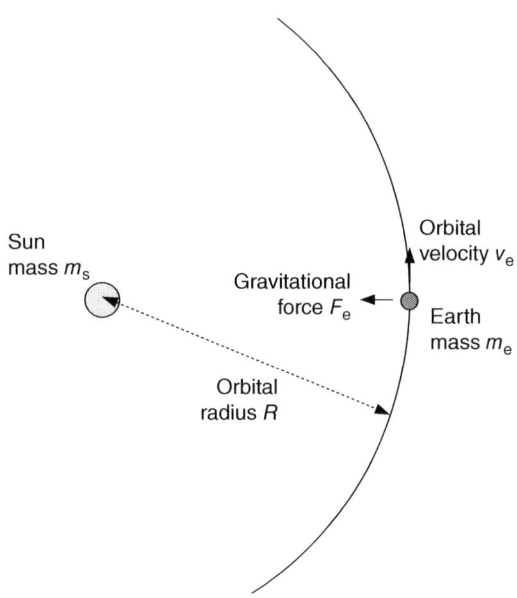

Figure 5.20 Earth in solar orbit.

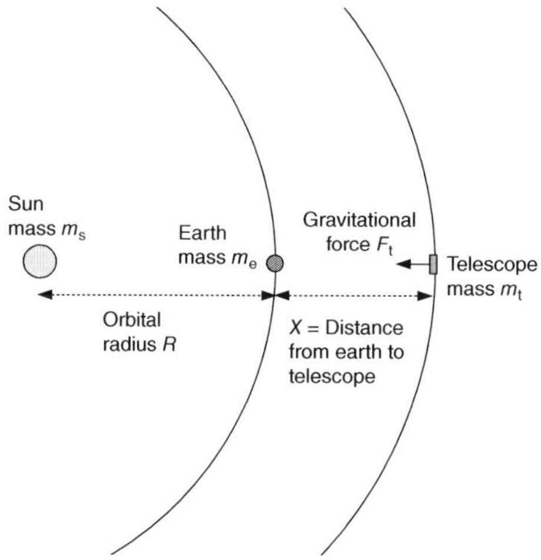

Figure 5.21 Earth and the JWST in solar orbit.

From this we can solve for the square of the orbital period

$$P_e^2 = \frac{4\pi^2 R^3}{Gm_s} \tag{5.25}$$

Now consider the JWST also in orbit about the Sun, at some point beyond the Earth's orbit by a distance x, as shown in Figure 5.21. The gravitational force on the telescope is

$$F_t = \frac{Gm_s m_t}{(R+x)^2} + \frac{Gm_e m_t}{x^2} \tag{5.26}$$

Assuming the telescope is also in a circular orbit of period P_t around the Sun, the centripetal force must be

$$F_t = \frac{4\pi^2(R+x)m_t}{P_t^2} \tag{5.27}$$

Equating these forces allows us to solve for the square of the telescope's orbital period

$$P_t^2 = \frac{4\pi^2(R+x)}{\frac{Gm_s}{(R+x)^2} + \frac{Gm_e}{x^2}} \tag{5.28}$$

The desired location for the JWST is the L2 Lagrangian point. This point is defined by the telescope's orbital period being equal to the Earth's period, or $P_t^2 = P_e^2$. So setting the two expressions above equal to each other

$$\frac{4\pi^2 R^3}{Gm_s} = \frac{4\pi^2(R+x)}{\frac{Gm_s}{(R+x)^2} + \frac{Gm_e}{x^2}} \tag{5.29}$$

We can turn this relation into a function of the distance x whose root is the distance to L2 simply with

$$f(x) = \frac{4\pi^2 R^3}{Gm_s} - \frac{4\pi^2(R+x)}{\frac{Gm_s}{(R+x)^2} + \frac{Gm_e}{x^2}} \tag{5.30}$$

However, before submitting this function to any numerical root solver, we must exercise some care regarding the particular form of the expressions we code in C and pass to a root solver function. All the algorithms detailed in Section 5.1 involve starting at some initial point or points in the function argument range and iteratively converging on the root, evaluating the object function at each step. That object function should not be expressed in a form that would allow any particular argument value encountered during the search to trigger any computational failures or result in large roundoff errors. Leaving factors containing a dependence on x in fractional denominators in Equation (5.30) would leave us vulnerable to these potential numerical issues. By multiplying through and moving these dependencies on x to the numerators, we eliminate the traps of singularities in the function that can easily cause any iterative algorithm to diverge. The result is a fifth-order polynomial function in x.

See the source code file `JWST.c` on the Student Companion website for the main program code that solves for the location of the JWST at the L2 Lagrangian point with the Secant algorithm, passing a function that will go to zero when its argument is the distance in meters from Earth to the JWST.

Compiling and running at a Linux shell prompt window:

```
$ gcc -Wall -Wshadow -O2 -o JWST JWST.c CPLibrary/root.c -lm
$ ./JWST
./JWST <a> <b>
$ ./JWST 1e9 2e9
JWST distance = 1.50330368e+09 meters
```

A variation of this problem to locate the Solar and Heliospheric Observatory (SOHO) telescope at the L1 Lagrangian point is available for the student as Exercise 5.4.

5.8 Application: Planck's Radiation Law

A seminal development in quantum mechanics at the very start of the twentieth century was the postulate of Planck's law for black body radiation [1, Chap. 1; 23, Chap.3; 24, Chap. 2; 25, Chap. 12]. The radiation energy density per unit of wavelength, as a function of wavelength, is given by

$$u(\lambda) = \frac{8\pi hc}{\lambda^5}\frac{1}{e^{\frac{hc}{\lambda kT}} - 1} \tag{5.31}$$

and this energy density spectrum for four black body temperatures is plotted in Figure 5.22. This plot also indicates the range of wavelengths corresponding to visible light.

A feature of this spectrum that is important to quantify is the wavelength at which the radiation spectrum reaches its maximum. We can follow the canonical method of finding the maximum by differentiating this energy density function, and then solving for the independent wavelength variable that makes the derivative zero.

$$\frac{du}{d\lambda}(\lambda_{max}) = 0 \tag{5.32}$$

However, while setting up the solution of Equation (5.32) we are met with a transcendental function that we must find the root of, and so we must submit the derivative function to a numerical root-finding routine. This coding problem is Exercise 5.5 for the student. But examination of the derivative function $\frac{du}{d\lambda}$ in Equation (5.32) will reveal that the wavelength

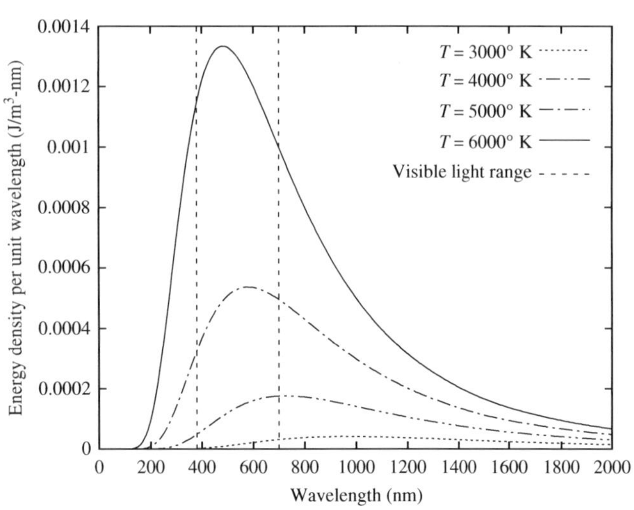

Figure 5.22 Radiation energy density per unit wavelength versus wavelength for T = 3000, 4000, 5000, and 6000 K.

λ only occurs in the transcendental expression multiplied by the absolute temperature T. This implies that the numerical solution of Equation (5.32) need only be performed once at one nominal temperature, and that all other values of λ_{max} at other temperatures will follow the inverse proportionality of Wien's displacement law

$$\lambda_{max} = \frac{\alpha}{T} \tag{5.33}$$

Solving for the constant α in Equation (5.33) is also left as an exercise for the student. We will revisit Planck's radiation law in conjunction with the cosmic microwave background radiation in Section 11.10.

5.9 Application: Radioactive Decay

This section examines some basic modeling of nuclear decay statistics and the appearance of transcendental equations that must be solved numerically. We first consider a simple case of a nuclide "A" decaying into a different nuclide "B," and that the decay events are governed by some constant probability λ_{AB} of a quantum mechanical tunneling event within some unit time interval. This implies that the expected value of reduction of the population N_A in each unit time interval is $\lambda_{AB}N_A$. Then in the limit, as the population becomes large and the unit time interval becomes infinitesimal, the population of A nuclides at some particular time is given by a function $N_A(t)$, and we can write the differential equation

$$\frac{dN_A}{dt} = -\lambda_{AB}N_A \tag{5.34}$$

This is a simple first-order differential equation, and assuming an initial population $N_A(0) = N_{A0}$ at some point in time, the straightforward solution for the population as a function of time is

$$N_A(t) = N_{A0}e^{-\lambda_{AB}t} \tag{5.35}$$

From Equation (5.35), we may make a corresponding characterization of the decay probability λ_{AB} by defining the *half life* $T_{1/2}$ of the process as the time required for half of the original population to decay.

$$T_{1/2} = \frac{\ln(2)}{\lambda_{AB}} \tag{5.36}$$

The population of B nuclides, $N_B(t)$, will grow at the same instantaneous rate at which the A nuclides decay, so

$$\frac{dN_B}{dt} = \lambda_{AB}N_A = \lambda_{AB}N_{A0}e^{-\lambda_{AB}t} \tag{5.37}$$

Assuming that the initial population $N_B(0) = 0$, we may find the B population as a function of time by straightforward integration of Equation (5.37) to be

$$N_B(t) = N_{A0}(1 - e^{-\lambda_{AB}t}) \tag{5.38}$$

If we were to define the half life of nuclide A to be instead the time required for the B population to reach half of the initial A population, we would find the same result as in Equation (5.36). The time dependence of these model equations for both nuclide populations is plotted in Figure 5.23.

Now suppose we have a more elaborate nuclear decay process, where nuclide A decays to B, but the B nuclides then spontaneously decay into another nuclide C. The decay probabilities for the two processes are unrelated, as are the two different values λ_{AB} and λ_{BC}. With the same understanding of the time rate of change of the populations of the three nuclides, in the continuous limit we may write the three differential equations

$$\begin{aligned}
\frac{dN_A}{dt} &= -\lambda_{AB}N_A \\
\frac{dN_B}{dt} &= \lambda_{AB}N_A - \lambda_{BC}N_B \\
\frac{dN_C}{dt} &= \lambda_{BC}N_B
\end{aligned} \tag{5.39}$$

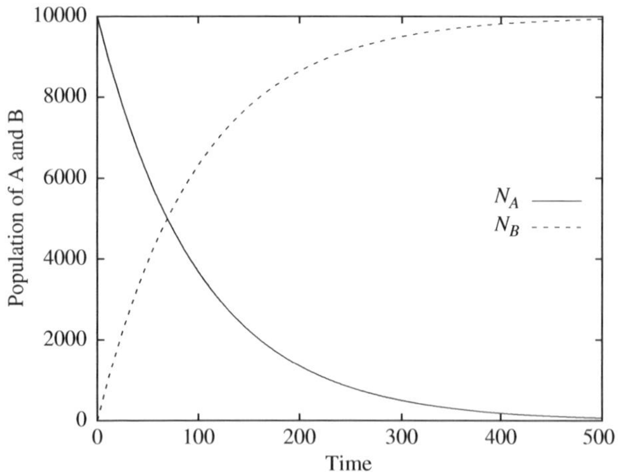

Figure 5.23 Population of A and B nuclides versus time, with $N_{A0} = 10^4$, $\lambda_{AB} = 0.01$.

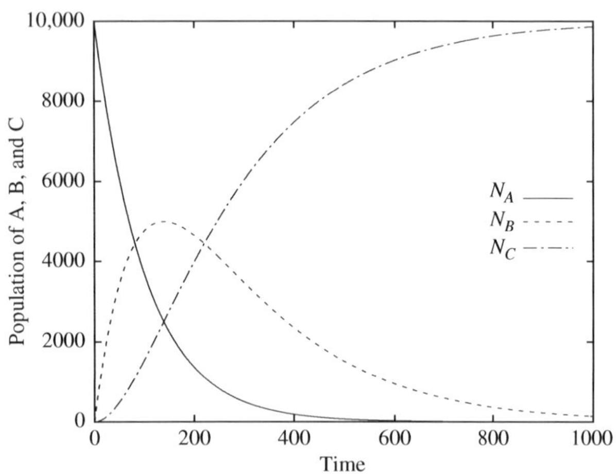

Figure 5.24 Population of A, B, and C nuclides versus time, with $N_{A0} = 10^4$, $\lambda_{AB} = 0.01$, $\lambda_{BC} = 0.005$.

The student may verify that with the initial conditions $N_A(0) = N_{A0}$, $N_B(0) = 0$, and $N_C(0) = 0$, the solution to this system is the set of three population functions

$$N_A(t) = N_{A0}e^{-\lambda_{AB}t} \tag{5.40}$$

$$N_B(t) = \frac{\lambda_{AB}}{\lambda_{BC} - \lambda_{AB}} N_{A0} \left(e^{-\lambda_{AB}t} - e^{-\lambda_{BC}t}\right) \tag{5.41}$$

$$N_C(t) = \frac{1}{\lambda_{BC} - \lambda_{AB}} N_{A0} \left[\lambda_{BC}(1 - e^{-\lambda_{AB}t}) - \lambda_{AB}(1 - e^{-\lambda_{BC}t})\right] \tag{5.42}$$

From the plots in Figure 5.24 of these three functions, over time we can visualize the initial population of A nuclides gradually decaying into B nuclides, and then the B nuclides finally decaying into C.

If we want to define an equivalent half life $T_{1/2}$ for this process, we may try to define this time constant in a way similar to Equation (5.36), as the time required for half of the original population of A nuclides N_{A0} to decay to B nuclides. In that case Equation (5.36) still applies. However, since the population of the B nuclides is only transitory, a more meaningful definition is to find the time required for the population of the C nuclides to reach half of the initial A population. That is, we find the solution of

$$N_C\left(T_{1/2}\right) = \frac{1}{2}N_{A0} \tag{5.43}$$

Inspecting Figure 5.24, we would expect the half life defined by Equation (5.43) to be considerably longer than that defined by Equation (5.36). When using the relation of Equation (5.42) to solve Equation (5.43), we will find an elaborate transcendental equation that must be approached numerically with the techniques of this chapter. The coding of the solution is available to the student as Exercise 5.6.

We will revisit the topic of radioactive decay in Section 8.7, where we explore pseudorandom techniques to model the stochastic processes of nuclear decay, and in Section 11.7, where we use the least squares fitting method to experimentally determine the probability parameter λ.

5.10 Coding: Finding Multiple Roots with Stepping

The premise of the coding practices outlined in Chapter 3 is that by structuring our coding work in modules with consistent interfaces, we are able to assemble existing and debugged code tools into more elaborate solutions. In this section, we look at an example of such a code reuse method applied to the problem of locating multiple roots of a nonlinear function. We will adapt the stepping code of Chapter 4 and plug in the root-finding tools of this chapter.

Consider a problem shown generally in Figure 5.25. Here a function with three roots within some region is plotted, and we would like to locate all three roots to some specified precision. We cannot rely on one particular initial point to converge to all three roots, so we must first make a survey pass through the argument space with a sequential step, searching for two consecutive stepped points giving opposite function value signs. At this point, the survey is paused and convergence on a root is started with either one or two bracketing argument values with the root-finding algorithm of choice. When the nearby root is converged on, the surveying step process is resumed until the entire range of interest is covered. The step increment value for the survey pass should be chosen carefully. Too fine an increment will consume needless computation load, but too coarse an increment risks that we will not bracket all function roots or initialize the convergence too far from the root. Four tool routines for locating an arbitrary number of roots in any target function with this sequential stepping method can be found in the source code file `step_roots.c`, together with its companion header file `step_roots.h`, part of the `CPLibrary` package on the Student Companion website.

There are four solver routines in the `step_roots.c` utility tool, `step_newton_raphson()`, `step_secant()`, `step_regula_falsi()`, and `step_bisection()`. The following excerpt is the `step_regula_falsi()` routine, and the reader may follow along in this example as we develop the solution strategy in the following paragraphs.

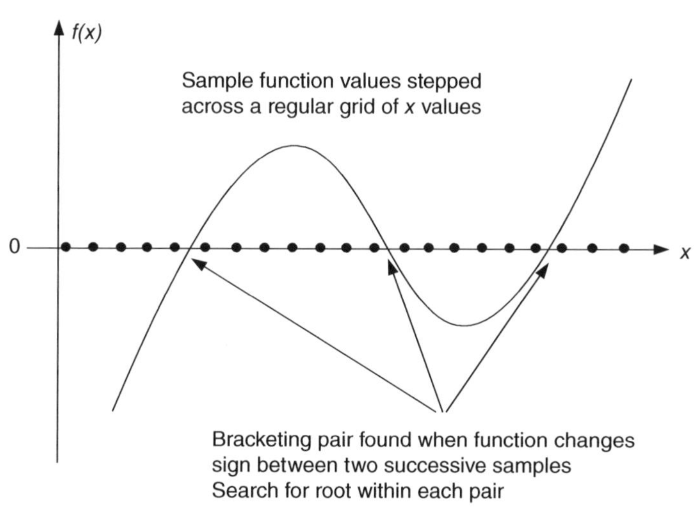

Figure 5.25 Finding multiple roots with stepping.

```
int step_regula_falsi(double (*func)(double,void *),
                      GRID *xgrid_ptr,void *p,int nmax,double tol,
                      ARRAY1D *roots_ptr) {
  double a,b,fa,fb,r;
  int i,n,n_alloc,retval;

  n_alloc = 0;
  roots_ptr->n_elems = 0;
  roots_ptr->elem = NULL;
  retval = STEP_ROOT_NOT_FOUND;
  n = NUM_GRID_PTS(*xgrid_ptr);
  a = xgrid_ptr->init;
  fa = (*func)(a,p);
  i = 1;
  while (i < n) {
    b = xgrid_ptr->init + ((double) i) * xgrid_ptr->inc;
    fb = (*func)(b,p);
    if ((fa * fb) < 0.0) {
      retval = regula_falsi(func,a,b,p,nmax,tol,&r);
      if (retval != ROOT_SUCCESS) return(retval);
      if (roots_ptr->n_elems >= n_alloc) {
        n_alloc += 10;
        roots_ptr->elem = realloc(roots_ptr->elem,n_alloc *
                                  sizeof(double));
      }
      roots_ptr->elem[roots_ptr->n_elems] = r;
      roots_ptr->n_elems++;
    }
    a = b;
    fa = fb;
    i++;
  }
  return(retval);
}
```

The task of a utility function to step through and find all the roots and report them back to a calling program has an interesting requirement for memory allocation. The size of the memory array needed to store all the converged root values is not only not known at compile time, but it is actually not even known until all the survey points have been numerically tested by the algorithm along a particular function. Only then is the number of roots of whatever object function that has been passed for analysis that are bracketed between stepped points known, and may range from zero roots to a large number for an extreme function with a number of oscillations. Almost all the solver routines developed in the `CPLibrary` for this text in general follow the recommended practice that any memory needed for the result of the routine is always allocated by the calling code, so that the complementary processes of memory allocation and freeing can be kept track of in one code module. For example, the `step()` routine in Section 4.1 must be called with an `ARRAY1D` result array that has already been allocated with the proper length to accommodate all the requested stepped samples. However, for the purpose of locating an unknown number of roots of a function being scanned numerically, we will need a variable dynamic memory allocation technique performed in the solver routine itself. The array where the routine stores the resulting roots initially has no memory allocated, but it can grow in length to an arbitrarily large memory array by a block at a time as more and more roots are numerically discovered. This capability was briefly introduced in Section 3.1.3, but we are now at a stage to examine some coding details. Only in this special case will the calling code invoke the solver tool with an `ARRAY1D` structure for its result that does not point to any allocated memory. The calling code, however, still has the responsibility to free the memory allocated by the solver function when no longer needed.

Figure 5.26 shows the plan for successively larger memory allocations. The routine that will search for an arbitrary number of roots keeps track of the current size of the dynamic allocation using an integer variable `n_alloc`. This variable is initialized at 0, and reflects the initial state of no array elements allocated. As roots are found by the algorithm, the allocated array size is increased by a block of 10 elements at a time in this example tool to accommodate any required number of elements to store the root values. That is, when the first root is found, an array of 10 elements is dynamically allocated, allowing for the first root and up to nine subsequent roots. If an eleventh root is found, the array size is increased from 10 to 20, and so on, as shown. The variable `n_alloc` will be updated with the count of double-precision floating point elements in the allocation.

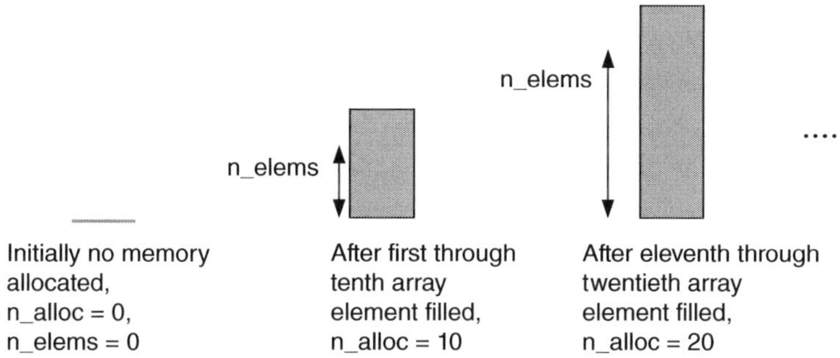

Figure 5.26 Scheme for arbitrary length arrays.

Changing the size of the allocated array is accomplished by the `realloc()` library function call. This function expects to be called with two arguments: a pointer to an existing block of dynamically allocated memory, and the new expanded size required. For each of the solving tools in `step_roots.c`, the last function argument expected is a pointer to an `ARRAY1D` structure. This will be the array that the tool will fill with an arbitrary number of roots. Note that at the beginning of each solver function, the `ARRAY1D` structure is initialized for zero size and a `NULL` pointer for the array element memory, along with setting `n_alloc = 0`. This scheme takes advantage of the property of the `realloc()` function that if the pointer to the existing memory block has an initial value of `NULL`, the function assumes there is no existing block, and essentially performs the same task as a `malloc()` call to allocate a new block.

The code to survey the entire span of function values is essentially the same in all four tools, and is copied from the function stepping code used in Chapter 4. When a root is found to be bracketed in the survey, a different root-finding routine from the `root.c` tool is called to converge on the root with the specified precision, such as the `regula_falsi()` routine in the example above. Once all roots are properly converged on, the function returns with the array pointed to in the `ARRAY1D` structure filled with an arbitrary number of roots located, and the structure element `n_elems` reporting their count. In most cases, there will be some unused array elements in the last allocated block, unless the number of roots happens to be a whole multiple of the reallocation block size. For finding roots of typical functions used in the physical sciences, a block size of 10 is usually a good compromise between minimizing calls to the `realloc()` system function and the small amount of wasted memory at the array end. The needed header file for this utility is `step_roots.h`.

5.11 Application: Quantum Energy Levels of Bound Particles

In this section, we work through an example physics problem that requires the multiple root tool described in Section 5.10. We wish to find the allowed quantum energies of an electron constrained in a rectangular potential well. For the following development, download two source code files, `quantum_rectwell_step.c` and `quantum_rectwell_roots.c`, from the Student Companion website. We will examine solutions of the time-independent Schrödinger equation in one dimension

$$\frac{d^2\psi(x)}{dx^2} = -\frac{2m}{\hbar^2}\,[E - V(x)]\,\psi(x) \tag{5.44}$$

for the abrupt rectangular potential profile shown in Figure 5.27. The wells of width a extend from an x coordinate of $-\frac{a}{2}$ to $+\frac{a}{2}$. The electron potential is 0 in the field area outside the well, and $-V_0$ inside the well. The electron energy is assumed to be at some negative level greater than $-V_0$ so that the electron is constrained in the well. For this simple potential profile, the one-dimensional solutions to the Schrödinger equation are found easily by analytical means, since in each region the potential is constant, and the general solution expressions are shown in the figure. The solution must be oscillatory in region 2 within the well, and because the potential is an even function of x, the standing wave solution must be symmetrical, and therefore the wave function can be expressed simply as a linear combination of the sine and cosine dependence. In contrast, the solution in regions 1 and 3 on either side of the well must be an exponential function of x. Because we are solving for a wave function of x that describes the probability density of the existence of the electron around a given x coordinate, any solution that has physical meaning must fall to zero outside the well region. This assertion allows the integration of the wave function over all x to be limited to a finite value, and the integral of the wave function squared may be normalized to unity. Therefore, the form of the solution must be an exponential of x with a positive coefficient for x values to the left of the well, and with a negative coefficient to the right of the well, and we can set both B_1 and A_3 in the figure to zero.

The general solution forms in Figure 5.27 are known within the multiplicative constants A_1, A_2, B_2, and B_3. We allow one of these constants, say A_1, to take on an arbitrary value for now that would be used for scaling the entire solution for normalization. The other constants are determined relative to the value of A_1 in order to fit the geometry of the potential well and the continuity requirement of the wave function. There are two classes of possible solutions and the further refinement of the forms of the solutions for the two cases of even and odd parity wave function solutions, shown in Figures 5.28 and 5.29, respectively.

We may also rely on the principle that both the wave function and its first derivative must be continuous functions of x for a physically meaningful solution. Starting with the even parity case, equating these two continuity conditions at the right-hand edge of the well, where $x = \frac{a}{2}$ gives

$$\begin{aligned} A_2 \cos\left(k_2 \frac{a}{2}\right) &= B_3 e^{-k_3 \frac{a}{2}} \\ -k_2 A_2 \sin\left(k_2 \frac{a}{2}\right) &= -k_3 B_3 e^{-k_3 \frac{a}{2}} \end{aligned} \tag{5.45}$$

Dividing the two equations to eliminate the exponential term gives one transcendental equation [1, App.G; 4, Chap.1] to solve numerically for allowed values of E. (Note that k_2 and k_3 are both functions of E.)

$$f_{\text{even}}(E) = k_2 \sin\left(k_2 \frac{a}{2}\right) - k_3 \cos\left(k_2 \frac{a}{2}\right) = 0 \tag{5.46}$$

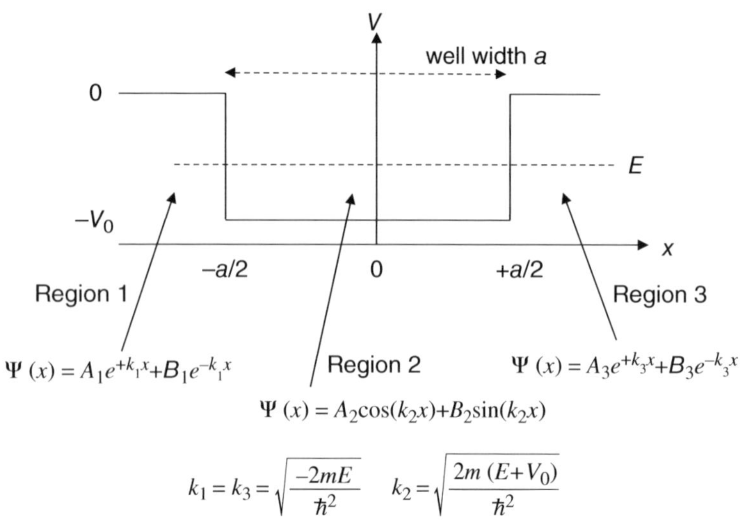

Figure 5.27 Quantum rectangular well potential and wave solution forms.

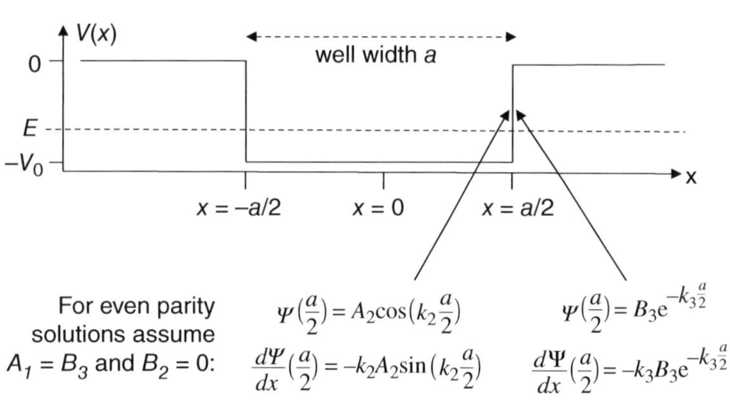

Figure 5.28 Even parity continuity requirements for a rectangular quantum potential well.

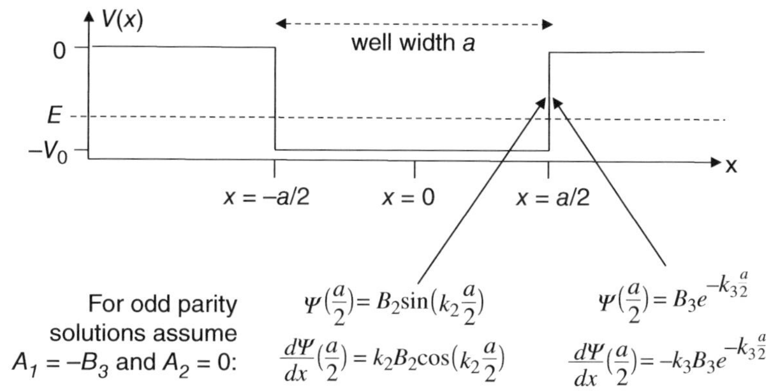

Figure 5.29 Odd parity continuity requirements for a rectangular quantum potential well.

Similarly, for the odd parity case, the continuity conditions imply

$$\begin{aligned} B_2 \sin\left(k_2 \frac{a}{2}\right) &= B_3 e^{-k_3 \frac{a}{2}} \\ k_2 B_2 \cos\left(k_2 \frac{a}{2}\right) &= -k_3 B_3 e^{-k_3 \frac{a}{2}} \end{aligned} \tag{5.47}$$

Again dividing these two equations to eliminate the exponential term gives a second transcendental equation for the odd parity case. Similar to the even parity case, the roots of this function of E (through the values of k_2 and k_3) are the allowed electron energy levels within a given potential well.

$$f_{\text{odd}}(E) = k_2 \cos\left(k_2 \frac{a}{2}\right) + k_3 \sin\left(k_2 \frac{a}{2}\right) = 0 \tag{5.48}$$

Note that the values of the functions in Equations (5.46) and (5.48) do not have any particular physical meaning, but the energy value E that make the functions zero, or the roots of the functions, are the allowed energy levels for the electron. These are transcendental equations, so we must use numerical methods to actually locate the roots. Moreover, we do not know a priori how many roots there will be, as their count will vary with changes in the well profile geometry, so the algorithm in Section 5.10 must be used to accommodate any number of roots.

What do these functions look like? Let's use the visualization tools of Chapter 4 to first survey the function shapes and gain an understanding of the ease or difficulty in converging on the multiple roots we are interested in. That is the task performed by the program `quantum_rectwell_step.c`.

At the top of the program, the two functions in Equations (5.46) and (5.48) are coded into C functions `quantum_rectwell_even()` and `quantum_rectwell_odd()`. Then the `step()` tool of Chapter 4 is called twice to step both functions and generate two datasets in the output stream. Running this program in a Linux shell prompt window

```
$ gcc -Wall -Wshadow -O2 -o quantum_rectwell_step quantum_rectwell_step.c CPLibrary/step.c
    ↪ CPLibrary/array.c -lm
$ ./quantum_rectwell_step
./quantum_rectwell_step <width in nm> <depth in eV> <mass * c^2 in Mev>
$ ./quantum_rectwell_step 1 10 0.5110 >| quantum_rectwell_step.dat
```

Starting an interactive gnuplot session and giving the commands

```
gnuplot> set xlabel 'Energy level (eV)'
gnuplot> set ylabel 'Root function'
gnuplot> plot 'quantum_rectwell_step.dat' i 0 u 1:2 w l t 'Even parity','' i 1 u 1:2 w l t 'Odd
    ↪ parity'
```

produces the plot seen in Figure 5.30.

This plot shows that, for this particular example, potential well profile, there are three roots each for the two parity classes. The first root for the even function is the so-called "ground state," the minimum possible energy level for a bound electron. Both functions appear smooth and well behaved, so we can expect a multiple root tool such as `step_regula_falsi()` described in Section 5.10 to converge well to all roots.

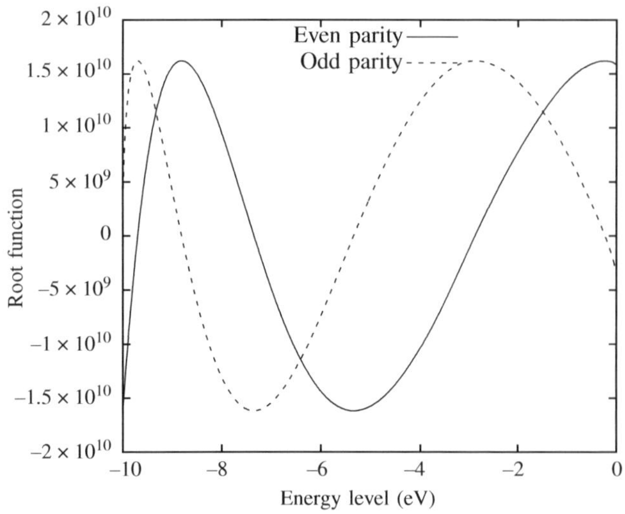

Figure 5.30 The two functions whose roots are the allowed energies of a bound electron for even and odd parities.

We would certainly like to obtain numerical solutions of the allowed energy levels, but to gain more understanding, we would also like to visualize the stationary wave function solution at each allowed energy level to see the way the solutions satisfy the required forms of the wave function with a resonance fitted into the potential profiles. So let's demonstrate the power and flexibility of our tool library by writing one comprehensive main program that makes use of both the `step_roots.c` tool for multiple root finding, combined with the `step.c` tool for tracing out the shape of the wave functions that correspond to each root. Once the root-finding algorithm has located allowed values of the energy E and the k_2 and $k_1 = k_3$ wave numbers for each eigenvalue, the wave functions are known from Figure 5.27 and the continuity condition of the wave function. Following the general solution forms of Figure 5.27 and setting the coefficient $A_1 = 1$ arbitrarily we may obtain unnormalized wave functions suitable for plotting and visualization as

$$\psi(x) = \begin{cases} e^{k_3 x} & x < -\frac{a}{2} \\ A_2 \cos(k_2 x) & -\frac{a}{2} < x < \frac{a}{2} \\ e^{-k_3 x} & \frac{a}{2} < x \end{cases} \tag{5.49}$$

where

$$A_2 = \frac{e^{-k_3 \frac{a}{2}}}{\cos\left(k_2 \frac{a}{2}\right)}$$

for the even parity case, and

$$\psi(x) = \begin{cases} e^{k_3 x} & x < -\frac{a}{2} \\ B_2 \sin(k_2 x) & -\frac{a}{2} < x < \frac{a}{2} \\ -e^{-k_3 x} & \frac{a}{2} < x \end{cases} \tag{5.50}$$

where

$$B_2 = -\frac{e^{-k_3 \frac{a}{2}}}{\sin\left(k_2 \frac{a}{2}\right)}$$

for the odd parity case.

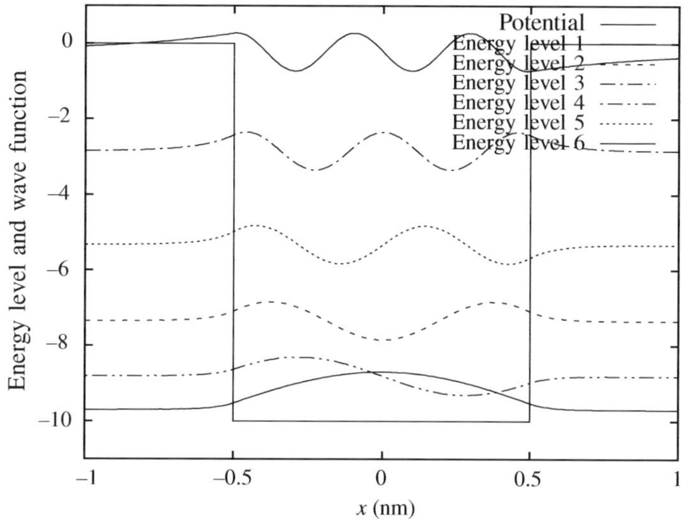

Figure 5.31 Stationary wave functions corresponding to the allowed energies of a bound electron, width = 1 nm, depth = 10 eV.

See the main program in `quantum_rectwell_roots.c`. After reading the specification of the potential profile from the command line, the program generates a set of data points in the standard output stream. When plotted by `gnuplot` with connecting lines, these points will graphically represent the potential. This program code copies the two object functions `quantum_rectwell_even()` and `quantum_rectwell_odd()` that compute the functions of Equations (5.46) and (5.48), and these are each passed to the `step_regula_falsi()` tool in turn to locate the two classes of allowed energies. But two additional object functions `quantum_rectwell_wave_even()` and `quantum_rectwell_wave_odd()` are added in this program. These functions compute the stationary wave function solutions of Equations (5.49) and (5.50), respectively. Once the available energy roots are found by calls to `step_regula_falsi()`, the program iteratively loops through the found roots and fills in a `QRWELLWAVE` parameter structure with the values corresponding to each energy level. Then the wave functions are passed to the `step()` tool to trace out the form for each wave solution.

The usual custom for these sorts of bound quantum mechanical problems is to plot a family of allowed wave function solutions offset vertically against the potential profile with each corresponding energy level, as in Figure 5.31. Note in the program some additional work is needed to scale the wave function solutions so that each wave can be plotted with a similar size curve. A call is made to the `minmax_array1d()` utility function in the `array.c` tool to allow this scaling operation. Then the final data set values sent to the standard output stream are the sum of the scaled wave function values and the corresponding energy level root. Running this program in a Linux shell prompt window sends the six allowed energy levels in units of electron volts to the standard error stream so that they will be displayed on the terminal while the standard output stream is redirected to a data file, as well as preparing an output file of multiple data sets to be plotted

```
$ gcc -Wall -Wshadow -O2 -o quantum_rectwell_roots quantum_rectwell_roots.c CPLibrary/step_roots.c
    ↪ CPLibrary/root.c CPLibrary/array.c -lm
$ ./quantum_rectwell_roots
./quantum_rectwell_roots <width in nm> <depth in eV> <mass * c^2 in Mev>
$ ./quantum_rectwell_roots 1 10 0.5110 >| quantum_rectwell_roots.dat
3 roots found for even parity:
-9.7023406 -7.3476269 -2.8541675
3 roots found for odd parity:
-8.8135057 -5.3360802 -0.2295175
```

Starting an interactive gnuplot session and giving the commands

```
gnuplot> set xrange [-1:1]
gnuplot> set xlabel 'X-coordinate (nm)'
gnuplot> set yrange [-11:1]
gnuplot> set ylabel 'Energy level and wave function'
gnuplot> plot 'quantum_rectwell_roots.dat' i 0 u ($1 * 1e9):2 w l lt -1 t 'Potential', for [i=1:6]
    ↪ '' i (i) u ($1 * 1e9):2 w l lt (i) t sprintf("Energy level %d",i)
```

produces the plot seen in Figure 5.31. The potential is plotted as the first dataset, so that the fit of each stationary wave solution within it can be visualized. Note the use of some powerful features of `gnuplot` in this example, such as setting up an iterative loop within the `plot` command to plot all six waves in sequence, with each curve titled with its corresponding solution index.

Abrupt rectangular potential profiles such as the one shown in Figure 5.27 are appealing for seeing the general tendency of bound quantum particles to be constrained to discrete energy levels, but are not particularly physically realistic. Abrupt changes in potential require infinite localized forces, which are clearly not expected in nature. The power of numerical analysis can be appreciated as we expand the types of potential profiles that can be treated in these types of quantum analyses. We will revisit these solutions for other more realistic potentials in Section 13.6.

5.12 Application: Ideal Single-slit Diffraction

The general problem of wave diffraction in two dimensions was treated in Section 4.7, where the numerical analysis allowed any geometrical dimensions relative to the radiation wavelength. We may imagine a more idealized geometry shown in Figure 5.32 with parallel plane waves incident from the bottom diffracting through a single slit of width a and observed at the top at a distance very far away relative to the wavelength of the waves.

An analytic expression for the relative wave intensity as a function of the observation deflection angle θ is derived in many introductory textbooks [17, Chap. 37; 19, Chap. 36; 20, Chap. 38]. For this ideal geometry, the one-dimensional power intensity pattern is

$$I(\theta) = \left(\frac{\sin \alpha}{\alpha}\right)^2 \tag{5.51}$$

where $I(\theta)$ is the power intensity as a function of the deflection angle relative to the peak intensity, and

$$\alpha = \frac{\pi a}{\lambda} \sin \theta \tag{5.52}$$

This characteristic is plotted in Figure 5.33 for the example case of $a = 5$ and $\lambda = 1$ in arbitrary units. The locations of the minima of the intensity pattern can be readily identified analytically, but solving for the deflection angles where the value of the intensity is at some specified fraction of the peak must be accomplished numerically. Since the number of crossing points of the various lobes of the intensity curve with some target level depends on the level chosen, this is a good application for the flexible `step_roots.c` tool. This solution is Exercise 5.9 for the student.

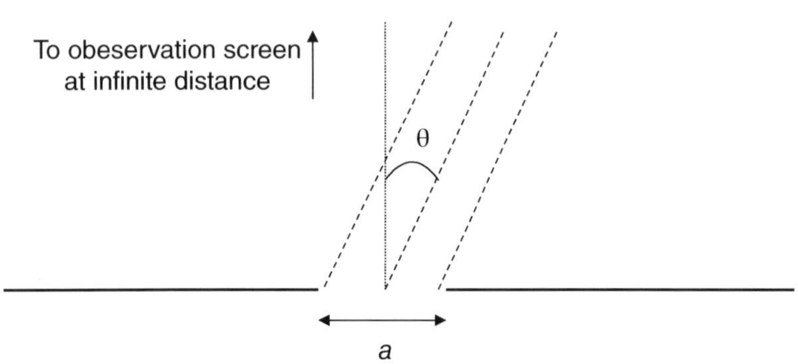

Figure 5.32 Geometry of ideal single-slit diffracting incident plane waves and observed at an infinite distance.

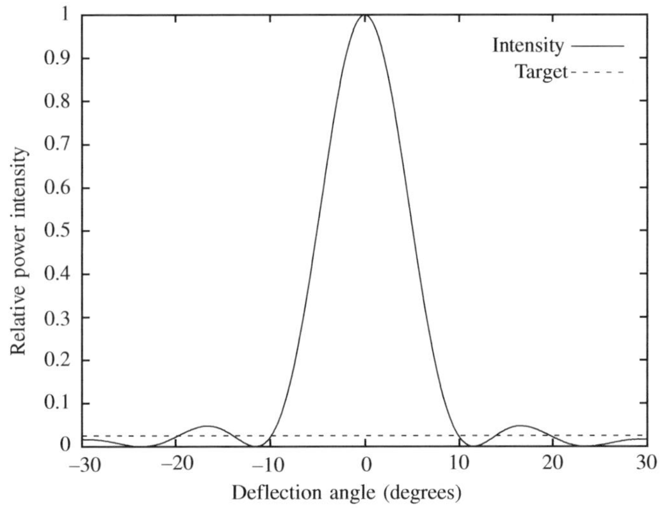

Figure 5.33 Power intensity one-dimensional diffraction pattern from a single slit, $a = 5$ and $\lambda = 1$.

5.13 Exercises

5.1 (a) Use the visualization methods of Chapter 4 with the `step.c` tool to write a program `LED_V_step.c` to generate the data for a plot of the LED voltage drop in the circuit of Figure 5.13 as a function of current flow. As discussed in the text, the series resistor limits the current flow to a maximum of 0.025 A. Use the example parameter values in Section 5.4 to verify that for a current flow between 0.001 and 0.025 A, we may reasonably expect the solution to involve an LED voltage drop of less than 1.5 V.

(b) Write a program `LED.c` to solve for the current flowing in the circuit of Figure 5.13. Use the approximate LED voltage drop from the plot in part (a) with Ohm's law to arrive at an initial guess and an initial bracket of the current for both the Newton–Raphson and Secant algorithms.

5.2 Write a program `stunt.c` to solve the car stunt problem, using the physical parameters in Section 5.5: $D_1 = 200$ m, $D_2 = 50$ m, $D_3 = 100$ m, $v_1 = 20$ m/sec, $v_2 = 25$ m/sec. The program should solve the governing nonlinear equation for the time of collision, and from that value calculate the angles θ and ϕ.

5.3 Use the coding techniques developed in Section 4.13 to adapt the `orbit_step.c` program from Section 5.6 to write a new program `orbit_step_gpanim.c` to produce animated graphics of a body in an elliptical orbit. Use an eccentricity $\epsilon = 0.9$ and an argument of periapsis $\omega = \frac{\pi}{4}$. Your animation should resemble that shown in the downloadable file `orbit_step_gpanim_e0p9.gif`.

5.4 Using the example solution program `JWST.c` for locating the JWST in Section 5.7 as a starting point, write a new program `SOHO.c` to solve for the distance to the L1 Lagrangian point where the SOHO telescope is located.

5.5 (a) Differentiate the energy density predicted by Planck's radiation law in Section 5.8 with respect to wavelength λ. As shown in Equation (5.33), the root of this derivative function is the wavelength at which the radiation power density is maximum λ_{max}. Write a program `planck_max_w.c` to find the root of this transcendental function with the Secant algorithm at a temperature of 3000 Kelvin.

(b) Using your root found in part (a), find the constant α in Wien's displacement law, Equation (5.33).

(c) Check the inverse proportionality of Wien's displacement law by comparing the prediction of Equation (5.33) using your value of α found in part (b) against the result of another run of the numerical solution in part (a), run at some other temperature, such as 4000, 5000, or 6000 K.

5.6 Write a program `decay2_step.c` to perform two functions: Step the population functions in Equations (5.40) through (5.42) for the two nuclide decay process of Section 5.9 by invoking the `step()` tool of Section 4.1 three times, and then solve Equation (5.43) with the Newton–Raphson method and report a half life for the process of nuclide A to C decay. This will all involve including five C functions in your program, the first three for each of the population functions to step, one for a function to find the root of and another for its derivative. Run the program with the parameters shown in Figure 5.24: $N_{A0} = 10^6$, $\lambda_{AB} = 0.01$, $\lambda_{BC} = 0.005$ and plot the time dependence of the three populations as in Figure 5.24.

5.7 Under the assumption that an ideal rectangular potential well has infinitely high walls, we can analytically find the allowed energies for a bound particle in the limit that the ends of the oscillatory portions of the wave function inside the well are anchored with boundary conditions at zero. Measured upward from the bottom of the well, these allowed energy levels are

$$E_n = \frac{\pi^2 \hbar^2 n^2}{2ma^2} \tag{5.53}$$

In Section 5.11, the program `quantum_rectwell_roots.c` was demonstrated that not only solved for allowed electron energy levels in a rectangular well of finite depth, but also generated the corresponding wave function values so that we may visualize the wave functions fitting both inside and outside the geometry of the well. Copy and modify the source code for `quantum_rectwell_roots.c` to make another program `quantum_rectwell_roots_bottom.c` that sends to the standard output only the height of the discrete allowed energy levels above the bottom of the well. Run your program for an electron bound in a 1-nm wide well with a depth of 100 eV. Create a plot that graphically compares the set of energy levels with those energy levels predicted by Equation (5.53) in the limit of an infinite depth.

5.8 **(a)** For a very lightly damped harmonic oscillator with an initial displacement of 1 unit and a natural period of 1 time unit, what set of times corresponding to the minima and maxima of the displacement time dependence do you expect? (The minimum displacement is defined as the maximum negative displacement, rather than the minimum absolute value of displacement.)

(b) Use the `damped_oscillator.c` tool from Section 4.2 to generate a plot of the time dependence of the damped harmonic oscillator described in part (a), but with a damping factor of 0.5.

(c) Write a new program `damped_oscillator_minmax.c` to use the `step_secant()` tool to find all the times between 0.1 and 5 time units of the displacement extrema for the oscillator modeled in part (b). Compare with your expected values for the lightly damped oscillator in part (a).

5.9 For the idealized single-slit diffraction geometry discussed in Section 5.12, write a program `diffraction_1slit_step_roots.c` to determine all the angles where the power intensity is $\frac{1}{40}$ of the peak value by calling the `step_secant()` tool. Run the program for the case of $a = 5$ and $\lambda = 1$ in arbitrary spatial units. Hint: Consider the guarding techniques discussed in Section 3.3.6 to avoid the condition of dividing by zero.

6

Systems of Linear Equations

In this chapter and in Chapter 7, we explore the modeling of physical systems that are characterized by simultaneous systems of both linear and nonlinear equations, respectively. Such a characterization often arises in systems composed of similar and interconnected components, such as interacting spring-mass oscillators or electrical networks. When we model physical systems with linear equations, often the strict linearity in the mathematical expression is an approximation that is valid only to some limited extent. For example, the force-compression characteristic of actual spring behavior, or the restoring forces between elements of a flexible membrane or between atoms in a lattice, can be somewhat nonlinear, but we may limit the oscillation displacement to cover only a range where the behavior is modeled linearly to an acceptable approximation.

In general, we may cast the system of linear equations modeling our system as shown as Equations (6.1). Customarily the subscript notation for a system of n equations in n unknowns range from 1 to n, but to be consistent with the array subscripts of the C programming language, we will write subscripts ranging from 0 to $n-1$.

$$\begin{aligned}
a_{0,0}x_0 + a_{0,1}x_1 \quad & +a_{0,2}x_2 + \cdots \quad + a_{0,n-1}x_{n-1} = b_0 \\
a_{1,0}x_0 + a_{1,1}x_1 \quad & +a_{1,2}x_2 + \cdots \quad + a_{1,n-1}x_{n-1} = b_1 \\
a_{2,0}x_0 + a_{2,1}x_1 \quad & +a_{2,2}x_2 + \cdots \quad + a_{2,n-1}x_{n-1} = b_2 \\
& \vdots \\
a_{n-1,0}x_0 + a_{n-1,1}x_1 \quad & +a_{n-1,2}x_2 + \cdots \quad + a_{n-1,n-1}x_{n-1} = b_{n-1}
\end{aligned} \tag{6.1}$$

Using the rules of matrix arithmetic, this system can be written as in Equation (6.2)

$$\begin{bmatrix}
a_{0,0} & a_{0,1} & a_{0,2} & \cdots & a_{0,n-2} & a_{0,n-1} \\
a_{1,0} & a_{1,1} & a_{1,2} & \cdots & a_{1,n-2} & a_{1,n-1} \\
a_{2,0} & a_{2,1} & a_{2,2} & \cdots & a_{2,n-2} & a_{2,n-1} \\
\vdots & & & & & \\
a_{n-2,0} & a_{n-2,1} & a_{n-2,2} & \cdots & a_{n-2,n-2} & a_{n-2,n-1} \\
a_{n-1,0} & a_{n-1,1} & a_{n-1,2} & \cdots & a_{n-1,n-2} & a_{n-1,n-1}
\end{bmatrix}
\begin{bmatrix} x_0 \\ x_1 \\ x_2 \\ \vdots \\ x_{n-2} \\ x_{n-1} \end{bmatrix}
=
\begin{bmatrix} b_0 \\ b_1 \\ b_2 \\ \vdots \\ b_{n-2} \\ b_{n-1} \end{bmatrix} \tag{6.2}$$

or, using the notation of linear algebra, the compact expression of Equation (6.3).

$$\mathbf{A}\vec{x} = \vec{b} \tag{6.3}$$

Given a matrix of constant coefficients $\mathbf{A}$ and a constant "right-hand side" vector $\vec{b}$, one class of solutions we often seek is finding a unique vector $\vec{x}$ that solves this system, if indeed a unique solution does exist. The mathematical method for finding the solution vector has been provided by the development of linear algebra for many years. The idealized method

Computational Physics Using C: Efficient Programming with Ease, First Edition. John W. Fattaruso. © 2026 John Wiley & Sons, Inc.
All rights reserved, including rights for text and data mining and training of artificial intelligence technologies or similar technologies.
Published 2026 by John Wiley & Sons, Inc. Companion Website: http://www.wiley.com/go/CompPhysUsingC

of Gaussian elimination will be reviewed in the next section, and then the adaptation of this method that allows numerical precision to be preserved for practical problems will be detailed in Section 6.2.

Another class of solutions we seek is in the case of the right-hand side vector $\vec{b}$ not constant, but rather of the form $\lambda\vec{x}$ for some unknown set of eigenvalues λ_i and eigenvectors $\vec{x}_i$. Solving for eigenvalues and eigenvectors for the special case of a symmetric coefficient matrix, common in physical systems, is shown in Section 6.3.

6.1 Algorithms: Gaussian Elimination

The solution process known as Gaussian elimination is an iterative process in which "row operations" are applied successively to the rows of both the coefficient matrix and the right-hand side vector in a linear system. The row operations derive from the fact that multiplying one row in a system of equations by any nonzero constant and adding it to another will not change the solution vector, if such a solution exists for the system. The goal of the row operations in Gaussian elimination is to strategically select the row multiplication factors so that the system is reduced to one with a coefficient matrix in upper diagonal form, as shown in Equation (6.4).

$$\begin{bmatrix} a'_{0,0} & a'_{0,1} & a'_{0,2} & \cdots & a'_{0,n-2} & a'_{0,n-1} \\ 0 & a'_{1,1} & a'_{1,2} & \cdots & a'_{1,n-2} & a'_{1,n-1} \\ 0 & 0 & a'_{2,2} & \cdots & a'_{2,n-2} & a'_{2,n-1} \\ \vdots & & & & & \\ 0 & 0 & 0 & \cdots & a'_{n-2,n-2} & a'_{n-2,n-1} \\ 0 & 0 & 0 & \cdots & 0 & a'_{n-1,n-1} \end{bmatrix} \begin{bmatrix} x_0 \\ x_1 \\ x_2 \\ \vdots \\ x_{n-2} \\ x_{n-1} \end{bmatrix} = \begin{bmatrix} b'_0 \\ b'_1 \\ b'_2 \\ \vdots \\ b'_{n-2} \\ b'_{n-1} \end{bmatrix} \tag{6.4}$$

From this point, the components of the solution vector may be found with the recursive back substitution steps in Equation (6.5).

$$\begin{aligned} x_{n-1} &= \frac{b'_{n-1}}{a'_{n-1,n-1}} \\ x_{n-2} &= \frac{b'_{n-2} - a'_{n-2,n-1}x_{n-1}}{a'_{n-2,n-2}} \\ &\vdots \\ x_i &= \frac{b'_i - \sum\limits_{j=i+1}^{n-1} a'_{i,j}x_j}{a'_{i,i}} \end{aligned} \tag{6.5}$$

Consider the task of choosing a row operation that will annihilate the $a_{1,0}$ coefficient in Equation (6.2) to zero. This can be accomplished by adding row 0 multiplied by the factor $-\frac{a_{1,0}}{a_{0,0}}$ to row 1. Then the $a_{2,0}$ coefficient may be annihilated to zero by adding row 0 multiplied by the factor $-\frac{a_{2,0}}{a_{0,0}}$ to row 2. This process is repeated for each row under the $a_{0,0}$ diagonal element. Then the coefficients under the $a_{1,1}$ diagonal element are annihilated to zero with a similar process, with the row multiplication factor being $-\frac{a_{i,1}}{a_{1,1}}$ for row i under the diagonal, and so on for all lower diagonal elements. At each step down the main diagonal of the matrix, the diagonal element under which we annihilate elements to zero, and whose value is used in the denominator of the row factor, is called the *pivot* element. Note that the b_i elements in the right-hand side vector must also be included in the row operations so that the solution of the linear system is preserved.

The case of a degenerate system for which no unique solution exists will be revealed in the elimination process by the prescription for a row multiplication factor turning out to be an undefined quantity, resulting from an attempt to divide by a zero diagonal pivot element. We classify coefficient matrices that have no corresponding unique system solution as *singular* matrices. Note that if the process completes successfully, the remaining nonzero components of the new coefficient matrix and right-hand side vectors are now modified from their original values, as notated by the primes in Equation (6.4).

6.2 Algorithms: Pivoting Strategies

The simple Gaussian elimination method discussed in the previous section can be a practical algorithm in some limited cases. To be confident it will result in meaningful and accurate results, the linear system that is specified in the coefficient matrix must be well formed with all coefficients nonzero and close to the same order of magnitude. This condition is, unfortunately, not always met in linear systems describing physical systems.

Consider, for example, the rather contrived system of Equation (6.6) with four linear equations in four unknowns.

$$\begin{bmatrix} 0 & 3 & 2 & 1 \\ 3 & 2 & 1 & 4 \\ 2 & 1 & 4 & 3 \\ 1 & 4 & 3 & 0 \end{bmatrix} \begin{bmatrix} x_0 \\ x_1 \\ x_2 \\ x_3 \end{bmatrix} = \begin{bmatrix} 16 \\ 26 \\ 28 \\ 18 \end{bmatrix} \tag{6.6}$$

Such a set of equations could plausibly result from modeling some physical system, and in this example, two of the important diagonal matrix elements turn out to have troublesome zero values. Nonetheless, a unique mathematical solution for this system does exist, and is the vector

$$\begin{bmatrix} x_0 \\ x_1 \\ x_2 \\ x_3 \end{bmatrix} = \begin{bmatrix} 1 \\ 2 \\ 3 \\ 4 \end{bmatrix} \tag{6.7}$$

In the prescription for Gaussian elimination, the zero values falling on the diagonal would mean an undefined value for the very first row multiplication factor, since reducing the column below the element $a_{0,0}$ requires a multiplication factor with $a_{0,0}$ in the denominator. Yet we know this coefficient matrix is non-singular since there does exist a unique solution. When solving such a system by hand with Gaussian elimination, the conventional workaround is to simply reorder the equations so that the coefficients do not start with a zero value in the first diagonal. So any general purpose linear equation solution routine must provide an equation reordering capability automatically.

Even with coefficient matrices with no diagonal elements identically zero, a mechanism for reordering equations is also necessary to preserve the greatest possible numerical precision with practical problems. Solving linear equations with Gaussian elimination is intrinsically an *order 3* problem, that is, the number of arithmetic operations required grows as the *cube* of the dimension of the coefficient matrix. This results in a large amount of computation for matrices of even modest dimension, and in a tremendous numerical churning of matrix element values. Potentially troublesome for preserving precision are the subtractions aimed at zeroing out matrix elements. To return to the example above, when zeroing out the $a_{1,0}$ element under the $a_{0,0}$ diagonal, all other elements in row 1 must be replaced by $a_{1,j} - \frac{a_{1,0}}{a_{0,0}} \cdot a_{0,j}$ for $j = 1, \dots, n-1$ as the row operation is carried out. The process by which numerical roundoff error is accumulated with all the churning of the Gaussian algorithm is quite complex, and much research over literally centuries has been devoted to optimizing its precision, but we can get a hint of the tendencies by considering the mechanism of loss of precision errors introduced in Section 2.3. From the expression above, we can see that the smaller the magnitude of the diagonal element $a_{0,0}$ and the larger the magnitude of the elements of row 1, then the greater the tendency for this subtraction to be operating on two relatively large numbers of close magnitudes. This is just the condition discussed in Section 2.3 for a significant numerical error to be introduced. We will be introduced to the Vandermonde matrix at the beginning of Chapter 9, which is a particularly troublesome matrix form for roundoff error. The general lesson to be learned is that the least numerical error is introduced when the rows are reordered so that the diagonal pivot elements are chosen to be as large as possible.

Even though the Gaussian elimination algorithm already involves a large amount of computation, it is very worthwhile from the standpoint of numerical precision to optimize the order of the remaining rows before processing each row according to the elimination algorithm. There have been various strategies proposed by numerical analysis research for how to choose an optimum pivot element at each step at the expense of some reasonable additional computation. The system solution tool developed in this text uses a method found to be a good compromise between extra calculation work and numerical performance. This method goes by the name of *scaled partial pivoting* [10, Chap. 2; 12, Chap. 7].

We will apply the Gaussian elimination algorithm directly to solve a selection of physics problems in this chapter, and also make use of it in Section 14.2 for the development of implicit methods for solving differential equations.

6.3 Algorithms: The Jacobi Eigenvalue Method

Another class of solutions for systems described by Equation (6.2) with a square matrix of coefficients is finding eigenvalues and eigenvectors of the coefficient matrix. In general, this is a challenging numerical task for arbitrary matrices of large order, and several numerical algorithms have been developed to address this problem. In this section we explore Jacobi's method [9, Chap. 4; 11, Chap. 11; 26, Chap. 7], perhaps the most straightforward of the numerical eigenvalue algorithms, but applicable only to symmetric matrices. Because finding eigenvalues and eigenvectors is a common numerical problem in the physical sciences, there are packages available that contain routines representing decades of refinement and improvement in efficiency, and applicable to more general matrices. But the Jacobi routine in textbook library code `lineq.c` provides a relatively simple routine to study as an initial case, provided we are working with a symmetric and real-valued coefficient matrix.

The goal of the Jacobi algorithm is to transform the original $\mathbf{A}_1$ coefficient matrix into a purely diagonal matrix after M iterations with successive *similarity transforms*

$$\begin{aligned}
\mathbf{A}_1 &= \mathbf{S}_1^{-1}\mathbf{A}_1\mathbf{S}_1 \\
\mathbf{A}_2 &= \mathbf{S}_2^{-1}\mathbf{A}_2\mathbf{S}_2 \\
&\vdots \\
\mathbf{A}_M &= \mathbf{S}_M^{-1}\mathbf{A}_M\mathbf{S}_M
\end{aligned} \tag{6.8}$$

Somewhat similar to Gaussian elimination, in each iteration we perform linear algebraic operations to annihilate a matrix element above the center diagonal to zero, and repeat the iterations with this annihilation scanning through the matrix. The target element for the annihilation is referred to as a *pivot* element a_{IJ} for the Jacobi algorithm. At the beginning of the iterations, we initialize the pivot element to the top row, just to the right of the diagonal, and scan to the right and downward in subsequent iterations. For the annihilation, we make use of the similarity transform because of its property of leaving unchanged the eigenvalues of the matrix being transformed. We then construct a similarity transformation matrix **S** that is an *orthonormal* rotation matrix for only row and column indices I and J. The rotation angle for the kth iteration is θ_k. The rotation matrix is constructed to place the $-\sin(\theta_k)$ element at the row and column coordinates of the target pivot element (I, J), and the $\sin(\theta_k)$ element at the diagonally symmetrical coordinate. The two $\cos(\theta_k)$ terms are on the diagonal, at coordinates (I, I) and (J, J).

$$\mathbf{S_k} = \begin{bmatrix}
1 & 0 & \cdots & \cdots & \cdots & 0 & 0 \\
0 & 1 & \cdots & \cdots & \cdots & 0 & 0 \\
 & & & \vdots & & & \\
0 & \cdots & \cos\theta_k & \cdots & -\sin\theta_k & \cdots & 0 \\
 & & & \vdots & & & \\
0 & 0 & \cdots & 1 & \cdots & 0 & 0 \\
 & & & \vdots & & & \\
0 & \cdots & \sin\theta_k & \cdots & \cos\theta_k & \cdots & 0 \\
 & & & \vdots & & & \\
0 & 0 & \cdots & \cdots & \cdots & 1 & 0 \\
0 & 0 & \cdots & \cdots & \cdots & 0 & 1
\end{bmatrix} \tag{6.9}$$

Since **S** is an orthonormal matrix, the inverse of **S** is simply the transpose of **S**, or $\mathbf{S}_k^{-1} = \mathbf{S}_k^T$. So for the *k*th iteration

$$\mathbf{A_{k+1}} = \mathbf{S_k^T A_k S_k} \tag{6.10}$$

and the sequence of $\mathbf{A_k}$ matrices will all be real and symmetric. We carefully select the rotation angle θ_k for the *k*th iteration to be that angle required to annihilate the pivot element $a_{I,J}$ to zero.

$$\theta_k = \frac{1}{2}\tan^{-1}\left(\frac{2a_{I,J}}{a_{I,I} - a_{J,J}}\right) \tag{6.11}$$

We may be hopeful that one pass through all the off-diagonal elements is sufficient to turn them all to zero, and leave the matrix in its desired strictly diagonal form, in a way reminiscent of Gaussian elimination, annihilating all the lower diagonal elements in one pass. Unfortunately, in the Jacobi algorithm, elements that have been turned to zero, in general, will not stay zero during subsequent iterations that pivot on different upper-diagonal elements. However, it can be shown that under the condition that the rotation angle is restricted to the range $-\frac{\pi}{4} \le \theta_k \le \frac{\pi}{4}$, then at least the off-diagonal elements are all reduced in magnitude each step. This implies that repeatedly running iterations will eventually be able to reduce all off-diagonal elements below some specified tolerance level. When we are able to scan through all the pivot elements for each iteration and not find an element with an absolute magnitude greater than our tolerance for an entire scan above the diagonal, we declare the algorithm successfully concluded. At this point, the eigenvalues of the original matrix $\mathbf{A}_1$ may be found along the matrix diagonal elements. They will appear in no particular numerical order, so if the eigenvalues need to be sorted, say, smallest to largest, in some application, an additional sorting operation may be applied.

And how do we find the eigenvectors that correspond to each eigenvalue? If we start with an identity matrix **I**, and successively multiply by the same transformation matrix $\mathbf{S}_k$ after each iteration, we will end up with a matrix **V** whose columns are the eigenvectors of the original matrix.

$$\mathbf{V} = \mathbf{I}\mathbf{S}_1\mathbf{S}_3\mathbf{S}_3 \cdots \mathbf{S}_M \tag{6.12}$$

The eigenvector in column *j* of the resulting **V** matrix will correspond to the eigenvalue on the diagonal element with coordinates (j, j) in the final transformed matrix $\mathbf{S}_M$.

6.4 Coding: The Systems of Linear Equations Tool

The code for the `lineq.c` tool, together with its header file `lineq.h`, capable of finding various classes of solutions of systems of linear equations, is part of the `CPLibrary` package that can be downloaded from the Student Companion website. Different Gaussian elimination routines are included for the cases of the coefficient matrix being described by both full and banded matrices. One important task in these routines is to test whether the coefficient matrix passed to it is singular or not. As discussed above, this decision is equivalent to testing if a diagonal matrix element has been annihilated to zero at some point in the elimination process. We must heed the lessons of Section 3.3.4 for this test, and instead of testing for a floating point value being identically zero, we establish a very small numerical region around the value of zero for this test that allows for some roundoff error in practical floating point computation hardware. However, the small width of this region must be chosen appropriately for each particular system of equations being solved, as the scale of the equation coefficients may vary over a wide range for different physical problems.

Therefore, each of the routines in this tool accepts a double-precision function argument named `min`. This is considered to be the minimum absolute value that a diagonal element may assume in the elimination process to be considered nonzero, and for the process to be allowed to continue. If the absolute value of a diagonal element is found to be less than this limit, the fraction that forms the row multiplication factor would result in a very large number, and at that point the routine is halted and the matrix is declared to be singular. In this case, a particular integer error code is returned to the calling program so it may take appropriate action.

The first routine in this tool is named `gauss()`, and implements the basic Gaussian elimination algorithm on a coefficient matrix and right-hand side vector passed to it. This routine can be useful in limited situations where we may know from our knowledge about the physical system being modeled that the coefficients are all nonzero and of approximately the same numerical order. In this case, the simple algorithm with its minimum amount of computation can generally be relied on. The basic code in the `gauss()` routine that implements the Gaussian elimination process is

```
1  // Iterate down the diagonal pivot elems
2  for (k = 0; k < n - 1; k++) {
3    if (fabs(coeff->elem[k][k]) < min) {
4      return(LINEQ_SINGULAR);
5    }
6  // Iterate down the rows under pivot row
7    for (i = k + 1; i < n; i++) {
8      factor = coeff->elem[i][k] / coeff->elem[k][k];
9  // Iterate over the columns to the right of the pivot column
10     for (j = k + 1; j < n; j++) {
11       coeff->elem[i][j] -= factor * coeff->elem[k][j];
12     }
13     rhs->elem[i] -= factor * rhs->elem[k];
14   }
15 }
```

A subtle technique is used in this routine to save a small but significant amount of unneeded computation [10, Chap. 2]. The technique grows from the observation that the matrix elements under the diagonal pivot element that are turned to zero by the Gaussian elimination method are never actually used again in any computation steps, in neither the diagonalization phase nor the back substitution phase. Therefore, there is no need to actually carry out the computation that results in zero for those elements under the pivot elements. We may safely skip that algebraic step and start the actual computation with the column *to the right* of the element under the pivot element. This economy is easily implemented by carefully choosing the loop iteration parameters for the j `for` statement in line 10 above.

Then once the coefficient matrix has been transformed into an upper diagonal form, the backsubstitution phase is implemented with the basic code

```
// Iterate up the rows for backsubstitution
for (i = n - 1; i >= 0; i--) {
  if (fabs(coeff->elem[i][i]) < min) {
    return(LINEQ_SINGULAR);
  }
  sum = rhs->elem[i];
// Iterate over the columns for summation
  for (j = i + 1; j < n; j++) {
    sum -= coeff->elem[i][j] * x->elem[j];
  }
  x->elem[i] = sum / coeff->elem[i][i];
}
```

These other routines in this tool build more features into this basic algorithm. The second routine in this tool is named `gauss_band()`, and implements the same basic Gaussian elimination algorithm, but for a *banded* linear system. The coefficient matrix is assumed to be stored in an `ARRAY2D_BAND` banded two-dimensional array as described in Section 3.1.4. This solution method is appropriate for physical systems with some sort of physical array of elements with limited interaction between them, so that the coefficient matrix to be transformed has nonlinear elements only within a narrow band of elements around the center diagonal. In this case, we may save the computation effort of trying to annihilate to zero the matrix elements that are already zero. For a large physical system, this can potentially save an enormous computational load. To implement this algorithm in the `gauss_band()` routine, the array indexing is changed to the system shown in Figure 3.29, and limits are placed on the iterative loops to process only the stored elements within the band.

The next routine, `gauss_pivotmax()`, implements a relatively simple scaled partial pivoting strategy for choosing the optimum row order as the elimination algorithm for a full square matrix of coefficients proceeds. When the routine is first called, the code scans through the coefficient matrix and develops a one-dimensional table of the maximum absolute coefficient value in each row. (The local memory space for this table is obtained by a dynamic memory allocation that is freed before the function returns.) This provides the algorithm with a measure of the scale of the coefficients in each linear equation in the system, which may vary over many orders of magnitude for some physical problems. Then during the elimination process, the algorithm will reorder the remaining rows so that the coefficient with the largest value relative to its equation's scale is used as the pivot element, that is, the coefficient that will appear in the denominator of the row multiplication factors. This method will also automatically reorder rows in the case of a diagonal element being identically 0 due to the physical system being modeled.

The actual mechanism for reordering of the rows here makes use of a very important facility in the storage mechanism for two-dimensional arrays developed in Section 3.1.4. To accomplish the reordering of rows in the coefficient matrix, we want to avoid the needless computational load of physically moving all the data elements of entire rows from one area of computer memory to another. We can accomplish the same reordering virtually by simply exchanging two pointer values in the `elem` array of row pointers in the `ARRAY2D` structure by lines 3 through 5 below.

```
1  // Swap in the best pivot row if necessary
2  if (imax != k) {
3    aswap = coeff->elem[imax];
4    coeff->elem[imax] = coeff->elem[k];
5    coeff->elem[k] = aswap;
6    dswap = rhs->elem[imax];
7    rhs->elem[imax] = rhs->elem[k];
8    rhs->elem[k] = dswap;
9    dswap = scale[imax];
10   scale[imax] = scale[k];
11   scale[k] = dswap;
12 }
```

This is yet another of the many examples of the significant computation efficiency afforded by making good use of pointer variables in the C language. We do need to take care, however, to also exchange the corresponding elements of the right-hand side vector, in lines 6 through 8 above, and the original scale measurement array, in lines 9 through 11 above, when exchanging virtual rows.

Further routines are provided in the `lineq.c` tool for other computations common in linear algebra. The routines `det()` and `det_pivotmax()` compute the determinant of a square matrix using the principle of linear algebra that the product of all the diagonal elements of a triangular matrix is the determinant. The Gaussian elimination process is carried out first to reduce the original coefficient matrix to an upper diagonal matrix, which will not change the determinant from its original value. The first of these routines will perform a simple Gaussian elimination, and the second uses scaled partial pivoting to preserve numerical precision for practical matrices.

The last routine in the tool, `jacobi()`, implements the iterative Jacobi algorithm for finding the eigenvalues and eigenvectors of a symmetric coefficient matrix. Since the Jacobi method is fundamentally applicable only to symmetric square matrices, we can save almost a factor of two in memory storage and processing cost by supplying and maintaining the argument coefficient matrix as an upper triangular matrix allocated by the `alloc_array2d_utri()` function, described in Section 3.1.4. The missing lower-diagonal elements are assumed to be diagonally symmetric to the stored elements. The calling program must, however, allocate a separate full square matrix with the `alloc_array2d()` tool and pass it as a second argument for storing the matrix of eigenvectors that this routine develops. Care is taken to calculate the rotation angle θ_k for each iteration efficiently by avoiding actual calls to C library trigonometric functions. Since almost all the elements in the $\mathbf{S_k}$ transformation matrices are simply 1, the two matrix products in the expression $\mathbf{S_k^T A_k S_k}$ are performed together with select arithmetic expressions so that the series of $\mathbf{A_k}$ matrices can all continue to be stored in an upper triangular array.

All these functions accept needed pointer arguments passed to them, pointing at one- and two-dimensional arrays to hold the coefficient matrix, the right-hand side vector, and the solution vector, as required. The Gaussian elimination routines assume that the calling program has allocated memory space in an `ARRAY1D` structure for the solution vector, and that vector elements may then be stored in this space in the back substitution phase. The Jacobi routine assumes that the calling program has allocated a square array for storing the found eigenvectors. Even for very large systems, the function calling procedure is simplified by using only a few structure pointers as arguments.

Note that all these functions are declared with a return type of `int`. This allows the success or failure of the corresponding linear algebraic operation to be reported to the calling program with the constants `LINEQ_SUCCESS` or `LINEQ_SINGULAR`, the latter in the case of the routine encountering a singular coefficient matrix. In addition, the `jacobi()` function may return a value of `LINEQ_MAX_ITERATIONS` if the maximum number of allowed iterations has been exceeded. These constants are defined in the header file `lineq.h`. This header should be included in any source file calling these routines, so that the return values may be tested and appropriate action taken in any case. With the algorithm status reported in the function return value, we rely on pointers passed as function arguments for memory locations where the computation results will be stored, whether they be vectors or scalers.

In the sections and exercises that follow, we will explore some physical problems that rely solely on the routines in this tool code for solutions. However, the capability of solving systems of linear equations will also appear as an essential component in other numerical algorithms in the following chapters, particularly Chapters 7 and 14.

6.5 Application: Modes of Coupled Oscillators

Physical systems that may be modeled as an array of spring-mass oscillators are common, and this model is almost universally analyzed in classical mechanics textbooks. However, the analysis typically stops at the point of deriving the discrete normal modes of oscillation. Here we extend this analysis to situations where the initial conditions excite oscillations that are a mixture of the normal modes, where a solution of a linear system is required.

We'll begin with a review of the derivation of normal oscillation modes for some simple coupled mechanical networks [27, Chap. 11; 28, Chap. 4; 29, Chap. 11]. A coupled oscillator with two identical masses with mass m and three identical linear springs with spring constants k is shown in Figure 6.1. Let the square of a reference frequency be $\omega_0^2 = \frac{k}{m}$ and we may derive the equations of motion from straightforward Newtonian laws as

$$\begin{aligned} \frac{d^2x_1}{dt^2} &= -2\omega_0^2 x_1 + \omega_0^2 x_2 \\ \frac{d^2x_2}{dt^2} &= \omega_0^2 x_1 - 2\omega_0^2 x_2 \end{aligned} \tag{6.13}$$

where x_i is the displacement of mass i. We make the assumption that the normal mode solutions exist with both masses oscillating at identical physical frequencies, and therefore can be written in a complex form as

$$\begin{aligned} x_1 &= a_1 e^{i\omega t} \\ x_2 &= a_2 e^{i\omega t} \end{aligned} \tag{6.14}$$

with ω being the physical frequency of a particular mode. Then substituting into the equations of motion gives

$$\begin{aligned} -\omega^2 a_1 e^{i\omega t} &= -2\omega_0^2 a_1 e^{i\omega t} + \omega_0^2 a_2 e^{i\omega t} \\ -\omega^2 a_2 e^{i\omega t} &= \omega_0^2 a_1 e^{i\omega t} - 2\omega_0^2 a_2 e^{i\omega t} \end{aligned} \tag{6.15}$$

Solving for the conditions of the coefficients a_1 and a_2 that will allow a nontrivial solution for all values of time t

$$\begin{aligned} -\omega^2 a_1 &= -2\omega_0^2 a_1 + \omega_0^2 a_2 \\ -\omega^2 a_2 &= \omega_0^2 a_1 - 2\omega_0^2 a_2 \end{aligned} \tag{6.16}$$

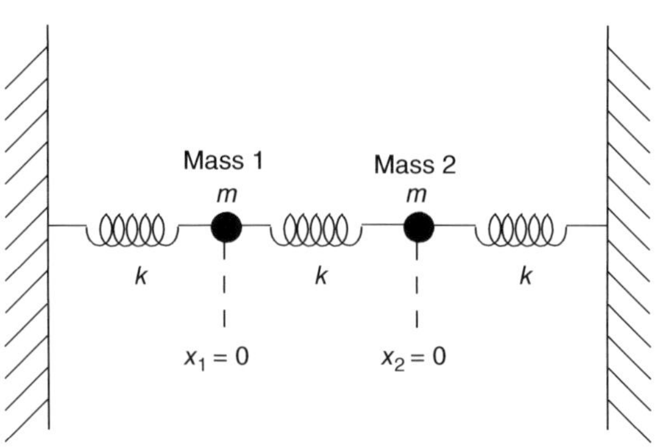

Figure 6.1 A simple coupled oscillator with two masses.

or, in matrix form

$$\begin{bmatrix} 2\omega_0^2 & -\omega_0^2 \\ -\omega_0^2 & 2\omega_0^2 \end{bmatrix} \begin{bmatrix} a_1 \\ a_2 \end{bmatrix} = \omega^2 \begin{bmatrix} a_1 \\ a_2 \end{bmatrix} \tag{6.17}$$

or, setting $\lambda = \frac{\omega^2}{\omega_0^2}$,

$$\begin{bmatrix} 2 & -1 \\ -1 & 2 \end{bmatrix} \begin{bmatrix} a_1 \\ a_2 \end{bmatrix} = \lambda \begin{bmatrix} a_1 \\ a_2 \end{bmatrix} \tag{6.18}$$

This is the classical eigenvalue problem, which for this small order of two, may be solved easily. We rearrange this into a simultaneous linear system

$$\begin{bmatrix} 2-\lambda & -1 \\ -1 & 2-\lambda \end{bmatrix} \begin{bmatrix} a_1 \\ a_2 \end{bmatrix} = 0 \tag{6.19}$$

For this system to have a nontrivial solution with nonzero values for the two amplitudes a_1 and a_2, the coefficient matrix must be singular, and therefore the determinant of the matrix must be zero.

$$\det \begin{bmatrix} 2-\lambda & -1 \\ -1 & 2-\lambda \end{bmatrix} = 0 \tag{6.20}$$

So we may find the two allowed eigenvalues λ_1 and λ_2 as the roots of the characteristic polynomial

$$\lambda^2 - 4\lambda + 3 = 0 \tag{6.21}$$

The two roots are the eigenvalues $\lambda_1 = 1$ and $\lambda_2 = 3$. Since $\omega^2 = \lambda\omega_0^2$, these eigenvalues correspond to the physical frequencies of the two normal modes of oscillation being $\omega = \pm\omega_0 = \pm\sqrt{\frac{k}{m}}$ and $\omega = \pm\sqrt{3}\omega_0 = \pm\sqrt{3\frac{k}{m}}$.

Since Equation (6.19) is a degenerate system of equations, we may find the solution of the two amplitudes a_1 and a_2 only within a multiplicative constant. By substituting the two allowed eigenvalues of λ back into either row of the matrix in Equation (6.19), we can find the allowed relationships between the amplitudes, or the system eigenvectors, to be $a_2 = a_1$ or $a_2 = -a_1$ corresponding to the eigenvalues of 1 and 3, respectively. For numerical convenience, we can arbitrarily choose $a_1 = 1$, making the possible eigenvector of coefficient values in Equation (6.14) that correspond to each eigenvalue

$$\lambda = 1 \Rightarrow \begin{bmatrix} a_1 \\ a_2 \end{bmatrix} = \begin{bmatrix} 1 \\ 1 \end{bmatrix} \qquad \lambda = 3 \Rightarrow \begin{bmatrix} a_1 \\ a_2 \end{bmatrix} = \begin{bmatrix} 1 \\ -1 \end{bmatrix} \tag{6.22}$$

Note that there is an algebraic relationship between the eigenvalues and the *square* of the physical frequencies, so each eigenvalue actually corresponds to two normal modes, one with a positive oscillation frequency and another with a negative frequency, that mathematically satisfy the eigenvalue constraints. Since the motion solution we are seeking is the solution of a linear differential equation, we know that a linear combination of these normal modes will also be a consistent solution. Moreover, because the normal modes are a complete basis for any motion, we can express any possible motion that may be observed as following the combined relationships

$$\begin{aligned} x_1(t) &= b_{1p}e^{i\sqrt{\lambda_1}\omega_0 t} + b_{1n}e^{-i\sqrt{\lambda_1}\omega_0 t} + b_{2p}e^{i\sqrt{\lambda_2}\omega_0 t} + b_{2n}e^{-i\sqrt{\lambda_2}\omega_0 t} \\ x_2(t) &= b_{1p}e^{i\sqrt{\lambda_1}\omega_0 t} + b_{1n}e^{-i\sqrt{\lambda_1}\omega_0 t} - b_{2p}e^{i\sqrt{\lambda_2}\omega_0 t} - b_{2n}e^{-i\sqrt{\lambda_2}\omega_0 t} \end{aligned} \tag{6.23}$$

The coefficients b_{1p}, b_{1n}, b_{2p}, and b_{2n} are arbitrary complex numbers that determine the contribution of each normal mode, including both positive and negative frequencies. We have included the relationship between the modes constrained by Equation (6.22) by the relative signs of each of the four b coefficients that appear in the expressions for $x_1(t)$ and $x_2(t)$

with the corresponding oscillatory terms at the normal mode frequencies. Since we seek to characterize the motion of a physical system with real-valued displacements, we can also restrict the resulting values of $x_1(t)$ and $x_2(t)$ to be real. With some algebraic manipulation, this constraint implies that the coefficients b_{1p} and b_{1n} must be complex conjugates of each other, as must b_{2p} and b_{2n}. In this case the real-valued displacements of the two masses may be expressed as

$$\begin{aligned} x_1(t) &= c_{1c}\cos(\sqrt{3}\omega_0 t) + c_{1s}\sin(\sqrt{3}\omega_0 t) + c_{2c}\cos(\omega_0 t) + c_{2s}\sin(\omega_0 t) \\ x_2(t) &= c_{1c}\cos(\sqrt{3}\omega_0 t) + c_{1s}\sin(\sqrt{3}\omega_0 t) - c_{2c}\cos(\omega_0 t) - c_{2s}\sin(\omega_0 t) \end{aligned} \tag{6.24}$$

where the real-valued coefficients c_{1c}, c_{1s}, c_{2c}, and c_{2s} are arbitrary constants to be evaluated from the real-valued initial displacement conditions.

In Sections 14.1 and 14.4, we will simulate the oscillations of this coupled oscillator directly from the equations of motion. We can see the time dependence of the two displacements for the two normal modes plotted in Figures 14.2 and 14.3.

Let us extend this same analysis to a simple coupled oscillator with three identical masses of mass m and four identical springs with constant k, as shown in Figure 6.2. As before, let the square of a reference frequency $\omega_0^2 = \frac{k}{m}$ and the equations of motion become

$$\begin{aligned} \frac{d^2x_1}{dt^2} &= -2\omega_0^2 x_1 + \omega_0^2 x_2 \\ \frac{d^2x_2}{dt^2} &= \omega_0^2 x_1 - 2\omega_0^2 x_2 + \omega_0^2 x_3 \\ \frac{d^2x_3}{dt^2} &= \omega_0^2 x_2 - 2\omega_0^2 x_3 \end{aligned} \tag{6.25}$$

Assume the solutions of the normal oscillation modes are in the form

$$\begin{aligned} x_1 &= a_1 e^{i\omega t} \\ x_2 &= a_2 e^{i\omega t} \\ x_3 &= a_3 e^{i\omega t} \end{aligned} \tag{6.26}$$

Then substituting the assumed solution forms into the equations of motion gives

$$\begin{aligned} -\omega^2 a_1 &= -2\omega_0^2 a_1 + \omega_0^2 a_2 \\ -\omega^2 a_2 &= \omega_0^2 a_1 - 2\omega_0^2 a_2 + \omega_0^2 a_3 \\ -\omega^2 a_3 &= \omega_0^2 a_2 - 2\omega_0^2 a_3 \end{aligned} \tag{6.27}$$

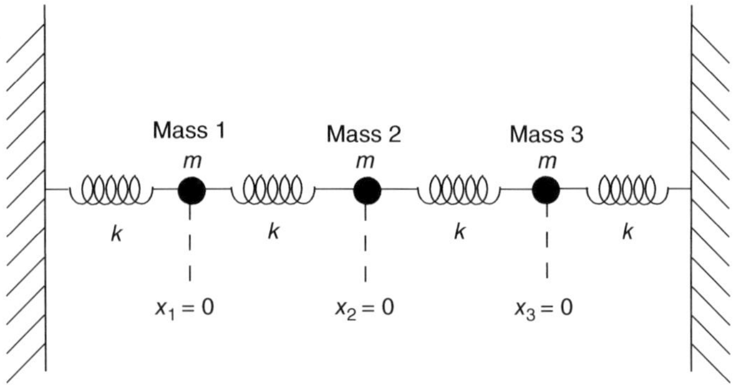

Figure 6.2 A simple coupled oscillator with three masses.

or, in matrix form

$$\begin{bmatrix} 2\omega_0^2 & -\omega_0^2 & 0 \\ -\omega_0^2 & 2\omega_0^2 & -\omega_0^2 \\ 0 & -\omega_0^2 & 2\omega_0^2 \end{bmatrix} \begin{bmatrix} a_1 \\ a_2 \\ a_3 \end{bmatrix} = \omega^2 \begin{bmatrix} a_1 \\ a_2 \\ a_3 \end{bmatrix} \tag{6.28}$$

or, with $\lambda = \frac{\omega^2}{\omega_0^2}$,

$$\begin{bmatrix} 2 & -1 & 0 \\ -1 & 2 & -1 \\ 0 & -1 & 2 \end{bmatrix} \begin{bmatrix} a_1 \\ a_2 \\ a_3 \end{bmatrix} = \lambda \begin{bmatrix} a_1 \\ a_2 \\ a_3 \end{bmatrix} \tag{6.29}$$

Following the same procedure as with the second-order case, we rearrange this into a system of three linear equations. We set the determinant of the coefficient matrix to zero for finding the conditions on the eigenvalues for nontrivial solutions. We then must find the roots of the characteristic polynomial

$$(2 - \lambda)(\lambda^2 - 4\lambda + 2) = 0 \tag{6.30}$$

One eigenvalue root is evident from the first factor as $\lambda_1 = 2$, and the other two eigenvalues are the quadratic roots $\lambda_2 = 2 - \sqrt{2}$ and $\lambda_3 = 2 + \sqrt{2}$. Of course, these eigenvalues correspond to the physical frequencies of the three normal modes of oscillation by $\omega_i = \pm\sqrt{\lambda_i}\omega_0 = \pm\sqrt{\lambda_i \frac{k}{m}}$.

Again, by substituting these values of λ back into the first two rows in the matrix of Equation (6.29), we can find the permissible eigenvectors corresponding to each eigenvalue within a multiplicative constant. One arbitrary choice of scaling is

$$\lambda = 2 \Rightarrow \begin{bmatrix} a_1 \\ a_2 \\ a_2 \end{bmatrix} = \begin{bmatrix} -1 \\ 0 \\ 1 \end{bmatrix} \quad \lambda = 2 - \sqrt{2} \Rightarrow \begin{bmatrix} a_1 \\ a_2 \\ a_2 \end{bmatrix} = \begin{bmatrix} 1 \\ \sqrt{2} \\ 1 \end{bmatrix} \quad \lambda = 2 + \sqrt{2} \Rightarrow \begin{bmatrix} a_1 \\ a_2 \\ a_2 \end{bmatrix} = \begin{bmatrix} 1 \\ -\sqrt{2} \\ 1 \end{bmatrix} \tag{6.31}$$

We extend the derivation of Equation (6.23) to the case of the three-mass system, including the mathematical existence of normal modes with both positive and negative oscillation frequencies. A full solution to the differential equation can be constructed as a linear combination of six normal modes as

$$\begin{aligned} x_1(t) &= -b_{1p}e^{i\sqrt{\lambda_1}\omega_0 t} - b_{1n}e^{-i\sqrt{\lambda_1}\omega_0 t} \\ &\quad + b_{2p}e^{i\sqrt{\lambda_2}\omega_0 t} + b_{2n}e^{-i\sqrt{\lambda_2}\omega_0 t} \\ &\quad + b_{3p}e^{i\sqrt{\lambda_3}\omega_0 t} + b_{3n}e^{-i\sqrt{\lambda_3}\omega_0 t} \\ x_2(t) &= \sqrt{2}b_{2p}e^{i\sqrt{\lambda_2}\omega_0 t} + \sqrt{2}b_{2n}e^{-i\sqrt{\lambda_2}\omega_0 t} \\ &\quad - \sqrt{2}b_{3p}e^{i\sqrt{\lambda_3}\omega_0 t} - \sqrt{2}b_{3n}e^{-i\sqrt{\lambda_3}\omega_0 t} \\ x_3(t) &= b_{1p}e^{i\sqrt{\lambda_1}\omega_0 t} + b_{1n}e^{-i\sqrt{\lambda_1}\omega_0 t} \\ &\quad + b_{2p}e^{i\sqrt{\lambda_2}\omega_0 t} + b_{2n}e^{-i\sqrt{\lambda_2}\omega_0 t} \\ &\quad + b_{3p}e^{i\sqrt{\lambda_3}\omega_0 t} + b_{3n}e^{-i\sqrt{\lambda_3}\omega_0 t} \end{aligned} \tag{6.32}$$

where, again, we have included the constraints of the relative amplitudes of the three normal modes with the signs and multiplicative constants of the b coefficients. We then follow the derivation of Equation (6.24) and formulate our solution for real values of displacement as

$$\begin{aligned}
x_1(t) &= -c_{1c}\cos(\sqrt{\lambda_1}\omega_0 t) - c_{1s}\sin(\sqrt{\lambda_1}\omega_0 t)\\
&\quad +c_{2c}\cos(\sqrt{\lambda_2}\omega_0 t) + c_{2s}\sin(\sqrt{\lambda_2}\omega_0 t)\\
&\quad +c_{3c}\cos(\sqrt{\lambda_3}\omega_0 t) + c_{3s}\sin(\sqrt{\lambda_3}\omega_0 t)\\
x_2(t) &= \sqrt{2}c_{2c}\cos(\sqrt{\lambda_2}\omega_0 t) + \sqrt{2}c_{2s}\sin(\sqrt{\lambda_2}\omega_0 t)\\
&\quad -\sqrt{2}c_{3c}\cos(\sqrt{\lambda_3}\omega_0 t) - \sqrt{2}c_{3s}\sin(\sqrt{\lambda_3}\omega_0 t)\\
x_3(t) &= c_{1c}\cos(\sqrt{\lambda_1}\omega_0 t) + c_{1s}\sin(\sqrt{\lambda_1}\omega_0 t)\\
&\quad +c_{2c}\cos(\sqrt{\lambda_2}\omega_0 t) + c_{2s}\sin(\sqrt{\lambda_2}\omega_0 t)\\
&\quad +c_{3c}\cos(\sqrt{\lambda_3}\omega_0 t) + c_{3s}\sin(\sqrt{\lambda_3}\omega_0 t)
\end{aligned} \tag{6.33}$$

This case of three coupled masses is also simulated in Section 14.1. We can see the time dependence of the three displacements plotted for the three normal modes in Figures 14.6–14.8. Intuitively, we can interpret these plots to understand the different natural frequencies associated with each of these modes. The slowest frequency is associated with the mode with the three masses oscillating back and forth mostly together. In that case, the only springs being substantially compressed and stretched during each cycle are the two springs anchoring the system to the walls, and with the small counteracting force of only two springs, the acceleration is slow. The highest frequency is associated with the mode with the two outer masses together in their motion, but with the central mass oscillating in opposition. This motion stretches all the springs, and the larger counteracting forces leads to faster accelerations. The case of a medium oscillation frequency comes from a mode where the central mass is stationary, and the motions of the two outer masses are counteracted with moderately stretched springs. This same intuitive reasoning lets us understand the variation in mode frequencies for more elaborate systems of higher order as well.

Let us consider the case where the initial conditions applied to the oscillator system take the form only of initial static displacements, with no initial velocities of the masses. In that case, we may set the coefficients of all the $\sin(\omega_i t)$ in Equation (6.33) to zero. This leaves the motion of the masses being described by

$$\begin{aligned}
x_1(t) &= -c_1\cos(\sqrt{\lambda_1}\omega_0 t) + c_2\cos(\sqrt{\lambda_2}\omega_0 t) + c_3\cos(\sqrt{\lambda_3}\omega_0 t)\\
x_2(t) &= \sqrt{2}c_2\cos(\sqrt{\lambda_2}\omega_0 t) - \sqrt{2}c_3\cos(\sqrt{\lambda_3}\omega_0 t)\\
x_3(t) &= c_1\cos(\sqrt{\lambda_1}\omega_0 t) + c_2\cos(\sqrt{\lambda_2}\omega_0 t) + c_3\cos(\sqrt{\lambda_3}\omega_0 t)
\end{aligned} \tag{6.34}$$

The components of the vectors in Equation (6.31) are the initial displacement values that will excite oscillation of this system purely in the corresponding normal mode, or only along one axis of the orthogonal basis. But what about the case of an arbitrary set of initial displacements, not in general one of these eigenvectors? Since the underlying equations of motion are linear, we expect that this will excite motion that is some linear combination of the trajectories of the three normal modes. How can we predict the motion of the masses for any arbitrary initial condition? The problem can be distilled down to finding the solution of a linear system of equations, where the component of each normal mode is resolved

$$\begin{bmatrix} -1 & 1 & 1 \\ 0 & \sqrt{2} & -\sqrt{2} \\ 1 & 1 & 1 \end{bmatrix} \begin{bmatrix} c_1 \\ c_2 \\ c_3 \end{bmatrix} = \begin{bmatrix} d_1 \\ d_2 \\ d_3 \end{bmatrix} \tag{6.35}$$

where the d_i are the three initial displacements of the masses at time $t = 0$, and the c_i are the components of the three normal modes required to support those initial conditions. Since the system is linear, the same components will subsequently be present in the same proportions for times $t > 0$. The columns of the coefficient matrix for this linear system are the eigenvectors corresponding to the eigenvalues for each normal mode, in the same order as the mode coefficients in the c_i solution vector.

Consider the main program in `coupled_osc3_step.c`, available for download from the Student Companion website, to model the displacement of all three masses as a function of time. We will recall the function stepping tool `step.c` of Chapter 4 to generate tables of displacement versus time in the output stream. That leaves the only task in this program to be writing three functions computing displacement as a function of time for each of the masses, assuming all three normal modes may be superposed. These three functions are then passed to the `step()` tool. This program accepts six command line arguments to set the physical parameters of the simulation. The first argument is the reference frequency expressed as a real frequency in cycles per second, that is, $\frac{\omega_0}{2\pi}$. The next three arguments specify the arbitrary initial displacement of the three masses, not necessarily only those displacements that are the mode eigenvectors. Finally, the last two command line arguments give the final and increment time of the time variable grid for the simulation.

The first section of the program sets up the linear system of Equation (6.35) and solves it for the three initial displacements provided on the command line. The solution of the linear system is stored in the `ARRAY1D` structure that is a member of the `OSC` structure. Then the `OSC` structure is passed as a parameter set to all three functions `mass1()`, `mass2()`, and `mass3()`. These are the three object functions that are stepped by the `step.c` tool to model the displacement for times $t > 0$. In order to superpose the solutions of motion for each of the normal modes in the proper proportions, these object functions make use of the solution of the mode-mixing linear system as parameters. Since the linear system being solved for this problem is only of order three, we allocate memory for the right-hand side and solution vectors statically, and fill in the `ARRAY1D` structure elements directly. The `ARRAY2D` coefficient matrix is allocated dynamically with a call to `alloc_array2d()`. Since the eigenvector coefficients are numerically similar, calling the simple `gauss()` routine is sufficient.

Compiling this main program with both the linear system solution and function stepping tool routines and running in a Linux shell prompt window with some example parameter values and the example initial mass displacements of −1, 1, and 0.5 gives

```
$ gcc -Wall -Wshadow -O2 -o coupled_osc3_step coupled_osc3_step.c CPLibrary/lineq.c
    ↪ CPLibrary/step.c CPLibrary/array.c -lm
$ ./coupled_osc3_step
./coupled_osc3_step <freq> <init x1> <init x2> <init x3> <tfinal> <tinc>
$ ./coupled_osc3_step 1 -1 1 0.5 10 0.01 >| coupled_osc3_step.dat
Mode 1:0.75   Mode 2:0.22855339   Mode 3:-0.47855339
```

The solution for this example set of initial displacements gives the components of the three modes of oscillations needed to support the initial conditions to be 0.75, 0.22855339, and −0.47855339, respectively.

Starting an interactive `gnuplot` session and giving the commands

```
gnuplot> set key default
gnuplot> set xlabel 'Time'
gnuplot> set ylabel 'Displacement'
gnuplot> plot 'coupled_osc3_step.dat' i 0 w l t 'Mass 1','' i 1 w l t 'Mass 2','' i 2 w l t 'Mass
    ↪ 3'
```

produces the plot seen in Figure 6.3. In this plot, we see the coupled oscillating motion of the masses in a superposition of the three modes in the solved proportions. The motion is complicated and intricate, and not at all apparent by inspection of the plot that three uniform normal modes form its basis.

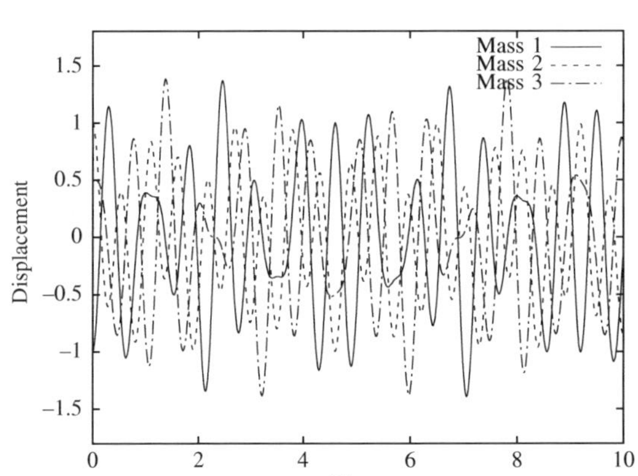

Figure 6.3 Plot of mixed-mode solution of the three-mass oscillator for initial displacements of −1, 1, 0.5.

Beyond third-order matrices, solving for eigenvalues for a particular matrix can easily become intractable, and numerical methods must be used. The `jacobi()` routine included in the `lineq.c` package for this textbook is applicable to the coupled oscillator system due to the symmetry of the coefficient matrix. The program to use this tool routine to calculate the eigenvalues corresponding to the cases of 2–19 masses is in `coupled_osc_eigenvalues.c` on the Student Companion website. Compiling this main program with the linear system routines and running in a Linux shell prompt window with an example parameter value of $\omega_0 = 1$ radians per second gives

```
$ gcc -Wall -Wshadow -O2 -o coupled_osc_eigenvalues coupled_osc_eigenvalues.c CPLibrary/lineq.c
    ↪ CPLibrary/array.c -lm
$ ./coupled_osc_eigenvalues
./coupled_osc_eigenvalues <w0>
$ ./coupled_osc_eigenvalues 1 >| coupled_osc_eigenvalues.dat
```

Starting an interactive `gnuplot` session and giving the commands

```
gnuplot> unset key
gnuplot> set xlabel 'Number of masses'
gnuplot> set xrange [0:20]
gnuplot> set ylabel 'Physical frequency of eigenvalue $\omega$'
gnuplot> set yrange [0:2]
gnuplot> plot 'coupled_osc_eigenvalues.dat' i 0 u 1:3 w p pt 7
```

produces the plot seen in Figure 6.4. In this plot, we see the frequencies associated with eigenvalues, assuming the reference frequency $\omega_0 = 1$ radian per second, for two masses at 1 and $\sqrt{3} \approx 1.7321$ radians per second, and those for three masses at $\sqrt{2} \approx 1.4142$, $\sqrt{2-\sqrt{2}} \approx 0.7654$, and $\sqrt{2+\sqrt{2}} \approx 1.8478$ radians per second, as derived above algebraically. Then past these two counts, as the network grows to larger counts of masses, we see more and more eigenvalues introduced that spread out and fill the frequency space between 0 and 2 radians per second in a way that approaches uniform coverage.

If we examine the eigenvectors associated with the first five and the last eigenvalues, we can see in Figure 6.5 plots of displacement versus the location coordinate of the mass along the line of springs.

We can interpret these normal mode eigenvector plots as either the set of initial displacements of the masses that, when released, will excite the system oscillating solely in the corresponding normal mode, or as a snapshot of the positions of the masses when the system is busy oscillating solely in one of the normal modes. The displacement of each mass will oscillate about zero, with all masses oscillating at the characteristic frequency of each particular mode. Note that there are "nodes" of oscillation, at the anchor end points and at certain internal masses, where there is no oscillation, and "antinodes" where the oscillation is at maximum amplitude. We know that this same behavior of oscillation nodes and antinodes is observed when standing, or stationary, waves are propagating in two opposite directions in a wave-supporting medium with reflecting discontinuities at each end. Examples are sound waves resonating in a pipe or along vibrating strings at certain frequencies. We can imagine the normal modes shown in Figure 6.5 for spring-mass oscillators being the equivalent of

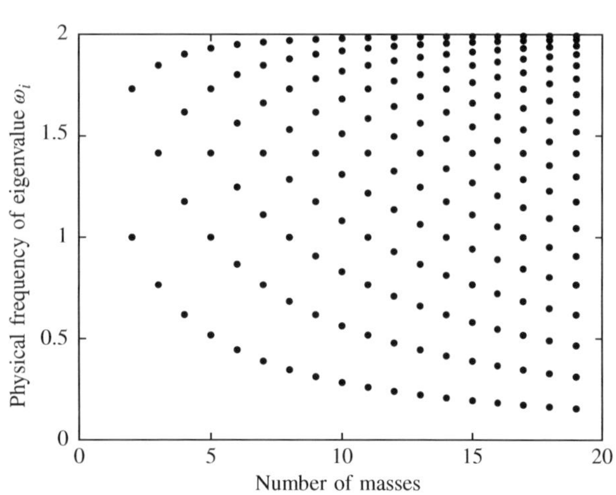

Figure 6.4 Physical frequencies ω_i corresponding to eigenvalues of coupled spring-mass oscillators with 2–19 masses.

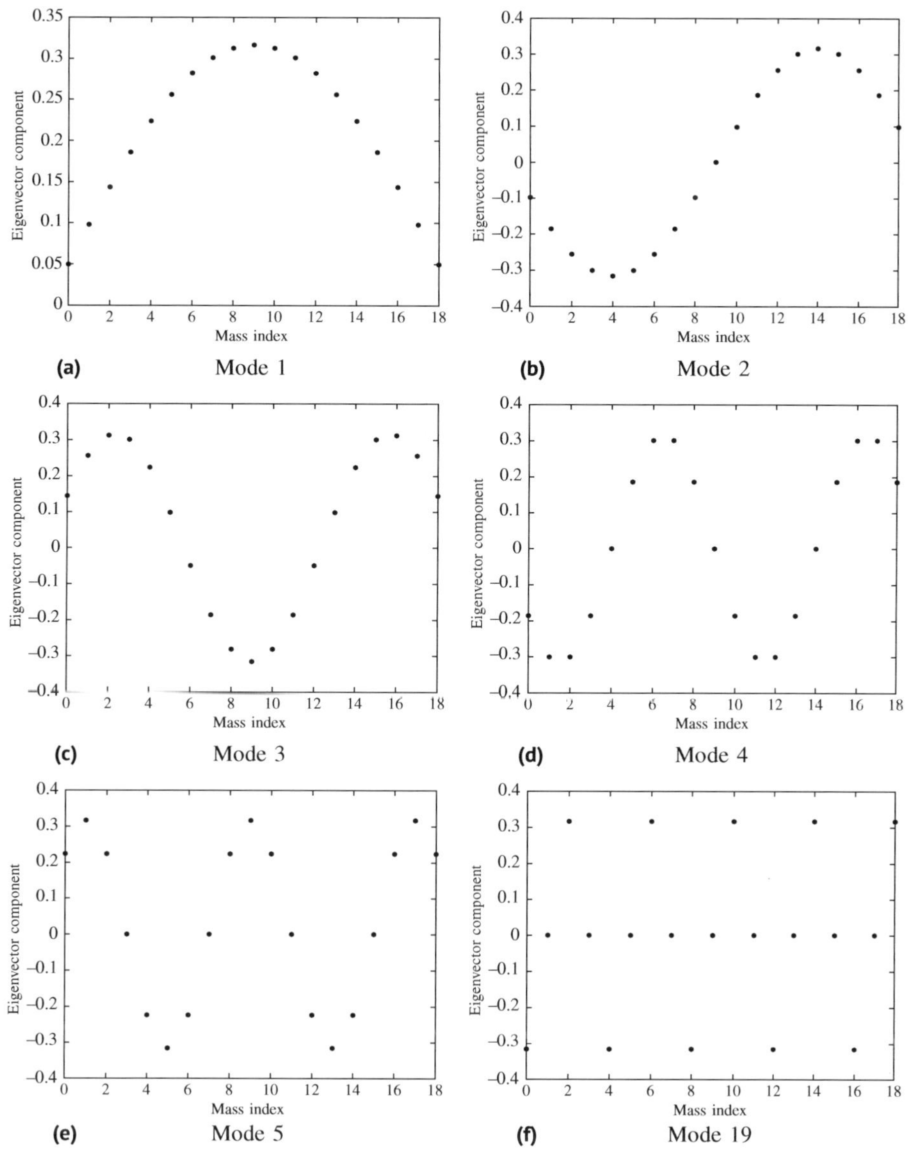

Figure 6.5 Six of the normal oscillation modes for 19 masses showing standing wave propagation.

standing longitudinal waves resonating in a wave medium between the anchor points of the outer springs. We will revisit this system in Section 6.9 and see how to generate animated graphics that let us clearly visualize the wave motion of the normal modes, and again in Section 14.5 where we numerically solve the wave function.

To see the coupled springs being essentially a discrete wave medium, we should reformulate the mass quantities and spring constants making up the system. For the distribution of eigenvalues in Figure 6.4, when we start with the two-mass oscillator and progress by including additional masses and springs, we assume all the masses in the system are identical to the original two masses, and we are making the system more and more massive by increasing the order. Similarly, we assume we are adding additional coupling springs just as stiff as the original three springs. For comparison with a wave medium, now let us consider growing the order of the system by keeping the total mass and the stiffness from anchor to anchor constant as we add additional masses. For a system with N masses, we now set the mass and spring constant in each segment to have the relationship with a reference mass and stiffness m_0 and k_0 given in Equation (6.36).

$$\begin{aligned} m_N &= \frac{m_0}{N} \\ k_N &= Nk_0 \end{aligned} \tag{6.36}$$

So we consider modeling and analyzing a particular length of a continuous longitudinal wave propagating medium, with a particular linear mass density ρ and Young's modulus Y for the medium material, and we seek to model this continuous medium with a larger and larger number of discrete coupled masses and springs. We know from continuous longitudinal wave solutions in solid materials [30, Chap. 8], the wave velocity is

$$v = \sqrt{\frac{Y}{\rho}} \tag{6.37}$$

For a given unit length of material that we will approximate as one of N discrete segments with a single mass and spring, this is equivalent to

$$v = \sqrt{\frac{k_N}{m_N}} \tag{6.38}$$

Since we consider the equivalent length of each segment to be a unit length in some systems of units, the delay time of the wave propagating through each segment is numerically $\frac{1}{v}$, so the delay time of propagation through the entire length of N segments is

$$D = N\sqrt{\frac{m_N}{k_N}} \tag{6.39}$$

We wish to compare the numerical solution of the eigenvalues of the discrete spring-mass system as the number of masses N grows large with the *equivalent* of the eigenvalues of the continuous wave system. When we solve for the eigenvalues of a vibrating mechanical system, what are we basically finding? We are discovering the set of discrete frequencies at which the system may oscillate in its normal modes, a set limited due to some resonance constraints. Let's approach a system with continuous wave propagation and reflections from the view of finding the set of frequencies of the waves that sustain a continuous and consistent oscillation. Consider the description of a wave system expressed as a signal flow graph of Figure 6.6.

A general complex form of an oscillation $Ae^{i\omega t}$ is assumed to be present at the left edge of a medium as shown. After propagating left-to-right through the medium, and subject to the propagation delay D, the oscillation $Ae^{i\omega(t-D)}$ will be present at the right side of the medium. At this point, we assume it is reflected without frictional losses on the right edge, and returns, now traveling right-to-left. Of course, in a linear wave medium, solutions of the wave equation may travel in either direction, with the total displacement being the superposition of both waves resulting in the standing or stationary wave behavior. The wave is delayed by another D units of time in its traversal back to the left edge, resulting in an oscillation at that point of $Ae^{i\omega(t-2D)}$. In order to sustain a consistent oscillation solution at one normal mode frequency, we must have the condition that these two waves result in the same displacement oscillation, or

$$Ae^{i\omega t} = Ae^{i\omega(t-2D)+i2\pi n} \tag{6.40}$$

where n is an arbitrary integer greater than zero. The additive term in the exponent on the right-hand side is appropriate because a reflected wave can return from the right and be consistent with the exciting wave at the left edge after any arbitrary amount of phase delay in multiples of 2π. Then what are the conditions on ω that a consistent solution exists, for a given value of n? After factoring out $Ae^{i\omega t}$ from Equation (6.40), we may solve this to find

$$\omega = \frac{\pi n}{D} \tag{6.41}$$

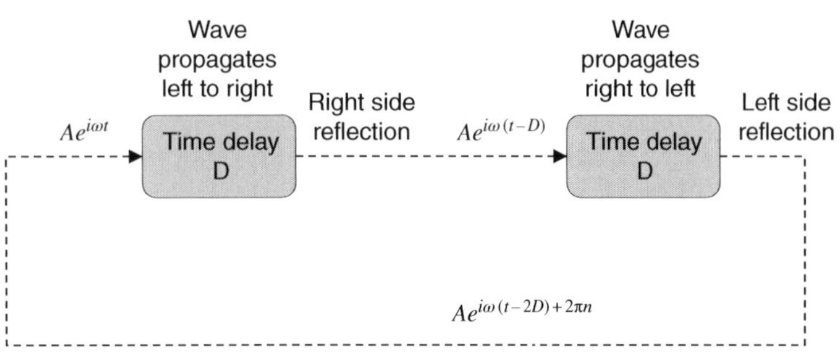

Figure 6.6 Signal flow graph of coupled spring-mass oscillator interpreted as a wave medium.

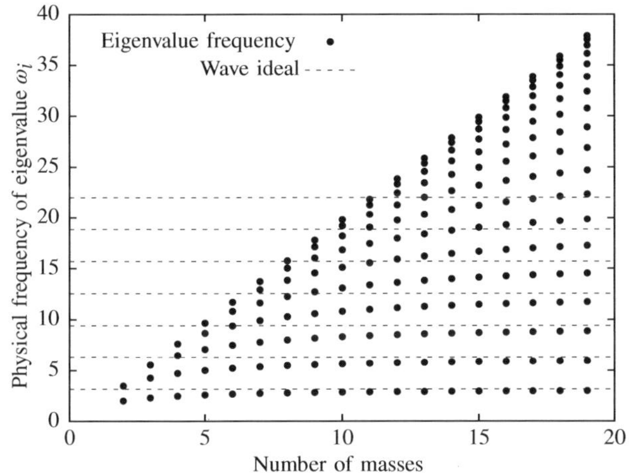

Figure 6.7 Physical frequencies ω_i corresponding to eigenvalues of coupled spring-mass oscillators with 2–19 masses, and scaled for discrete approximation for a wave medium.

This relationship gives the allowed frequencies that will support a normal mode of oscillation in a system of reflected linear wave propagation [29, Chap. 11]. The set is generated by an arbitrary quantum number n, from 1 to any large integer, although in practical systems the oscillation frequency will always be limited by second-order time and phase delays. How does this set of frequencies compare to those we solve numerically for a discrete system of spring-mass oscillators? To answer this, consider a modified program in `coupled_osc_wave_eigenvalues.c`. This program is essentially the same as the program that generated the eigenvalues plotted in Figure 6.4, except that as a growing number of spring-mass segments are introduced, the individual masses and spring constants are scaled by Equation (6.36). Compiling this main program with the linear system routines and running in a Linux shell prompt window with an example parameter value of $\omega_0 = 1$ radians per second gives

```
$ gcc -Wall -Wshadow -O2 -o coupled_osc_wave_eigenvalues coupled_osc_wave_eigenvalues.c
    ↪ CPLibrary/lineq.c CPLibrary/array.c -lm
$ ./coupled_osc_wave_eigenvalues
./coupled_osc_wave_eigenvalues <w0>
$ ./coupled_osc_wave_eigenvalues 1 >| coupled_osc_wave_eigenvalues.dat
```

Now consider the plot of frequencies corresponding to the eigenvalues found by the `jacobi()` tool, found in Figure 6.7. The allowed frequencies are plotted as points, and also appearing as horizontal lines are the first seven ideal frequencies in the continuous wave set of Equation (6.41). Note that for given values of n, the discrete coupled oscillator frequencies gradually approach the ideal values as N, the number of spring-mass segments in our discrete approximation, increases. The higher the allowed frequency, larger numbers of discrete segments are needed to make our modeling approximation close to the ideal continuous wave value. Of course, with a spring-mass system with as numerous a number of elements as the number of air molecules and springy molecular interactions in a typical sample of air, fluid, or rigid solid, the continuous wave treatment is an excellent model. So with this numerical work, we are able to graphically visualize the transition between discrete and continuous wave transmission media.

6.6 Application: The Laplace Equation

With the tools for solving systems of linear equations, we are now in a position to make use of the estimates from Equations (2.10) and (2.11) of the first and second derivatives of a function to solve the Laplace equation within a region. This example provides us with an introduction to the important technique of numerical solutions known as finite element analysis. The Laplace equation finds common application in many areas of physics, including electrostatics, fluid dynamics, and heat flow. For example, finding the electrostatic potential V in a charge-free region where the boundaries are fixed at known potentials requires solving the Laplace equation (6.42) with Dirichlet boundary conditions.

$$\nabla^2 V = 0 \tag{6.42}$$

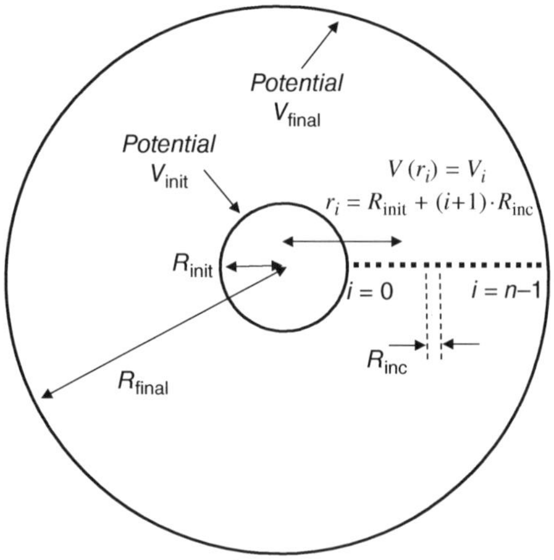

Figure 6.8 Indexing scheme for a grid of points in a cylindrical region of interest where the Laplace equation is to be solved.

To start with a simple example in one dimension, we consider a system with cylindrical symmetry, with a cross-section shown in Figure 6.8. We will assume that the potentials vary only in the radial coordinate, in which case solving the Laplace equation becomes an ordinary differential equation boundary value problem. The potential of the inner cylindrical conductor of radius R_{init} is fixed at V_{init}, and the potential of the outer conductor of radius R_{final} is fixed at V_{final}. The potential in the insulating region between the conductors will be solved at a set of n discrete radial points, indexed by the scheme in the Figure 6.8. The potential V_i is the solution of the continuous potential function sampled at a uniform grid of radii with step size r_{inc}, where $V_i = V(r_i)$, and where $r_i = R_{init} + (i+1) \cdot r_{inc}$. This implies that the unknown sampled potentials V_i are all in the interior region, and that the first and last radial points r_0 and r_{n-1} are separated from the inner and outer conductors, respectively, by the step size r_{inc}. We will see later in the C code that the required count of sample points n is calculated from the statement

```
nr = NUM_GRID_PTS(param.rgrid) - 2;
```

using the `NUM_GRID_PTS` macro from `grid.h`.

The continuous solution of the Laplace equation for this simple system is analytically solvable. In cylindrical coordinates with cylindrical symmetry

$$\frac{d^2V(r)}{dr^2} + \frac{1}{r} \cdot \frac{dV(r)}{dr} = 0 \tag{6.43}$$

We can solve this ordinary differential equation by integrating it twice, accumulating two unknown constants of integration, and we find the general solution as

$$V(r) = C_1 \ln(r) + C_2 \tag{6.44}$$

The constants of integration C_1 and C_2 can be found by setting $V(r)$ to the boundary potentials $V(r) = V_{init}$ and $V(r) = V_{final}$ at the boundary radii $r = R_{init}$ and $r = R_{final}$, and are

$$C_1 = \frac{V_{final} - V_{init}}{\ln\left(\frac{R_{final}}{R_{init}}\right)} \qquad C_2 = V_{final} - C_1 \ln(R_{init}) \tag{6.45}$$

Although we know the continuous solution to this particular problem, let's arrive at the same solution by sampled numerical means, as this will provide a detailed example of preparing a system of linear equations in a way to solve more involved boundary value problems, including those with no analytical solution. In Section 2.5, we examined the error characteristics of the estimates of first and second derivatives of a function calculated from values of the function sampled on a uniform grid of argument values. For our sampled numerical solution, we make use of the estimates of the first and second derivatives from Equations (2.10) and (2.11) to turn the continuous differential equation (6.43) into the discrete approximation

$$\frac{V(r-h) - 2V(r) + V(r+h)}{h^2} + \frac{1}{r} \cdot \frac{V(r+h) - V(r-h)}{2h} = 0 \tag{6.46}$$

where $h = r_{inc}$. Collecting terms, we put this in a form that identifies the elements that may be filled into a coefficient matrix and right-hand side vector that together define a system of linear equations governing the potentials at the interior sampling points. First define the convenient constants

$$a_i = \left(\frac{1}{h^2} - \frac{1}{2r_i h}\right) \quad b = -\left(\frac{2}{h^2}\right) \quad c_i = \left(\frac{1}{h^2} + \frac{1}{2r_i h}\right) \tag{6.47}$$

Then the set of potentials at the interior sample points for $i > 0$ and $i < n-1$ must satisfy

$$a_i V_{i-1} + bV_i + c_i V_{i+1} = 0 \tag{6.48}$$

Note that the right-hand side vector components for these equations are all zero. However, at the two ends of the sample points we must modify this relationship. At the first point with $i = 0$, there is no unknown potential V_{i-1} to appear in this linear combination, and similarly at the last point with $i = n-1$, there is no unknown potential V_{i+1} being solved for. Both of those potentials land on the boundary conductors, where the potential is a known constant. For these two relationships, we will algebraically swing the constant terms to the right side of Equation (6.48) in order to leave the left-hand side solely a linear combination of unknown system variables. Therefore, adjacent to the inner radial boundary, we fill only two coefficients into the matrix and set a nonzero component of the right-hand side vector with the equation

$$bV_0 + c_0 V_1 = -a_0 V_{\text{init}} \tag{6.49}$$

Similarly, adjacent to the outer radial boundary, we also fill in two coefficients and a nonzero component of the right-hand side vector with

$$a_{n-1} V_{n-2} + bV_{n-1} = -c_{n-1} V_{\text{final}} \tag{6.50}$$

Of course, we must have some nonzero right-hand side vector components in any linear system with a nonsingular matrix in order to have a unique solution. The system can now be written as

$$\begin{bmatrix} b & c_0 & 0 & 0 & 0 & \cdots & 0 \\ a_1 & b & c_1 & 0 & 0 & \cdots & 0 \\ 0 & a_2 & b & c_2 & 0 & \cdots & 0 \\ 0 & 0 & a_3 & b & c_3 & \cdots & 0 \\ \vdots & & & & & & \\ 0 & 0 & 0 & \cdots & a_{n-2} & b & c_{n-2} \\ 0 & 0 & 0 & \cdots & 0 & a_{n-1} & b \end{bmatrix} \begin{bmatrix} V_0 \\ V_1 \\ V_2 \\ V_3 \\ \vdots \\ V_{n-2} \\ V_{n-1} \end{bmatrix} = \begin{bmatrix} -a_0 V_{\text{init}} \\ 0 \\ 0 \\ 0 \\ \vdots \\ 0 \\ -c_{n-1} V_{\text{final}} \end{bmatrix} \tag{6.51}$$

See the source code in the program file `laplace_cyl.c` for an example program to solve this problem. The `ARRAY2D` structure to store the coefficient matrix and the two `ARRAY1D` structures to hold the right-hand side vector and the solved potential vector are all allocated according to the geometry and boundary potentials specified on the command line. An excerpt from this program shows an iterative loop that will fill in the values to represent the system of Equation (6.51).

```
zero_array2d(&a);
zero_array1d(&rhs);
b = -2.0 / (param.rgrid.inc * param.rgrid.inc);
for (i = 0; i < nr; i++) {
  r = param.rgrid.init + ((double) (i + 1)) * param.rgrid.inc;
  ai = 1.0 / (param.rgrid.inc * param.rgrid.inc) -
       0.5 / (r * param.rgrid.inc);
  ci = 1.0 / (param.rgrid.inc * param.rgrid.inc) +
       0.5 / (r * param.rgrid.inc);
  a.elem[i][i] = b;
  if (i == 0) rhs.elem[i] -= ai * param.Vinit;
  else a.elem[i][i-1] = ai;
  if (i == nr - 1) rhs.elem[i] -= ci * param.Vfinal;
  else a.elem[i][i+1] = ci;
}
```

We can examine the solution of a simple example problem by compiling and running this program with

```
$ gcc -Wall -Wshadow -O2 -o laplace_cyl laplace_cyl.c CPLibrary/lineq.c CPLibrary/step.c
    ↪ CPLibrary/array.c -lm
$ ./laplace_cyl
./laplace_cyl <rinit> <rfinal> <rinc> <V1> <V2>
$ ./laplace_cyl 1 10 0.1 10 0 >| laplace_cyl.dat
Elapsed user time = 0.000 seconds,
C1 = -4.3429448    C2 = 10
```

This example is for a geometry with an inner conductor radius of 1 unit where the potential is fixed at 10 units, and an outer conductor at a radius of 10 units with a potential of 0 units. We may take these distance and potential units as meters and volts in the SI system, if desired, but the actual units are arbitrary. The interior region is sampled with a step size of 0.1 units, providing 89 sample points between the boundaries. Also included in the program is a measurement of the processor time required by the Gaussian elimination tool `gauss_pivotmax()` by using the `times()` routine provided by the Linux operating system. Finally, the program also computes the analytical solution of Equation (6.44) for reference.

Inspection of the matrix representation of Equation (6.51) shows that by treating this linear system with the full Gaussian elimination algorithm, we are churning a great deal of unnecessary computation. Note that all but one of the array coefficients under the central diagonal are already zero, and the annihilation steps that compute how to reduce an element to zero that is already zero constitute wasted instructions in the CPU. As has been discussed in Section 6.2, the computation required for the full Gaussian elimination algorithm grows as the *cube* of the equation count. We can test this first attempt at a solution program with a large matrix by specifying a geometry with an outer radius of 100 units, still with a step size of 0.1, on the command line with

```
$ ./laplace_cyl 1 100 0.1 10 0 >| laplace_cyl.dat
Elapsed user time = 2.700 seconds,
C1 = -2.1714724    C2 = 10
```

The wasted computation effort of the full Gaussian elimination method is now evident by the long execution time of well over 2 seconds running on an example processor. Therefore, taking the opportunity to treat this coefficient matrix with the tools for a banded matrix as in Figure 3.29 can save a tremendous amount of computation load, as well as memory requirements.

To see an example of greatly improving computation efficiency, examine the example source code of the program `laplace_cyl_band.c` from the Student Companion website. In this code we use the banded matrix representation detailed in Section 3.1.4. We see that all the utility routines in `array.c` for allocating, zeroing, and freeing matrices have been changed to the variety coded for the banded matrix scheme of Figure 3.29. Since we have only one diagonal in the matrix on each side of the central diagonal with nonzero coefficients, we allocate the banded matrix with a half band width argument $w = 1$. Then the iterative code segment to fill in coefficients is changed from indexing the coefficient matrix with `[i][j]` rectangular row and column indices to use instead the diagonal indices `[p][q]` with the transformation $p = j - i + w$ and $q = i$, as detailed in Section 3.1.4.

```
zero_array2d_band(&a);
zero_array1d(&rhs);
b = -2.0 / (param.rgrid.inc * param.rgrid.inc);
for (i = 0; i < nr; i++) {
  r = param.rgrid.init + ((double) (i + 1)) * param.rgrid.inc;
  ai = 1.0 / (param.rgrid.inc * param.rgrid.inc) -
       0.5 / (r * param.rgrid.inc);
  ci = 1.0 / (param.rgrid.inc * param.rgrid.inc) +
       0.5 / (r * param.rgrid.inc);
  a.elem[1][i] = b;
  if (i == 0) rhs.elem[i] -= ai * param.Vinit;
  else a.elem[0][i] = ai;
  if (i == nr - 1) rhs.elem[i] -= ci * param.Vfinal;
  else a.elem[2][i] = ci;
}
```

Let's compile and run this new program for the example large geometry of $R_{\text{init}} = 1$, $R_{\text{final}} = 100$, and $R_{\text{inc}} = 0.1$.

```
$ gcc -Wall -Wshadow -O2 -o laplace_cyl_band laplace_cyl_band.c CPLibrary/lineq.c CPLibrary/step.c
    ↪ CPLibrary/array.c -lm
$ ./laplace_cyl_band
./laplace_cyl_band <rinit> <rfinal> <rinc> <V1> <V2>
$ ./laplace_cyl_band 1 100 0.1 10 0 >| laplace_cyl.dat
Elapsed user time = 0.000 seconds,
C1 = -2.1714724   C2 = 10
```

This dialog reports the impressive results that the `gauss_band()` function executes so rapidly that its processor time is shorter than the minimum time increment that the Linux system is able to measure. This is a powerful example of the importance of the special treatment of diagonal coefficient matrices.

Now running the banded program with the original geometry of $R_{\text{init}} = 1$ and $R_{\text{final}} = 10$, this program generates a solution that matches the analytical solution of Equation (6.44) with the coefficients C_1 and C_2 also computed by the program, as shown in Figure 6.9. The limited graphical resolution of the full plot does not allow any discrepancy between the discrete sampled solution and the analytical continuous solution to be seen, but a small level of truncation error is apparent with the zoomed-in view of the inset plot. There we see the truncation error that comes from using only a second-order estimate of the derivatives from Equations (2.10) and (2.11).

Now that we have worked through an initial example of solving the Laplace equation in one dimension, we may turn to a more interesting problem in a region spanned by two Cartesian dimensions. The Laplacian operator in Equation (6.42) then involves two partial second derivatives, and the Laplace equation becomes the partial differential equation

$$\frac{\partial^2 V(x,y)}{\partial x^2} + \frac{\partial^2 V(x,y)}{\partial y^2} = 0 \tag{6.52}$$

Applying the second-order estimate of the second derivative of Equation (2.11), and extending it to the case of partial derivatives, we arrive at a discrete approximation to the two-dimensional Laplace equation

$$\frac{V(x-h_x,y) - 2V(x,y) + V(x+h_x,y)}{h_x^2} + \frac{V(x,y-h_y) - 2V(x,y) + V(x,y+h_y)}{h_y^2} = 0 \tag{6.53}$$

where h_x is the step size in the x direction and h_y is the step size in the y direction. With the definition of three convenient constants for computation

$$c_x = \frac{1}{h_x^2} \qquad c_y = \frac{1}{h_y^2} \qquad c_{xy} = -\frac{2}{h_x^2} - \frac{2}{h_y^2} \tag{6.54}$$

Equation (6.53) then becomes

$$c_{xy}V(x,y) + c_xV(x-h_x,y) + c_xV(x+h_x,y) + c_yV(x,y-h_y) + c_yV(x,y+h_y) = 0 \tag{6.55}$$

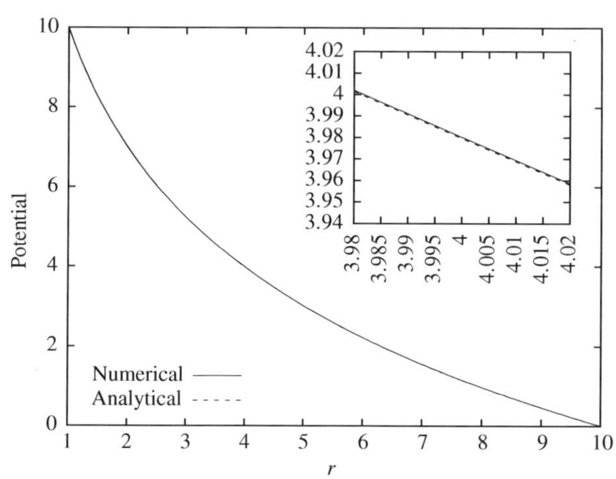

Figure 6.9 Solution of the Laplace equation showing electric potential as a function of radius in a cylindrically symmetric region.

Note that this equation is simply a linear combination of the unknown sampled potential values, and will allow us to configure a system of linear equations that may be solved for the unknown potentials. We will construct a system where the potential function $V(x, y)$ is governed by a separate equation in the form of Equation (6.55) for each coordinate (x, y) in the region of interest. This establishes a system with the same number of equations as variables, and given some nonzero boundary conditions, provides a unique solution.

We assume here that the potentials at all grid points in our region of interest are unknown, but that the region is bounded by forced and known potentials completely around its circumference. For convenience, we choose to start and end our array of grid points of unknown potentials one grid step distance separated from the boundary lines of fixed potentials. Figure 6.10 shows an indexing scheme that will be convenient to convert the partial differential equation problem into a system of linear equations to be solved. The physical region where the solution is to be found is sampled by a two-dimensional grid of points, with each point having coordinates (x_j, y_i). The grid of columns of x_j values is indexed by j, and the grid of rows of y_i values is indexed by i. The physical coordinate values at the grid points are $x_j = x_{\text{init}} + (j + 1) \cdot x_{\text{inc}}$, and similarly $y_i = y_{\text{init}} + (i + 1) \cdot y_{\text{inc}}$. By selecting the proper number of grid points in each direction n_x and n_y, we create a spacing of x_{inc} and y_{inc} between the first and last grid point and the boundary lines of fixed potentials. Similar to the cylindrical case, in the C code we will carefully compute the required count of points in the two directions with the statements

```
nx = NUM_GRID_PTS(xgrid) - 2;
ny = NUM_GRID_PTS(ygrid) - 2;
```

again using the `NUM_GRID_PTS` macro from `grid.h`. However, while the rows and columns in Figure 6.10 reflect the physical discretization of the region of interest, it is important to keep in mind that they *are not* the rows and columns of the coefficient matrix of the required system of equations. We will construct a matrix of coefficients and a right-hand side vector where each row corresponds to an equation resembling Equation (6.55), and the rows of the solution vector will represent all the unknown potentials in the two-dimensional physical array. So for indexing the rows of equations, right-hand side constants, and variables in our system, we will need a third index, k. As shown in Figure 6.10, the index k starts at 0 in the lower-left corner, increments by one as each physical row is traversed, and then grows to larger values as the region is scanned toward the top-right corner. We can always find the value of k corresponding to the given physical grid indices of (i, j) with $k = in_x + j$ as shown in Figure 6.10. Then we will construct our system coefficient matrix and right-hand side vector by letting the k index range from 0 to $n_x n_y - 1$ to address all the required rows of equations and unknown variables in our system. With this scheme, the Laplace equation approximation in Equation (6.55) becomes indexed as

$$c_{xy} V_k + c_x V_{k-1} + c_x V_{k+1} + c_y V_{k-nx} + c_y V_{k+nx} = 0 \tag{6.56}$$

Extending the one-dimensional example of cylindrical symmetry above, for the vast majority of equations in our two-dimensional system centered on a physical grid point that has all four adjacent grid points still within the region of interest, we would need simply to fill the coefficients from Equation (6.56) into the row in our coefficient matrix indexed by the

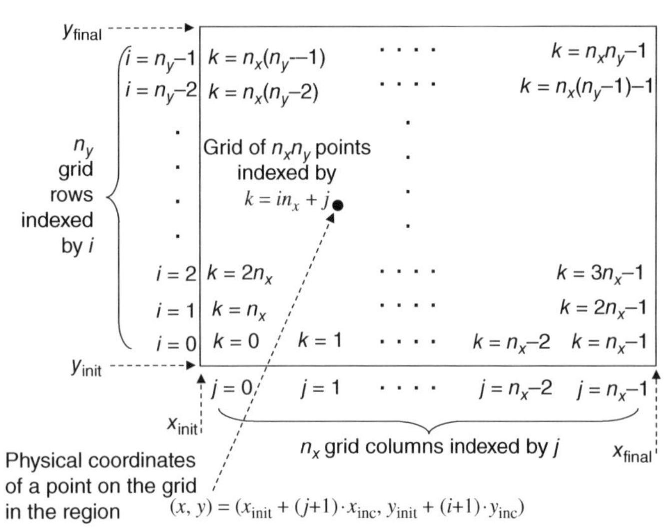

Figure 6.10 Indexing scheme for a grid of points in a rectangular region of interest where the Laplace equation is to be solved.

corresponding value of k, and set the corresponding kth element of the right-hand side vector to zero. Since the constant c_{xy} is generally the largest magnitude of the three constants in Equation (6.54), and c_{xy} will end up on the diagonal of the coefficient matrix, these matrices generally will turn out to be well conditioned for limited roundoff error. However, just as in the cylindrical case, we will need to be more careful with the equations corresponding to physical grid points at the edges of the region. For those combinations of the i and j indices, we will substitute a known potential from the boundary condition for one (or two, for corner locations) of the terms of Equation (6.56) whose index would otherwise land it outside the region. Since this known boundary potential is a constant, we then algebraically move the corresponding term to the right side of the equation and enter it as a nonzero component k of the right-hand side vector. For example, in the case of a grid point with $j = 0$ along the left edge of the region in Figure 6.10, the linear equation for the corresponding kth row becomes

$$c_{xy}V_k + c_xV_{k+1} + c_yV_{k-nx} + c_yV_{k+nx} = -c_xV_{\mathrm{left}}(y_i) \tag{6.57}$$

with a nonzero right-hand side, where $V_{\mathrm{left}}(y_i)$ is the value of the function that describes the fixed boundary potential outside the left edge of the region evaluated at a sampled y coordinate of y_i. For grid points along the right, top, and bottom edges, we make a similar adjustment to its corresponding linear equations.

Equation (6.57) describes coefficients for a banded system, although the bands of nonzero coefficients around the center diagonal are, in general, wide and sparse. For a matrix row describing the neighboring relationships around the potential indexed by k, we will have a large (in magnitude) coefficient c_{xy} on the diagonal at the matrix row and column indices of (k, k), and then we can have smaller coefficients c_x both one column to the left and right, and coefficients c_y both located n_x columns to the left and right. The total number of equations in our linear system is the product of the counts $n_x \cdot n_y$. With only a modest number of sampled grid points along each of the two dimensions of the solution space n_x and n_y, this total can grow to a large number. So learning from the tremendous improvement in efficiency we enjoyed by applying the banded matrix tools in the one-dimensional cylindrical case, we also use the banded matrix routines for solving this two-dimensional problem, but now we allocate the banded coefficient matrix with a half band width of n_x.

However, it should be mentioned here that the numerical methods presented in this section are only appropriate for a modest sampling density. If a higher resolution solution is required, which would typically be advisable since the fundamental derivative estimate we are using is only of the second order, and especially if the geometry is extended to include a third dimension, other more efficient algorithms for solving sparse linear systems must be used. There are iterative algorithms that are more widely used for this type of application in practice, but these are beyond the scope of this text.

Let's consider a particular physical example to solve. Let us solve a problem where a rectangular region of interest ranges inside of ± 5 units in the x direction and ± 3 units in the y direction. Along the boundaries aligned in the y direction, at $x = \pm 5$, we set the potential to 0. Along the top boundary aligned in the x direction, at $y = +3$, we arrange for the potential to follow the quadratic contour between 0 and 1 volt as

$$V(x) = \frac{(x-c)^2}{a^2} \tag{6.58}$$

where c is set to the center of the x range and a is half the width of the x range. The potential along the bottom boundary at $y = -3$ is similar but of the opposite polarity. For a main program solving this example see the source code file `laplace_rect.c` available for download from the Student Companion website. As in the case of the `laplace_cyl.c` program, the numerical analysis of the program does not need to assume any particular system of units. Compiling and running the example

```
$ gcc -Wall -Wshadow -O2 -o laplace_rect laplace_rect.c CPLibrary/lineq.c CPLibrary/array.c -lm
$ ./laplace_rect
./laplace_rect <xinit> <xfinal> <xinc> <yinit> <yfinal> <yinc>
$ ./laplace_rect -5 5 0.2 -3 3 0.2 >| laplace_rect.dat
```

Plotting with `gnuplot` generates the plot in Figure 6.11, with arbitrary units.

```
gnuplot> set hidden3d
gnuplot> set xlabel 'x'
gnuplot> set ylabel 'y'
gnuplot> set zlabel 'Potential' rotate by 90
gnuplot> set ticslevel 0.25
gnuplot> splot 'laplace_rect.dat' w l
```

Once we have found the sampled values of the electrostatic potential from the Laplace equation, we know from the definition of potential that the electric field vector may be found from the negative gradient of the potential

$$\vec{E} = -\nabla V \tag{6.59}$$

Since we have computed samples of the potential on a uniform grid, we extend the double-sided estimate of the first derivative from Equation (2.10) to the case of partial derivatives and find the x and y components of the $\vec{E}$ field vector in Cartesian coordinates with

$$\begin{aligned} E_x(x, y) &= \frac{V(x + h_x, y) - V(x - h_x, y)}{2h_x} \\ E_y(x, y) &= \frac{V(x, y + h_y) - V(x, y - h_y)}{2h_y} \end{aligned} \tag{6.60}$$

Now we revisit the graphical technique first developed in Section 4.11 for plotting vectors that follow the direction of the local $\vec{E}$ field vector, with the magnitude encoded in the vector length, in order to visualize the contours of the field. In Figure 6.12, we may see the field lines originate on the positive charges that must be present along the top edge to establish its positive potential, and stream toward the negative charges beyond the bottom edge.

The coding for generating the graphics of the $\vec{E}$ field vectors appears as Exercise 6.3 for the student. This is most easily accomplished by extending the downloaded source code file `laplace_rect.c` with the graphical techniques from Section 4.11.

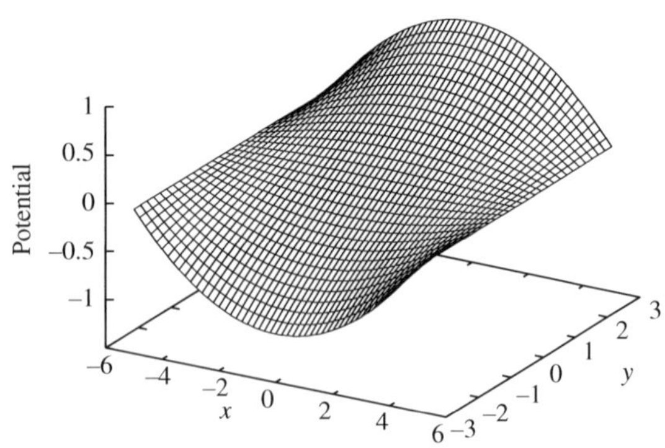

Figure 6.11 Solution of the Laplace equation showing electric potential over the two-dimensional electrostatic region.

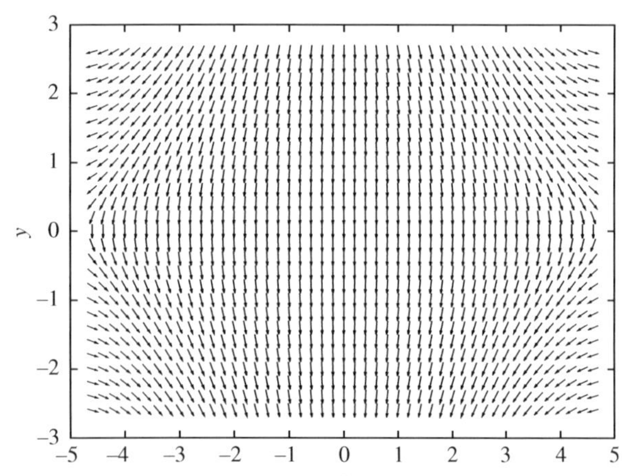

Figure 6.12 Electric field vector directions in the two-dimensional electrostatic region.

6.7 Application: Kirchoff's Current Law

In Section 5.4, we explored an electrical network with a topology of a single loop of electric current, and solved for the current flowing around the loop with a numerical solution of Kirchoff's Voltage Law (KVL). In this section, we examine the case of electrical networks with topologies forming multiple current loops and interconnection nodes. We may solve for the currents flowing in various loops with multiple simultaneous KVL equations, but for completeness we look here at a solution method applying the corresponding law of lumped electrical networks, Kirchoff's Current Law, or KCL.

See the diagram of a portion of a general electrical network in Figure 6.13. For a network of electrical components interconnected at the network nodes, the electric potential at node i is V_i relative to a reference node called "ground" located at the bottom right in this diagram. We will make the assumption that the electrical reference potential of zero at the ground node is achieved by an ideally infinite reservoir of electrical charge. KCL, an expression of the conservation of charge, asserts that the sum of all currents flowing into each network node is 0. The jth element has current I_j flowing through it and V_j potential drop across it, where V_j is the difference between two node potentials, or between a source voltage and a node voltage. If the relationships between I_j and V_j for each circuit element are linear, that is, the elements are linear resistors that obey Ohm's law, then the node voltages V_i can be found by solving a system of linear equations.

As an example practical electrical network to analyze, consider the circuit shown in Figure 6.14. This network is known as the "Wheatstone bridge," and is commonly used in experimental measurements. The potential at the top node is fixed at some reference level, in this case 10 V, above the ground potential by an ideal voltage source. Resistances R_1 through R_4 are arranged in ring, with resistance R_5 "bridging" two intermediate nodes of the ring. Typically, three of the four elements R_1 through R_4 are fixed resistor components, and the fourth is a measurement sensor of some sort whose electrical resistance varies with respect to some physical property, such as temperature, displacement, force, or pressure. Resistance R_5 is the effective resistance of an instrument recording the measurements. The bridge network is devised so that the electric current I_5 through the instrument resistance R_5 will be zero when the bridge is "balanced," that is, when the potentials V_1 and V_2 at the two intermediate nodes are equal. We can show numerically that this balanced condition is achieved when the ratios $\frac{R_1}{R_2}$ and $\frac{R_3}{R_4}$ are equal. Typically this balance is designed to correspond to some nominal physical condition of the sensor resistance. When the sensor resistance changes with physical measurements, the current I_5 will differ from zero and can be sensed and recorded. The potentials V_1 and V_2 are determined by the relationship between all the resistances, and we may solve for them using KCL and the `lineq.c` tool. Once these potentials have been found, the current I_5 is determined by Ohm's law across R_5. Setting up the KCL equations, asserting that the sum of currents flowing into the two nodes with unknown potentials is zero, we write

$$
\begin{aligned}
\frac{10 - V_1}{R_1} + \frac{0 - V_1}{R_2} + \frac{V_2 - V_1}{R_5} &= 0 \\
\frac{10 - V_2}{R_3} + \frac{0 - V_2}{R_4} + \frac{V_1 - V_2}{R_5} &= 0
\end{aligned}
\tag{6.61}
$$

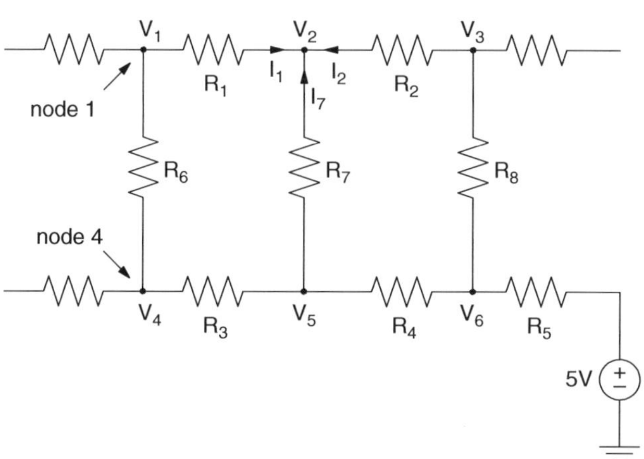

Figure 6.13 Kirchoff's Current Law with a resistive network.

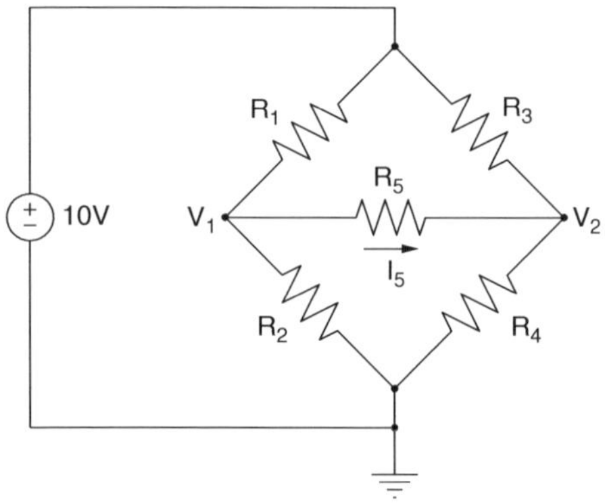

Figure 6.14 Schematic diagram of the Wheatstone bridge electrical network.

We may then arrange these relationships into a system of two linear equations in the two unknown potentials V_1 and V_2 in the form of $\mathbf{A}\vec{v} = \vec{b}$, where the elements of the coefficient matrix $\mathbf{A}$ and the right-hand side vector $\vec{b}$ are

$$\begin{aligned} a_{0,0} &= \frac{1}{R_1} + \frac{1}{R_2} + \frac{1}{R_5} \\ a_{0,1} &= -\frac{1}{R_5} \\ a_{1,0} &= -\frac{1}{R_5} \\ a_{1,1} &= \frac{1}{R_3} + \frac{1}{R_4} + \frac{1}{R_5} \\ b_0 &= \frac{10\ \text{V}}{R_1} \\ b_1 &= \frac{10\ \text{V}}{R_3} \end{aligned} \tag{6.62}$$

An example program `wheatstone.c` to solve this network is available for download from the Student Companion website. The program accepts the values of the resistances R_1 through R_5 from its command line, assembles the `ARRAY2D` and `ARRAY1D` structure element values according to Equation (6.62), and calls the `gauss_pivotmax()` tool. The solution of the potentials V_1 and V_2 will be found as the $\vec{v}$ vector, and the measurement current I_5 is then calculated and reported to the output stream. An example command shell dialog compiling and running this program is

```
$ gcc -Wall -Wshadow -O2 -o wheatstone wheatstone.c CPLibrary/lineq.c CPLibrary/array.c -lm
$ ./wheatstone
./wheatstone <R1> <R2> <R3> <R4> <R5>
$ ./wheatstone 1e3 1e3 10e3 10e3 1e6
Current I5 = 0
$ ./wheatstone 1e3 1e3 20e3 10e3 1e6
Current I5 = 1.6548072e-06
$ ./wheatstone 1e3 1e3 20e3 20e3 1e6
Current I5 = 0
```

Let us consider the configuration where resistances R_1 and R_2 are equal at 1000 ohms each. The R_3 resistance represents some physical sensor, with a nominal resistance of 10000 ohms. A reference resistance R_4 is set to this same value to balance the bridge for the nominal condition with equal resistance ratios. The instrument resistance R_5 is known to be 10^6 ohms. In the first run of the program, we can see a measurement current of zero reported. In the second run the sensor resistance has changed to 20000 ohms in response to some environmental change to be measured, and the instrument current is now nonzero. In the third run, the reference resistance R_4 has been adjusted to bring the bridge back into balance for this new condition.

Now let us consider the more elaborate resistive network shown in Figure 6.15. R_0 is some nonzero reference resistance, and the physical resistances in the network have a value of either R_0 or $2R_0$, in the pattern shown. With this prescribed arrangement of resistance values, this network is capable of performing a function known as a digital-to-analog conversion, in this case converting a 6-bit binary digital number into a corresponding analog voltage. This function is a necessary

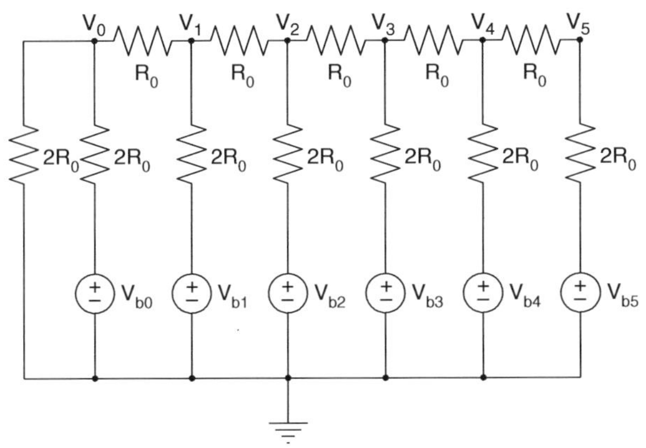

Figure 6.15 A simple 6-bit digital-to-analog converter.

component in, for example, systems that play back multimedia streams that have been digitized into an analog form suitable for human viewing or hearing. The six digital input bits are assumed to be represented by the state of the six voltage sources in the network, V_{b0} through V_{b5}. We assume the state of V_{b5} will be a 5-V source if the most significant bit in the binary number, or the bit with a significance of $2^5 = 32$, is a 1, and a 0-V source if the bit is 0. This same mapping of binary input bits to source voltage ranges down to V_{b0}, which is assumed to be 5 or 0 V for the two cases of the least significant bit, or the bit with a significance of $2^0 = 1$, is a 1 or a 0, respectively. The electric potential at node 5, or V_5, is considered to be the converted output voltage. The "ground" symbol in the diagram represents the network connection assumed to be at a reference potential of zero. The general R_0 and $2R_0$ pattern of this network, known as an "R-2R ladder," may be extended to larger counts of digital bits, until limited by practical considerations of component matching.

For this general network topology to function as a linear *n*-bit digital-to-analog conversion with "1" digital bits represented by 5-V sources and "0" digital bits represented by 0-Vsources, we would expect the potential at node *n* to follow the relationship

$$V_n = \left(\frac{D_{in}}{2^n}\right) \cdot 5\ \text{V} \tag{6.63}$$

where D_{in} is the numerical value of the digital binary input bits, ranging from 0 to $2^n - 1$. We may write equations forming the sum of all the currents flowing into each network node by using Ohm's law for each of the network branches consisting of a linear resistor. For the *j*th network branch, which is assumed to connect between nodes *i* and *k*, we may write

$$I_j = \frac{(V_k - V_i)}{R_j} \tag{6.64}$$

where I_j is the current flowing into node *i* through the branch resistor *j*; R_j is the resistance of the branch resistor *j*; V_i is the electric potential at node *i*; and V_k is the electric potential at node *k* at the other end of branch *j*. The node voltages may be among the set of unknown potentials we are solving for, or may be fixed by a physical voltage source. For each node *i*, we must form a KCL equation that asserts that the sum of all the I_j currents flowing into the node according to the network topology is zero. We may solve for all the unknown node voltages V_i by assembling a system of all these linear equations and finding a unique solution for all V_i.

Since the interconnection pattern of this topology is limited to adjacent nodes in a relatively sparse network, the coefficient matrix formed for this solution will be in a limited banded structure. So this problem provides a good motivation for using the `gauss_band()` routine in the `lineq.c` tool, with the coefficients stored in the banded matrix by diagonals with the `ARRAY2D_BAND` structure. This analysis is available as Exercise 6.4 for the student.

6.8 Application: Determinate Structures

As we look at the designed things in the world around us, in order to understand how buildings retain structural stability during storms, or how bridges carry loads over rivers, or how airplane wings remain rigid, we may start with a study of static forces. According to Newton's laws of motion, in static situations where there is no motion, the forces and torques must balance to zero. In order that structures maximize space for their use and minimize their weight and cost, we build structures out of interconnected members of material that may support forces in tension and compression. If a structure has just enough members at their nominal lengths so that the system of member tensions and supporting forces constitute a system of linear equations with a unique solution, we call such a structure *determinate*.

In this section, we study a ubiquitous fundamental determinate substructure known as a *truss*. One of the most admired structures in the world, with its stability dependent on trusses and truss-like designs, is shown in Figure 6.16.

We will seek to compose a system of linear equations that we will solve with the numerical tools in this chapter. The solution of this system will describe a consistent set of tensions in the structure members and reaction forces at supporting points that all together simultaneously enforce a balanced condition of forces at pivot points. For a determinate structure, the total number of member tensions and supporting forces will equal the total number of equations describing the net force in all dimensions of the force vectors. In order to ensure that the overall structure will not fail, we must solve this sort of system to test whether or not the tension (or compression) in any particular member is within the allowable strength of the material used to build it. To simplify our task, we will consider only trusses in two dimensions, and the need to balance only the x and y components of the forces at the pins that interconnect members.

Refer to Figure 6.17 for an indexing scheme we can use to systematically construct our system of linear equations. We may imagine this diagram being a schematic representation of a bridge, and may categorize this class of truss, called the Warren Truss, by an integer n, which is the number of horizontal members arranged along the top row. The angle that all the diagonal elements form with the horizontal is a design parameter θ. We assume that there will be n loading weights applied in the $-y$ direction on the internal pins that model the load on the structure. We also assume that the truss is supported by the equivalent of a hinge on the bottom-left end that will hold that end stationary by supplying a two-dimensional reaction force with both x and y components. At the bottom-right end, we assume the support is the equivalent of a roller bearing that is capable of supplying a vertical reaction force in the y direction only, but will allow slippage in the x direction to account for some small material expansion or contraction. We must have n of at least 1, but for extensive structures we would like to be able to analyze trusses with a large number of members, and, therefore, need a system to programmatically generate a system of linear equations of any arbitrary order. The variables in this system that we will solve for will be the set of unknown tensions in all the members, with a negative tension signifying compression, and the unknown supporting reaction forces. The coefficients in the linear equations will be determined by the geometry of the interconnections at the joining pins, and describe the proportions of the linear combination of member tensions that make up both the x and y components of the force acting on a particular pin. These forces must either sum to zero for the top row and corner pins, or sum to the load weights for the pins along the bottom row. These load weights will constitute the right-hand sides in the system of equations.

(a) View of the entire structure **(b)** Trusses in the tower design

Figure 6.16 Eiffel Tower, Paris, France.

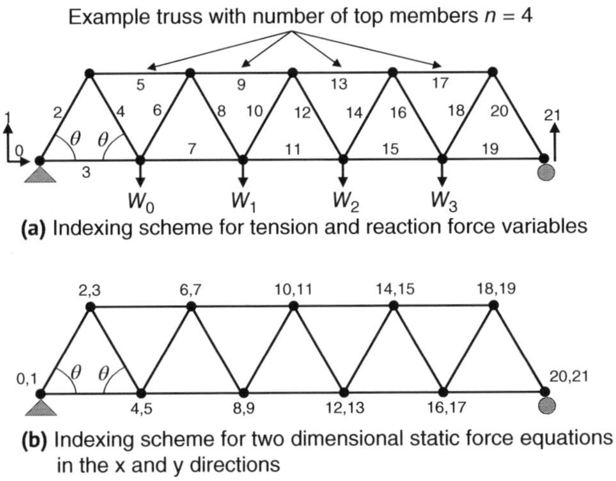

Figure 6.17 Indexing schemes for an example truss with $n = 4$.

In order to create an algorithm for generating an arbitrary number of linear equations for a truss of arbitrary length, a convenient choice of a scheme for indexing the variables and equations characterizing the structure is shown in Figure 6.17 for an example case of a truss with $n = 4$. For any truss there will be, in general, $4n + 6$ equations and variables, and in this particular case of $n = 4$ we will have a system with 22 equations and variables, indexed 0 through 21. In Figure 6.17a, variables 0 and 1 are the x and y components, respectively, of the reaction force at the hinge on the left side. Variable 21 is the y-directed supporting force of the roller on the right side. Then variables 2 through 20 refer to the tensions in all the truss members in a meandering path through the structure from left to right as shown. When we submit our system to the `gauss_pivotmax()` tool for solution, these index numbers will refer to the element in the solution `ARRAY1D` structure that will hold the tension or reaction force solution. In Figure 6.17b, the equations that enforce the static condition at all the pins are indexed. These range from equations 0 and 1 for the x and y components at the pin at the left end to equations 20 and 21 for the x and y components at the pin at the right end. These indices will locate the row in the coefficient matrix where we will assemble the coefficients for each equation in our linear system.

Of course, Figure 6.17 shows only one particular case of a truss with $n = 4$. In the general case, the last tension variable, in the diagonal member on the right side, will be with index $4n + 4$, and the right side supporting the reaction force will be with index $4n + 5$. Similarly, the equations for the last static condition at the pin on the right side will be with equation indices $4n + 4$ and $4n + 5$.

How do we design a general algorithm that will fill in the values of the coefficient matrix and right-hand side vector for our linear system of arbitrary order? The utility of the indexing scheme of Figure 6.17 will become clear if we now consider using a primary index k as a parameter for this algorithmic generation task. For an arbitrary truss, we will let k range iteratively over $2n + 3$ values from 0 to $2n + 2$, inclusive, with an increment of 1, and in each step create the linear equations indexed as $2k$ and $2k + 1$. For $k = 0$, we fill in the coefficient values for the x and y components at the left end supported pin. For $k = 2n + 2$, we construct the two component equations at the right end supported pin. For the intermediate values of k, for k odd we create coefficients in the two component directions for the top row pins, and for k even we create coefficients and right-hand side values for the loaded bottom row pins.

With this primary index k defined, Figure 6.18 shows how we may construct the force balancing equations in two dimensions for each value of k. The tensions in member i are T_i; the loading weight i is W_i; the left-hand supporting reaction force has components F_{lx} and F_{ly}; and the vertical right-hand supporting reaction force is F_{ry}. We can see that with the indexing scheme of Figure 6.17, algorithmically filling in the elements of the coefficient array and the right-hand side vector is possible by following this consistent pattern of indices over the iterative values of k.

The coding of this tool program is left as Exercise 6.5 for the student. The program should accept the value of n and the angle θ on its command line, and then a sequence of an appropriate number of loading weights for the load values W_i in Figure 6.17. A dialog to compile and run this program for three example cases is

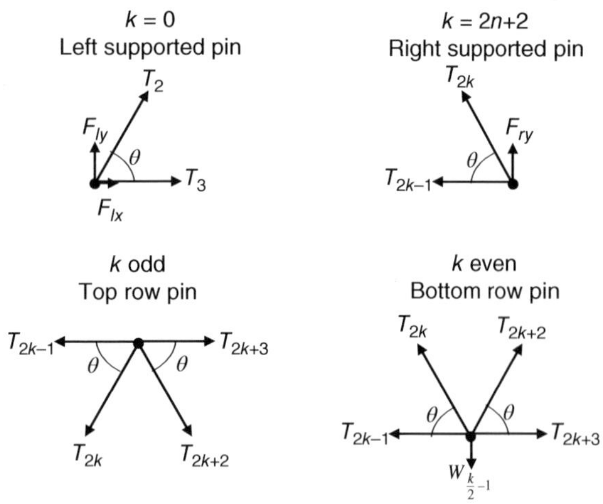

Figure 6.18 Static force balancing diagrams for any.

```
$ gcc -Wall -Wshadow -O2 -o truss truss.c CPLibrary/lineq.c CPLibrary/array.c -lm
$ ./truss
./truss <n> <theta> <w0> <w1> ... <wn-2> <wn-1>
$ ./truss 1 60 100
Left hand hinge reaction force: 0,50
Right hand roller reaction force: 0,50
Member  2 tension:   -57.735027
Member  3 tension:    28.867513
Member  4 tension:    57.735027
Member  5 tension:   -57.735027
Member  6 tension:    57.735027
Member  7 tension:    28.867513
Member  8 tension:   -57.735027
$ ./truss 3 60 0 50 0
Left hand hinge reaction force: 7.1054274e-15,25
Right hand roller reaction force: 0,25
Member  2 tension:   -28.867513
Member  3 tension:    14.433757
Member  4 tension:    28.867513
Member  5 tension:   -28.867513
Member  6 tension:   -28.867513
Member  7 tension:     43.30127
Member  8 tension:    28.867513
Member  9 tension:   -57.735027
Member 10 tension:    28.867513
Member 11 tension:     43.30127
Member 12 tension:   -28.867513
Member 13 tension:   -28.867513
Member 14 tension:    28.867513
Member 15 tension:    14.433757
Member 16 tension:   -28.867513
$ ./truss 5 60 0 0 50 0 0
Left hand hinge reaction force: -1.7763568e-15,25
Right hand roller reaction force: 0,25
Member  2 tension:   -28.867513
Member  3 tension:    14.433757
Member  4 tension:    28.867513
Member  5 tension:   -28.867513
Member  6 tension:   -28.867513
Member  7 tension:     43.30127
Member  8 tension:    28.867513
Member  9 tension:   -57.735027
Member 10 tension:   -28.867513
```

```
Member 11 tension:    72.168784
Member 12 tension:    28.867513
Member 13 tension:    -86.60254
Member 14 tension:    28.867513
Member 15 tension:    72.168784
Member 16 tension:   -28.867513
Member 17 tension:   -57.735027
Member 18 tension:    28.867513
Member 19 tension:     43.30127
Member 20 tension:   -28.867513
Member 21 tension:   -28.867513
Member 22 tension:    28.867513
Member 23 tension:    14.433757
Member 24 tension:   -28.867513
```

In the first example run, a simple truss with only one top member and structure angle θ of 60° is loaded at its one bottom row pin with a weight of 100 in some arbitrary system of units. In the second example, we extend this truss to three top row members, and with a load of 50 located on the center bottom row pin. The third example extends this truss to $n = 5$, again, with a load of 50 at the center pin. The maximum tension or compression we see in any member is approximately 72.2 units, on the same order as the loading weights, even as the span of this "bridge" covers longer and longer distances and the truss structure distributes this load over the span. Note that the only locations where a vertical load is able to ultimately be supported are on the hinge support at the left end and on the roller support at the right end. In all three of these example cases, we note that even though the loads are distributed as tensions and compressions throughout the structure, the sum of the vertical components of the loads at these two supports is the sum of the loads placed on the bottom row pins, as required for a static solution.

6.9 Coding: Animated Modes of Coupled Oscillators

Let us return to the analysis of coupled spring-mass oscillators begun in Section 6.5. In that section, we referred to the static plots in Figures 14.6–14.8 describing the three normal modes of oscillation for a system with three masses. Generally, we can imagine the motion in different modes from static plots for just a few masses, but for more complex systems, an animated representation can be very effective. In this section, we will use the methods of generating animated graphics that were detailed in Section 4.13 applied to spring-mass systems of any order.

First work through the source code in the file `coupled_oscN_eigenvalues_step.c` from the Student Companion website. The first two command line arguments that the main program accepts are the number of masses in the system and the real reference frequency in hertz, $\frac{\omega_0}{2\pi}$. The final two arguments specify a grid of timestep values. For any number of masses, the program constructs an appropriate coefficient matrix and calls the `jacobi()` function in the `lineq.c` tool to find all the eigenvalues and corresponding eigenvectors for the normal oscillation modes. Then the program uses the `step()` tool to call the `mass()` function to model the motion of each of the masses in each of the modes. The stepped displacement samples are stored in an aggregation of `ARRAY1D` tables, as allocated by the methods shown in Section 3.1.4. Then a `gen_output()` function is called to write an output stream listing displacement time samples for all the modes. The motion of each mode is described in an output file data set, and the lines in each data set have data samples arranged in columns for each of the mass displacements. With this tool, static plots resembling Figures 14.6–14.8 could be prepared for systems of any order.

Consider next the modified program in the file `coupled_oscN_eigenvalues_step_gpanim.c` from the Student Companion website. We follow the method in Section 4.13 of replacing the `gen_output()` function with a `gen_output_gpanim()` function that writes lines of `gnuplot` plotting commands to the standard output instead of raw data lines. The plotting commands are designed to animate the motion of all the masses in all the modes into a sequence of frame image files. For each mode, a horizontal line of dots represents the instantaneous position of each of the masses, and the different modes are represented in a vertical arrangement of lines. See, for an example system with nine masses, the animated GIF file `coupled_oscN_eigenvalues_step_gpanim.gif` in the Student Companion website. This file was rendered with the following compilation and execution sequence, again using a Linux pipe to channel the programmatically generated plotting commands into a `gnuplot` process.

```
$ gcc -Wall -Wshadow -O2 -o coupled_oscN_eigenvalues_step_gpanim
   ↪ coupled_oscN_eigenvalues_step_gpanim.c CPLibrary/step.c CPLibrary/lineq.c CPLibrary/array.c
   ↪ -lm
$ ./coupled_oscN_eigenvalues_step_gpanim
./coupled_oscN_eigenvalues_step_gpanim <N masses> <freq> <tfinal> <tinc>
$ ./coupled_oscN_eigenvalues_step_gpanim 9 1 10 0.02 | gnuplot
$ ffmpeg -f image2 -i frame%05d.png -vf 'split[v],palettegen,[v]paletteuse' -r 30 -y
   ↪ coupled_oscN_eigenvalues_step_gpanim.gif
$ rm -f frame*
```

Recall that the `jacobi()` function reports eigenvalues along the diagonal of the transformed coefficient matrix, but in no particular order. So the mode number labels along the vertical axis of this animated plot identify the unique modes, but are not in any particular order of frequency. The motion across the oscillator with some masses remaining stationary clearly illustrates the standing waves resonating at different frequencies across the structure.

6.10 Exercises

6.1 Historically a method to model the variation of refractive index with wavelength for a glass in a prism has been used, modeling the index n as a function of the wavelength λ by the Cauchy's equation:

$$n = c_0 + \frac{c_1}{\lambda^2} + \frac{c_2}{\lambda^4} \tag{6.65}$$

where c_0, c_1, and c_2 are constants selected to fit the function to the measured points. Three measured indices of refraction for a certain type of glass are in the following table:

Table 6.1 Index of refraction versus wavelength for a certain glass.

Wavelength (nm)	Index of refraction
400	1.5
550	1.43
700	1.4

These three measured data points provide three equations in the form of the Cauchy equation that relate n to λ. Note that the resulting equations are linear with respect to the unknown c_i coefficients. Write a C program `cauchy.c` calling the `gauss_pivotmax()` routine that solves for the c_i coefficients using the measured wavelengths in Table 6.1. Accept on the command line of your program a different test wavelength λ_{test}. The program should also include an evaluation of the model of Equation (6.65) at the test wavelength with the coefficients solved for with the linear system. Run the program specifying $\lambda_{test} = 600$ nm to find the corresponding index of refraction.

6.2 Using the programs `coupled_osc3_step.c` and `coupled_oscN_eigenvalues_step.c` from Section 6.5 as starting examples, write a program `coupled_oscN_step.c` that uses the `jacobi()` tool to solve for the eigenvalues and eigenvectors of a system of coupled spring-mass oscillators with N identical masses and identical springs. As in the `coupled_osc3_step.c` program, the new program must resolve some particular mixed mode initial conditions specified on the command line into components of the normal modes with a solution of linear equations using the `gauss()` tool. After a command line argument specifying the number of masses, additional arguments specifying all initial displacements must follow.

(a) Test your program by simulating the case of three masses, with the initial conditions of Figure 14.9, and compare with the results in the figure.

(b) Run a simulation for the case of nine masses with the initial conditions 1, 0, 1, 0, 1, 0, 1, 0, 1, and plot the results. These results will be referred to in Exercise 14.1.

6.3 Extend the source code file `laplace_rect.c` described in Section 6.6 that solves the Laplace equation for the electrostatic potential in a rectangular area to write a new program `laplace_rect_E.c`. The program must generate the data to plot vectors following the direction of the electric field and the magnitude encoded as lengths as described in Section 4.11. Use the same boundary conditions as in Section 6.6.

6.4 Write a program `dac.c` to solve for the node potentials in the resistive network for a general n-bit R-2R ladder digital-to-analog converter (DAC) described in Section 6.7. The program should accept the number of bits as the first command line argument, followed by the voltages that correspond to the sequence of digital bit inputs. Demonstrate that for converting a 6-bit digital code of `100101` with the digital input potential of 5 V corresponding to each `1` bit, the output voltage follows the ideal relationship of Equation (6.63).

6.5 Write the code for the `truss.c` program to solve for the member tensions in a truss described in Section 6.8. The program should accept the value of n and the angle θ on its command line, and then a sequence of an appropriate number of loading weights for the load values W_i in Figure 6.17. Compile the program as shown in Section 6.8 and run it, for the example, case of 5 top row members with a 60° structure angle. Place a load of 20 force units at each of the 3 leftmost bottom row pin joints, but no load at the 2 bottom row pins on the right side. Report the highest level of tension or compression experienced by a member. Verify that the vertical support loads at the two ends sum to the expected value.

7

Systems of Nonlinear Equations

In Chapter 5, we explored algorithms to converge on roots of nonlinear functions, and in Chapter 6, we studied the linear algebraic methods for solving systems of multiple linear equations. With this foundation, we are now in a position to consider the more challenging case of systems of nonlinear equations. As with single nonlinear functions, we start with an algorithm attributed to Newton and Raphson, a method that requires a starting guess for the vector root, and use a slope-based method to hopefully iteratively converge on the desired root. The development of this method will also make use of linear algebraic concepts to extend the dimensions of the functions and slopes guiding the convergence.

To illustrate the principles involved, consider a system of two simultaneous nonlinear equations in two unknown argument values, as in Equation (7.1). Figure 7.1 graphically illustrates an example problem where there may be multiple roots of the system to converge on. Each of the two contours in the two-dimensional argument plane represent the contours over which each of the functions in the system independently go to zero. Our task in converging on the roots of this system is to locate any points in the argument space where these contours intersect, if such points even exist. If there are multiple roots to locate, we will require some knowledge of the function space so that the solver algorithm may be guided to the desired roots with different initial guesses in different vicinities.

$$\begin{aligned} f(x, y) &= 0 \\ g(x, y) &= 0 \end{aligned} \tag{7.1}$$

General systems with n nonlinear functions in n unknowns would conceptually involve the same task extended to an n-dimensional space.

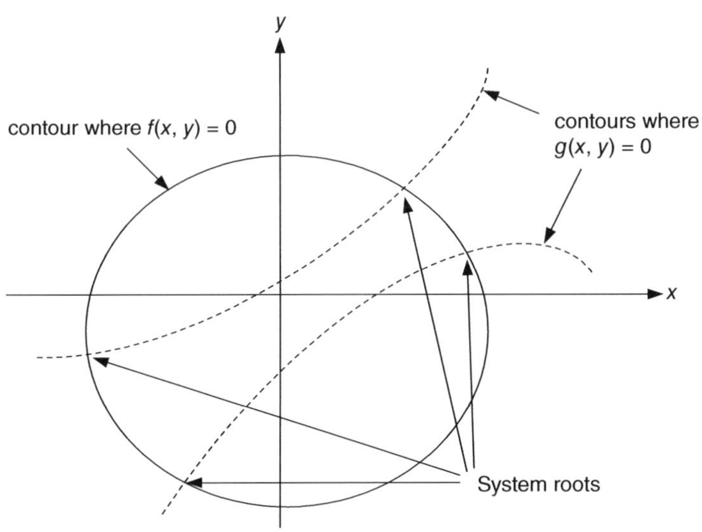

Figure 7.1 Example problem of finding multiple roots of a system of functions.

Computational Physics Using C: Efficient Programming with Ease, First Edition. John W. Fattaruso. © 2026 John Wiley & Sons, Inc.
All rights reserved, including rights for text and data mining and training of artificial intelligence technologies or similar technologies.
Published 2026 by John Wiley & Sons, Inc. Companion Website: http://www.wiley.com/go/CompPhysUsingC

7.1 Algorithms: Multidimensional Newton–Raphson Method

Let's review the one-dimensional method of Chapter 5. Equations (5.1) and (5.2) are a simple derivation of Newton's method of converging on a root in one dimension, that is, a root of a function of a single argument variable. This derivation may then be readily extended to multiple dimensions, for a system of multiple functions of multiple variables. Customarily, the subscript notation for a system of n equations in n unknowns range from 1 to n, but to be consistent with the array subscripts of the C programming language, we will write subscripts ranging from 0 to $n-1$. First we define

$$\vec{x} = \begin{bmatrix} x_0 \\ x_1 \\ x_2 \\ \vdots \\ x_{n-2} \\ x_{n-1} \end{bmatrix} \tag{7.2}$$

as a column vector in the n-dimensional function argument space of the n functions in the system, and another column vector

$$\vec{f}(\vec{x}) = \begin{bmatrix} f_0(\vec{x}) \\ f_1(\vec{x}) \\ f_2(\vec{x}) \\ \vdots \\ f_{n-2}(\vec{x}) \\ f_{n-1}(\vec{x}) \end{bmatrix} \tag{7.3}$$

of the n function values evaluated at that argument vector. Now define the Jacobian matrix of first partial derivatives as

$$\mathbf{J}(\vec{x}) = \begin{bmatrix} \frac{\partial f_0(\vec{x})}{\partial x_0} & \frac{\partial f_0(\vec{x})}{\partial x_1} & \cdots & \frac{\partial f_0(\vec{x})}{\partial x_{n-2}} & \frac{\partial f_0(\vec{x})}{\partial x_{n-1}} \\ \frac{\partial f_1(\vec{x})}{\partial x_0} & \frac{\partial f_1(\vec{x})}{\partial x_1} & \cdots & \frac{\partial f_1(\vec{x})}{\partial x_{n-2}} & \frac{\partial f_1(\vec{x})}{\partial x_{n-1}} \\ \frac{\partial f_2(\vec{x})}{\partial x_0} & \frac{\partial f_2(\vec{x})}{\partial x_1} & \cdots & \frac{\partial f_2(\vec{x})}{\partial x_{n-2}} & \frac{\partial f_2(\vec{x})}{\partial x_{n-1}} \\ & & \vdots & & \\ \frac{\partial f_{n-2}(\vec{x})}{\partial x_0} & \frac{\partial f_{n-2}(\vec{x})}{\partial x_1} & \cdots & \frac{\partial f_{n-2}(\vec{x})}{\partial x_{n-2}} & \frac{\partial f_{n-2}(\vec{x})}{\partial x_{n-1}} \\ \frac{\partial f_{n-1}(\vec{x})}{\partial x_0} & \frac{\partial f_{n-1}(\vec{x})}{\partial x_1} & \cdots & \frac{\partial f_{n-1}(\vec{x})}{\partial x_{n-2}} & \frac{\partial f_{n-1}(\vec{x})}{\partial x_{n-1}} \end{bmatrix} \tag{7.4}$$

From the definition of partial derivatives, a first-order truncated multidimensional Taylor series approximation, a multidimensional extension of the single variable Taylor series of Section 2.4, can be formed as

$$\vec{f}(\vec{x}_{i+1}) = \vec{f}(\vec{x}_i) + \mathbf{J}(\vec{x}_i) \cdot (\vec{x}_{i+1} - \vec{x}_i) \tag{7.5}$$

where the dot operator denotes the matrix product of a matrix with a vector, and the subscripts i and $i+1$ on the vector $\vec{x}$ denote an iteration index rather than a vector component index. Following the derivation of the single-dimensional Newton iteration formula Equation (5.2), we set $\vec{f}(\vec{x}_{i+1}) = \vec{0}$ at the multidimensional root we aspire to converge on, and solve for $\vec{x}_{i+1}$, the next argument vector in the iteration sequence. This procedure gives

$$\vec{x}_{i+1} = \vec{x}_i - \mathbf{J}^{-1}(\vec{x}_i) \cdot \vec{f}(\vec{x}_i) \tag{7.6}$$

as the multidimensional Newton algorithm iteration formula. Note the similarity between this iteration formula and the single dimensional case of Equation (5.2). Now we observe that solving any system of linear equations $\mathbf{A}\vec{y} = \vec{b}$ for the unknown vector $\vec{y}$ is equivalent to finding the matrix product $\vec{y} = \mathbf{A}^{-1} \cdot \vec{b}$. Therefore in the iteration formula of Equation (7.7), we can identify the term $\mathbf{J}^{-1}(\vec{x}_i) \cdot \vec{f}(\vec{x}_i)$ as simply the vector solution of a system of linear equations. That linear system is formed with a coefficient matrix consisting of the Jacobian matrix **J** (with its partial derivatives evaluated at the current argument vector $\vec{x}_i$) and with a right-hand side vector of $\vec{f}(\vec{x}_i)$. This system solution can be found with the numerical Gaussian elimination methods of Chapter 6. So the multidimensional Newton–Raphson iteration formula becomes the two-step process

$$\begin{aligned} &\text{solve } \mathbf{J}(\vec{x}_i)\vec{d} = \vec{f}(\vec{x}_i) \text{ for the displacement vector } \vec{d} \\ &\vec{x}_{i+1} = \vec{x}_i - \vec{d} \end{aligned} \tag{7.7}$$

As with the one-dimensional Newton method of Chapter 5, the root search algorithm must be initialized at some guess value of the argument vector $\vec{x}$. The success or failure of the convergence to an appropriate root, if indeed a root even exists, may depend strongly on the closeness of the initial guess to the actual root.

We will apply the Newton–Raphson algorithm directly to solve a selection of physics problems in this chapter, and also make use of it in Section 14.2 for the development of implicit methods for solving differential equations.

7.2 Coding: The Systems of Nonlinear Equations Tool

The code for the `sys_root.c` tool, together with its header file `sys_root.h`, capable of solving systems of nonlinear equations can be found as part of the `CPLibrary` download package. This code file contains one function, `sys_newton_raphson()`.

Recall from Chapter 5 that the `newton_raphson()` function in the `root.c` tool must be called with two pointers to C functions, one calculating the function whose root is to be found, and another calculating that function's derivative. What coding techniques can we use to extend these calling arguments to multiple dimensions needed for `sys_newton_raphson()`? We revisit the techniques introduced in the code for extending the `step()` function to the `step2d()` function for the function stepper tool in Section 4.1, and used in the diffraction pattern visualization of Section 4.7.

The functions in the systems solved in this chapter must accept an argument vector, and we already have the enhanced array structure `ARRAY1D` that will provide the memory for the components of this vector. So we may pass a pointer to such a structure to a C function to supply the function's argument. For the sake of coding efficiency, we will want to make one call to a single C function and have it evaluate the values of all the mathematical functions in our system, and essentially return all the components of the vector $\vec{f}(\vec{x})$ of Equation (7.3). The most efficient technique for a function to return a vector to the calling program is passing another pointer argument to a `ARRAY1D` structure that the C function will use to store the n computed values into. Since the dimension of this array is known in the calling routine, the function will assume that the calling code has allocated the `ARRAY1D` memory for this storage. Finally, we will also want to continue the practice of passing a `void` pointer to the C function to point to some arbitrary memory block of physical parameter values. So using all these elements, the C function that computes the vector of system functions will be coded in the form

```
void func(ARRAY1D *x,void *p,ARRAY1D *f) {
  ...
  f->elem[0] = ... ;  // Compute first vector component
  f->elem[1] = ... ;  // Compute next vector component
  ...
  return;
}
```

Since the return values are vector components which the function fills into an array, we do not need to make use of the formal function return value mechanism provided by the C language, and we may simply declare these functions with a `void` return type. Likewise the C function that computes the Jacobian matrix must return a two-dimensional matrix of first derivative values. So we code this in the form

```
void jacobian(ARRAY1D *x,void *p,ARRAY2D *a) {
  ...
  a->elem[0][0] = ...;
  a->elem[0][1] = ...;
  a->elem[1][0] = ...;
  a->elem[1][1] = ...;
  ...
  return;
}
```

Pointers to these two functions above form the first two arguments to the `sys_newton_raphson()` tool in the `sys_root.c` code. We assume that, as in the single-dimensional case, calculation of the first derivatives requires the same physical parameter block as does the calculation of the function values. Other extensions from the one-dimensional root solver tool are extending the initial function argument, the convergence tolerance limit, and the converged root return value from double precision scalars to vectors. The `sys_newton_raphson()` tool assumes that the calling program has allocated the memory in the `ARRAY1D` structure `root` to store the converged root values into. Since the multidimensional Newton–Raphson algorithm requires a solution to a linear system in its iteration formula, a coefficient matrix minimum value must also be passed. All these extensions make the function definition for the system solver

```
int sys_newton_raphson(void (*func)(ARRAY1D *,void *,ARRAY1D *),
                       void (*jacobian)(ARRAY1D *,void *,ARRAY2D *),
                       ARRAY1D *init,void *param,
                       ARRAY1D *tol,double min,int n_iters,
                       ARRAY1D *root_ptr) {
  ...
}
```

The remainder of the code implements iterations around the formula of (7.7), until successful convergence or the iteration limit is met. The displacement vector for each iteration, which moves the current function argument vector to the next vector in the iteration sequence, is found by creating and solving a linear system of equations using the `gauss_pivotmax()` function from Chapter 6. This sequence of statements in each iteration that compute the current displacement vector is

```
(*jacobian)(root_ptr,param,&a);
(*func)(root_ptr,param,&rhs);
if (gauss_pivotmax(&a,&rhs,&d,min) == LINEQ_SINGULAR) {
  ...
}
```

Note that for a successful convergence to be declared, all n components of the displacement vector must have converged within the tolerance specified by the argument vector `tol`. Since this test requires iteratively testing each component of the displacement vector in sequence, we can efficiently combine this iterative loop with the loop required to also update the current argument vector components. An integer "flag" variable `done` is used to store the condition that at least one component of the displacement vector has failed the tolerance test

```
done = 1;
for (i = 0; i < n; i++) {
  root_ptr->elem[i] -= d.elem[i];
  if (fabs(d.elem[i]) > tol->elem[i]) done = 0;
}
if (done) break;
```

7.3 Application: Statics Problems

See Figure 7.2 for a statics problem with two masses on three flexible strings. At first glance it may appear to be comparable to a simple statics problem with a determinant structure, but note here the support members between the masses are assumed to be flexible strings with an attachment angle at each mass that can pivot freely. Therefore, unlike the case of a fixed structure, the angle the members will make relative to horizontal and vertical are unknown variables, and again we

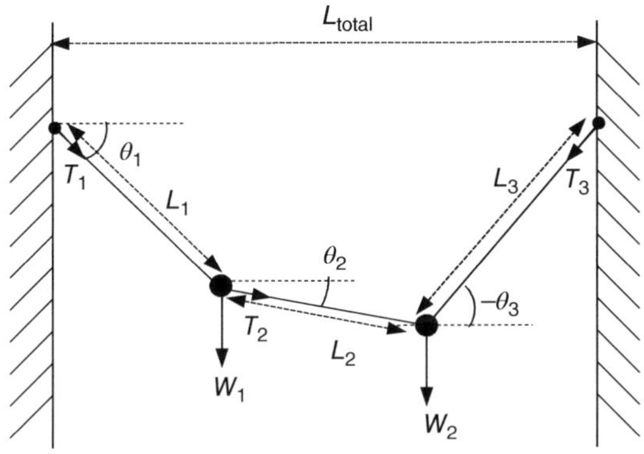

Figure 7.2 Nonlinear statics problem of strings and two masses.

are in the realm of transcendental equations that must be solved numerically. Two weights are connected to the joints in the strings, with weights W_1 and W_2. The lengths of the three strings are L_1, L_2, and L_3; the distance between the two walls that the outer strings are tied to is L; and the two strings are tied to the walls at the same altitude. We must solve for the unknown tensions in the three strings T_1, T_2, and T_3, and the unknown angles the three strings make with the horizontal, θ_1, θ_2, and θ_3. This produces a system of nonlinear equations that must be solved with the multidimensional Newton–Raphson algorithm.

The main program that calls the `sys_newton_raphson()` function should accept six command line arguments, for the initial guesses for the six argument vector components. Writing the C function for the Jacobian computation with its 36 partial derivatives is somewhat cumbersome, but each derivative is either relatively simple or identically zero. We may have the convenience of expressing the angles θ_1, θ_2, and θ_3 in degrees by using the multiplicative conversion constants `RADTODEG` and `DEGTORAD` in the header file `constants.h`. For this example problem, we may use arbitrary units for the dimensions and the forces. From straightforward static force analysis,

$$
\begin{aligned}
f_0 &= L_1 \cos\theta_1 + L_2 \cos\theta_2 + L_3 \cos\theta_3 - L_{\text{total}} = 0 \\
f_1 &= L_1 \sin\theta_1 + L_2 \sin\theta_2 + L_3 \sin\theta_3 = 0 \\
f_2 &= T_1 \sin\theta_1 - T_2 \sin\theta_2 - W_1 = 0 \\
f_3 &= -T_1 \cos\theta_1 + T_2 \cos\theta_2 = 0 \\
f_4 &= T_2 \sin\theta_2 - T_3 \sin\theta_3 - W_2 = 0 \\
f_5 &= -T_2 \cos\theta_2 + T_3 \cos\theta_3 = 0
\end{aligned}
\tag{7.8}
$$

Deriving the partial derivatives of each of these functions to place in the Jacobian matrix is straightforward, although a bit laborious, but only a fraction of these derivatives are nonzero. The main program in the file `strings2.c`, which can be found on the Student Companion website, includes the functions to compute Equations (7.8) and the corresponding Jacobian matrix. Note that since the dimension of the vectors in this problem is fixed at six, we can make use of static allocation for the vector elements as discussed in Section 3.1.4. For example, to save a call to the operating system for dynamic allocation, we simply declare a static C array to hold the six initial root guess components, translate their numerical values from the text command line arguments, and then fill in the members of the `ARRAY1D init` vector with

```
double init_state[6];
init_state[0] = atof(argv[1]);
init_state[1] = atof(argv[2]);
init_state[2] = atof(argv[3]);
init_state[3] = atof(argv[4]);
init_state[4] = atof(argv[5]);
init_state[5] = atof(argv[6]);
ARRAY1D init = {.elem = init_state,.n_elems = 6};
```

Once this and other vectors have been prepared, the multidimensional Newton–Raphson solver routine is called with

```
if (sys_newton_raphson(strings2,jacobian,&init,&param,&tol,1.0e-12,
                       100,&root) == SYS_ROOT_SUCCESS) {
  print_array1d(&root,"%.8g ");
  exit(0);
}
else {
  fprintf(stderr,"Convergence failure\n");
  exit(1);
}
```

to test whether the initial guesses on the command line are sufficient to successfully converge on a solution. If convergence is successful, the solution vector is printed with the `print_array1d()` library function.

When we compile this main program, we of course, must include the tool code `sys_root.c` to link the code to the `sys_newton_raphson()` solver, but we also must include the `lineq.c` code that is needed by the multidimensional Newton–Raphson algorithm. Note that the main program includes setting some example physical parameters of the system as $L_{total} = 30$ units, $L_1 = 12$, $L_2 = 10$, $L_3 = 15$, $W_1 = 100$ units, and $W_2 = 150$ units. Building and running the code with the initial guesses on the command line converges to the solution

```
$ gcc -Wall -Wshadow -O2 -o strings2 strings2.c CPLibrary/sys_root.c CPLibrary/lineq.c
    ↪ CPLibrary/array.c -lm
$ ./strings2
./strings2 <initial T1> <T2> <T3> <theta1> <theta2> <theta3>
$ ./strings2 100 100 100 30 30 30
189.02886 142.84744 186.85721 42.060226 10.745143 41.316878
```

Exercise 7.1 will let the student extend this example statics solution to more strings and masses.

7.4 Application: Nonlinear Circuits

In Section 6.7, we used Kirchoff's Current Law to solve for the unknown node potentials in a straightforward linear electrical circuit. Electrical networks composed of solely linear resistive components can be useful in some particular applications, but including the complexity of components that behave with nonlinear relationships between their terminal potentials and current flow opens the possibilities of wide and rich electronic functions. In particular, on the subject of digital computation, designing electronic networks that perform the arithmetic, logical, and memory functions involved in the central processing unit (CPU) system of Figure 1.6 requires components with significant nonlinear behavior. Over the developmental history of the twentieth century, these components have been electromagnetic relays, vacuum tubes, bipolar junction transistors (BJTs), and metal-oxide-semiconductor field effect transistors (MOSFETs). The modeling of real semiconductor devices is a very involved field of study, but we can gain an introductory understanding of the network topology that is capable of storing a single binary bit in a CPU register by assuming a simple model of a hypothetical transistor device. The electrical behavior of this component is shown in Figure 7.3.

$$g(V_G) = \frac{e^{\frac{V_G - V_T}{V_X}}}{1 + e^{\frac{V_G - V_T}{V_X}}} \tag{7.9}$$

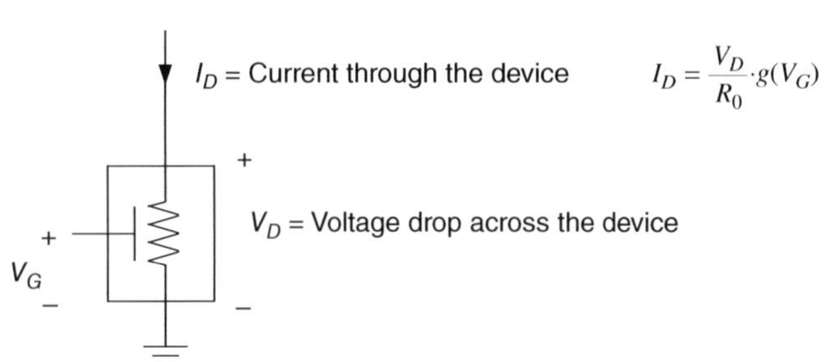

Figure 7.3 Hypothetical simple nonlinear transistor device.

We'll assume this hypothetical electronic device is functionally a linear resistor between the top and bottom current conduction terminals, but with the resistance modulated by the side gate potential. The current through the device is linearly proportional to the device voltage across the device. However, this proportionality constant is the nonlinear sigmoid gate function plotted in Figure 7.4 and described by Equation (7.9) divided by the nominal resistance of R_0.

$$I_D = \frac{V_D}{R_0} g(V_G) \tag{7.10}$$

In a conceptual sense, we think of this device being an electric valve with a certain intrinsic linear resistance to current flow through it, and with a gate terminal that can gradually "open" or "close" the current valve by applying higher or lower gate potentials, respectively. We also assume that there is no current being drawn into the gate terminals of the transistor device, similar to an actual MOSFET.

We will work with the sigmoid function more closely in Section 12.1 as we study numerical solutions to an example differential equation. With two of these devices and two linear resistors, we can assemble a simple network known as a "flip-flop," or with more historical accuracy, the Eccles–Jordan bistable multivibrator, as shown in Figure 7.5. Circuits of this type are at the heart of any computation hardware and are capable of storing the state of a single binary arithmetic bit represented in electric potentials.

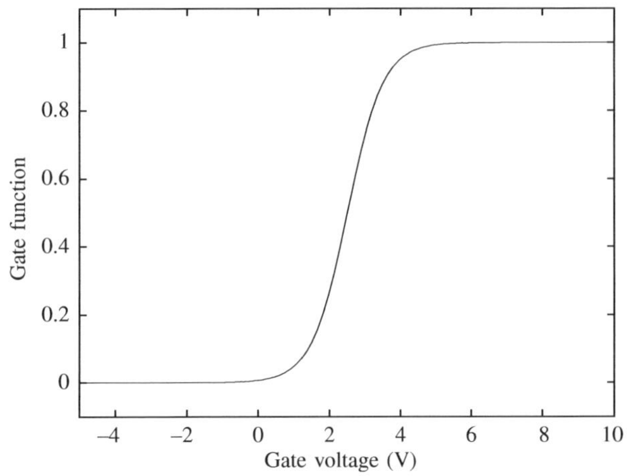

Figure 7.4 Sigmoid gating function of hypothetical transistor device with V_T = 2.5 V, V_X = 0.5 V.

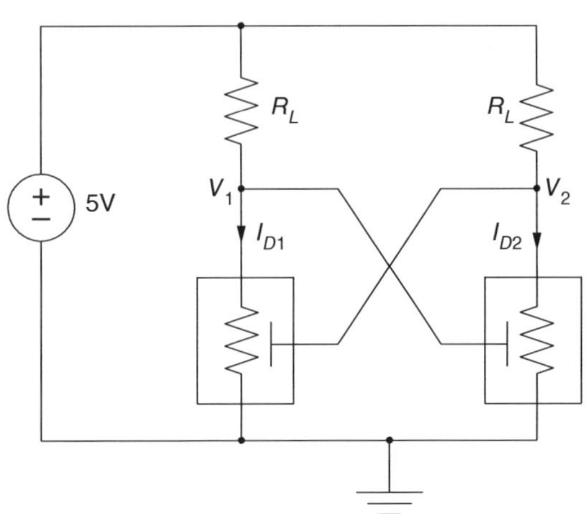

Figure 7.5 Nonlinear circuit network for a 1-bit binary storage flip-flop.

There are two circuit network nodes with unknown potentials, V_1 and V_2. We may set up a system of two simultaneous equations that express the constraints of Kirchoff's Current Law at each circuit node.

$$\begin{aligned} I_{D1} - \frac{5V - V_1}{R_L} &= 0 \\ I_{D2} - \frac{5V - V_2}{R_L} &= 0 \end{aligned} \tag{7.11}$$

When we include the device characteristics from Equation (7.10) in the system of (7.11) it becomes a system of nonlinear equations we may submit to the Newton–Raphson algorithm as

$$\begin{aligned} f_0(V_1, V_2) &= \frac{V_1}{R_0} g(V_2) - \frac{5V - V_1}{R_L} = 0 \\ f_1(V_1, V_2) &= \frac{V_2}{R_0} g(V_1) - \frac{5V - V_2}{R_L} = 0 \end{aligned} \tag{7.12}$$

Note that the gate function of each device is dependent on the potential at the other side of the circuit. This dependence reflects the cross-connected topology of the network. This system may be solved for the two-node voltages with the Newton–Raphson algorithm. As we will see in Section 12.1 in the context of solutions of ordinary differential equations, if a sigmoid function

$$y(x) = \frac{e^{kx}}{e^{kx} + c} \tag{7.13}$$

for any constant c, then its first derivative can be written implicitly as

$$\frac{dy}{dx} = ky(1 - y) \tag{7.14}$$

Once a simple C function has been written to compute the gating function $g(V_G)$, then this implicit derivative can be used to easily code the partial derivatives in the Jacobian matrix for the circuit solution as

$$\begin{aligned} \frac{\partial f_0(V_1, V_2)}{\partial V_1} &= \frac{1}{R_0} g(V_2) + \frac{1}{R_L} \\ \frac{\partial f_0(V_1, V_2)}{\partial V_2} &= \frac{V_1}{R_0} \frac{1}{V_X} g(V_2)\,(1 - g(V_2)) \\ \frac{\partial f_1(V_1, V_2)}{\partial V_1} &= \frac{V_2}{R_0} \frac{1}{V_X} g(V_1)\,(1 - g(V_1)) \\ \frac{\partial f_1(V_1, V_2)}{\partial V_2} &= \frac{1}{R_0} g(V_1) + \frac{1}{R_L} \end{aligned} \tag{7.15}$$

and by calling the sigmoid gating function code.

The coding of the program simulating this system is available as Exercise 7.2 for the student.

7.5 Application: Hyperbolic Radio Navigation

Figure 7.6 shows a three-dimensional navigation system based on radio beacon transmitters at fixed, known locations. The user carries a portable receiver, and needs to solve for his unknown location at coordinates (x, y, z). A primary transmitter sends out pulses of electromagnetic waves at regular intervals, and triggers the three secondary transmitters to send out their coded pulses. There is no way for the receiver to know any absolute time relative to the clocks on the transmitters, but the receiver may measure the *time delays* between the primary pulse arriving at the receiver and each of the secondary pulses. We let the time delay between receiving the primary pulse and secondary pulse 1 be t_{10}, the delay between the primary pulse and pulse 2 be t_{20}, and the delay to pulse 3 to be t_{30}. We assume that the radio pulses propagate from each transmitter to the receiver at the speed of light, and that the measured time delays have already been corrected for the triggering time delay between the primary and each secondary transmitters, since their locations are fixed and known. This system is the general basis for the operation of the historic LORAN navigation system, as well as the widely used GPS system of current day. If the three-dimensional coordinates of each of the transmitters are fixed and known as shown in the figure, how may we solve for the unknown receiver coordinates (x, y, z) from the measured time delays t_{10}, t_{20}, and t_{30}?

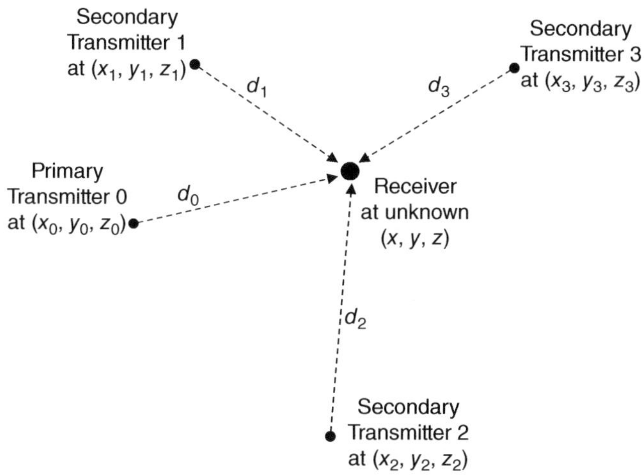

Figure 7.6 Hyperbolic radio navigation in three dimensions.

If the unknown distances between transmitter i and the receiver is d_i as shown in Figure 7.6, then from the geometry we know

$$\begin{aligned}
d_0^2 &= (x - x_0)^2 + (y - y_0)^2 + (z - z_0)^2 \\
d_1^2 &= (x - x_1)^2 + (y - y_1)^2 + (z - z_1)^2 \\
d_2^2 &= (x - x_2)^2 + (y - y_2)^2 + (z - z_2)^2 \\
d_3^2 &= (x - x_3)^2 + (y - y_3)^2 + (z - z_3)^2
\end{aligned} \tag{7.16}$$

From the propagation of radio waves we may write

$$\begin{aligned}
\sqrt{d_1^2} - \sqrt{d_0^2} &= ct_{10} \\
\sqrt{d_2^2} - \sqrt{d_0^2} &= ct_{20} \\
\sqrt{d_3^2} - \sqrt{d_0^2} &= ct_{30}
\end{aligned} \tag{7.17}$$

where c is the speed of light. From this, and using the definitions of the d_i^2 in Equation (7.16), we may define a system of three nonlinear equations to be solved simultaneously as

$$\begin{aligned}
f_0(x, y, z) &= \sqrt{d_1^2} - \sqrt{d_0^2} - ct_{10} = 0 \\
f_1(x, y, z) &= \sqrt{d_2^2} - \sqrt{d_0^2} - ct_{20} = 0 \\
f_2(x, y, z) &= \sqrt{d_3^2} - \sqrt{d_0^2} - ct_{30} = 0
\end{aligned} \tag{7.18}$$

We know from Section 7.1 that to solve this system numerically with the Newton–Raphson algorithm, we will need to derive the Jacobian matrix with nine partial derivatives

$$\mathbf{J} = \begin{bmatrix}
\frac{\partial f_0}{\partial x} & \frac{\partial f_0}{\partial y} & \frac{\partial f_0}{\partial z} \\
\frac{\partial f_1}{\partial x} & \frac{\partial f_1}{\partial y} & \frac{\partial f_2}{\partial z} \\
\frac{\partial f_2}{\partial x} & \frac{\partial f_2}{\partial y} & \frac{\partial f_2}{\partial z}
\end{bmatrix} \tag{7.19}$$

For these nine derivations, it will be convenient to make use of the chain rule and the definitions in Equation (7.16); for example

$$\frac{\partial f_0}{\partial x} = \frac{\partial \sqrt{d_1^2}}{\partial (d_1^2)} \cdot \frac{\partial (d_1^2)}{\partial x} - \frac{\partial \sqrt{d_0^2}}{\partial (d_0^2)} \cdot \frac{\partial (d_0^2)}{\partial x} \tag{7.20}$$

The coding work for solving this system of nonlinear equations is available as Exercise 7.3.

7.6 Algorithms: Numerical Estimates of the Jacobian Partial Derivatives

For a general n-dimensional system of nonlinear equations, the task of deriving and coding mathematical expressions for all the elements of an $n \times n$ dimensional Jacobian matrix can become daunting for a large order. For many problems, such as the strings-and-masses statics problem of Section 7.3, the large number of Jacobian elements that may be identically zero helps to simplify the task considerably. But for the general case, we can propose a method to numerically derive estimates of the first partial derivatives in the Jacobian matrix that will vastly reduce the work required to prepare a solution using the multidimensional Newton–Raphson method.

We have actually already seen a numerical algorithm that replaces a first derivative with a discrete estimate, in the derivation of the Secant algorithm in Section 5.1.2. This method can be seen as basically similar to the one-dimensional Newton–Raphson algorithm for root finding, but with the approximation of a secant line inside the curve of a function replacing the adjacent tangent line on the outside of the curve. The secant line is constructed in the algorithm as the slope between two adjacent samples of the function, a slope drawn between the current function evaluation point and the point in the previous iteration. As such, this line forms an estimate of the tangent line that comes from a one-sided estimate of the function's derivative. Even with this crude estimate, we know the Secant algorithm converges accurately on the roots of functions. So in this section we extend this general idea to the multidimensional root-finding method, but also move to an improved derivative estimate.

In our exploration of truncation error in numerical algorithms in Section 2.5, we examined the two-sided second-order estimate of the first derivative in Equation (2.10). Using this estimation method to numerically fill in all the partial derivatives in a Jacobian matrix will give us a second-order multidimensional slope to guide the convergence of the multidimensional version of the Newton–Raphson method [13, Chap. 2].

Consider as an example the program in the file `strings2_numjac.c` on the Student Companion website. This code solves the same statics problem of Figure 7.2, but without requiring any derivation or coding of the partial derivatives of the system functions. As before, this program will call `sys_newton_raphson()` to converge on the root of the system functions, but will now use a pointer to the function `jacobian_numerical()` as its second argument. The `jacobian_numerical()` numerical function will systematically fill in all 36 elements in the Jacobian matrix with a second-order estimate of each partial derivative. The estimate is formed by generating a test vector in the function argument space that displaces one component at a time in the argument space by both a positive and negative increment value, and calling the C function that evaluates the nonlinear system function vector once at each displacement. The estimate for an entire column in the Jacobian matrix is derived as

$$\frac{\partial \vec{f}(\vec{x})}{\partial x_j} \approx \frac{\vec{f}(\vec{x}_+) - \vec{f}(\vec{x}_-)}{2h_j} \tag{7.21}$$

where the test vector $\vec{x}_+$ is the current $\vec{x}$ argument vector with component j displaced by $+2h_j$, the test vector $\vec{x}_-$ is the current $\vec{x}$ argument vector with component j displaced by $-2h_j$, and h_j is a step size for component j.

For the convenience of avoiding the task of deriving and coding all n^2 partial derivatives, and letting the processor numerically compute all our estimates for us, we do have to provide an extra parameter in this process, namely the step size h_j for each of the dimensions in the argument space. This we must do using our physical understanding of the system being solved. In general, we would want to specify a "small but significant" displacement distance for each vector coordinate. For example, in the two-mass statics problem, we could specify increment values of 1.0 units for each of the three string tension components, and 0.5 degrees for each of the three angle components.

And how are these increment values communicated to the numerical Jacobian function in our code? We may simply add an array of these increment values to the existing structure we already pass as physical parameters. These modifications can be seen in the source code `strings2_numjac.c`. Compiling and running this program reports the same solution as did the code with the analytic Jacobian elements of Section 7.3.

```
$ gcc -Wall -Wshadow -O2 -o strings2_numjac strings2_numjac.c CPLibrary/sys_root.c
    ↪ CPLibrary/lineq.c CPLibrary/array.c -lm
$ ./strings2_numjac
./strings2_numjac <initial T1> <T2> <T3> <theta1> <theta2> <theta3>
$ ./strings2_numjac 100 100 100 30 30 30
189.0289 142.8474 186.8572 42.0602 10.7451 41.3169
```

The code for the `jacobian_numerical()` function in this example can be easily copied and edited for other solution problems, and we see an example application in Section 7.7.

7.7 Application: The Covalent Bond

In Section 5.11, we tackled the problem of finding the allowed energy levels for an electron bound in a single one-dimensional rectangular quantum potential well. Mathematically this problem could be solved by locating the multiple roots along the oscillating contours of two nonlinear functions of a single dimension, as plotted in Figure 5.30.

Let us now explore a similar physical problem, still solving for allowed energy levels of an electron, but now we assume the particle is bound in an expanded one-dimensional potential consisting of two adjacent and identical rectangular wells. This geometry is shown in Figure 7.7. The field potential is assumed to be zero, and the potential at the bottom of each well is $-V_0$. Each well has the same width w, and the two wells are separated by a potential step at the field potential of width s. Allowed wave functions that satisfy the second-order spatial differential equation of Schrödinger, Equation (5.44), have an arbitrary multiplicative constant used for normalization. But for the purpose of solving for the allowed energy levels for an electron bound in this structure, we can simply find wave functions that are "normalizable," that is, the square of the wave function will integrate over the entire spatial coordinate x range to some finite number. Normalizable wave functions supported by this potential must follow a growing exponential for x values to the left of the wall at $x = -\frac{s}{2} - w$, and a decaying exponential for x values to the right of the wall at $x = +\frac{s}{2} + w$. If we make the arbitrary assumption that the exponential term has a positive coefficient on the left side, then the sign of the coefficient on the right side will be positive for even parity wave functions and negative for odd parity solutions. We also know that the Schrödinger equation solution must follow a sinusoidal form within the potential wells for bound electrons. Relying on the condition of the continuity of the wave function, the solution through this entire structure will be a piecewise stitching together of exponential and sinusoidal regions. We need only to solve for the wave function for $x > 0$, and then the solution for $x < 0$ will be either a

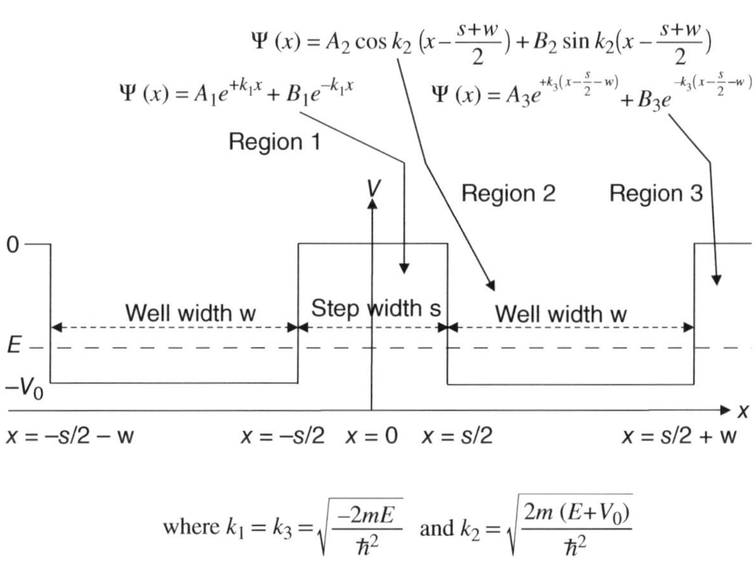

Figure 7.7 Two adjacent quantum rectangular potential wells.

symmetric or anti-symmetric extension. Since we will be asserting the continuity of both the wave function and its first derivative at the two points shown in Figure 7.7, mathematically, this stitching problem will be resolved by finding the simultaneous roots of a system of four nonlinear equations in four unknowns.

The right-hand potential well runs from $x = \frac{s}{2}$ to $x = \frac{s}{2} + w$. In region 3, we make the assertion that $A_3 = 0$ for any normalizable wave function with physical meaning, and it is convenient to make the remaining right-hand exponential have a zero argument at $x = \frac{s}{2} + w$

$$\psi(x) = B_3 e^{-k_3\left(x - \frac{s}{2} - w\right)} \tag{7.22}$$

The center of the right-hand well is at $x = \frac{s+w}{2}$, so for convenience in region 2, we make the possible sin() and cos() component functions centered around that x value where their arguments go to zero

$$\psi(x) = A_2 \cos k_2 \left(x - \frac{s+w}{2}\right) + B_2 \sin k_2 \left(x - \frac{s+w}{2}\right) \tag{7.23}$$

Then the possible center exponential in region 1 can be written with arguments around $x = 0$

$$\psi(x) = A_1 e^{+k_1 x} + B_1 e^{-k_1 x} \tag{7.24}$$

First, solving for energy levels of the *even parity* states, we set $B_1 = A_1$. Then applying continuity of $\psi(x)$ at $x = \frac{s}{2} + w$ we have

$$B_3 = A_2 \cos k_2 \left(\frac{w}{2}\right) + B_2 \sin k_2 \left(\frac{w}{2}\right) \tag{7.25}$$

Applying continuity of $\psi'(x)$ at $x = \frac{s}{2} + w$ gives

$$-k_3 B_3 = -k_2 A_2 \sin k_2 \left(\frac{w}{2}\right) + k_2 B_2 \cos k_2 \left(\frac{w}{2}\right) \tag{7.26}$$

Applying continuity of $\psi(x)$ at $x = \frac{s}{2}$ gives

$$A_1 e^{\frac{k_1 s}{2}} + A_1 e^{\frac{-k_1 s}{2}} = A_2 \cos k_2 \left(-\frac{w}{2}\right) + B_2 \sin k_2 \left(-\frac{w}{2}\right) \tag{7.27}$$

Applying continuity of $\psi'(x)$ at $x = \frac{s}{2}$ gives

$$k_1 A_1 e^{\frac{k_1 s}{2}} - k_1 A_1 e^{\frac{-k_1 s}{2}} = -k_2 A_2 \sin k_2 \left(-\frac{w}{2}\right) + k_2 B_2 \cos k_2 \left(-\frac{w}{2}\right) \tag{7.28}$$

Since we are looking for only a normalizable solution, we may arbitrarily set $B_3 = 1$ and, together with $k_1 = k_3$, construct a system of four nonlinear equations in four unknowns to solve for the coefficients A_1, A_2, B_2, and the energy level E for the even parity case. The electron energy E determines the wave numbers k_1 and k_2 outside and inside the wells, respectively, as shown in Figure 7.7.

$$\begin{aligned}
f_0(A_1, A_2, B_2, E) &= B_3 - A_2 \cos k_2 \left(\frac{w}{2}\right) - B_2 \sin k_2 \left(\frac{w}{2}\right) = 0 \\
f_1(A_1, A_2, B_2, E) &= -k_1 B_3 + k_2 A_2 \sin k_2 \left(\frac{w}{2}\right) - k_2 B_2 \cos k_2 \left(\frac{w}{2}\right) = 0 \\
f_2(A_1, A_2, B_2, E) &= A_1 e^{\frac{k_1 s}{2}} + A_1 e^{\frac{-k_1 s}{2}} - A_2 \cos k_2 \left(-\frac{w}{2}\right) - B_2 \sin k_2 \left(-\frac{w}{2}\right) = 0 \\
f_3(A_1, A_2, B_2, E) &= k_1 A_1 e^{\frac{k_1 s}{2}} - k_1 A_1 e^{\frac{-k_1 s}{2}} + k_2 A_2 \sin k_2 \left(-\frac{w}{2}\right) - k_2 B_2 \cos k_2 \left(-\frac{w}{2}\right) = 0
\end{aligned} \tag{7.29}$$

Moving on to solving for energy levels of *odd parity* states, we set $B_1 = -A_1$. Applying continuity of $\psi(x)$ at $x = \frac{s}{2} + w$ we have

$$B_3 = A_2 \cos k_2 \left(\frac{w}{2}\right) + B_2 \sin k_2 \left(\frac{w}{2}\right) \tag{7.30}$$

Applying continuity of $\psi'(x)$ at $x = \frac{s}{2} + w$ gives

$$-k_1 B_3 = -k_2 A_2 \sin k_2 \left(\frac{w}{2}\right) + k_2 B_2 \cos k_2 \left(\frac{w}{2}\right) \tag{7.31}$$

Applying continuity of $\psi(x)$ at $x = \frac{s}{2}$ gives

$$A_1 e^{\frac{k_1 s}{2}} - A_1 e^{\frac{-k_1 s}{2}} = A_2 \cos k_2 \left(-\frac{w}{2}\right) + B_2 \sin k_2 \left(-\frac{w}{2}\right) \tag{7.32}$$

Applying continuity of $\psi'(x)$ at $x = \frac{s}{2}$ gives

$$k_1 A_1 e^{\frac{k_1 s}{2}} + k_1 A_1 e^{\frac{-k_1 s}{2}} = -k_2 A_2 \sin k_2 \left(-\frac{w}{2}\right) + k_2 B_2 \cos k_2 \left(-\frac{w}{2}\right) \tag{7.33}$$

Again we arbitrarily set $B_3 = 1$ and construct a different system of four nonlinear equations in four unknowns to solve for the coefficients A_1, A_2, B_2, and the energy level E for the odd parity case

$$\begin{aligned}
f_0(A_1, A_2, B_2, E) &= B_3 - A_2 \cos k_2 \left(\frac{w}{2}\right) - B_2 \sin k_2 \left(\frac{w}{2}\right) = 0 \\
f_1(A_1, A_2, B_2, E) &= -k_1 B_3 + k_2 A_2 \sin k_2 \left(\frac{w}{2}\right) - k_2 B_2 \cos k_2 \left(\frac{w}{2}\right) = 0 \\
f_2(A_1, A_2, B_2, E) &= A_1 e^{\frac{k_1 s}{2}} - A_1 e^{\frac{-k_1 s}{2}} - A_2 \cos k_2 \left(-\frac{w}{2}\right) - B_2 \sin k_2 \left(-\frac{w}{2}\right) = 0 \\
f_3(A_1, A_2, B_2, E) &= k_1 A_1 e^{\frac{k_1 s}{2}} + k_1 A_1 e^{\frac{-k_1 s}{2}} + k_2 A_2 \sin k_2 \left(-\frac{w}{2}\right) - k_2 B_2 \cos k_2 \left(-\frac{w}{2}\right) = 0
\end{aligned} \tag{7.34}$$

The two sets of four nonlinear functions in Equations (7.29) and (7.34) involve quite a challenging assortment of trigonometric and exponential terms. If we imagine the task of deriving and coding the 16 partial derivatives associated with each of these systems, we would foresee a significant level of work. So this problem is an excellent example of one where application of the numerical Jacobian coding technique of Section 7.6 would be very advantageous.

Assembling the code for this twin well problem with the system functions in Equations (7.29) and (7.34), together with a copy of the numerical Jacobian routine adapted from the `strings2_numjac.c` code, gives us the code in the file `quantum_2rectwell_root.c`, available for download from the Student Companion website. The compilation step and "help" line display at a command terminal is

```
$ gcc -Wall -Wshadow -O2 -o quantum_2rectwell_root quantum_2rectwell_root.c CPLibrary/sys_root.c
    ↪ CPLibrary/lineq.c CPLibrary/step.c CPLibrary/array.c -lm
$ ./quantum_2rectwell_root
./quantum_2rectwell_root <well width> <step width> <well depth> <mass> <initial A2> <B2> <A1> <E>
    ↪ e|o
```

The program expects the physical parameters of well width, separation width, and well depth on the command line, followed by initial guesses for the four system variables. The last argument that this program expects on its command line is not a numerical value, but rather a single character code that selects the parity of the solution being solved for. The character may be `'e'` or `'o'`, and with the following conditional coding structure

```
if (argv[9][0] == 'e') {
  retval = sys_newton_raphson(quantum_2rectwell_even,
              jacobian_numerical_even,&init,&rparam,&tol,
              1.0e-12 * CHARGE,100,&rwparam.root);
  ...
}
else {
    retval = sys_newton_raphson(quantum_2rectwell_odd,
              jacobian_numerical_odd,&init,&rparam,&tol,
              1.0e-12 * CHARGE,100,&rwparam.root);
  ...
}
```

the program will select among two possible calls to `sys_newton_raphson()` appropriate for solving for even or odd parity wave functions, respectively. If the Newton–Raphson algorithm has successfully converged on a system root, the program outputs not only the allowed energy E, but also the three coefficients A_2, B_2, and A_1. From these values the program goes on to use the `step.c` tool of Chapter 4 to send a data set of the sampled piecewise wave function stitched together from the expressions in Figure 7.7 to the standard output stream for plotting.

These sets of nonlinear functions represent a tricky numerical challenge, with the presence of exponential functions able to swing the function values over many orders of magnitude. Making reasonable guesses for initial values of the four-dimensional argument vector in the neighborhood of a desired root is essential. In particular, we will need to make a close guess of the allowed energy levels E to be able to successfully converge on desired roots. So using a little physical reasoning, we expect the levels to be not too far from those energy levels for a single well, since we will have only a small likelihood of electron tunneling and interaction through the step barrier potential between the wells. So we return to the `quantum_rectwell_roots.c` program developed in Section 5.11 and run the program solving for the allowed electron energies in a single rectangular well. These wave function solutions are plotted in Figure 5.31. Picking the same example dimensions of a 10-eV deep well of 1 nm width,

```
$ ./quantum_rectwell_roots
./quantum_rectwell_roots <width in nm> <depth in eV> <mass * c^2 in Mev>
$ ./quantum_rectwell_roots 1 10 0.5110 >| quantum_rectwell_roots.dat
3 roots found for even parity:
-9.7023952 -7.3481053 -2.8553652
3 roots found for odd parity:
-8.8137221 -5.3369033 -0.23062891
```

From these results we expect that for the twin well profile in this section, for energy levels near the bottom of the wells, the lowest four levels should not be too far from −9.7, −8.8, −7.3, and −5.3 eV. For the twin well program to converge successfully, we will also need to prepare a set of reasonable initial guesses for the coefficients A_2, B_2, and A_1 in Figure 7.7. Let us visualize stitching together wave function solutions at various energy levels from Figure 5.31 to form stationary wave solutions that are of even or odd parity over the space of two potential wells. We understand that if the wave functions as plotted in Figure 7.8 are solutions to the linear Schrödinger equation with the rectangular well boundary conditions, then the negative of these wave functions will also be viable solutions. We can then make some approximate sketches that resemble the shapes shown in Figure 7.8 that satisfy the continuity relationships at the stitching points shown in Figure 7.7.

From these sketches for the even and odd parity wave functions for the first and third energy levels at approximately −9.7 and −7.3 eV, we can visualize the signs of the coefficients that would correspond to each of these wave function modes. Recall from Figure 7.7 that A_2 is the coefficient of the cosine term in the center of the right-hand well; the coefficient B_2 scales the sine term in the right-hand well; and the A_1 coefficient is the factor for the exponential term between the wells. Matching the shapes of the curves in Figure 7.8, we can determine the signs of these coefficients. With the signs matching the proper contours, we find that some approximate magnitudes of the coefficients will initialize the Newton–Raphson algorithm to converge readily to the desired solutions. To match the assumed value of the coefficient $B_3 = 1$, and observing that both of these first and third energy levels require *primarily a cosine dependence* inside the potential wells, let us guess the oscillatory coefficients to be $A_2 = \pm 1$ and $B_2 = 0$, and the exponential coefficient A_1 to be ± 0.1, according to Table 7.1.

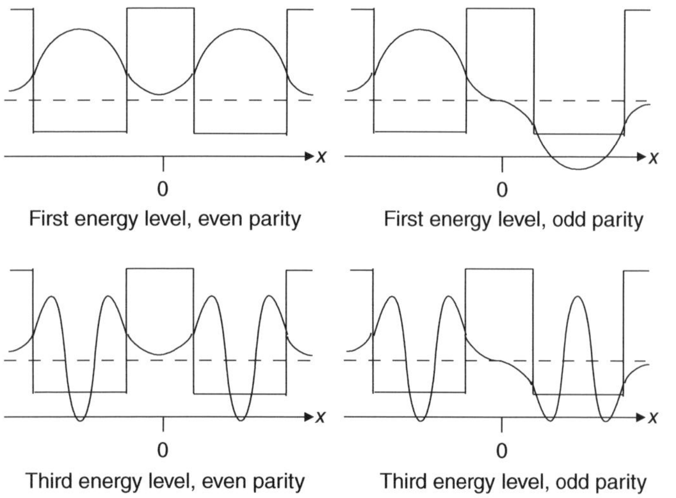

Figure 7.8 Wave function sketches for electron bound in two rectangular potential wells.

Table 7.1 Initial guesses for wave function solution system variables.

Wave function	A_2	B_2	A_1	E
First energy level, even parity	1	0	0.1	−9.7
First energy level, odd parity	−1	0	−0.1	−9.7
Third energy level, even parity	−1	0	0.1	−7.3
Third energy level, odd parity	1	0	−0.1	−7.3

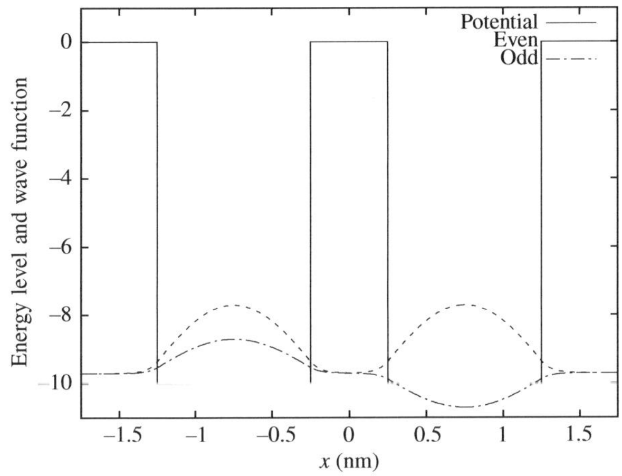

Figure 7.9 Even and odd wave functions for electron bound in two rectangular potential wells; well width = 1 nm, step width = 0.5 nm, depth = 10 eV, energy level 1.

The following dialog shows the runs of our solution program with the appropriate initial guess values for the first and third levels. For the first energy level with a wide potential step of 0.5 nm separating the wells, as plotted in Figure 7.9

```
$ ./quantum_2rectwell_root 1 0.5 10 0.5110 1 0 0.1 -9.7 e >| quantum_2rectwell_root.dat
5.7963725481 -0.000337855282934 0.0185231999803 -9.70236265859
$ ./quantum_2rectwell_root 1 0.5 10 0.5110 -1 0 -0.1 -9.7 o >| quantum_2rectwell_root.dat
-5.79594299368 -0.000337642808366 -0.0185114250792 -9.70231853939
```

Following the `quantum_2rectwell_root.c` source code, the list of converged system variables are displayed on the terminal line in the order A_2, B_2, A_1, and E. Both energy levels, the final value displayed on the output lines, for the even and odd parity solutions are quite close to the level for a single well reported by the run of `quantum_rectwell_roots.c` listed above. With a wide separation between the wells, there is very little probability of any sharing of the electron probability between the wells, so the solution is essentially that of isolated wells. Then again for the first energy level but with the width of the potential step separating the wells narrowed to 0.1 nm, as plotted in Figure 7.10

```
$ ./quantum_2rectwell_root 1 0.1 10 0.5110 1 0 0.1 -9.7 e >| quantum_2rectwell_root.dat
5.94766328285 -0.246178027014 0.554439263105 -9.71779554867
$ ./quantum_2rectwell_root 1 0.1 10 0.5110 -1 0 -0.1 -9.7 o >| quantum_2rectwell_root.dat
-5.68707915135 -0.167629101594 -0.37779659481 -9.69108122574
```

With this narrow well separation, we can see a small difference emerging between the energy levels of the even and odd parity solutions as the probability of interaction between the wells increases.

Now moving up to the third energy level with a wide potential step of 0.5 nm separating the wells, as plotted in Figure 7.11

```
$ ./quantum_2rectwell_root 1 0.5 10 0.5110 -1 0 0.1 -7.3 e >| quantum_2rectwell_root.dat
-1.94187582989 0.000827879607852 0.0310840917676 -7.3481008222
$ ./quantum_2rectwell_root 1 0.5 10 0.5110 1 0 -0.1 -7.3 o >| quantum_2rectwell_root.dat
1.94152896053 0.000827446343603 -0.0310628731865 -7.3471531721
```

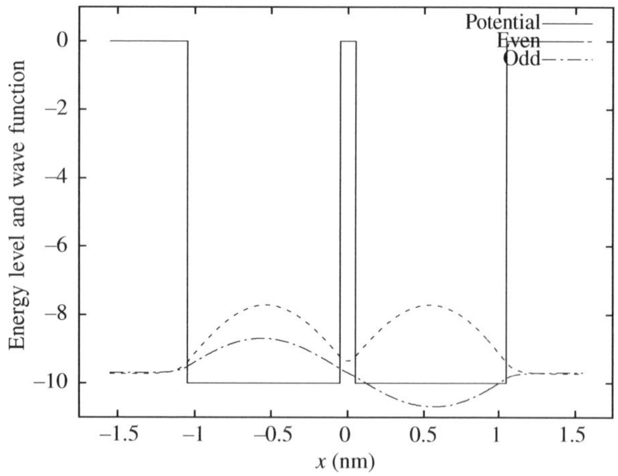

Figure 7.10 Even and odd wave functions for electron bound in two rectangular potential wells; well width = 1 nm, step width = 0.1 nm, depth = 10 eV, energy level 1.

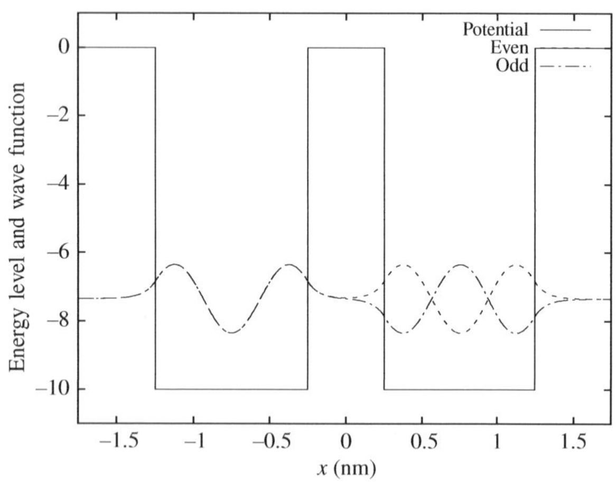

Figure 7.11 Even and odd wave functions for electron bound in two rectangular potential wells; well width = 1 nm, step width = 0.5 nm, depth = 10 eV, energy level 3.

Again there is a small separation of the energy levels for the even and odd parity solutions in this case. Narrowing the well separation for the third energy level to 0.1 nm, as plotted in Figure 7.12

```
$ ./quantum_2rectwell_root 1 0.1 10 0.5110 -1 0 0.1 -7.3 e >| quantum_2rectwell_root.dat
-1.97740536663 0.237502893173 0.549342667037 -7.47891084228
$ ./quantum_2rectwell_root 1 0.1 10 0.5110 1 0 -0.1 -7.3 o >| quantum_2rectwell_root.dat
1.89237998005 0.189828896451 -0.44224965654 -7.23538256824
```

We now see that for an energy level closer to the top of the potential wells, narrowing the separation distance causes a significant split in the energy levels between the even and odd parity solutions.

Plotting the wave function solutions using the data stream output from our program produces the following plots. These four plots are for the first and third energy levels, and for the two separation distances of 0.5 and 0.1 nm. The curves shown together on each graph are for the even and odd parity cases. The larger magnitudes of the wave function at the center of the step barrier for the narrow separation cases give us the indication that bringing the wells into closer vicinity greatly increases the likelihood of the electron tunneling between potential wells, and the electron probabilistically being shared between the wells. We use the usual graphical convention in these plots: the vertical coordinate that corresponds to the zero value of the wave function $\psi(x)$ for a given state is the coordinate that corresponds to the allowed electron energy level for that state plotted against the potential energy.

Another important observation to make with these plots and the program text output is the *splitting* of the energy levels from the single well case to those associated with the even and odd parity solutions in the twin well case. This energy

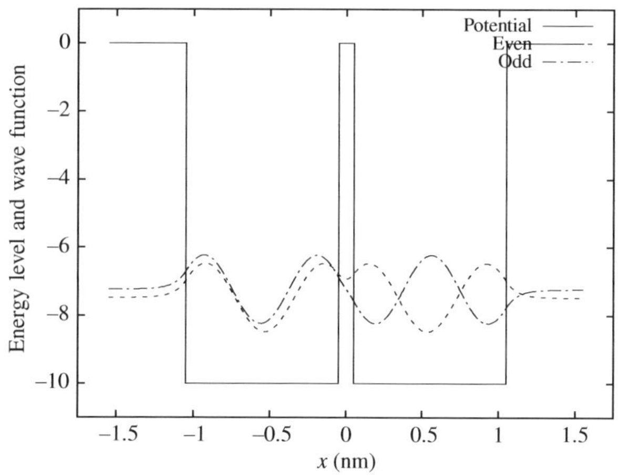

Figure 7.12 Even and odd wave functions for electron bound in two rectangular potential wells; well width = 1 nm, step width = 0.1 nm, depth = 10 eV, energy level 3.

level splitting becomes clearly apparent, as the well separation is small. Since the even parity solutions support the highest magnitude wave function probabilities in the center of the potential barrier, and, therefore, the highest probability of electron exchange between the wells, it is the even parity solution where we see the allowed energy level lowering, and the odd parity solution the energy level rising.

We can hardly consider the simple physical problem of Figure 7.7 to be a very rigorous model of two ionic atomic sites. This geometry is of only one dimension and not three; there is only one free electron being considered as bound in the potential wells, and the potential profiles have unrealistic abrupt transitions. Nonetheless, we can see the general tendency of the mathematical solutions for these adjacent potential wells to allow the electron to share probability between the wells and to split two allowed energy levels. This tendency increases markedly as the separation between the wells decreases.

Looked at from an energy level perspective, the level splitting will drop the even parity allowed energies below that for widely separated wells. From a force perspective, a dropping energy level with a decreasing separation implies that there is an attractive force between the two potential well sites. This tendency for the solutions lets us have a glimpse at the quantum mechanical principles of what is known as a covalent bond between atoms in a molecule [2, Chap. 9; 3, Chap. 4].

The first and third energy levels analyzed above required primarily a cosine function dependence of the wave functions inside the wells to be consistent with the continuity requirements. On the other hand, continuity of the solutions for the second and fourth energy levels will require *primarily sine function dependence* inside the wells. To complete the analysis, the process of sketching the general shapes of the wave function solutions for these energy levels, guessing corresponding coefficient values, and running the `quantum_2rectwell_root.c` program to find actual solutions remains as Exercise 7.4 for the student. The solutions at these levels will demonstrate similar energy level splitting and consequent bonding energy, but even more dramatically for the fourth level, owing to its higher probability of the electron being shared statistically between the wells. After the development of direct numerical solutions to the Schrödinger equation in Section 13.6, we will return to this simple model of the covalent bond with more realistic potential profiles in Exercise 13.7.

7.8 Exercises

7.1 Extend the statics system with two masses on three strings analyzed in Section 7.3 to comprise three masses on four flexible strings, as shown in Figure 7.13.

Using the example program `strings2.c` as a starting point, write a new program `strings3.c` to solve for the angles and string tensions shown in Figure 7.13 with the physical parameters $L_{total} = 10$ units, $L_1 = 4$, $L_2 = 4$, $L_3 = 4$, $L_4 = 4$, $W_1 = 100$ units, $W_2 = 200$ units, and $W_3 = 150$ units.

7.2 Write a program `flipflop.c` to solve the Kirchoff's Current Law problem of the nonlinear flip-flop circuit from Section 7.4. Use the physical parameters $R_L = 15 \times 10^3$ ohms, $R_0 = 1 \times 10^3$ ohms, $V_T = 2.5$ V, and $V_X = 0.5$ V in a parameter structure to pass to the object functions. As in Section 5.5, you may use static allocation for the required two-dimensional vectors. Compiling and running the main program with two sets of initial conditions should show

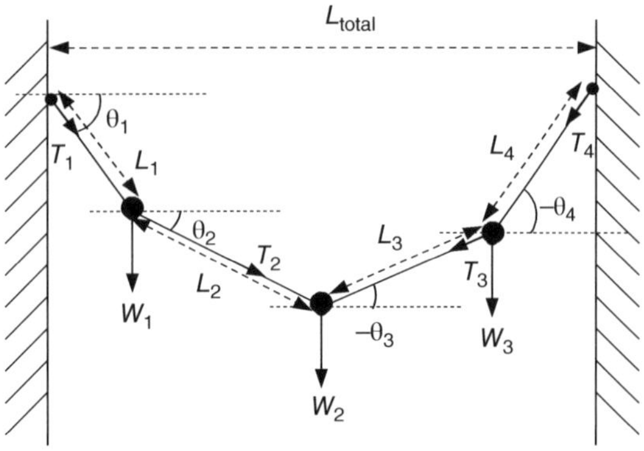

Figure 7.13 Nonlinear statics problem of strings and three masses.

the two stable states the circuit can settle in, each state corresponding to a root of the nonlinear system. These two circuit states represent the storage of the state of a single binary bit, 0 or 1, in the bit map diagrams of Figures 2.4, 2.6, and 2.7. To formulate your initial guesses for this solution, make use of the knowledge that for this circuit network composed solely of resistive elements that only dissipate energy, the potentials of internal nodes must lie between 0 and 5 V.

7.3 Write a program `navigation.c` to solve the hyperbolic navigation triangulation problem described in Section 7.5. Place the primary transmitter at the origin of a three-dimensional Cartesian coordinate system. The three secondary transmitters are located at example coordinates $(1 \times 10^6, 00)$, $(0, 1 \times 10^6, 0)$, and $(0.0, 1 \times 10^6)$, respectively, in meters. The `navigation.c` should accept six command line arguments. The first three are the measured delay times t_{10}, t_{20}, and t_{30} in seconds, and the last three are initial guess values of the unknown position coordinates (x, y, z) that are to be solved for. Test your program with example, delay times of 0.1, 0.2, and 0.3 ms, respectively.

7.4 Following the development in Section 7.7, sketch the general shapes of the four wave function solutions for the second and fourth energy levels for the same twin well geometry of 10-eV deep wells of 1 nn widths. From the solution of the single well potential profile, we know that the corresponding energy levels will be at approximately −8.8 and −5.3 eV. Construct a table of good guess values for the problem of finding the allowed electron energies for the system solution vector following Table 7.1. Compile and run the example program `quantum_2rectwell_root.c` for two well separation distances of 0.5 and 0.1 nm to demonstrate the splitting of the even and odd parity allowed energy levels, and plot the resulting wave functions in a manner similar to Figures 7.9 to 7.12.

8

Monte Carlo Simulation

The extensive literature devoted to Monte Carlo simulation is essentially a computer simulation of what happens in the building pictured in Figure 8.1. This building is the Grand Casino of Monte Carlo, located in the Principality of Monaco. In this chapter, we explore the branch of numerical analysis that is heavily dependent on the laws of statistics, and is, therefore, named after an institution that owes its opulence to the application of the same laws of statistics in order to extract wealth from its patrons over a large number of trials.

In previous chapters, we have made a careful study of algorithms designed to bring numerical data in programs as close as possible to ideal values from mathematical relationships. In this chapter, we turn to techniques to generate and make use of long sequences of numerical data designed to conform to statistical distributions.

8.1 Algorithms: Applications of Pseudorandom Numbers

What sort of computation applications require generating random numbers? First, there are, of course, certain physical processes such as statistical mechanics and quantum mechanics that are intrinsically characterized by statistical laws. We can naturally simulate such processes and systems by generating numbers that represent fundamentally random physical parameters, such as quantum mechanical tunneling events. Second, some analyses involve integrations over multidimensional surfaces and volumes, and systematically covering multidimensional spaces with uniform sampling can involve a prohibitive amount of computation if the dimension is high. In these cases, statistical sampling of a multidimensional space can be a relatively efficient way to approximate surface or volume integrals.

What is a "random" number in the context of digital computation? Imagine throwing truly "random" darts at a real number line between two limiting values. The sequence of x coordinates where each dart lands represents an ideal random variable. We would expect the sequence of x values to exhibit the three statistical properties:

1. Uniform sampling: All x values within a given range are equally likely.
2. Uncorrelated values: The past x values from previous darts have no effect on the next value.
3. Unpredictable samples: There is no way to predict the likelihood of a given x value.

The fundamental computation hardware diagrammed in Figure 1.6 is based on the completely deterministic design of digital electronic networks. If the same computation procedure is run on this machine with the same numerical input, we would expect the same numerical results. In fact, if results differ each time identical computation is repeated, we would conclude the hardware is faulty. Therefore, strictly speaking, achieving property 3 above is not possible in the context of conventional digital hardware. (There are contemporary extensions of computing hardware that do include special electronic circuits that are capable of generating truly random bits derived from the statistics of quantum mechanical limits of electronic devices.) So algorithms that run on conventional processors are capable of generating sequences of numbers that exhibit only properties 1 and 2 above, and these are called "pseudorandom" sequences. However, in Section 8.5 we will introduce a facility of the Linux operating system that is designed to make use of the practical complexity of a contemporary shared process operating system and hardware to offer an excellent approximation to property 3.

Let us assume that we have available an algorithm that makes use of the system of Figure 1.6 to generate a sequence of integer values that exhibit properties 1 and 2 above. That is, as we inspect a large number of integers that are produced in sequence, we will find the coverage of a span of numbers between some limits to be uniform within a good approximation, and the likelihood of a particular value coming next in the sequence is never dependent on the history of past samples. Since

Computational Physics Using C: Efficient Programming with Ease, First Edition. John W. Fattaruso. © 2026 John Wiley & Sons, Inc.
All rights reserved, including rights for text and data mining and training of artificial intelligence technologies or similar technologies.
Published 2026 by John Wiley & Sons, Inc. Companion Website: http://www.wiley.com/go/CompPhysUsingC

Figure 8.1 The Grand Casino in Monte Carlo.

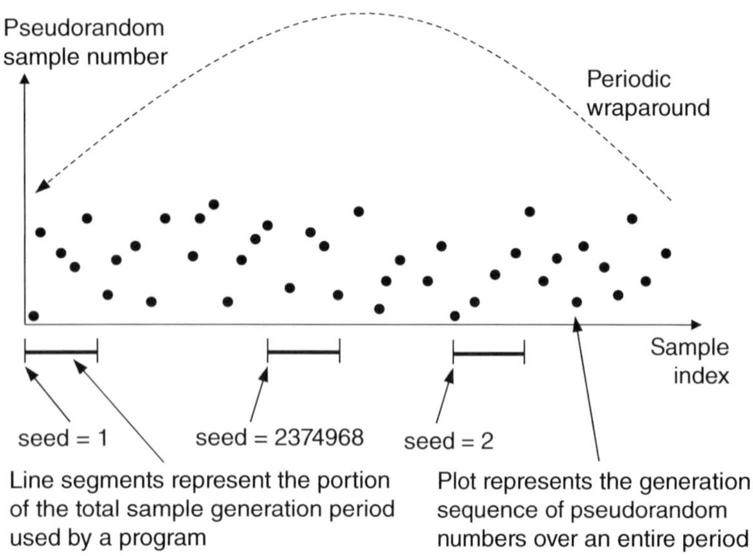

Figure 8.2 Period and seeding of a pseudorandom number generator.

integers may be represented by a finite number of binary bits, there will be some finite limit in the length of the sequence of pseudorandom values before the sequence will deterministically repeat and cover exactly the same sequence again and again. This process is illustrated in Figure 8.2.

As we seek the best algorithms for pseudorandom number generation for our modeling needs, our interest is always in finding a method with the longest possible period before its sequence repeats, for a given number of bits in an integer representation. If, in the example shown in Figure 8.2, a particular modeling task requires a sequence of pseudorandom integers of a length represented by the short line segments, we should ensure that we are using a generator with a period that is at least as long as the sequence needed for a complete computation execution. Otherwise, the statistical assumptions we make in our modeling efforts may easily be violated by using repeated numbers. We also may wish to simulate modeling our system with a pseudorandom sequence that approximates the unpredictability property 3 above. We can do this by making the initial value of our generator algorithm, called the "seed" value, available as a programmatic parameter. This is illustrated in Figure 8.2. When we initialize the sequence at some seed value, we will locate the current position of the algorithm at whatever place that value deterministically occurs, and subsequent pseudorandom samples must follow along by the intrinsic order of the algorithm. By supplying different seed values for different executions of our modeling program, we can approximate a completely random process. In Section 8.5, we will see how to automatically seed a pseudorandom algorithm with a number provided by the Linux operating system, and achieve, to a very good approximation, unpredictable random numbers.

8.2 Algorithms: Linear Congruential Method

Much study has been devoted to numerical algorithms that generate a pseudorandom integer value governed by a prescribed sequence. Since we generally consume a very large number of these values in a typical application program, computational efficiency for the task of calculating the next integer in the sequence is important. One of the most popular general purpose algorithms of this type is the Linear Congruential pseudorandom number algorithm [10, Chap. 10; 11, Chap. 7]. This class of generator is based on the iteration formula

$$r_{i+1} = (a \cdot r_i + c) \quad \text{mod } M \tag{8.1}$$

where r_i is the current pseudorandom value and the formula uses it to calculate the next value r_{i+1}. This formula requires very few arithmetic operations in the central processing unit (CPU). To design a practical algorithm, proper values of the parameters a, c, and M must be found to make the sample period as long as possible, up to the limit of $M - 1$ imposed by the mod M operation. Popular choices for the modulus M are the so-called *Mersenne primes*, in the form $M = 2^n - 1$ for some integer exponent n. This choice will allow some values of the multiplier a so that the generator will have the maximum possible period of $M - 1 = 2^n - 2$ with an offset $c = 0$. This saves one arithmetic operation in each generation cycle, with the tradeoff that we must eliminate the state of 0 being seeded. Three examples for different values of n are

$$r_{i+1} = (11 \cdot r_i) \text{ mod } 2^5 - 1 = (11 \cdot r_i) \text{ mod } 31 \tag{8.2}$$

$$r_{i+1} = (46 \cdot r_i) \text{ mod } 2^7 - 1 = (46 \cdot r_i) \text{ mod } 127 \tag{8.3}$$

$$r_{i+1} = (16807 \cdot r_i) \text{ mod } 2^{31} - 1 = (16807 \cdot r_i) \text{ mod } 2147483647 \tag{8.4}$$

Other choices that allow efficient arithmetic are using a power of two for the modulus, or $M = 2^n$ for some integer n. This choice, along with an additive offset $c \neq 0$, will allow the modulus operation to be essentially performed by simply overflowing the arithmetic and logic unit (ALU) register that is limited to n physical bits, and also allows state 0 to be present in the cycle.

To try some example generators, let us seed the generator in Equation (8.2), with the parameters $a = 11$, and $M = 2^5 - 1 = 31$, with the initial value of 1 and look at the first 80 samples of the resulting sequence of r_i:
1 11 28 29 9 6 4 13 19 23 5 24 16 21 14 30 20 3 2 22 25 27 18 12 8 26 7 15 10 17 **1** 11 28 29 9 6 4 13 19 23 5 24 16 21 14 30 20 3 2 22 25 27 18 12 8 26 7 15 10 17 **1** 11 28 29 9 6 4 13 19 23 5 24 16 21 14 30 20 3 2 22 ...

The initial state of 1 is in boldface type above. We can see by inspection the cycle of samples repeats and cycles back to the state of 1 after each period of $m = 2^5 - 2 = 30$ samples.

Figure 8.3 graphically shows the r_i sequence generated by the iteration formula in Equation (8.3). The samples begin at the seed value of 1 in the lower-left corner, and subsequent sample indices i are the x coordinate for this plot. Careful inspection of the dot pattern will show the pattern repeating after a sequence of 126 integers, plotted on the y axis.

These first two pseudorandom generators have periods and bit resolution that are too small to be of much practical use, but are useful to visualize how these linear congruential iteration formulas work. On the other hand, the generator in Equation (8.4) has the resolution to be of good general use, and we can test out this algorithm with the following simple program

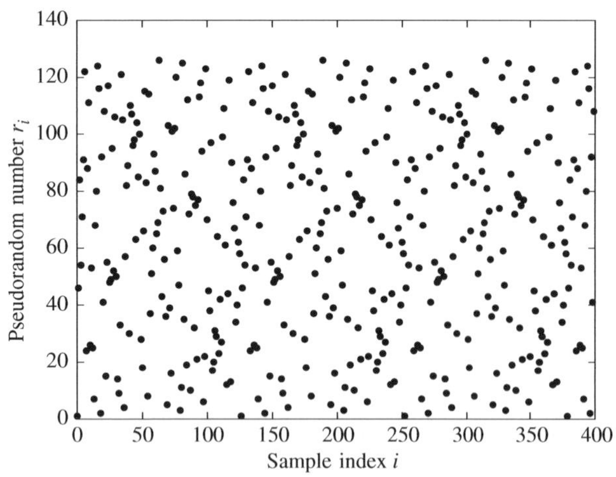

Figure 8.3 Four hundred samples of a linear congruential pseudorandom number generator, modulus = 127.

```
1  #include <stdio.h>
2  #include <stdlib.h>
3  #include <math.h>
4
5  int main(int argc,char *argv[]) {
6    long long unsigned int random_state;
7    int i_sample,n_samples;
8
9    if (argc != 3) {
10     fprintf(stderr,"%s <number of samples> <seed>\n",argv[0]);
11     exit(1);
12   }
13   n_samples = atoi(argv[1]);
14   random_state = atoll(argv[2]);
15   for (i_sample = 0; i_sample < n_samples; i_sample++) {
16     printf("%llu\n",random_state);
17     random_state = (16807LLU * random_state) % 2147483647LLU;
18   }
19   exit(0);
20 }
```

The modulus M for this algorithm in line 17 is $2^{31} - 1$, or 2 147 483 647. The value of the multiplier a in line 17 is carefully chosen at 16 807, and this gives a repetition period of the maximum 2 147 483 646. This is an enormous number of states to sequence through before repeating, so it is not practical to print out or plot the entire period. However, the program does output the first 1000 samples, after an example seed value of 1, which are plotted in Figure 8.4. Note that there are visible clusters and sparse areas of samples, which is the actual behavior of independent, uniform sampling, but there are no apparent patterns among the samples.

There are a few important details about the coding of this program that are instructive. In line 6 we use the data type `long long unsigned int` for the `random_state` variable, which will allocate a 64-bit integer for its storage. This is required because the r_i state value will range from 1 to $2^{31} - 1$, and require 31 bits of integer precision. Multiplying by $a = 16807$ in line 17 for many samples will result in a temporary product requiring many more bits than 31, before the mod operator is applied. The size of integers declared as `long unsigned int` is somewhat dependent on particular implementations of the C language on various processors, but is often only 32 bits, which can overflow after the multiplication by c. Note the corresponding conversion factor of `%llu` in the call to `printf()` in line 16 and the constant suffix designations `LLU` in line 17. Even with the relatively limited sampling of only the first 1000 samples in Figure 8.4, we can see a good coverage of the range from 1 to 2 147 483 646.

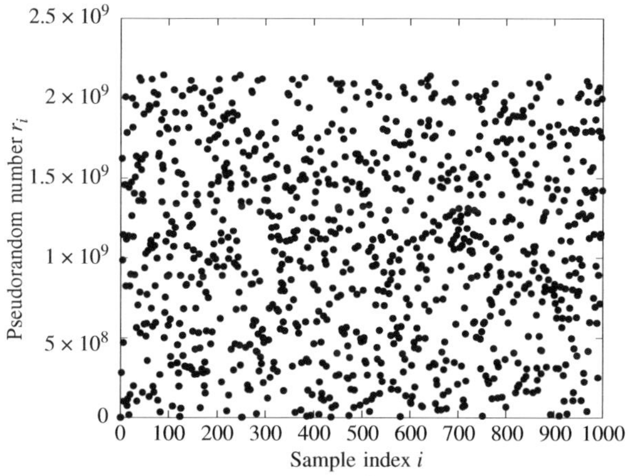

Figure 8.4 One thousand samples of a linear congruential pseudorandom number generator, modulus = 2 147 483 647.

8.3 Coding: The Pseudorandom Number Generator Tool

The code for a practical tool for generating a sequence of pseudorandom numbers with either uniform or approximately Gaussian distributions can be found in the files `random.c` and `random.h`, part of the `CPLibrary` package from the Student Companion website. There are many pseudorandom generators in easily available libraries for more exacting requirements, such as the standard `drand48()` routine. But this tool has reasonable properties for the example code in this text, and gives us the opportunity to work with, and possibly modify, a generator tool from basic code.

The `random.c` tool contains three C functions: `random_seed()`, `random_gen()`, and `random_normal_gen()`. `random_seed()` is used to seed the algorithm at any unsigned integer greater than 0 and less than 2147483647. The simple demonstration program above reported the sequence of generated samples in their fundamental integer form, but in practice this is not the most useful form of pseudorandom numbers for scientific applications. Generally we would like to have available an approximation to random numbers distributed on a real number line to simulate physical processes. Note in the source code for the `random_gen()` function that the double precision return value is formed by multiplying the current integer state by $\frac{1}{2147483647}$, which ensures a return value uniformly distributed ≥ 0 but < 1. Also note that the `long long unsigned int` internal variable that holds the current r_i state of the generator is named `random_state`. This must be declared as a `static` variable in the C code, which will allocate memory for its storage in a location that will remain undisturbed for the entire time the program is executing. The default case for variables declared within program blocks is that their values remain defined only while a function is executing.

How do we generate a uniform distribution for a particular application over some arbitrary range other than 0 to 1? We can see three example operations in Figure 8.5 that illustrate an easy technique. By combining multiplication and addition operations on the value returned from `random_gen()` in the C application code, we may spread the uniform distribution to run between any desired limits.

We have seen how the basic pseudorandom generator has been designed to simulate samples with a uniform random distribution over some limits. However, there may be other distributions that are required by a particular simulation task. For applications that, for example, require a pseudorandom number sequence whose values are distributed in a normal, or Gaussian, distribution we may call the `random_normal_gen()` function instead of `random_gen()`. This routine makes use of central limit theorem of statistics, that the sum of a large number of random variables that share the same distribution will approach a normal distribution. The random variable S_n formed as

$$S_n = \sqrt{\left(\frac{12}{n}\right)} \sum_{i=1}^{n} U_i - \frac{n}{2} \tag{8.5}$$

from the sum of n variables U_i uniformly randomly distributed between 0 and 1 will approximate a standard normal distribution, with mean of 0 and standard deviation of 1. By choosing $n = 12$, we conveniently eliminate some arithmetic in Equation (8.5), and this sum is implemented in the `random_normal_gen()` function. With only 12 variables summed,

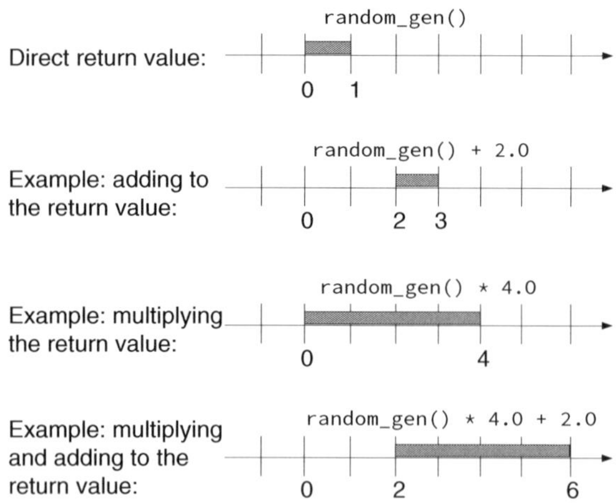

Figure 8.5 Transforming samples from a pseudorandom number generator to a uniform distribution between any arbitrary limits.

the conformance to a normal distribution is only approximate, but is of good enough fidelity for many applications. Other values of mean and variance are generated by the function with simple scale and offset operations as the function's return value is formed.

8.4 Application: Monte Carlo Simulation of π

As a first example application of Monte Carlo methods, we look at an exercise in estimating the value of π, which is in essence an example of performing a multidimensional integration. Consider the areas of a unit circle, with radius 1 and centered on the origin, and a square extending from -1 to $+1$ in both the x and y directions. The area of the circle is $\pi r^2 = \pi$, and the area of the square is 4. If we were to generate a large number N_{total} of pseudorandom two-dimensional point coordinates within the square, and then simply count the number of points N_{inside} that happen to lie inside the unit circle, since we assume our pseudorandom generator exhibits uniform sampling, the ratio of the area of the circle to that of the square must tend to be $\frac{N_{\text{inside}}}{N_{\text{total}}}$ as N_{total} grows large. This implies that the area of the circle must be $4\frac{N_{\text{inside}}}{N_{\text{total}}}$, and that area will be an estimate of the value of π. Figure 8.6 shows graphically how 5000 example pseudorandom points are distributed by an example Monte Carlo run. Random points inside the unit circle are shown in black, and those outside the unit circle are plotted in gray.

To generate pseudorandom points within the square, we may simply call the `random_gen()` function twice, and use a transformation of the type illustrated in Figure 8.5 to spread the returned samples to cover the range of -1 to 1. One sample is assigned to the x coordinate of the point, and the other is assigned to its y coordinate. As we write the code to test whether or not a candidate point is inside or outside the unit circle, we may first consider calculating the distance from the origin to the point $r = \sqrt{x^2 + y^2}$ and comparing r with 1. However, note that we can save the processing cost of calling the `sqrt()` function by simply testing if $r^2 \leq 1$ instead. The program `pi_MonteCarlo.c` is available for download from the Student Companion website. Compiling and running the main program for simulating the value of π

```
$ gcc -Wall -Wshadow -O2 -o pi_MonteCarlo pi_MonteCarlo.c CPLibrary/random.c -lm
$ ./pi_MonteCarlo
./pi_MonteCarlo <N trials> <seed>
$ ./pi_MonteCarlo 10000 1
3.12120000
$ ./pi_MonteCarlo 100000 1
3.14256000
$ ./pi_MonteCarlo 1000000 1
3.14335200
```

For all three runs, we have used the example seed value of 1, but we could choose any integer less than $2^{31} - 1$. With a larger number of random trials made, we can see that the accuracy of the estimate of π is improved. Three successive runs with the same seed will return the same sequence of pseudorandom numbers, and, therefore, give the same estimate of the value of π

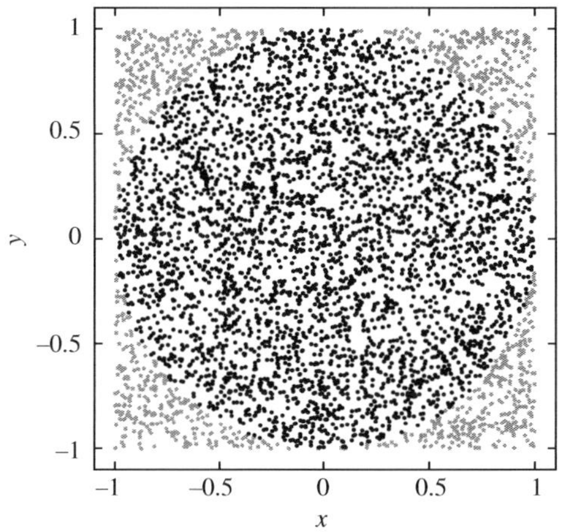

Figure 8.6 Monte Carlo simulation of the value of π with 5000 samples.

```
$ ./pi_MonteCarlo 1000000 1
3.14335200
$ ./pi_MonteCarlo 1000000 1
3.14335200
$ ./pi_MonteCarlo 1000000 1
3.14335200
```

It should be noted here that generating patterns of points over a two-dimensional space as illustrated in Figure 8.6 is actually a simple initial test of the conformance of a candidate pseudorandom number algorithm to properties 1 and 2 above. The human visual system is often able to spot patterns in such a plot that reveal either a lack of uniform coverage or a correlation between successive samples. More sophisticated statistical tests are necessary to fully characterize candidate algorithms, but if our pseudorandom points are being transformed for a particular application, a simple graphical test can often provide a first confirmation that we are achieving the desired statistical properties.

8.5 Coding: The Linux /dev/random Device

When operating under the Linux operating system, a special software device is available for reading by programs that is capable of generating a sequence of bytes with properties that are an excellent approximation to the third property listed in Section 8.1. One of the many tasks the operating system assumes is collecting measurements from the dozens of processes running in the background of a Linux system and assembling them in a memory storage buffer to be read by other programs when needed. The data is dependent on physical processes such as network traffic and the operation of peripheral hardware devices, and is very close to being unpredictable. We can access this sequence of bytes by using calls to operating system functions that are equivalent to reading bytes from an apparent data file by the name of `/dev/random` in the Linux file system. The demonstration program in the file `pi_MonteCarlo_devrandom.c` from the Student Companion website, which is a modification of the basic π simulation program of `pi_MonteCarlo.c`, shows an example of reading eight binary bytes from this source and using it to form a seed value for the pseudorandom number generator. We first use an array of bytes in memory addressed as an array of `char` types as required by the call to the `read()` system function to fill a memory buffer with the random bytes. Then we use the cast operator to assign the constant `char` pointer `buffer` to the pointer variable `ptr`, which has been declared as a pointer to `long unsigned int`. This allows us to refer to the buffer of eight memory bytes alternatively as a `long unsigned int` type to call the `random_seed()` function. This is illustrated in Figure 8.7, with an example array of random byte values in hexadecimal, and is yet another example of the power we have available with the facility of pointer variables. A manually chosen seed value is no longer required on the command line. Running this program three times gives us three different Monte Carlo estimates for the value of π:

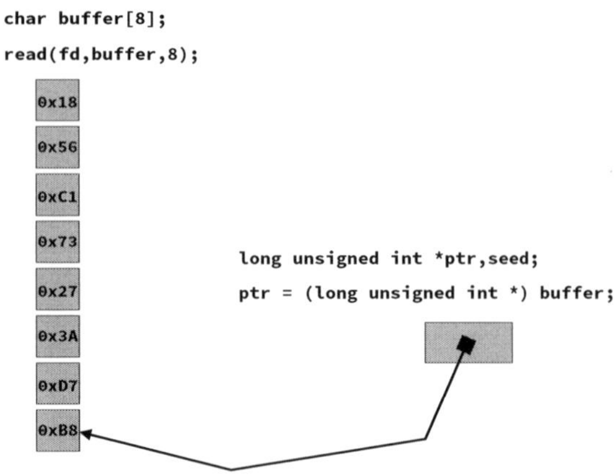

Figure 8.7 Casting a pointer to an array of bytes to a pointer to an integer.

```
$ gcc -Wall -Wshadow -O2 -o pi_MonteCarlo_devrandom pi_MonteCarlo_devrandom.c CPLibrary/random.c
    ↪ -lm
$ ./pi_MonteCarlo_devrandom
./pi_MonteCarlo_devrandom <N trials>
$ ./pi_MonteCarlo_devrandom 1000000
3.14294400
$ ./pi_MonteCarlo_devrandom 1000000
3.14310800
$ ./pi_MonteCarlo_devrandom 1000000
3.14110400
```

8.6 Application: Random Walks

We now look at modeling a stochastic physical process that is the basis for phenomena such as diffusion of one gas-like molecular species within another gas-like substance, as well as early universe behavior. The basic model we will examine is called a "random walk," invoking an image of a person setting off for a walk, but choosing a random direction for each step taken. For the ease of plotting and visualization, we'll work in two dimensions, which is sufficient to gain an understanding of real-world three-dimensional processes.

See Figure 8.8 for an illustration of a particle subjected to randomly directed forces in two dimensions once each time step. We use this simple model to approximate, for example, molecules in a gas being subjected to the random battering of Brownian motion. We initialize the coordinates of the particle $x_0 = 0$ and $y_0 = 0$. Then at each time step we generate Δx and Δy as pseudorandom coordinates on the unit circle and update $x_i = x_{i-1} + \Delta x$ and $y_i = y_{i-1} + \Delta y$. The resulting sequence of particle coordinates will follow a two-dimensional random walk. We may also keep track of the quantity D_n as the total distance traveled in the walk after n steps.

$$D_n^2 = (\Delta x_1 + \Delta x_2 + \Delta x_3 + \cdots + \Delta x_n)^2 + (\Delta y_1 + \Delta y_2 + \Delta y_3 + \cdots + \Delta y_n)^2 \tag{8.6}$$

After n steps, the distance D_n will simply be the distance to a point reached by the sum of all the individual steps

$$\begin{aligned} D_n^2 &= \Delta x_1^2 + \Delta x_2^2 + \Delta x_3^2 + \cdots + \Delta x_n^2 + 2\Delta x_1 \Delta x_2 + 2\Delta x_1 \Delta x_3 + \cdots \\ &\quad + \Delta y_1^2 + \Delta y_2^2 + \Delta y_3^2 + \cdots + \Delta y_n^2 + 2\Delta y_1 \Delta y_2 + 2\Delta y_1 \Delta y_3 + \cdots \end{aligned} \tag{8.7}$$

For large n, the expected value of all the cross terms will tend toward zero, and so on average

$$\begin{aligned} D_n^2 &\approx \langle \Delta x_1^2 + \Delta x_2^2 + \Delta x_3^2 + \cdots + \Delta x_n^2 + \Delta y_1^2 + \Delta y_2^2 + \Delta y_3^2 + \cdots + \Delta y_n^2 \rangle \\ &= n \cdot \langle r^2 \rangle \end{aligned} \tag{8.8}$$

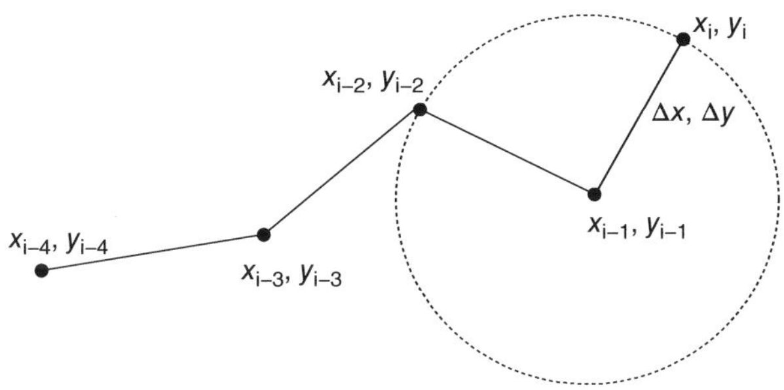

Figure 8.8 A two-dimensional random walk.

Therefore,

$$D_n \approx \sqrt{n} \cdot r_{\text{rms}} \tag{8.9}$$

Suppose, for instance, we were to open a bottle of perfume in the center of a room in still air. Equation (8.9) is a simple model for the expectation of the distance from the bottle where the smell of perfume may be perceived after diffusing through the air for some number of uniform time steps.

Consider the main program simulating a random walk, with any number of particles traversing any number of steps, in the file `walk_path.c` from the Student Companion website. The sequence of position coordinates for each particle in the ensemble is sent to the output stream, with each particle in a separate data set for selectively plotting with `gnuplot`.

As we consider coding strategies for generating pairs of pseudorandom points on a unit circle for the Δx and Δy required at each step, we may at first imagine generating a random angle uniformly between 0 and 2π radians and calling `cos()` and `sin()` functions to land a coordinate pair directly on the unit circle, as shown in the source code

```
while (i_step < n_steps) {
  angle = random_gen() * TWO_PI;
  dx = cos(angle);
  dy = sin(angle);
  x += dx;
  y += dy;
  printf("%.8f %.8f\n",x,y);
  i_step++;
}
```

This method does indeed work well, and does distribute points uniformly around the unit circle as desired for the concept of a random walk. But consider an alternative, and more efficient, scheme is shown in the source code

```
while (i_step < n_steps) {
  dx = 2.0 * (random_gen() - 0.5);
  dy = 2.0 * (random_gen() - 0.5);
  radius = sqrt(dx * dx + dy * dy);
  if (radius > 1.0) continue;
  dx /= radius;
  dy /= radius;
  x += dx;
  y += dy;
  printf("%.8f %.8f\n",x,y);
  i_step++;
}
```

This method borrows from our experience with the simulation of the value of π in Section 8.4. Trial points are generated so that they are distributed uniformly throughout the square area from $(-1, -1)$ to $(+1, +1)$, and then are accepted or rejected by a comparison of the radius of the point from the origin with 1. If the point is on or within the unit circle, it is moved radially out to the unit circle by dividing both coordinates by the radius. At the cost of a small minority of rejected points, illustrated by the gray points in Figure 8.6, we avoid calling C library trigonometric functions, which may be expensive in computation time. We may compile and run the main program, for an example, simulation of the paths of six particles over random walks of 100 steps with

```
$ gcc -Wall -Wshadow -O2 -o walk_path walk_path.c CPLibrary/random.c -lm
$ ./walk_path
./walk_path <N steps> <N particles> <seed>
$ ./walk_path 100 6 2 >| walk_path.dat
```

Starting an interactive gnuplot session and giving the commands

```
gnuplot> unset key
gnuplot> set size square
gnuplot> set xlabel "X-coordinate"
gnuplot> set xrange [-15:15]
gnuplot> set ylabel "Y-coordinate"
gnuplot> set yrange [-15:15]
gnuplot> plot 'walk_path.dat' i 0 u 1:2 w l,'' i 1 u 1:2 w l,'' i 2 u 1:2 w l,'' i 3 u 1:2 w l,''
    ↪ i 4 u 1:2 w l
```

produces the plot seen in Figure 8.9 tracing the paths of all six particles over their walks.

Let us build from this example program another tool to numerically visualize the simple model of Equation (8.9). We write the code only to keep track of the final positions of all the particles participating in their independent random walks,

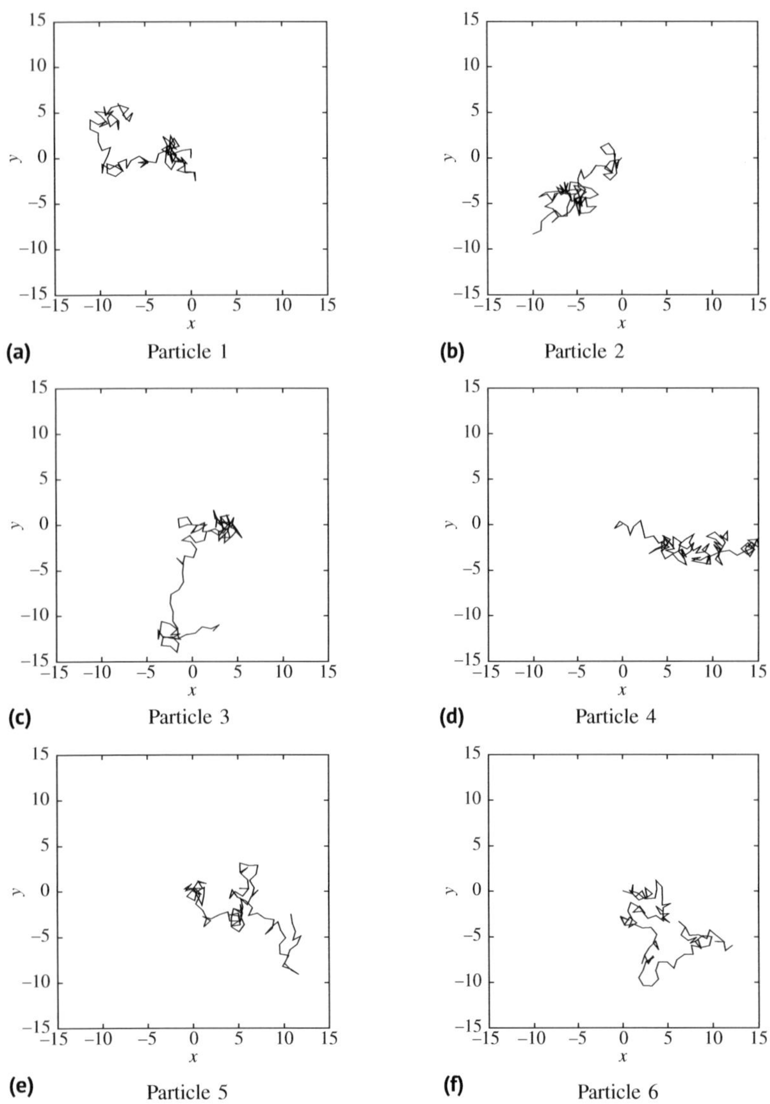

Figure 8.9 Paths of six example particles in two-dimensional random walks of 100 steps.

after a simulation lasting for a specified number of time steps. The first data set sent to the output stream is a collection of the final coordinates of all particles, so by running the program a number of times with different numbers of time steps, a scatter plot of all final coordinates provides a snapshot of how we expect the diffusion process to develop. In addition, the program computes a mean distance traveled from the origin across the particle ensemble at each step time point. These mean values are sent to the output, along with the square root of the time step index, as a function of the step index so that Equation (8.9) may be graphically tested. The main program for this simulation is in the file `walk.c` from the Student Companion website. Compiling and running the main program for simulating the final positions of the particles subjected to three random walks of varying durations

```
$ gcc -Wall -Wshadow -O2 -o walk walk.c CPLibrary/random.c CPLibrary/array.c -lm
$ ./walk
./walk <N steps> <N particles> <seed>
$ ./walk 100 100 1 >| walk_100.dat
$ ./walk 1000 100 1 >| walk_1000.dat
$ ./walk 10000 100 1 >| walk_10000.dat
```

Starting an interactive gnuplot session and giving the commands

```
gnuplot> set linetype 1 linecolor -1 linewidth 2 dashtype 1
gnuplot> set linetype 2 linecolor -1 linewidth 2 dashtype 2
gnuplot> unset key
gnuplot> set size square
gnuplot> set xlabel "X-coordinate"
gnuplot> set xrange [-150:150]
gnuplot> set ylabel "Y-coordinate"
gnuplot> set yrange [-150:150]
gnuplot> $data << EOD
gnuplot> 0 0 10
gnuplot> EOD
gnuplot> plot 'walk_100.dat' i 0 u 1:2 w p pointtype 7 pointsize 1,'$data' u 1:2:3 with circle
gnuplot> $data << EOD
gnuplot> 0 0 31.623
gnuplot> EOD
gnuplot> plot 'walk_1000.dat' i 0 u 1:2 w p pointtype 7 pointsize 1,'$data' u 1:2:3 with circle
gnuplot> $data << EOD
gnuplot> 0 0 100
gnuplot> EOD
gnuplot> plot 'walk_10000.dat' i 0 u 1:2 w p pointtype 7 pointsize 1,'$data' u 1:2:3 with circle
```

produces the plots seen in Figures 8.10 through 8.12.

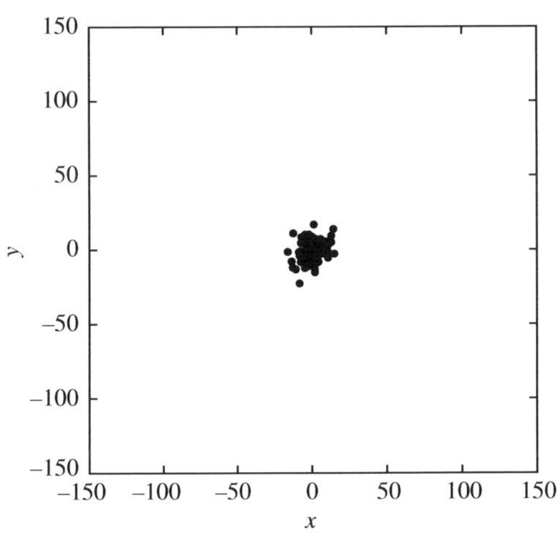

Figure 8.10 Two-dimensional random walk of 100 particles, 100 steps.

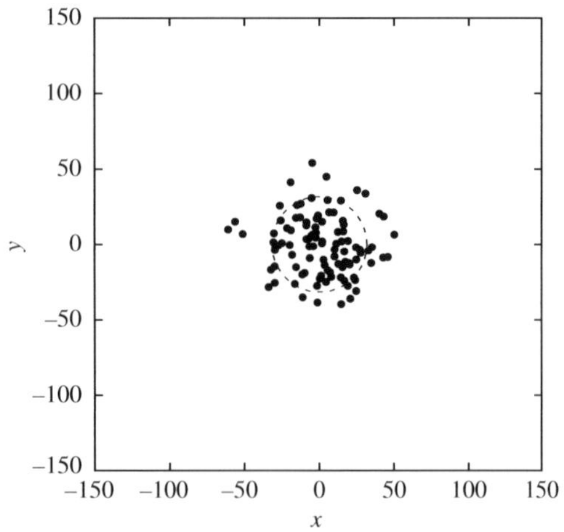

Figure 8.11 Two-dimensional random walk of 100 particles, 1000 steps.

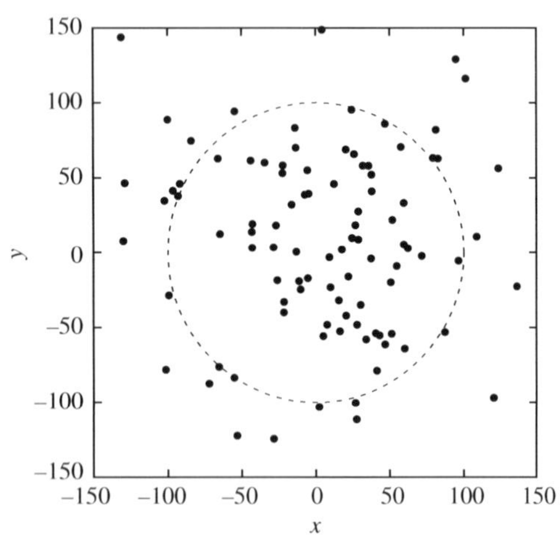

Figure 8.12 Two-dimensional random walk of 100 particles, 10 000 steps.

Each `plot` command above to `gnuplot` renders a scatter plot of the extent of the wandering of each particle in the ensemble and draws a circle with a dashed line at a radius set to the square root of the number of steps. We can graphically imagine the circle to be at about the root-mean-square (RMS) value of the final distances across the ensemble. Let us see this dependence more clearly with the final plot of Figure 8.13. This plot shows the expected ensemble average of the progression of walk distances over a large number of independent 10 000 step walks, extended to cover three dimensions, as a function of the step index. The coding work for generating this plot, which can be seen as a satisfying graphical representation of the model of Equation (8.9), is available as Exercise 8.1.

With some understanding of the basic properties of random walks that these simulations provide, the reader can now look back at Figures 2.14a and 2.14b and see how the process of the accumulation of roundoff errors in fundamental arithmetic operations that are limited to a finite number of binary bits is also basically another example of a one-dimensional random walk.

8.7 Application: Radioactive Decay Revisited

It is also natural to apply Monte Carlo simulation to physical processes that are dependent on quantum mechanical probabilities, such as nuclear decay. This section returns to the topic of radioactive nuclear decay that we first explored in Section 5.9 and examines Monte Carlo simulation of decay statistics.

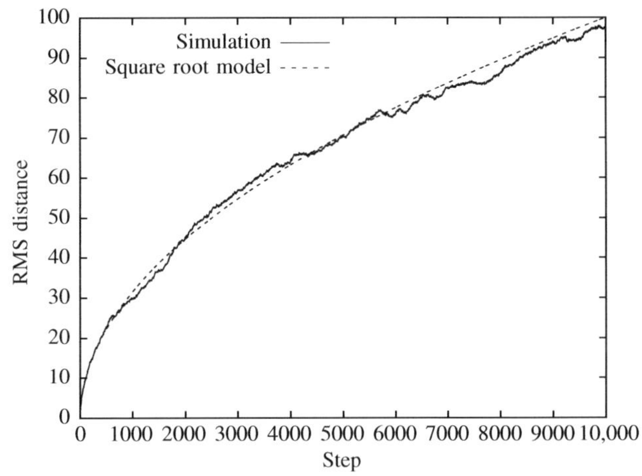

Figure 8.13 RMS distance of 100 particles in a three-dimensional random walk.

From Section 5.9, we know that for the simple decay process of a nuclide "A" to nuclide "B" the solution for the continuous limit of the population of A as a function of time is

$$N_A(t) = N_{A0}e^{-\lambda_{AB}t} \tag{8.10}$$

where N_{A0} is the initial population of the A nuclide and λ_{AB} is the probability that an A nuclide will decay to a B nuclide in a unit time interval.

However, instead of treating the continuous limit of the nuclide population functions, in this chapter we develop a simple model for the stochastic decay events and keep track of populations with accounting of actual integer nuclide counts. See the source code file `decay.c` from the Student Companion website. Assuming the integer variables `NA` and `NB` that hold the current populations of A and B nuclides have been initialized to N_{A0} and zero, respectively, for each simulated time step, we can run a sequence of random trials that essentially tests whether or not each of the A nuclides in the current population is selected to decay into B nuclides during the current step, according to the prescribed probability λ_{AB}. We execute the following iterative loop and generate a pseudorandom number between 0 and 1 in each iteration. An `if` statement conditionally tests whether the pseudorandom sample is less than the probability. If so, the current A nuclide is counted as having decayed, and another integer variable keeping track of decay events `NAB` is incremented. After the iterations through all remaining A nuclides are completed, the current populations of each nuclide are updated. In this way, we may simulate and send to the output stream the current population at each time step.

```
n_trial = NA;
NAB = 0;
while (n_trial > 0) {
  test = random_gen();
  if (test <= lambdaAB) NAB++;
  n_trial--;
}
NA -= NAB;
NB += NAB;
```

However, typically what is observable in a laboratory setting is the decay rate, as released decay products can be measured by instruments. So with an initial decay rate of R_{AB0}, we would expect a rate that also exhibits a similar time dependence

$$R_{AB}(t) = -\frac{dN_A}{dt} = \lambda_{AB}N_{A0}e^{-\lambda_{AB}t} = R_{AB0}e^{-\lambda_{AB}t} \tag{8.11}$$

This decay count is easily accessible from the simulation code above also by printing out the current value of the `NAB` variable each time step. Since the observed decay rate can vary over orders of magnitude due to the exponential factor, often a more convenient quantity to plot graphically is the logarithm of the decay rate. The probability of the nuclear decay can then be modeled as a simple linear slope with

$$\log(R_{AB}(t)) = \log(R_{AB0}) - \lambda_{AB}t \tag{8.12}$$

and this measurement can be used to infer the probability λ_{AB} and equivalently the half life of Equation (5.36). We will treat this problem in detail in Section 11.7.

Compiling and running the `decay.c` program, for an example, case of an initial population of 10^4 and a decay probability of 10^{-2} per time step

```
$ gcc -Wall -Wshadow -O2 -o decay decay.c CPLibrary/random.c -lm
$ ./decay
./decay <NA initial> <lambdaAB> <N steps> <seed>
$ ./decay 10000 0.01 500 123 >| decay.dat
```

Starting an interactive gnuplot session and giving the commands

```
gnuplot> set xlabel 'Time step'
gnuplot> set ylabel 'Population of A and B'
gnuplot> plot 'decay.dat' u 1:2 w l lt 2 t '$N_A$','' u 1:4 w l lt 3 t '$N_B$',10000*exp(-0.01*x)
    ↪ w l lt 1 t 'Model',10000*(1-exp(-0.01*x)) w l lt 1 notitle
```

will plot the A and B populations versus time. Also drawn on the plot seen in Figure 8.14 is the model of Equation (8.12), which shows good agreement with this Monte Carlo simulation.

Following the solution developed in Section 5.9, the continuous limit solution of a two-step radioactive nuclear decay process, from nuclide A to B and then from nuclide B to C, is

$$N_A(t) = N_{A0}e^{-\lambda_{AB}t} \tag{8.13}$$

$$N_B(t) = \frac{\lambda_{AB}}{\lambda_{BC} - \lambda_{AB}} N_{A0} \left(e^{-\lambda_{AB}t} - e^{-\lambda_{BC}t}\right) \tag{8.14}$$

$$N_C(t) = \frac{1}{\lambda_{BC} - \lambda_{AB}} N_{A0} \left[\lambda_{BC}(1 - e^{-\lambda_{AB}t}) - \lambda_{AB}(1 - e^{-\lambda_{BC}t})\right] \tag{8.15}$$

where N_{A0} is the initial population of the A nuclide, and the probabilities λ_{AB} and λ_{BC} govern the decay process from A to B and from B to C, respectively, in a unit time interval. By applying Monte Carlo techniques for modeling events with particular probabilities, a simulation of this two-nuclide decay sequence generates the plot shown in Figure 8.15.

The three populations plotted in Figure 8.15 are the integer counts in each time step accounted for with a similar process as used in the simulation for Figure 5.9. The characteristics in this plot agree quite closely with the solution of the differential equations of Figure 5.24, also plotted in the figure. The coding of this modeling program is Exercise 8.2 for the student. Furthermore, by applying the coding techniques of this section and Exercise 8.2, more elaborate chains of events governed by statistical probabilities may be simulated.

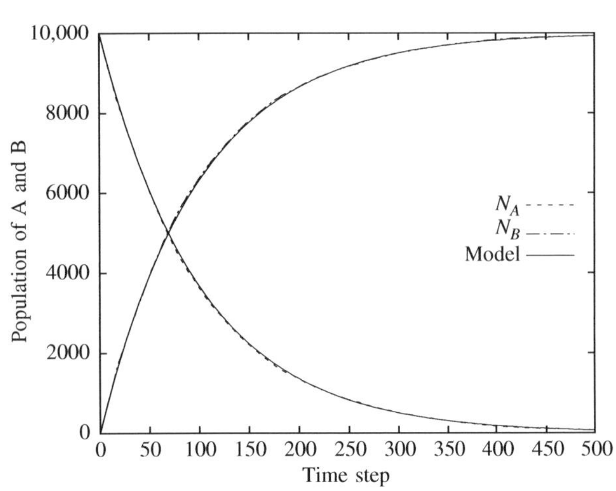

Figure 8.14 Monte Carlo simulation of radioactive decay, $N_{A0} = 10^4$, $\lambda_{AB} = 0.01$.

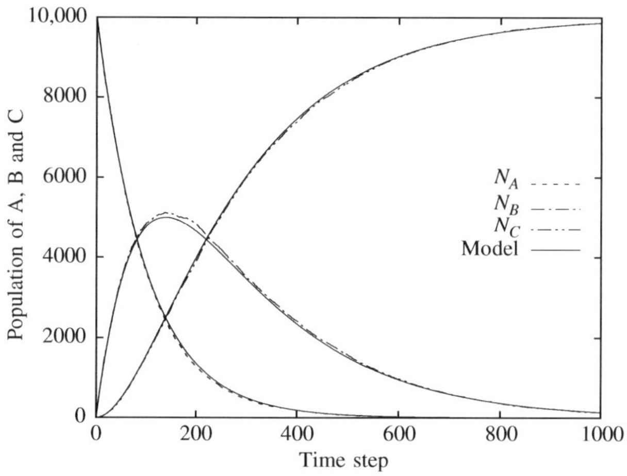

Figure 8.15 Monte Carlo simulation of two-step radioactive decay, $N_{A0} = 10^4$, $\lambda_{AB} = 0.01$, $\lambda_{BC} = 0.005$.

8.8 Application: Classical Scattering

In Section 4.8, we developed a model for a collision between two ideally hard spherical balls. Now combined with the tools for Monte Carlo simulation in this chapter, we are able to model the statistics governing the bombardment of a target by a large number of incident scattering objects.

Consider the arrangement shown in Figure 8.16. Ball 1 is incident from the left, initially located at a position vector $\vec{r}_{10}$ and with initial velocity $\vec{v}_{10}$. The target is ball 2, initially stationary and located at the origin.

Collisions are modeled by initially setting the y coordinate of $\vec{r}_{10}$ at some constant negative distance, but randomly varying the initial x and z coordinates to uniformly sample a square area to provide a random impact parameter for the collision. The range variation of the x and z coordinates is from $-(R_1 + R_2)$ to $(R_1 + R_2)$, where R_1 and R_2 are the radii of the two balls. This range ensures that some sort of collision must occur, even if only at a glancing angle. Repeating collision simulations with these initial coordinates varying gives a Monte Carlo simulation of the range of scattering trajectories that are physically possible, given the parameters of the ball masses and radii.

The main program `scattering.c`, available for download from the Student Companion website, gives the user the option of selecting the masses and radii of the two balls. As we first saw in Section 4.8, we make use of the `sscanf()` function to translate a command line argument with several initial condition parameters into the numeric variables for the simulation with the lines

```
sscanf(argv[1],"%lf,%lf,%lf,%lf",&r10y,&v10y,&R1,&m1);
sscanf(argv[2],"%lf,%lf",&R2,&m2);
tfinal = atof(argv[3]);
```

Compiling and running the program for simulating 100 random trials, each of two instances of classical scattering, is shown below. The first case is for a symmetrical collision of the ball masses being identical at 1, and the second case for an asymmetrical arrangement of $m_1 = 1$ and $m_2 = 100$. In both cases, we set the ball radii to 1 unit, the initial incident velocity to 1 at an initial y coordinate of -10.

```
$ gcc -Wall -Wshadow -O2 -o scattering scattering.c CPLibrary/collide_balls.c CPLibrary/ball.c
    ↪ CPLibrary/random.c -lm
$ ./scattering
./scattering <r10y,v1by,R1,m1> <R2,m2> <tfinal> <num trials>
$ ./scattering -10,1,1,1 1,1 15 100 >| scattering_1.dat
$ ./scattering -10,1,1,1 1,100 15 100 >| scattering_100.dat
```

Starting an interactive gnuplot session and giving the commands

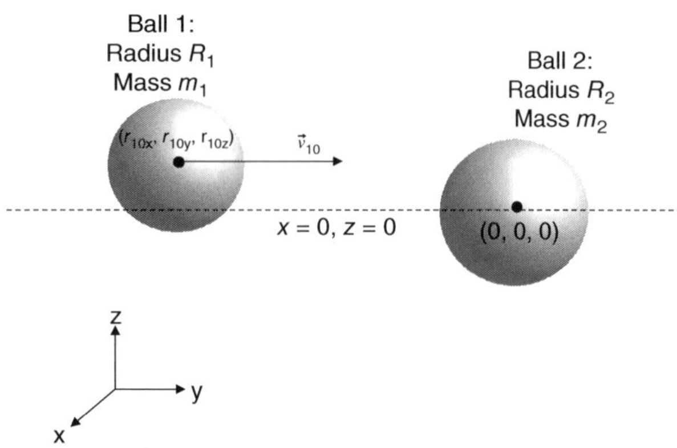

Figure 8.16 Scattering model with two ideal spherical balls.

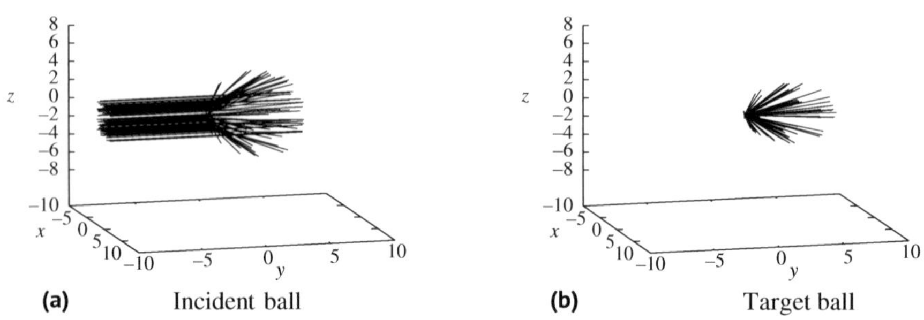

Figure 8.17 Monte Carlo simulation of scattering trajectories with $m_1 = 1$, $m_2 = 1$.

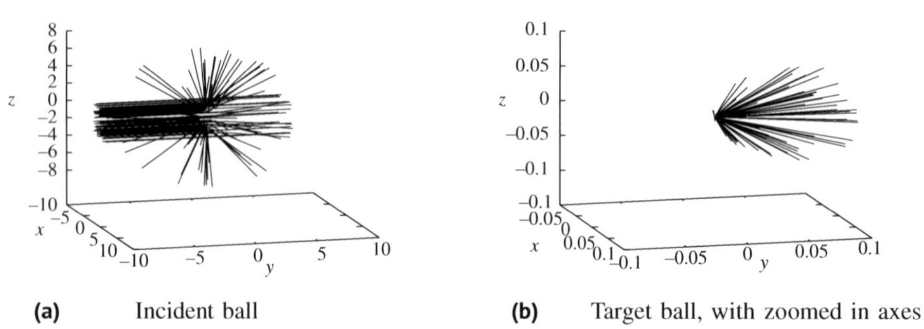

Figure 8.18 Monte Carlo simulation of scattering trajectories with $m_1 = 1$, $m_2 = 100$.

```
gnuplot> set ticslevel 0.25
gnuplot> set xlabel 'X'
gnuplot> set ylabel 'Y'
gnuplot> set zlabel 'Z'
gnuplot> set view 75,75
gnuplot> splot 'scattering_1.dat' u 2:3:4 w l t 'Incident','' u 5:6:7 w l t 'Target'
gnuplot> splot 'scattering_100.dat' u 2:3:4 w l t 'Incident','' u 5:6:7 w l t 'Target'
```

produces the plots seen in Figures 8.17 and 8.18. With the balls of equal mass in Figure 8.17 the angles at which Ball 1 rebounds are all in the positive y direction, and Ball 2 is displaced at a high velocity to the right.

The contrasting case is with a much more massive Ball 2 in Figure 8.18. Some rebound angles for Ball 1 include those traveling to the left in the negative y direction. At first glance at Figure 8.18b, it may appear that Ball 2 is being displaced as much as in the first case, but note that for clarity this plot has been drawn with much smaller scales than in Figure 8.17b, and in actuality, the more massive Ball 2 is barely being moved.

We can consider these two scattering scenarios to be the classical limits of two historically famous models of atomic scattering. The Monte Carlo simulation in Figure 8.17 represents the bombardment of nuclei assuming the original nineteenth-century postulate of the so-called "plum pudding" model of an atomic structure. The simulation in Figure 8.18 represents the observations in the famous Rutherford scattering experiment (the proverbial "cannon ball bouncing off a sheet of tissue paper") that proposed the current model of atomic structure, with the heavy nuclear mass concentrated in a small region.

8.9 Application: Corner Pocket Shots

Some physical processes that involve a large number of interacting objects can be treated effectively with statistical distributions, even though the mechanics governing the interactions are entirely deterministic. Consider, as an example, the distribution of balls around a pool table after a good break shot and other plays. Let's make use of a pseudorandom distribution of balls to be able to visualize the challenge of successfully making a corner pocket shot in the game of pool. Suppose that a player will choose to attempt a corner pocket shot if the cue ball and a target ball end up somewhere within the rectangles indicated in Figure 8.19.

See Figure 8.20 for the geometry applicable to this problem. The dimensions shown are those of a standard pool table, and the standard balls used have diameters of 2.25 inches. The cue ball, Ball 1 with radius R, is centered initially at a point (x_1, y_1) on the pool table coordinate system. It is struck with a cue stick by the player to send it at some trajectory angle θ_1 with respect to the x axis. The target ball, Ball 2 with radius R, is initially centered at a point (x_2, y_2). The impact of the collision with the cue ball must send Ball 2 to some target point (x_t, y_t), aimed at falling into the pocket. The center of the pocket is at (x_p, y_p), but the nominal width of the pocket is twice the radius of the balls, and the sides of the pocket are slanted to give some guidance to a ball approaching the pocket. So for simplicity, we assume that the target point (x_t, y_t) of Ball 2 can range from $(x_p - R, y_p)$ to $(x_p, y_p - R)$ for the shot to fall successfully into the pocket.

How difficult is this shot? In other words, how precisely must the trajectory angle θ_1 be realized by the player in order to send the target ball into the pocket? To answer this question, let us solve for the angle θ_1 that corresponds to the target point being at the two edges of the pocket.

In order to calculate the angle θ_1 corresponding to a target point, we make the same assumption about the ideal collision as was used in Section 4.8, namely that the impulse vector of the collision may only be directed normal to the surface of the colliding balls, and that its action takes place over an infinitesimal time duration. For the simple purpose here of only determining the linear path of the target ball after collision, we need not be concerned with any velocities or momenta, but rather only the path geometries. With these assumptions, and with the special case of Ball 2 being stationary before the impact, we know that at the point of impact, the center coordinates of Ball 1 must be colinear with the target point and the center of Ball 2. Therefore, we find the common slope between the three points as

$$\alpha = \frac{y_2 - y_c}{x_2 - x_c} = \frac{y_t - y_2}{x_t - x_2} \tag{8.16}$$

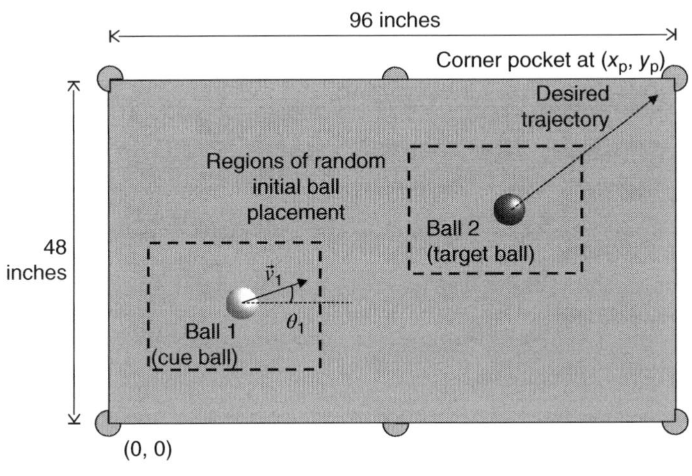

Figure 8.19 Schematic of a corner pocket shot in pool.

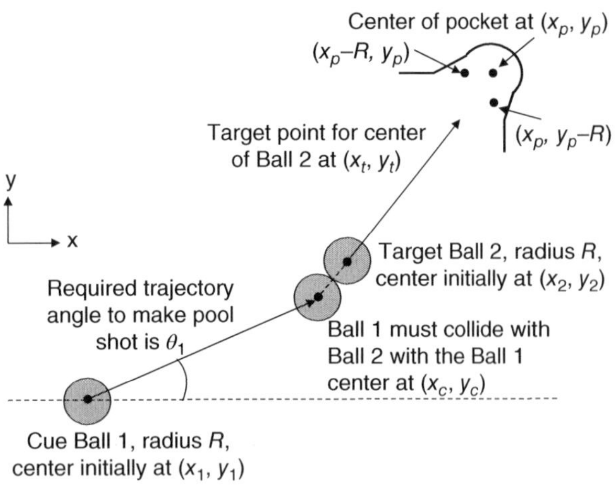

Figure 8.20 Geometry for solving for the cue ball trajectory angle to make a corner pocket shot.

or

$$(y_2 - y_c) = \alpha(x_2 - x_c) \tag{8.17}$$

The collision occurs when the separation distance between the two ball centers is exactly twice the ball radius, or

$$(x_2 - x_c)^2 + (y_2 - y_c)^2 = 4R^2 \tag{8.18}$$

Equations (8.17) and (8.18) may be solved simultaneously to find

$$\begin{aligned} x_c &= x_2 - \frac{2R}{\sqrt{1+\alpha^2}} \\ y_c &= y_2 - \alpha\frac{2R}{\sqrt{1+\alpha^2}} \end{aligned} \tag{8.19}$$

We then find the required cue ball trajectory to be

$$\theta_1 = \tan^{-1}\left(\frac{y_c - y_1}{x_c - x_1}\right) \tag{8.20}$$

This solution may be coded into a C program that makes a number of pseudorandom locations for the two balls according to the layout in Figure 8.19. The required trajectory angles are computed twice according to Equations (8.16) through (8.20), the first time for the Ball 2 target point being at the left edge of the pocket at $(x_p - R, y_p)$, and a second time for the target point being at the right edge at $(x_p, y_p - R)$. Then by subtracting the two solved angles, we may find the precision with which a player must line up the cue impact trajectory, over a large sampling of possible shot layouts.

Let us start our study with a shot covering a relatively short distance on the pool table. The pseudorandom points assigned in a sample Monte Carlo run for 500 trials over some restricted ranges limited to one end of the table are plotted in Figure 8.21. Then the allowable tolerance angles of the cue ball trajectory over each of the 500 trials are plotted in Figure 8.22. Most good pool players would consider trajectories that allow a tolerance of 0.25–0.5° to be of only moderate difficulty, and potential shots in this area would be considered a good strategic choice.

Next we consider a less fortunate ball placement, with pseudorandom placements over 1000 trials shown in Figure 8.23. The required distances of the ball displacements are considerably longer than in the previous example.

From the plot of Figure 8.24 we may visualize that an accomplished pool player must maintain very high precision of impact angles, typically less than a fifth of a degree! For most players, this would be considered a difficult shot, with less likelihood of success.

The coding of this simulation is Exercise 8.3 at the end of this chapter.

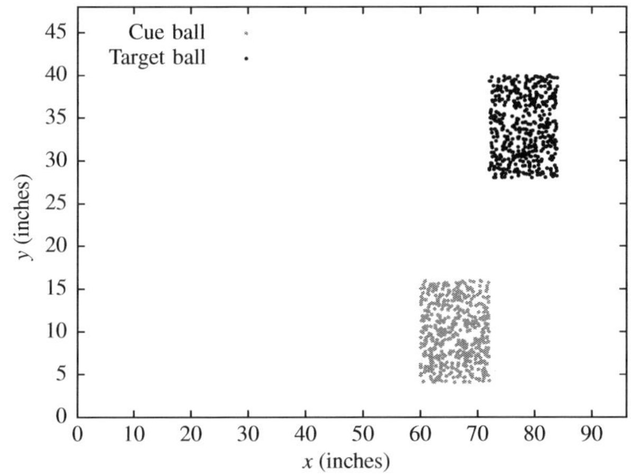

Figure 8.21 Pseudorandom ball placements for a short corner pocket shot.

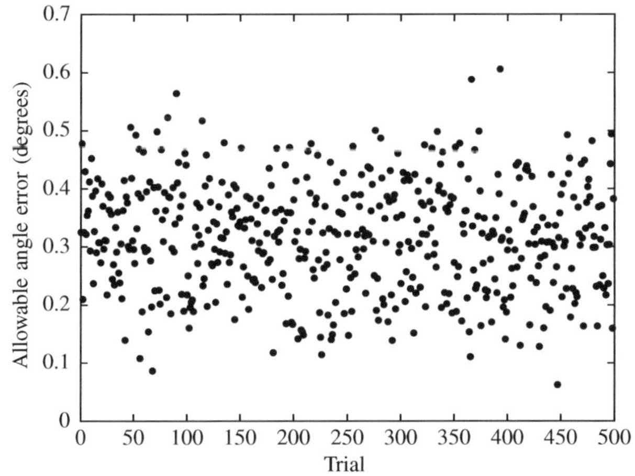

Figure 8.22 Allowable cue ball trajectory angle errors for a short corner pocket shot with pseudorandom ball placement.

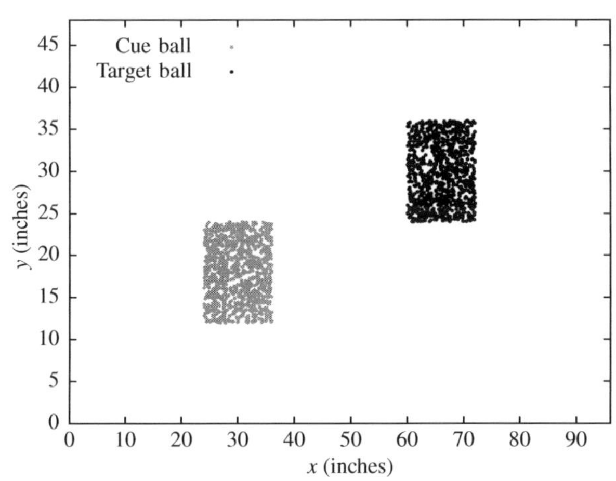

Figure 8.23 Pseudorandom ball placements for a long corner pocket shot.

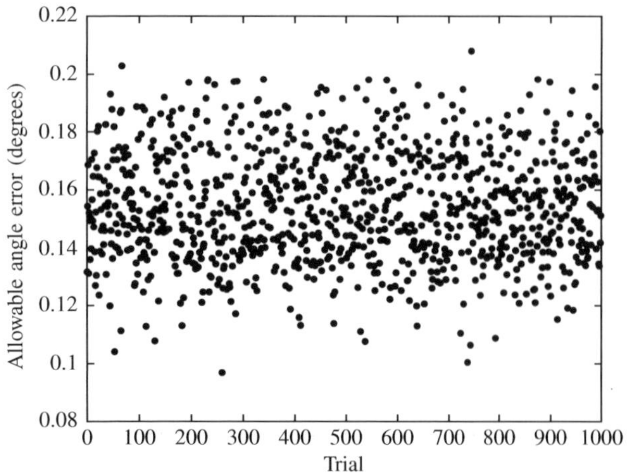

Figure 8.24 Allowable cue ball trajectory angle errors for a long corner pocket shot with pseudorandom ball placement.

8.10 Application: Olbers' Paradox

In Section 8.8, we explored modeling that gave us insight on an atomic scale, but in this section we turn to that which bears on questions on a cosmological scale. When the basic structure of the universe is pondered, a paradox emerges that is generally attributed to Heinrich Olbers. When we look up at the night sky under the condition of low light pollution, we see a primarily dark sky populated with specks of light from stars. Is that observation consistent with the postulate that the universe is of infinite extent and infinitely old? Figure 8.25 shows the geometry of an observation on Earth of two identical stars located at different distances. The apparent disc of more distant stars is smaller than that of nearby stars. We know that the light intensity from a star at a distance R from the Earth decreases as $\frac{1}{R^2}$, but that the circular disc of the sky area obscured by the star also decreases as $\frac{1}{R^2}$. Imagining then the line of site at every angle from an observer in the sky, would every line not terminate in a star at some distance? So why is the night sky primarily dark instead of uniformly filled with a light brightness equivalent to the average star?

The basic geometrical question to answer is: "Do the observed discs from the stars distributed with uniform average density throughout the volume of a large universe cover all the observed sky area?" If we model the geometry for convenience in two dimensions instead of three, this is equivalent to a problem of shooting arrows in a forest of infinite extent consisting of circular tree cross-sections. So we ask is it possible for any arrows to emerge from the forest without getting stuck somewhere in a tree? We can design a Monte Carlo experiment to answer this question.

Imagine shooting a large number of arrows inside the forest at a random angle and testing whether any arrows successfully emerge from the forest as the extent of the forest grows larger. For simplicity, we can limit our model of the forest to a single quadrant from the shooting source. We will plan our Monte Carlo experiment to fill a single quadrant of two-dimensional space with circles of uniform radii centered at random coordinates. We will then "launch" lines from the origin at one corner of the forest in a random direction with an angular heading between 0 and $\frac{\pi}{2}$. We will test if any of the linear flight paths are not obstructed by a circle, and test if the number of unobstructed lines goes to zero as we grow the forest to large dimensions. As we grow the forest, we keep the density of trees constant but increase the number of arrows launched until a clear limit is observed. Figure 8.26 shows the geometry to test whether an arrow will be caught in a tree.

The arrow has been launched into the forest at a random heading $0 < \theta < \frac{\pi}{2}$ in the direction of a circular tree cross-section centered at (x_0, y_0). To test if this arrow hits this tree, we may algebraically test if we can find real solutions to points common to both the line and the circle at some unknown coordinates (x, y). So to find the x coordinate of a possible collision point, we can use the equations of both the line and the circle and set $(x - x_0)^2 + (mx - y_0)^2 = r^2$, and then solve for x. We can rearrange this into a canonical quadratic equation in x, where the coefficients will be known in terms of the geometry.

$$Ax^2 + Bx + C = 0 \tag{8.21}$$

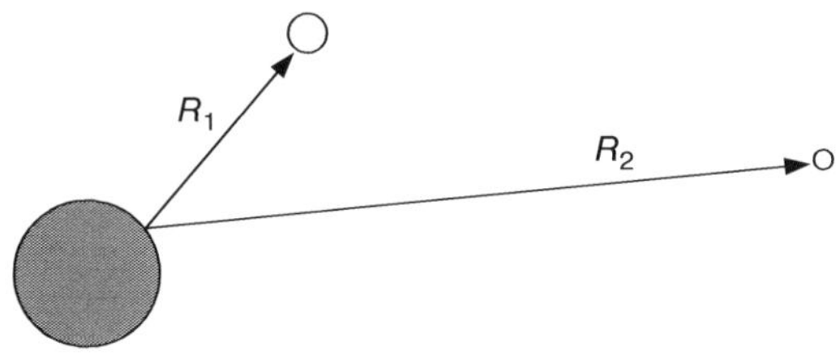

Figure 8.25 Olbers' paradox.

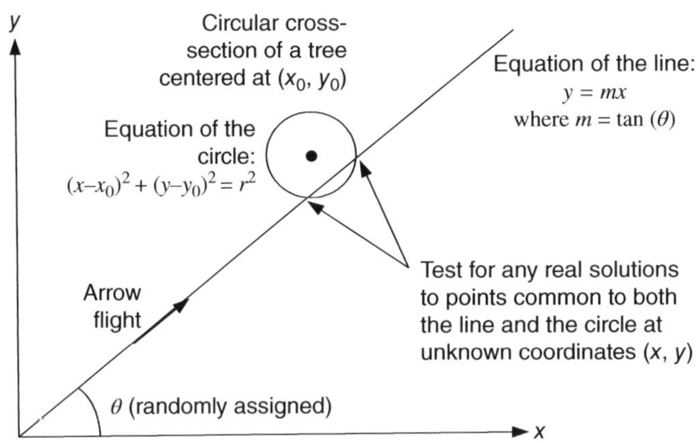

Figure 8.26 Analytic geometry for Olbers' paradox.

where the constants are

$$\begin{aligned} A &= m^2 + 1 \\ B &= -2(x_0 + my_0) \\ C &= x_0^2 + y_0^2 - r^2 \end{aligned} \tag{8.22}$$

The standard quadratic formula can then be used to find the collision x coordinate, under the condition that a collision actually occurs. Here we use the form of the quadratic formula that allows us to cancel out the factor of two in the expression for B in Equation (8.22) and saves a bit of computation.

$$x = \frac{-\left(\frac{B}{2}\right) \pm \sqrt{\left(\frac{B}{2}\right)^2 - AC}}{A} \tag{8.23}$$

We test for the condition that a collision occurs in Figure 8.26 by testing the sign of the discriminant $D = \left(\frac{B}{2}\right)^2 - AC$. If D is positive, then a real solution for x exists.

The coding work for the program `forest.c` is Exercise 8.4 for the student. This program will run Monte Carlo simulations that let us statistically verify the central premise of Olbers' paradox, that is, all linear sight lines in a static universe would eventually terminate in a star given a sufficiently large universe. The program should demonstrate statistics similar to those graphically illustrated in Figures 8.27 through 8.30.

In order to be confident of the statistics from our geometrical model, we will perform the Monte Carlo experiments over an ensemble of a large number of independent forests in which trees have been randomly distributed. We can start with an ensemble of small forests of only 100 size units on each side, with experiments launching 10 arrows into each forest of the ensemble. One example experiment with a particular forest in the ensemble is illustrated in Figure 8.27, where four arrows have been blocked by trees, but six have escaped the bounds of the forest. The program should run a Monte Carlo simulation over an ensemble of 100 randomly constructed forests, and report to the standard output stream how many forests in the ensemble allowed at least one arrow to escape. As we increase the size of the forests in our experiments, we will eventually

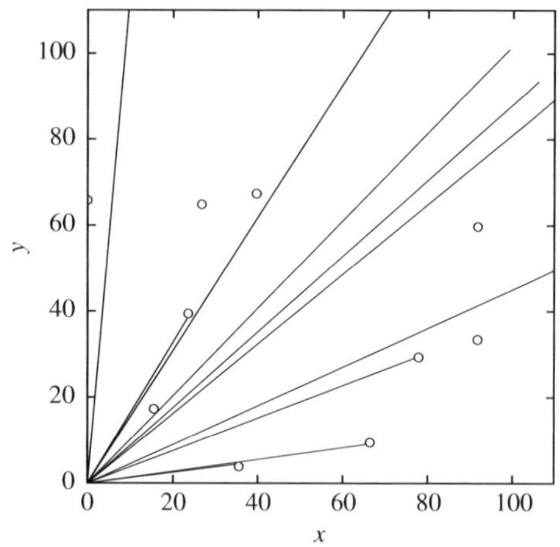

Figure 8.27 Monte Carlo simulation of Olbers' paradox, size = 100, density = 10^{-3}, 10 arrows, hits 4/10, escapes 6/10.

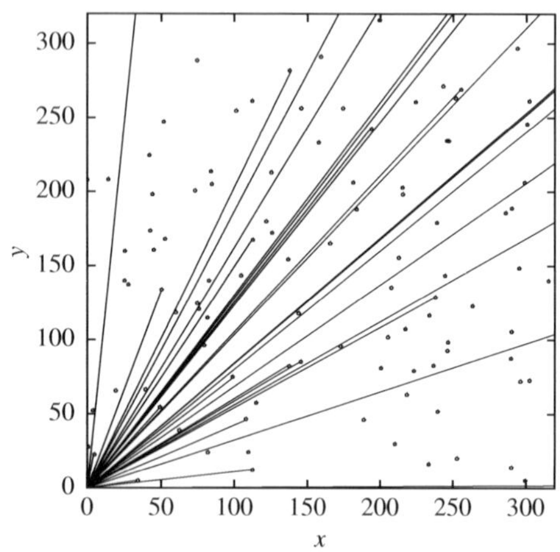

Figure 8.28 Monte Carlo simulation of Olbers' paradox, size = 316.2, density = 10^{-3}, 30 arrows, hits 15/30, escapes 15/30.

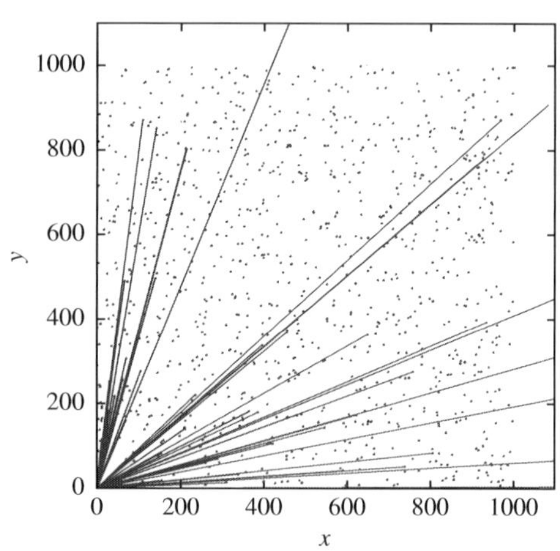

Figure 8.29 Monte Carlo simulation of Olbers' paradox, size = 1000, density = 10^{-3}, 100 arrows, hits 93/100, escapes 7/100.

observe that over all the experiments in the ensemble, no arrow will emerge from all the forests. One example experiment for a forest of sufficient size is illustrated in Figure 8.30.

With our simulations we will verify that, indeed, if a universe is large enough, all sight lines should be blocked eventually by a star. Some centuries after Olbers' observation, measurements were developed indicating that our universe is not static, but expanding, and this resolved the paradox of the dark night sky.

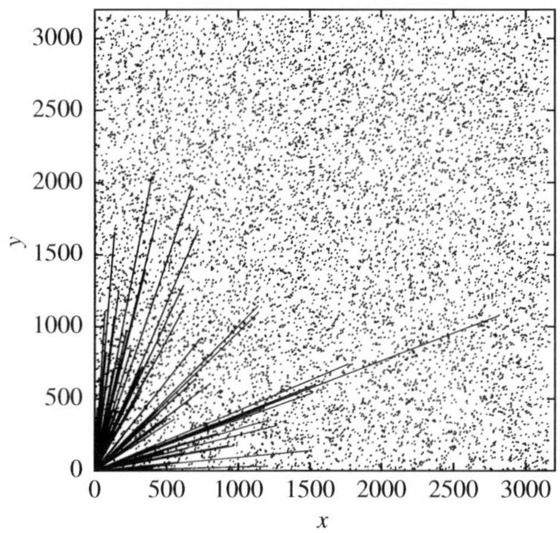

Figure 8.30 Monte Carlo simulation of Olbers' paradox, size = 3162.3, density = 10^{-3}, 300 arrows, hits 300/300, escapes 0/300.

8.11 Application: Ideal Gas Simulation

Developments in the field of statistical mechanics stemmed from the assumption that the number of interacting molecules in a gas was so enormous that we may use statistical laws to characterize the distributions of mechanical properties among microscopic particles. Making use of numerical tools, we can perform Monte Carlo simulations that allow us to visualize the transition between macroscopic mechanics and microscopic distributions in an ideal gas. We return to the ideal two-body collision model of Section 4.8 as a fundamental element.

As we plan how to construct an effective model for an ideal gas, we should, as always, keep computational practicality in mind. One possible approach would be to keep track of the instantaneous position and velocity of potentially many thousands of modeled molecules, as restrained by rebounds off the inner walls of some defined container. Having to keep track of the motion of a very large number of particles, and having to simulate many time steps to encounter the collisions that give rise to the statistics we are aiming at capturing, could easily involve a tremendous amount of computation. So we take a simplified approach here, intended only to model the particle collisions in isolated scenarios covering small regions of the gas volume. We may use Monte Carlo techniques to select and initialize the collision mechanics with random impact parameters that mimic the interaction of particles in a free gas region. This requires keeping track of only the three-dimensional velocities of all the particles in our ideal modeled gas.

Figure 8.31 shows an example scenario that represents the initialization of particle motions leading to collisions. Each time step, a subset of the particles in the gas are randomly selected in pairs to participate in collisions. The velocities of the two particles selected for each collision are used to establish initial particle positions only in the context of a local isolated scenario. The two particles are placed in initial positions calculated so that with their particular velocities at the current time step in the simulation, their centers would collide at the scenario origin. Then the initial position of one of the particles is displaced randomly, similar to the method used in the scattering collisions of Section 8.8, so that a range of random impact parameters will be encountered. The collision is then analyzed with the ideal model of Section 4.8 to arrive at a pair of new particle velocities after this staged collision. These new velocities are stored for possible participation in future time steps.

Refer now to the main program `gas.c`, from the Student Companion website. This relatively complex simulation requires a large number of parameters to be specified when the program is run. The first three arguments that this program expects on its command line are the mass of each particle in the gas, the simulated temperature of the gas, and the radius of each particle. Then the total number of particles in the gas and the number of pairs of particles randomly selected to participate in collisions during each time step are specified. Finally, the number of time steps to run the simulation, the histogram output interval, and the random seed finish the command line. To save on computation effort and output file size, the program computes and outputs a histogram of the speed distribution throughout all the gas particles, but only after a given periodic interval of time steps.

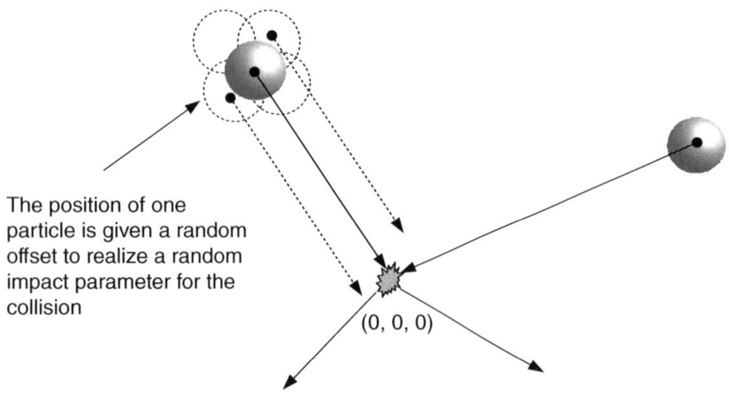

Figure 8.31 Simulated collision between two particles in an ideal gas.

Each particle in the simulated gas is represented by an instance of the structure

```
typedef struct gas_particle {
  double vx,vy,vz;  // x,y,z components of particle velocity
// flag variable =1 if the particle has been paired for a collision in
// current time step
  int paired;
} PART;
```

This structure holds the values of the three components of the velocity vector for each particle in each simulation time step. In addition, an integer flag variable is allocated that will be used in the random selection of pairs of particles to participate in a collision in each time step. The memory for the entire gas is allocated with the line

```
particle = malloc(n_particles * sizeof(PART));
```

with the variable `particle` having been declared to be of type `PART *`. With the expectation that a simulation of a large number of interacting particles will eventually lead to a speed distribution that will conform to the Maxwell–Boltzmann distribution [17, Chap. 20; 23, Chap. 8; 24, Chap. 9; 25, Chap. 11], we use that theoretical result to find a speed to initialize the motion of all of our gas particles for a particular temperature. In the lines

```
  init_energy = 1.5 * BOLTZMANN * T;
  init_speed = sqrt(2.0 * init_energy / m);
  for (i_ball1 = 0; i_ball1 < n_particles; i_ball1++) {
    do {
      particle[i_ball1].vx = 2.0 * (random_gen() - 0.5);
      particle[i_ball1].vy = 2.0 * (random_gen() - 0.5);
      particle[i_ball1].vz = 2.0 * (random_gen() - 0.5);
      radius = sqrt(particle[i_ball1].vx * particle[i_ball1].vx +
                    particle[i_ball1].vy * particle[i_ball1].vy +
                    particle[i_ball1].vz * particle[i_ball1].vz);
    } while (radius > 1.0);
    particle[i_ball1].vx *= init_speed / radius;
    particle[i_ball1].vy *= init_speed / radius;
    particle[i_ball1].vz *= init_speed / radius;
  }
```

we make use of a gas temperature specified on the command line to find an initial average energy and RMS speed for the particles. Then the iterative loop runs through all the particles in the gas and initializes their velocity vectors to be all uniform speed, but directed in a randomly chosen direction in three dimensions. To assign a vector heading uniformly distributed around in a sphere, we extend the method used in Section 8.4 for assigning random points on a unit circle. The random assignments here are repeated until a vector point inside a unit sphere is found, and then the point is moved to the surface of the sphere and scaled to the uniform initial speed.

The main iteration loop for all the time steps of the simulation is controlled by the statement

```
for (i_step = 0; i_step < n_steps; i_step++) {
  ...
}
```

Then within each time step, there is an inner iterative loop that executes for each simulated collision between pairs of particles controlled by

```
for (i_collision = 0; i_collision < n_collisions; i_collision++) {
}
```

For each collision, two candidate pairs of particles are randomly chosen by generating pseudorandom indices into the array of `PART` structures holding the current state of the gas. As random candidate particles are identified for each collision in each iteration of a `do {...} while()` loop, unfortunate pairs are rejected if either the particle has already participated in a collision previously in the current time step, or if the initial positions of the particles determined by the scenario of Figure 8.31 would end up placing the two particles so that they overlap. This last condition is tested with the lines

```
do {
  ...
  d = sqrt((ball1.rx - ball2.rx) * (ball1.rx - ball2.rx) +
           (ball1.ry - ball2.ry) * (ball1.ry - ball2.ry) +
           (ball1.rz - ball2.rz) * (ball1.rz - ball2.rz));
} while (particle[i_ball2].paired != 0 || d < 2.0 * R);
```

But statistically, almost all trials will result in two acceptable particles being selected, and their collision is modeled by the ideal `collide_balls()` tool from Section 4.8. That tool predicts the two particle velocities after the ideal collision, and these velocities are stored back in the array of `PART` structures for future time steps.

The count of particle speeds in each distribution bin is written to the `int` array `hist` after an interval of time steps has passed. This array is allocated at the beginning of the program with the statement

```
hist = malloc(n_hists * NUM_BINS * sizeof(int));
```

so that it allows memory for all the bin counts for all histograms. Then after all the time steps have been processed, the probability histograms of particle speeds are calculated from the bin counts in the `gen_output()` function.

```
void gen_output(int *hist,int n_hists,int n_particles,
                double init_speed) {
  int i_hist,i_bin;
  for (i_hist = 0; i_hist < n_hists; i_hist++) {
    for (i_bin = 0; i_bin < NUM_BINS; i_bin++) {
      printf("%.8g %.9f\n",init_speed * ((double) i_bin) / INIT_BIN,
             ((double) hist[i_hist * NUM_BINS + i_bin]) * INIT_BIN /
             (((double) n_particles) * init_speed));
    }
    printf("\n\n");
  }
}
```

Compiling this program and displaying its "help" line to remind us of the order of expected command line arguments shows

```
$ gcc -Wall -Wshadow -O2 -o gas gas.c CPLibrary/collide_balls.c CPLibrary/random.c
    ↪ CPLibrary/ball.c -lm
$ ./gas
./gas <mass> <temperature> <radius> <N particles> <N collisions> <N timesteps> <histogram
    ↪ interval> <random seed>
```

For some example runs of this simulator, we choose a gas comprising approximate helium atoms at 300 K. The kinetic model in `gas.c` includes three degrees of freedom from three-dimensional particle translation, but no particle rotation, and this model corresponds to the assumed behavior in a monatomic gas such as helium. So the first three command line arguments are a mass of 6.64647×10^{-27} kg, a temperature of 300 K, and a radius of approximately 1×10^{-9} m. By extracting

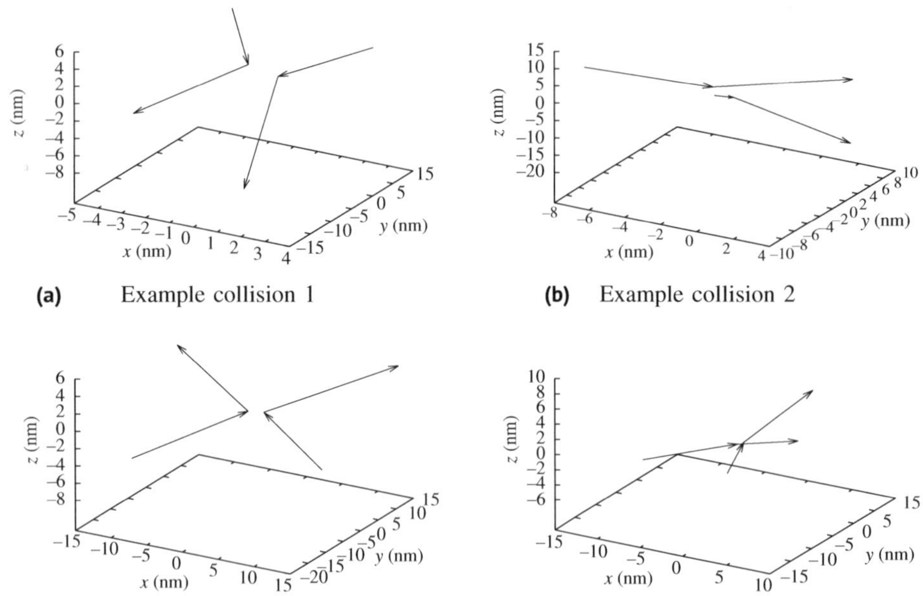

(a) Example collision 1
(b) Example collision 2
(c) Example collision 3
(d) Example collision 4

Figure 8.32 Example of random particle collision trajectories in ideal gas simulation.

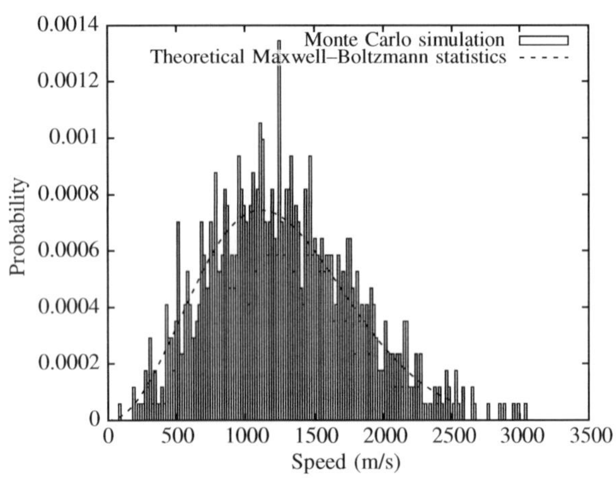

Figure 8.33 Simulated histogram of speed distribution for ideal He gas, 10^3 particles, T = 300 K.

the initial and final coordinates of a representative sample of some collisions (in a special version of the program code), we can visualize the trajectories that give rise to the statistical partitioning of the particle velocities in three dimensions shown in Figure 8.32.

The histograms of the distribution of particle speeds plotted in Figures 8.33 through 8.35 were the final distributions reported after three simulations of 500 time steps with 10^3, 10^4, and 10^5 particles. As more particles in the gas are included, the plots show the histogram approaching the Maxwell–Boltzmann distribution, which is plotted along with the simulated histograms for comparison using the code developed in Section 4.5 and Exercise 4.1. The progression of these plots shows the transition between a collection of mechanically interacting particles and a macroscopic gas that may be treated meaningfully statistically.

Finally, Figure 8.36 shows the evolution of several histograms of particle speeds throughout the simulation of the ideal gas with 10^5 particles. Since we initialized the gas with all particles having identical speeds at what the theoretical RMS speed should be, the initial histogram has only a single speed bin with a nonzero probability. We can visualize the process of the statistical distribution coming into equilibrium as the simulation proceeds. A clearer view of this equilibrium process can be seen in the animated GIF file `gas_gpanim.gif` on the Student Companion website. This animation was generated by substituting a different version of the `gen_output()` function for animated plot graphics, following the practice shown in Section 4.13.

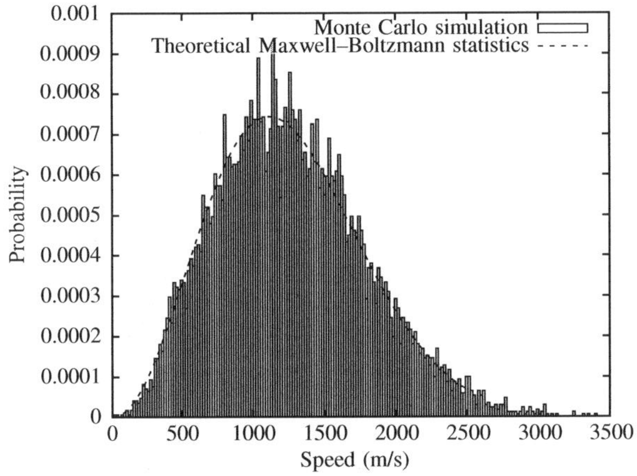

Figure 8.34 Simulated histogram of speed distribution for ideal He gas, 10^4 particles, T = 300 K.

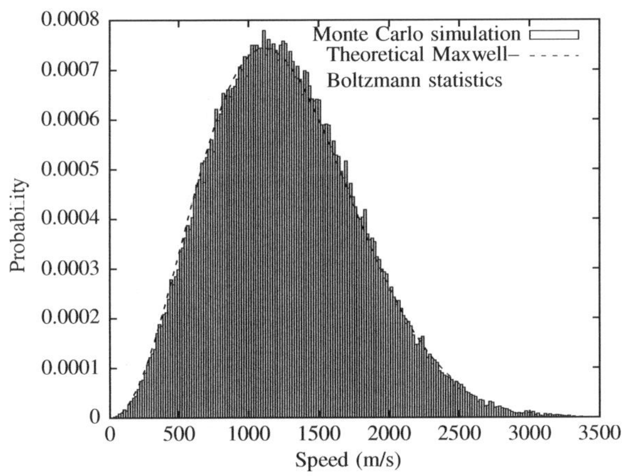

Figure 8.35 Simulated histogram of speed distribution for ideal He gas, 10^5 particles, T = 300 K.

8.12 Application: Integration of Gauss' Law

One of the fundamental relationships of electrostatics is the integral form of Gauss' Law of electrostatics

$$q = \epsilon_0 \oint_S \vec{E} \cdot d\vec{A} \tag{8.24}$$

where q is the charge enclosed inside a Gaussian surface S, $\vec{E}$ is the electric field vector on a surface element, $d\vec{A}$ is the outward unit vector normal to the surface element, and the "·" operator denotes a vector dot product. For some geometries, such as a charge exactly at the center of a symmetric volume, the integral is straightforward. However, we look here at another example of using the Monte Carlo method for estimating a multidimensional integration for other geometries for which the surface integral may be quite difficult.

Consider a spherical surface of radius 1 m. Suppose we choose three arbitrary locations, now not necessarily at the center, but elsewhere within the sphere, to locate three electrostatic charges of arbitrary strengths. For example, we may choose one charge to be of 0.75 coulombs at (x, y, z) coordinates of (0.5, 0.2, 0.6) meters.

To estimate the surface integral, we may use the `random_gen()` function to generate a large number of random points on the unit sphere. To each random point we associate a small patch of uniform surface area, as illustrated in Figure 8.37. The contribution each small area patch makes to the overall surface integral is easily calculated as a vector dot product with a single $\vec{E}$ field vector computed with Coulomb's Law, as shown in Figure 8.38. Since for a large number of random points, we assume the points are uniformly distributed on the surface, we can associate an average incremental area for each point dA_0 to be simply the total surface area of the sphere divided by the number of points.

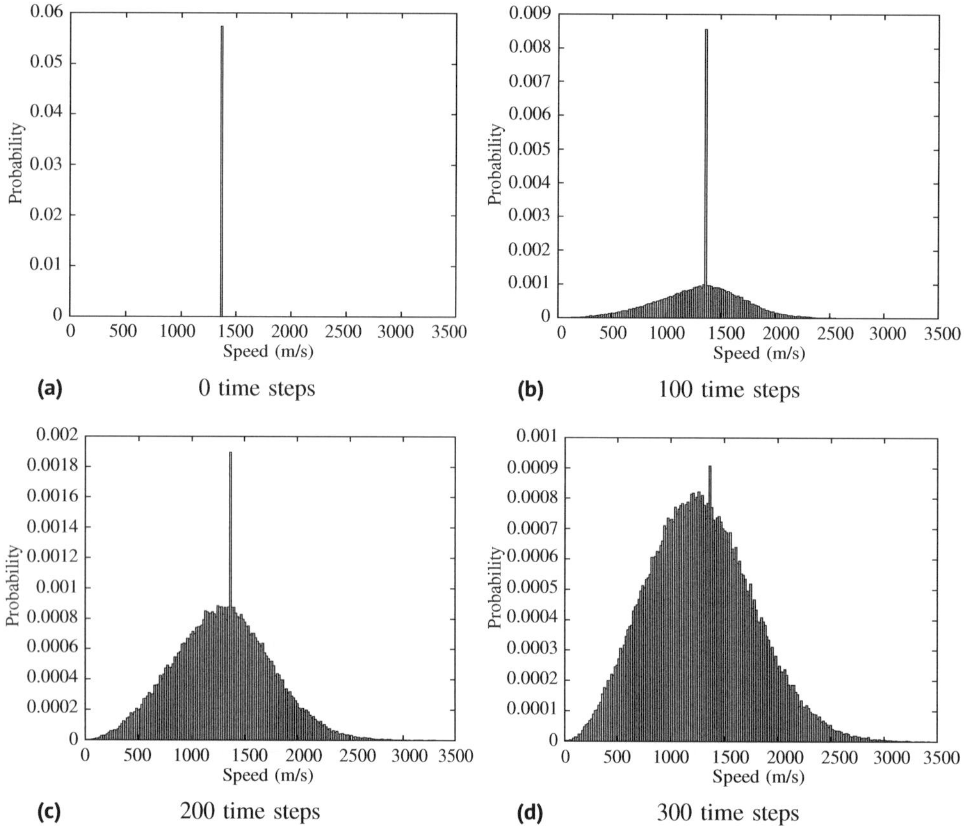

Figure 8.36 Evolution of the gas histogram through the simulation with 10^5 particles.

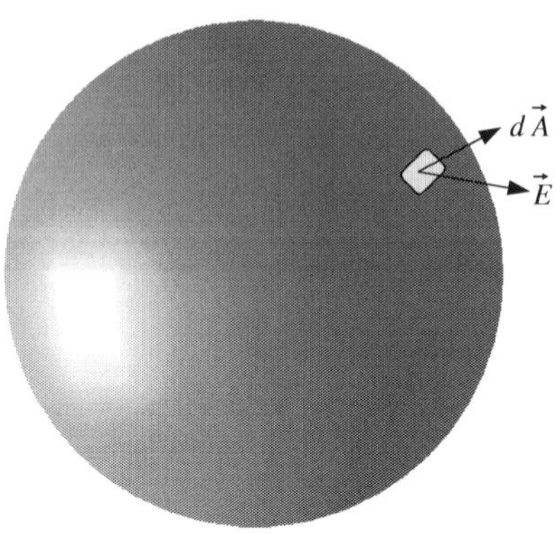

Figure 8.37 Gauss' Law for electric charges within a surface region.

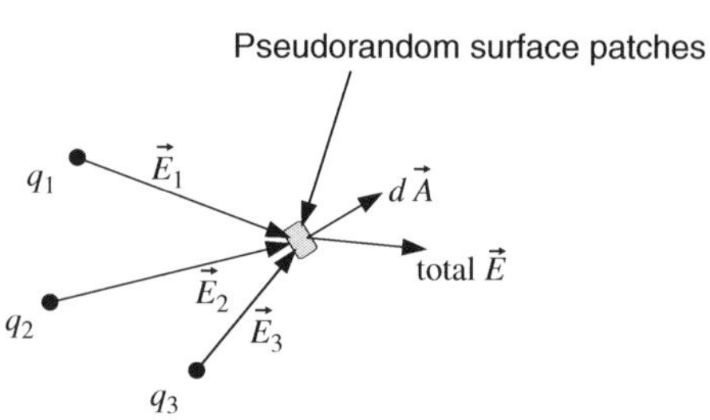

Figure 8.38 Computing the surface integral of Gauss' Law.

However, we must use care in selecting a sampling algorithm for generating random points on a sphere. For two-dimensional sampling on the path of a circle in Section 8.4, using an angle-based method was statistically equivalent to the method of distributing points in a square and moving them to the circle. But for three dimensions, we must be very careful with angle-based methods. Attempting, for example, to use spherical coordinates and setting the inclination angle θ to a uniform random variable between 0 and π and azimuthal angle ϕ to a uniform random variable between 0 and 2π, and then using standard conversion formulas between spherical and Euclidean coordinates with unit radius

$$x = \sin(\theta) \cdot \cos(\phi)$$

$$y = \sin(\theta) \cdot \sin(\phi)$$

$$z = \cos(\theta)$$

would distort the sampling due to the nonlinear products and cluster samples toward the poles of the sphere.

Therefore, we must use the method of spreading random points through a cube and moving them to the sphere surface if they are on or within the sphere, as used in Section 8.11 to initialize the velocities of the simulated gas particles uniformly in three dimensions. For each point on the unit sphere located at radial vector $\vec{r}$, we calculate the dot product of $\vec{E} \cdot d\vec{A} = \vec{E} \cdot dA_0\vec{r}$, where the $\vec{E}$ field is the sum of the fields from each of three interior charges at the surface point. Then we form a sum of the vector dot product values over all the random points on the sphere, and use that to estimate the surface integral. We will leave the coding of this simulation program as Exercise 8.5 for the student.

8.13 Exercises

8.1 Copy the `walk.c` program for the random walk in Section 8.6 to a file `walk_dist.c` and modify the code to produce a plot similar to the walk distance plot of Figure 8.13. First extend the walks to involve random displacements on the unit sphere in three dimensions instead of two, using the same approach used in the gas simulation of Section 8.11 and the Gauss' law integration of Section 8.12. Then add to the code a calculation at each time step of the RMS value of the distance walked over all the particles in the ensemble. Produce an output file with columns tabulating both the simulated RMS distance and the ideal square root model of Equation (8.9) for graphical comparison. Run the simulation for walks of 10 000 steps with an ensemble of 100 particles, and with three different random generator seed values.

8.2 Copy the `decay.c` source code to a file `decay2.c` and modify it for simulating the two-step nuclear decay process detailed in Section 8.7. Also run the `decay2_step.c` program from Exercise 5.6 to step the ideal model in the limits of very large populations. Running your programs and plotting the results together should reproduce the plot in Figure 8.15.

8.3 Write a program `corner_pocket.c` for the corner pocket pool shot simulations from Section 8.9. Include in the code a quantification of the difficulty of the short and long range shots by computing the mean and standard deviation of the tolerance angle over the trials. Use your C function solutions to Exercise 3.6 for these statistical calculations.

8.4 Write the program `forest.c` that simulates shooting arrows outward from the origin point of a forest for modeling a two-dimensional version of Olbers' paradox as discussed in Section 8.10. You need not generate the graphics shown in Figures 8.27 through 8.30, but simply run simulations and numerically keep track of the statistics. Hint: The general program structure should be in three nested loops

```
for (i_forest = 0; i_forest < n_forests; i_forest++) {
  ...
  for (i_arrow = 0; i_arrow < n_arrows; i_arrow++) {
    ...
    for (i_tree = 0; i_tree < n_trees; i_tree++) {
      ...
    }
  }
}
```

The output loop iterates through multiple experiments held in independently constructed forests with randomly distributed trees, all with a radius of 1 unit. The middle loop iterates over all the arrows launched at random angles during an experiment in a particular forest. Then the inner loop iterates through all the trees in the forest to check a particular arrow for a collision.

There should be command line arguments for the number of forests to simulate, the side dimension size of the square forest, the density with which trees are randomly distributed, the number of arrows to launch, and the pseudorandom seed value. The program must keep track of how many arrows emerge unobstructed from each forest trial, and then report the number of forests in the experiment out of which at least one arrow emerges unobstructed to the terminal. Use this program to observe that a sufficiently large forest is capable of blocking all arrows launched. As you grow the size of the forests, keep a tree density of 10^{-3} trees per unit area. Compile and run your program, growing the area of the forest by a factor of 10 at each step, as shown in the example cases of Figures 8.27 through 8.30. Depending on the particular pseudorandom seed you use, you may not duplicate exactly the statistics shown in the figures, but you should be able to demonstrate a similar limit.

8.5 Write the program `gausslaw.c` to use Monte Carlo surface integration from Section 8.12 with Gauss' Law to compute the total charge contained in a unit sphere. Make use of the point charge tool from Chapter 4. The point charge header file and utility code, `point_charges.h` and `point_charges.c`, have already been discussed in Section 4.11, and are available for download from the Student Companion website. Demonstrate the program with two arbitrarily chosen collections of three point charges within the sphere.

Hint: You probably will need quite a large number of random points to get a meaningful result, such as 10^6.

9

Interpolation of Sparse Data Points

In Chapters 4, 5, and 7, our numerical analysis of the equations of physical systems assumed that we have knowledge of the mathematical functions that govern behavior, and that those functions may be coded effectively in C and applied to various algorithms for visualization or solution. However, in some cases, we don't have knowledge of such a function a priori for the behavior we seek to understand, and instead are limited only to some discrete sampled data points. Examples of this situation are cases where we have data points found from experimental measurements, possibly expensive and sparse, or a set of data points that have been obtained from other computations, such as the Monte Carlo simulations of Chapter 8, and again possibly expensive and sparse. We assume that the samples of data points will describe physical behavior that is governed by some continuous dependence from physical laws, and that same dependence governs the behavior continuously between the sampled points as well. So in these cases, we often need numerical techniques to fill in data values between the known data points that follow plausible physical behavior. This process is called interpolation.

Exactly what constitutes "plausible physical behavior" can vary according to the particular problem at hand, but a good general approximation is to arrange for the intermediate interpolated data points to follow a smooth polynomial function that passes exactly through the known data points. We interpolate behavior between actual known data points not with the assumption that our interpolated points necessarily represent an exact physical solution, but with the understanding that we are obtaining at least some plausible solution for the intermediate behavior that is considerably better than no knowledge at all. Interpolation will also play a fundamental role in the methods for numerical integration we will explore in Chapter 10.

Figure 9.1 shows the fitting of a polynomial through a number of sampled points that follow some underlying nonlinear dependence. Mathematically, we know that there is one unique polynomial of degree $n - 1$ with n coefficients that exactly passes through n data points, although that polynomial may be expressed in a variety of algebraic formulations.

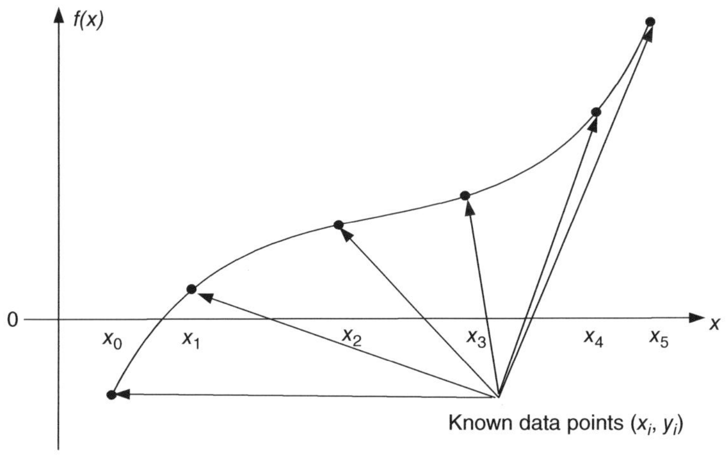

Figure 9.1 Fitting a nonlinear dependence with an interpolating polynomial.

Computational Physics Using C: Efficient Programming with Ease, First Edition. John W. Fattaruso. © 2026 John Wiley & Sons, Inc.
All rights reserved, including rights for text and data mining and training of artificial intelligence technologies or similar technologies.
Published 2026 by John Wiley & Sons, Inc. Companion Website: http://www.wiley.com/go/CompPhysUsingC

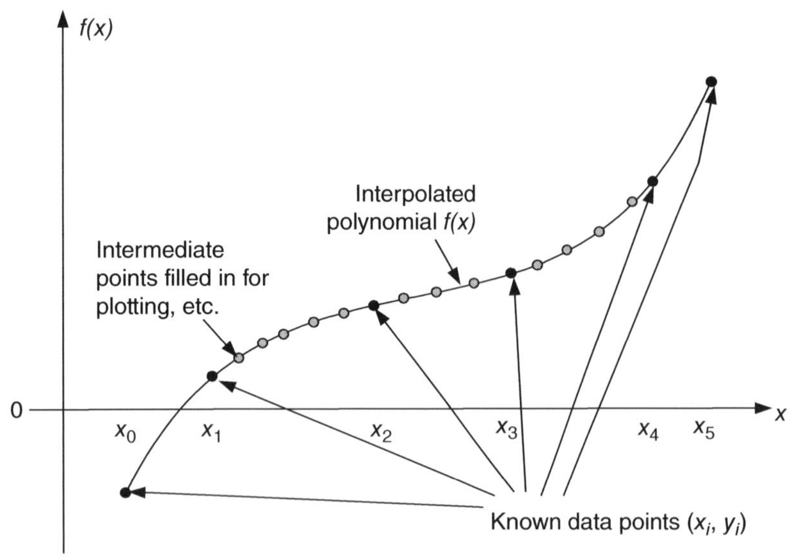

Figure 9.2 Filling in intermediate points on a curve with an interpolating polynomial.

Once the interpolating polynomial is found that exactly passes through the sampled points, any number of intermediate points along the interpolation curve can be computed by evaluating the polynomial function, as shown in Figure 9.2. These intermediate points may be used for plotting, solving for roots, integration, predicting, and other purposes.

To approach the task of solving for the polynomial that fits the data points, we might imagine beginning with the canonical formulation of a general polynomial

$$f(x) = a_{n-1}x^{n-1} + a_{n-2}x^{n-2} + \cdots + a_3x^3 + a_2x^2 + a_1x + a_0 \tag{9.1}$$

While the value of $f(x)$ depends on the powers of x, it is dependent *linearly* on the coefficients a_i. So we may imagine setting up a system of linear equations with the a_i as the unknown variables, and the coefficient matrix is a sequence of powers of the x coordinates of the known data values. So if the n known data points are at the coordinates (x_0, y_0), (x_1, y_1), (x_2, y_2), ..., (x_{n-1}, y_{n-1}), we would prepare the system

$$\begin{bmatrix} 1 & x_0 & x_0^2 & \cdots & x_0^{n-2} & x_0^{n-1} \\ 1 & x_1 & x_1^2 & \cdots & x_1^{n-2} & x_1^{n-1} \\ 1 & x_0 & x_2^2 & \cdots & x_2^{n-2} & x_2^{n-1} \\ \vdots & & & & & \\ 1 & x_{n-2} & x_{n-2}^2 & \cdots & x_{n-2}^{n-2} & x_{n-2}^{n-1} \\ 1 & x_{n-1} & x_{n-1}^2 & \cdots & x_{n-1}^{n-2} & x_{n-1}^{n-1} \end{bmatrix} \begin{bmatrix} a_0 \\ a_1 \\ a_2 \\ \vdots \\ a_{n-2} \\ a_{n-1} \end{bmatrix} = \begin{bmatrix} y_0 \\ y_1 \\ y_2 \\ \vdots \\ y_{n-2} \\ y_{n-1} \end{bmatrix} \tag{9.2}$$

and solve it with the methods of Chapter 6. This form of coefficient matrix is known as a *Vandermonde matrix* and, unfortunately, is particularly troublesome in linear systems. Since the powers of the x_i coordinates potentially grow to a very large range of coefficient values across this matrix, this set of coefficients leads to a linear system with a very high susceptibility to roundoff errors. Section 9.1 describes two alternative formulations for the fitting polynomial that not only allow solving for their coefficients with higher precision, but also avoid the order 3 Gaussian elimination algorithm. The numerical effort needed for the two required steps of first solving for the polynomial and then computing intermediate points can be traded off in different ways by the application of the two interpolation algorithms described in the next section.

9.1 Algorithms: Interpolation Methods

9.1.1 Newton Polynomial

Equations (9.3) through (9.5) show the Newton formulation [10, Chap. 4; 12, Chap. 12] of the interpolation polynomial of first, second, and third order.

$$f_1(x) = b_0 + b_1 \underbrace{(x - x_0)}_{P_1(x)} \tag{9.3}$$

$$f_2(x) = b_0 + b_1 \underbrace{(x - x_0)}_{P_1(x)} + b_2\underbrace{(x - x_0)(x - x_1)}_{P_2(x)} \tag{9.4}$$

$$f_3(x) = b_0 + b_1 \underbrace{(x - x_0)}_{P_1(x)} + b_2\underbrace{(x - x_0)(x - x_1)}_{P_2(x)} + b_3\underbrace{(x - x_0)(x - x_1)(x - x_2)}_{P_3(x)} \tag{9.5}$$

These can be more generally expressed for any order as

$$\begin{aligned} f(x) &= b_0 + \sum_{i=1}^{n-1} b_i \cdot P_i(x) \\ &\text{where } P_i(x) = \prod_{j=0}^{i-1} (x - x_j) \end{aligned} \tag{9.6}$$

As we consider an interpolation problem fitting an increasing number of data points, we would select a polynomial of increasing order with enough free coefficients to pass through the data. As we move to higher orders, note that the extra term we add on the right sides of Equations (9.3) through (9.5) is the only term with the higher order dependence on x. Also note that each term of this sum is composed of polynomials of increasing order whose roots are designed to be at the data point x coordinates.

With a polynomial structure designed in this manner, we may simply determine the interpolation polynomial $f(x)$ by solving for the set of Newton polynomial coefficients b_i with a straightforward set of recursive steps, without any system of simultaneous linear equations required. The first step is setting the value of the polynomial function evaluated at the x coordinate of the first known point equal to the y coordinate of the first known point

$$\begin{aligned} &f(x_0) = y_0 = b_0 \\ &\text{so set } b_0 = y_0 \end{aligned} \tag{9.7}$$

Once the coefficient b_0 is known, we may use it to solve for b_1 by a similar relation

$$\begin{aligned} &f(x_1) = y_1 = b_0 + b_1(x_1 - x_0) \\ &\text{so set } b_1 = \frac{y_1 - b_0}{(x_1 - x_0)} \end{aligned} \tag{9.8}$$

Now that the coefficients b_0 and b_1 are both known, we may make use of both of these to solve for b_2 with

$$\begin{aligned} &f(x_2) = y_2 = b_0 + b_1(x_2 - x_0) + b_2(x_2 - x_0)(x_2 - x_1) \\ &\text{so } b_2 = \frac{y_2 - b_0 - b_1(x_2 - x_0)}{(x_2 - x_0)(x_2 - x_1)} \end{aligned} \tag{9.9}$$

Continuing with this pattern, the coefficients b_0, b_1, and b_2 are now all known, and with these we may solve for b_3.

$$\begin{aligned} &f(x_3) = y_3 = b_0 + b_1(x_3 - x_0) + b_2(x_3 - x_0)(x_3 - x_1) + b_3(x_3 - x_0)(x_3 - x_1)(x_3 - x_2) \\ &\text{so } b_3 = \frac{y_3 - b_0 - b_1(x_3 - x_0) - b_2(x_3 - x_0)(x_3 - x_1)}{(x_3 - x_0)(x_3 - x_1)(x_3 - x_2)} \end{aligned} \tag{9.10}$$

This recursive procedure may be extended to any number of Newton polynomial coefficients b_i. Once the values of all the b_i coefficients are solved for, the interpolation polynomial is completely specified, and we may locate any number of points along it at any arbitrary x coordinates by evaluating Equation (9.6). For an application where many calculated points along the polynomial are needed, evaluating Equation (9.6) involves only a modest amount of computation, and by performing up front the recursive calculation of the coefficients with the continuing pattern starting with Equations (9.7) through (9.9), the simple evaluations of Equation (9.6) are made possible.

9.1.2 Lagrange Polynomial

Another alternative formulation for the interpolating polynomial is in Equations (9.11) and (9.12). These equations show the progression of a first-, second-, and third-order Lagrangian formulation[10, Chap. 4; 12, Chap. 12] of the unique interpolation polynomial.

$$f_1(x) = \underbrace{\frac{(x-x_1)}{(x_0-x_1)}}_{L_0(x)}c_0 + \underbrace{\frac{(x-x_0)}{(x_1-x_0)}}_{L_1(x)}c_1 \tag{9.11}$$

$$f_2(x) = \underbrace{\frac{(x-x_1)(x-x_2)}{(x_0-x_1)(x_0-x_2)}}_{L_0(x)}c_0 + \underbrace{\frac{(x-x_0)(x-x_2)}{(x_1-x_0)(x_1-x_2)}}_{L_1(x)}c_1 + \underbrace{\frac{(x-x_0)(x-x_1)}{(x_2-x_0)(x_2-x_1)}}_{L_2(x)}c_2 \tag{9.12}$$

$$\begin{aligned} f_3(x) = {} & \underbrace{\frac{(x-x_1)(x-x_2)(x-x_3)}{(x_0-x_1)(x_0-x_2)(x_0-x_3)}}_{L_0(x)}c_0 + \underbrace{\frac{(x-x_0)(x-x_2)(x-x_3)}{(x_1-x_0)(x_1-x_2)(x_1-x_3)}}_{L_1(x)}c_1 \\ & + \underbrace{\frac{(x-x_0)(x-x_1)(x-x_3)}{(x_2-x_0)(x_2-x_1)(x_2-x_3)}}_{L_2(x)}c_2 + \underbrace{\frac{(x-x_0)(x-x_1)(x-x_2)}{(x_3-x_0)(x_3-x_1)(x_3-x_2)}}_{L_3(x)}c_3 \end{aligned} \tag{9.13}$$

These can be more generally expressed for any order as

$$\begin{aligned} & f(x) = \sum_{i=0}^{n-1} L_i(x) \cdot c_i \\ & \quad \text{where } L_i(x) = \prod_{j=0, j\neq i}^{n-1} \frac{x-x_j}{x_i-x_j} \end{aligned} \tag{9.14}$$

The numerator of each of the terms in this formulation is a polynomial of the full $(n-1)$th order needed to fit n data points. The denominator of each term is a constant designed to equal the numerator when the latter is evaluated at one of the known data x coordinates. Note that with this construction, the polynomial terms are designed so that all but one of the terms will be zero when x takes the value of one of the fixed data points x_i, and the one nonzero term will simply equal the corresponding coefficient c_i. This means that conveniently the desired Lagrange interpolation polynomial coefficients $c_0, c_1, c_2, \ldots, c_{n-1}, c_n$ are simply the function values $f(x_0), f(x_1), f(x_2), \ldots, f(x_{n-1}), f(x_n)$, without any computation required. The drawback of this approach is that evaluating Equation (9.14) generally requires more computational effort than Equation (9.6).

Of course, when a given set of n fixed data points is used to solve for a set of polynomial coefficients, whether b_i for the Newton polynomial or c_i for the Lagrangian formulation, the two formulations will describe *exactly the same polynomial*. It is also exactly the same polynomial that would be described by the canonical form of 9.1 if the set of a_i coefficients could be accurately found by the linear system of Equation (9.2). The only difference between the formulations is the ease of computing the coefficients and the computational cost of calculating any intermediate points lying on the polynomial.

Moreover, there is no requirement in either the Newton or Lagrange formulation for the data points to be indexed in any particular order. In Figure 9.1 the points are indexed with increasing indices from left to right in the plot, with increasing x coordinate values. But this ordering is simply for graphical clarity; the very same polynomial will be fit with either formulation with any arbitrary ordering among the data points. The only constraint these interpolation methods enforce is that the polynomial will pass through the data points with the pairs of (x, y) coordinates defining each point.

9.1.3 Cubic Spline Interpolation

There are many examples of data points that are arrived at through experiment or calculation from physical systems through which polynomial interpolation provides an excellent method for filling in intermediate points. If the underlying physical mechanism that governs the variation of the data points is described by functions that have Taylor series approximations with only a few low-order terms that are primarily significant, then we can expect meaningful interpolation from polynomials. An example of this case is oscillatory behavior where a displacement is at least approximately sinusoidal in time. The Taylor series representation of the sine function is

$$\sin(x) = x - \frac{x^3}{3!} + \frac{x^5}{5!} - \frac{x^7}{7!} + \cdots \tag{9.15}$$

The significance of the terms in this series falls off rapidly due to the factorial in the denominators, so this series may be truncated at relatively few terms and still lead to meaningful approximations. Since a truncated Taylor series is simply a polynomial, we could expect a polynomial interpolation of any dependence that is not too far from a sinusoid to closely follow the true dependence. An example is shown in Figure 9.3, where nine points that follow a cosine function

$$f(x) = \cos(2\pi x) \tag{9.16}$$

are interpolated with an eighth-order polynomial of Newton's formulation.

Consider, however, another example of an unfortunate function that does not conform well to polynomial interpolation. The Runge function is

$$f(x) = \frac{1}{1 + x^2} \tag{9.17}$$

and is plotted in Figure 9.4. This function is not particularly abrupt or discontinuous, yet the poor result of an attempt to fit nine points along its contour to a polynomial is shown in Figure 9.5. In this example, we can see that the fit polynomial does indeed pass through the nine data points as constructed, but that trying to force an interpolation with polynomials leads to intermediate points that exhibit wide fluctuations, and are almost certainly not physically meaningful.

We may imagine that simply supplying more data points along a Runge function would improve the polynomial fit. See Figure 9.6 for the result with 13 points, showing even wider fluctuations. So to provide meaningful interpolation for a wider class of underlying dependence of data points, we must develop other interpolation methods. One very popular choice is that of fitting points with cubic splines.

Cubic spline interpolation is the method of using a third-order polynomial for the intermediate curves between the data points shown in Figure 9.1. But rather than fitting one polynomial function $f(x)$ of order $n-1$ through all n data points, a cubic spline fit defines a sequence of different third-order polynomial functions $s_i(x)$ for each segment of the curve between the fixed points. For n data points there will be $n-1$ different cubic polynomials defined, and the overall interpolated curve is obtained by a piecewise stitching together of all the $s_i(x)$ segment cubic polynomials. The segment polynomials are

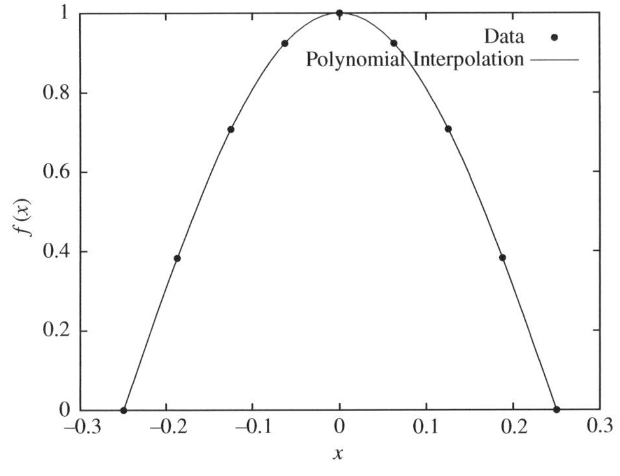

Figure 9.3 Polynomial interpolation of data points along the cosine function.

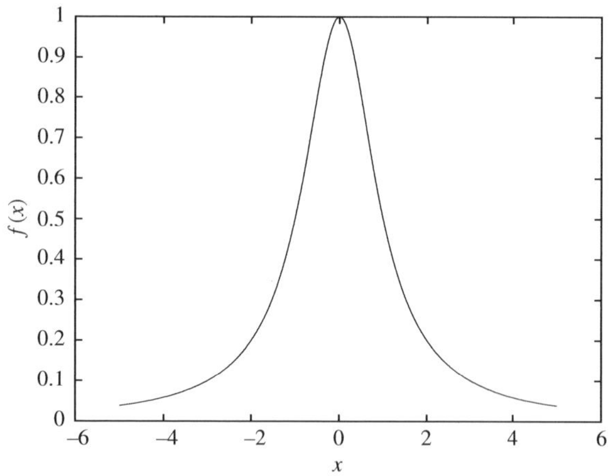

Figure 9.4 Plot of the Runge function.

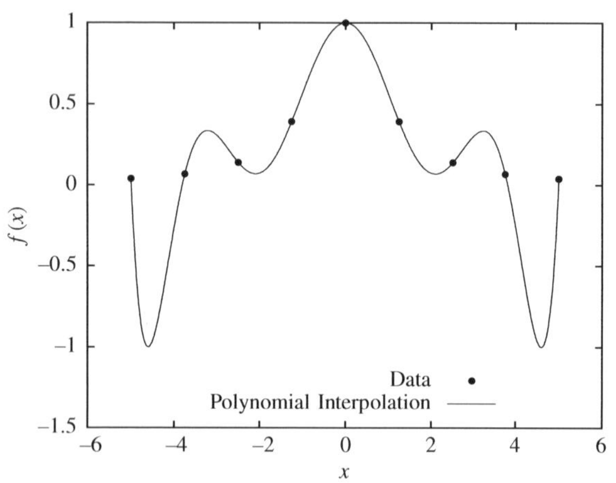

Figure 9.5 Polynomial interpolation of nine data points along the Runge function.

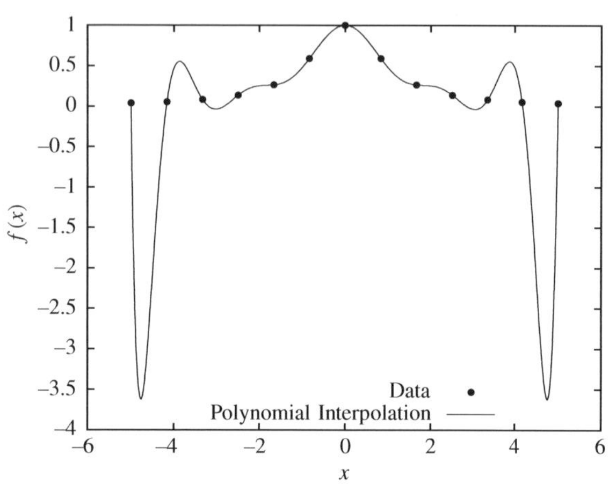

Figure 9.6 Polynomial interpolation of 13 data points along the Runge function.

aligned so that at the data points, which are the border points where one polynomial is handed off to the next, continuity constraints are established on the value and derivatives of the segment polynomials. To align all the segment functions in a continuous manner, we will need to solve a system of linear equations that expresses all the continuity relationships simultaneously. The name "spline" comes from a mechanical drafting technique where a thin flexible strip is arranged on a drafting surface with tension between anchoring pins at the known data points [12, Chap. 12].

To solve for all the segment polynomial coefficients, we follow here the similar derivations in [10, Chap. 6] and [13, Chap. 3]. We set up a system of equations that are written in terms of n variables z_i. The values of the z_i, for $i = 0$ through $n - 1$, are the second derivatives of the spline fit at the n data points in Figure 9.1. The second derivatives at the interior points, z_1 through z_{n-2}, are the shared continuous second derivatives of the $s_i(x)$ polynomials on either side of the transition points between segments. The second derivative at the two ends, z_0 and z_{n-1}, are subject to an arbitrary choice, and making the assignment that $z_0 = z_{n-1} = 0$, which models a constructed spline strip that straightens out and reaches its ends as inflection points, generally gives a smooth contour. This choice for z_0 and z_{n-1} is referred to as forming "natural" cubic splines. This assignment leaves $n - 2$ unknown values of the z_i interior second derivatives to be solved for with a system of equations.

To establish the continuity of the second derivative of the spline fit at the data points, and to linearly transition from one second derivative to another across each segment, we write

$$\frac{d^2 s_i(x)}{dx^2} = \frac{z_{i+1}}{h_i}(x - x_i) + \frac{z_i}{h_i}(x_{i+1} - x) \tag{9.18}$$

where $h_i = x_{i+1} - x_i$ for $i = 0$ to $n - 2$. These piecewise stitching constraints may be integrated twice to give third-degree polynomials for $s_i(x)$, and accumulate two arbitrary constants of integration that may be solved for with the constraints of the continuity of the $s_i(x)$ across boundary points. The form of the segment polynomial then becomes the cubic polynomial

$$\begin{aligned} s_i(x) = {} & \frac{z_{i+1}}{6h_i}(x - x_i)^3 + \frac{z_i}{6h_i}(x_{i+1} - x)^3 \\ & + \left(\frac{y_{i+1}}{h_i} - \frac{h_i}{6} z_{i+1}\right)(x - x_i) + \left(\frac{y_i}{h_i} - \frac{h_i}{6} z_i\right)(x_{i+1} - x) \end{aligned} \tag{9.19}$$

for $i = 0$ to $n - 2$. We must still apply the continuity constraint on the first derivatives of the $s_i(x)$ polynomials in order to arrive at a system of linear equations that we may solve for the z_i values for $i = 1$ to $n - 2$.

We define

$$\begin{aligned} u_i &= 2(h_{i-1} + h_i) \\ b_i &= \frac{1}{h_i}(y_{i+1} - y_i) \\ v_i &= 6(b_i - b_{i-1}) \end{aligned} \tag{9.20}$$

and the system to be solved becomes

$$\begin{bmatrix} u_1 & h_1 & 0 & 0 & 0 & \cdots & 0 \\ h_1 & u_2 & h_2 & 0 & 0 & \cdots & 0 \\ 0 & h_2 & u_3 & h_3 & 0 & \cdots & 0 \\ 0 & 0 & h_3 & u_4 & h_4 & \cdots & 0 \\ \vdots & & & & & & \\ 0 & 0 & 0 & \cdots & h_{n-4} & u_{n-3} & h_{n-3} \\ 0 & 0 & 0 & \cdots & 0 & h_{n-3} & u_{n-2} \end{bmatrix} \begin{bmatrix} z_1 \\ z_2 \\ z_3 \\ z_4 \\ \vdots \\ z_{n-3} \\ x_{n-2} \end{bmatrix} = \begin{bmatrix} v_1 \\ v_2 \\ v_3 \\ v_4 \\ \vdots \\ v_{n-3} \\ v_{n-2} \end{bmatrix} \tag{9.21}$$

The coefficient matrix for this linear system is in a banded form, as shown in Figure 3.29, and may be efficiently solved by the `gauss_band()` routine discussed in Section 6.4. In general, a cubic spline interpolation makes useful fits to any number of data points. The spline fit can give meaningful intermediate points for underlying function dependencies with Taylor series equivalents that would be troublesome for a single high order polynomial fit over a wide range. As an example, if we apply a cubic spline interpolation to the 13 points along the Runge function above, the intermediate points follow a much more natural contour, as plotted in Figure 9.7.

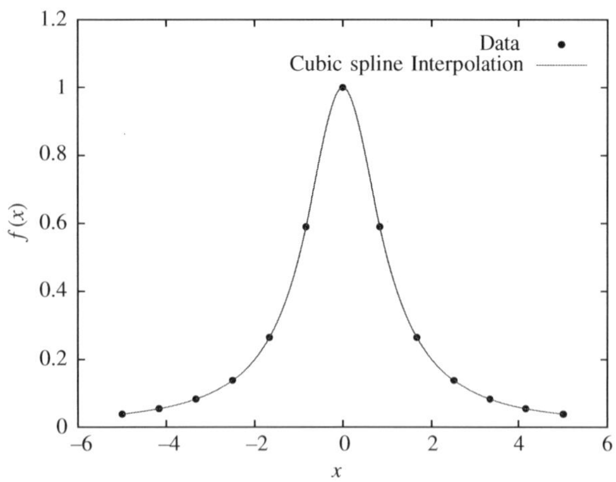

Figure 9.7 Cubic spline interpolation of data points along the Runge function.

9.2 Coding: The Data File Reading Tool

In this chapter, as in other chapters of this text, we would like to prepare a set of library routines to form polynomial and spline fits over any number of data points. Typically, those data points may originate from experimental measurements or observations, but in some cases they may be coordinates that are generated by other numerical analysis procedures. To make our interpolation routines generally applicable, we will code them to accept two separate lists of x and y coordinates stored in corresponding `ARRAY1D` structures. (In fact, we shall see later in Section 9.7 how useful it is to be able to swap which list is considered the "x," or independent variable, with the list that is considered the "y," or dependent variable.) In our coding design we want to include the case in which we would want to "plug in," in a modular coding sense, coordinates that originate from experiments quite independent from our computations. In that case, we'll assume the coordinates that will anchor our interpolation calculations will be stored by that external source into a columnar data file in our file system, in a format similar to the file streams our programs have been writing out, intended to be read by `gnuplot`. So we will now need to develop a utility routine capable of reading two- and three-column data files and filling two or three `ARRAY1D` structures with tables of coordinate values.

When a program that processes files that list data point coordinates is either compiled or run, there is no a priori knowledge of how many data points the program will find in the file, and that then must be stored in memory arrays. So we will code our file reading routines to use the same method for accumulating any number of data coordinates in the `ARRAY1D` one-dimensional tables as was used in Section 5.10 to accumulate an unknown number of function roots in a table. Memory for the tables is allocated in size increments of a quantization argument passed to the function. Data coordinates are read from a file in combinations needed to specify a point, two or three at a time. When the existing array memory allocations are full, the next point read from a file will cause the arrays to be reallocated up to the next larger size increment with the `realloc()` function. This file reading tool is another special case where the calling code should invoke the function with an `ARRAY1D` structure for its result that does not point to any allocated memory.

We also make use of the standard C library functions to open, read, and close a data file of a specified name in the file system. Before we read a file, we must call the `fopen()` function to ask the operating system to locate the file in the system. The first argument to this function is a character string holding the name of the desired file. This string may originate as an option on the program's command line, or perhaps be a fixed name coded in the program. The `fopen()` function returns a *stream pointer* or *file pointer* of the data type "pointer to `FILE`," where `FILE` is a structure holding the state of a file stream defined in the `stdio.h` system header file. (In this way, the definition and use of custom data types for structures such as `ARRAY1D` and function parameter sets in this textbook code library mirror how the C standard library is written.) The program must store the stream pointer returned by the `fopen()` function and subsequently use it as arguments to calls to the library functions `fscanf()` and `fprintf()` for reading and writing, respectively, from and to the open file. After all input or output is done, the program must call the `fclose()` function to terminate and close the file properly.

The tool code for reading two- and three-column data files can be found in the files `data_file.c` and `data_file.h`, part of the `CPLibrary` package available for download from the Student Companion website. The first function `read_2col_data_file()`, reads the text characters from a file specified by the first character string argument, interprets the words found as forming two columns, and fills the two arrays whose pointers are supplied as the next two arguments with an arbitrary number of elements. The values in the first column encountered in the file are filled into the first array argument, and the second column values are stored in the second array argument. The numerical values in these arrays may have any significance, but the formal arguments of this utility function are named `xtab_ptr` and `ytab_ptr`. These array names suggest the typical use of reading a file in which a sequence of x and y coordinates of data points are listed in the first and second columns, respectively. The last argument is the quantization of the increments of array size that accumulate an arbitrary number of points. If we have the knowledge that the data file may contain perhaps dozens of lines, but not hundreds, we might call this function with a quantization of 10, for instance, to save a large amount of unused memory. On the other hand, if we know we may be reading files with hundreds or thousands of points, perhaps we would call the function with a quantization of 100 to minimize the number of calls to the `realloc()` function.

The overall structure of this utility function is

```
int read_2col_data_file(char *file_name,ARRAY1D *xtab_ptr,
                        ARRAY1D *ytab_ptr,int quant) {
  ...
}
```

The file named in the first argument is opened for reading by the lines

```
FILE *stream;
stream = fopen(file_name,"r");
if (stream == NULL) {
  fprintf(stderr,"Can't open file '%s' for reading\n",file_name);
  return(DATA_FILE_ERROR);
}
```

If the operating system cannot open the named file for reading, due to, for example, an error in the file name or some file system permission prohibition, the `fopen()` function will return a `NULL` pointer. This condition is tested for and the error reported by the above lines, and the function will return with an error code defined in the `data_file.h` header file. Otherwise, the stream pointer is saved in the variable `stream` for subsequent file reading operations.

The integer variable `n_alloc` holds the current allocated size of the two `ARRAY1D` structures, and is initialized to zero. The array structures are initialized to a state corresponding to no allocation with the lines

```
n_alloc = 0;
xtab_ptr->n_elems = 0;
xtab_ptr->elem = NULL;
ytab_ptr->n_elems = 0;
ytab_ptr->elem = NULL;
```

By initializing the element pointer in the array structures to the `NULL` pointer, we prepare the first call to the `realloc()` function to operate equivalently as an original `malloc()` call.

Then words are read from the file iteratively by the loop

```
  while (fscanf(stream,"%lf %lf",&x,&y) == 2) {
    if (xtab_ptr->n_elems >= n_alloc) {
      n_alloc += quant;
      xtab_ptr->elem = realloc(xtab_ptr->elem,
                               n_alloc * sizeof(double));
      ytab_ptr->elem = realloc(ytab_ptr->elem,
                               n_alloc * sizeof(double));
    }
    xtab_ptr->elem[xtab_ptr->n_elems] = x;
    ytab_ptr->elem[ytab_ptr->n_elems] = y;
    xtab_ptr->n_elems++;
    ytab_ptr->n_elems++;
  }
  fclose(stream);
  return(DATA_FILE_SUCCESS);
```

For every pair of words successfully read by the `fscanf()` function and assigned to temporary `double` variables x and y, the current allocation is checked for sufficient room to store an additional pair of coordinates in the arrays. If the current arrays are full, their allocated length is incremented with a call to `realloc()`. In any event, the new values are stored in the arrays and their new sizes updated. When the supply of words from the file is exhausted, the file is properly closed and the function returns a code indicating a successful file read.

The second routine in the file `data_file.c`, named `read_3col_data_file()`, is a similar utility function, but applicable to three-column data files, filling three `ARRAY1D` tables with coordinate values. The two utility functions `read_2col_data_file()` and `read_3col_data_file()` may be "plugged in" to any solution code we write when columns of data point coordinates need to be read from a file. We will first make use of this utility in the interpolation tasks described later in this chapter.

9.3 Coding: The Interpolation Tools

The tool code for generating and evaluating both Newton and Lagrange interpolation polynomials that fit any number of fixed data points can be found in the files `interpolate.c` and `interpolate.h`, part of the `CPLibrary` package available for download from the Student Companion website.

Likewise, the files `cubic_spline.c` and `cubic_spline.h`, are also part of the `CPLibrary` package available for download. As the source file names imply, these routines will perform a cubic spline fit to any number of data points. All of these routines expect tables of data point coordinates to be passed as arguments in `ARRAY1D` structures. After carrying out the appropriate computation described in Section 9.1, the routines will either return to the calling program a direct interpolated value, or sets of coefficients that describe interpolation polynomials in other `ARRAY1D` tables. These cubic spline routines must be compiled with the `lineq.c` source code file to provide the solution to the constructed system of linear equations.

Six routines are available in the `interpolate.c` source code file, with arguments as shown here

```
double lagrange(ARRAY1D *x_ptr,ARRAY1D *y_ptr,double x) {
   ...
}

void find_newton_coeffs(ARRAY1D *x_ptr,ARRAY1D *y_ptr,
                        ARRAY1D *newton_coeff) {
   ...
}

double newton(ARRAY1D *x_ptr,ARRAY1D *newton_coeff,double x) {
   ...
}

double lagrange_grid(GRID *xgrid_ptr,ARRAY1D *y_ptr,double x) {
   ...
}

void find_newton_coeffs_grid(GRID *xgrid_ptr,ARRAY1D *y_ptr,
                             ARRAY1D *newton_coeff) {
   ...
}

double newton_grid(GRID *xgrid_ptr,ARRAY1D *newton_coeff,double x) {
   ...
}
```

The `lagrange()` routine accepts two pointers to `ARRAY1D` structures, with the two arrays containing the lists of the "x," or independent variable, coordinates, and the "y," or dependent variable, coordinates, respectively. The (x_i, y_i) coordinate pairs of the known data points are split up into these two arrays, and the coordinate values may come from other computations, or be filled in by the `read_2col_data_file()` utility routine as described in Section 9.2. The third argument is a single value of the independent variable anywhere along the intermediate points of the interpolation polynomial. The polynomial is evaluated at this x value, and the corresponding y value forms the function's return value. The body of the function is a straightforward evaluation of Equation (9.14) in two nested loops. As discussed in Section 9.1.2, the y values in the ARRAY1D structure pointed to by the second argument form the Lagrange polynomial coefficients without any computation required.

For the Newton polynomial formulation, the interpolation task is performed by two routines in concert, the first called once to compute the polynomial coefficients, and the second called any number of times to evaluate points along that polynomial. The find_newton_coeffs() routine accepts three pointers to ARRAY1D structures. Similar to the lagrange() routine, the first two arguments point to arrays containing the lists of the "x" and "y" coordinates, respectively. As before, the (x_i, y_i) data point coordinates may flexibly come from either prior calculations or a data file. The function body computes the Newton polynomial coefficients from the recursive method starting with Equation (9.7). The calculated coefficients are stored in the ARRAY1D structure pointed to by the third argument. Memory for the coefficient array must be allocated by the code that calls the find_newton_coeffs() routine, and the number of elements pointed to by the ARRAY1D structure must equal the number of data points.

Then the newton() routine makes use of the Newton polynomial coefficients computed by find_newton_coeffs() to calculate the value along the polynomial. Note that evaluation of Equation (9.6) requires the original "x" coordinate values as well as the solved Newton coefficients. Therefore, the first two arguments are pointers to the "x" coordinate array and the coefficient array, and the third argument is the x value to evaluate the polynomial at. As with the lagrange() routine, the return value of this function is the y value at the evaluated point.

The next three routines, lagrange_grid(), find_newton_coeffs_grid(), and newton_grid(), are essentially the same as the first three functions, except for a different type of first argument to specify the x coordinate values of the data points. These last three routines accept a pointer to a GRID structure as their first argument, instead of a pointer to an ARRAY1D array. These last three routines are useful for cases where the data points forming the basis for the interpolation process have come from prior calculations that have been performed on a uniform grid of x values, such as those from the step() tool. These functions are called with the assumption that the sequence of y coordinate values in the ARRAY1D table pointed to by the ytab_ptr argument correspond to the x grid values, index-by-index. An example of this application will be shown in Section 9.5.

Finally, two routines are available in the cubic_spline.c source code file, with arguments as shown here

```
int find_spline_coeffs(ARRAY1D *x_ptr,ARRAY1D *y_ptr,double min,
                       ARRAY1D *csz) {
  ...
}

double spline(ARRAY1D *x_ptr,ARRAY1D *y_ptr,ARRAY1D *csz,double x) {
  ...
}
```

Similar to the Newton interpolation routines described above, cubic spline interpolation is performed in two steps. The pairs of data point coordinates are passed in the first two ARRAY1D structure arguments to find_spline_coeffs(). Since solving for spline coefficients involves solving a system of linear equations, these routines must be compiled along with the lineq.c tool of Chapter 6. The double argument min to the find_spline_coeffs() routine is passed on to the gauss_band() solution routine, and the return value from gauss_band() is simply passed on as the return value from find_spline_coeffs() in order to trace back any unlikely matrix singularity errors. The solved spline fit coefficients z_i are stored in the ARRAY1D type csz, and this is then passed on to the spline() function that performs the interpolation at the x coordinate passed as its last argument. The memory for the array pointed to by the csv variable must be allocated by the calling code, with a size equal to *two less than* the number of data points.

The following sections describe making use of these routines for solving example problems. To provide starting code examples for calling these utility routines, download and study the main programs interpolate_step.c, interpolate_solve.c, and spline_step.c from the Student Companion website. The interpolate_step.c and spline_step.c programs read a data file named in an argument on its command line, form an interpolation polynomial, and then send to the standard output stream a sequence of coordinates evaluated as finely spaced intermediate points along the polynomial using both Newton and Lagrange formulations. The evaluation is performed on a uniform grid of x values also specified on the command line. Of course, since the two interpolation methods arrive at the very same polynomials, the two output data sets will be identical. These were the programs used to generate the plots shown in Figures 9.3 and 9.7, along with the data files interpolate_data_cosine9.dat and interpolate_data_runge13.dat that list the data point coordinates that are read by the programs. The following example command shell dialog shows compiling and running these programs to generate two data files for plotting

```
$ gcc -Wall -Wshadow -O2 -o interpolate_step interpolate_step.c CPLibrary/interpolate.c
    ↪ CPLibrary/array.c CPLibrary/data_file.c -lm
$ ./interpolate_step
./interpolate_step <input file> <xinit> <xfinal> <xinc>
$ ./interpolate_step interpolate_data_cosine9.dat -0.25 0.25 0.01 >| interpolate_step.dat
$ gcc -Wall -Wshadow -O2 -o spline_step spline_step.c CPLibrary/cubic_spline.c CPLibrary/lineq.c
    ↪ CPLibrary/data_file.c CPLibrary/array.c -lm
$ ./spline_step
./spline_step <input file> <xinit> <xfinal> <xinc>
$ ./spline_step interpolate_data_runge13.dat -5 5 0.01 >| spline_step.dat'
```

The `interpolate_solve.c` program performs inverse interpolation and solves the interpolated curve for some target y value. This application will be described in Sections 9.7 and 9.8.

9.4 Application: Estimating an Orbital Period

We suppose now that a new dwarf planet has been discovered at the outer range of our solar system, and telescopic images reveal a small moon in orbit around it. With observation time on high-power telescopes being a precious commodity, we assume that only a few images of the moon's orbital position on some successive nights are available. How can we use several images separated in time to estimate the moon's orbital period? Assume that, when viewed through a telescope, the diameter of the dwarf planet spans eight seconds of arc. The position of its moon relative to either the left or right edge of the planet is recorded at the same time each night for several nights. The moon is initially to the left of the planet, as shown in Figure 9.8.

Table 9.1 lists the observed angular displacements from an edge of the planet on successive nights. On the first four nights, in observations taken at the same time of night, the moon is seen to the left of the planet. On the fifth and sixth nights, the moon passes behind the planet, and then on nights seven, eight and nine, again at the same time, the moon is observed to the right of the planet. The orbital path is in the plane of the planet's equator. How may we use an interpolation routine to estimate the orbital period of the moon?

We first reform the data points so that they are on a continuous angle scale that we may then use for interpolation. We may also choose an arbitrary point for a zero angle reading, and arbitrarily choose either angular readings to the right or to the left as positive or negative, but in a way that accounts for the angular width of the planet. So arbitrarily choosing

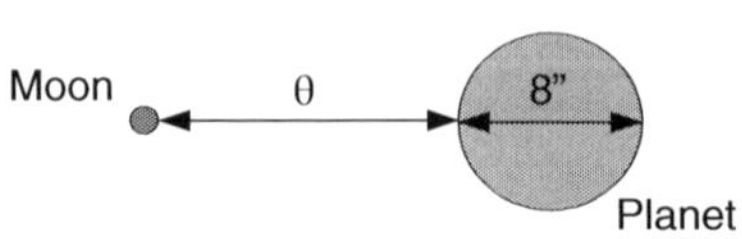

Figure 9.8 Observed moon orbiting a dwarf planet.

Table 9.1 Observations of a moon of a dwarf planet.

Night	Side of planet	Angular separation from edge (minutes of arc)
1	Left	2.282
2	Left	5.491
3	Left	5.687
4	Left	2.808
5	–	Obscured
6	–	Obscured
7	Right	4.219
8	Right	5.994
9	Right	4.596

readings to the right as positive, and choosing the center of the planet as a reference point for a zero angle, we subtract four seconds of arc from the negative of the readings on the left side of the planet and add four seconds of arc to the readings on the right. We then put our adjusted data into a file `interpolate_data_moon.dat`, with the time index in the first column and the angle in the second column

```
1 -6.282
2 -9.491
3 -9.687
4 -6.808
7 8.219
8 9.994
9 8.596
```

The tool code for evaluating an interpolation polynomial fitting the data points from the file can be found in the file `interpolate_step.c`, available for download from the Student Companion website. First the data file is read with the `read_2col_data_file()` utility routine. Then the Newton polynomial coefficients are found from the data points read from a file by calling `find_newton_coeffs()`. The interpolation polynomial is evaluated iteratively at all points specified by a grid of x values using the `step()` utility of Chapter 4. The object function called by the `step()` tool calls the `newton()` routine to evaluate the polynomial. Since the interpolation polynomial is evaluated many times after the polynomial coefficients are found just once, this is an example of a situation where we can make good use of the computation efficiency of the Newton formulation.

Running

```
$ gcc -Wall -Wshadow -O2 -o interpolate_step interpolate_step.c CPLibrary/interpolate.c
    ↪ CPLibrary/array.c CPLibrary/data_file.c -lm
$ ./interpolate_step
./interpolate_step <input file> <xinit> <xfinal> <xinc>
$ ./interpolate_step interpolate_data_moon.dat 0 10 0.01 >| interpolate_step.dat
```

provides a set of points stepped along the interpolation polynomial suitable for plotting or inspection. Plotting the original seven data points along with the sixth-order polynomial passing through them gives an image of the good fit to the interpolation polynomial in Figure 9.9. This is to be expected, as we know that orbital trajectories will be elliptical, so the coordinate of an orbit viewed on its plane will follow a dependence like sine and cosine functions with easily truncated Taylor series approximations.

How would we estimate the orbital period measured in days, to perhaps two significant fractional digits? We may use a text editor tool to examine the written data file `interpolate_step.dat`, or for expedience, we could also use a Linux pipe with

```
$ ./interpolate_step interpolate_data_moon.dat 1 9 0.01 | less
```

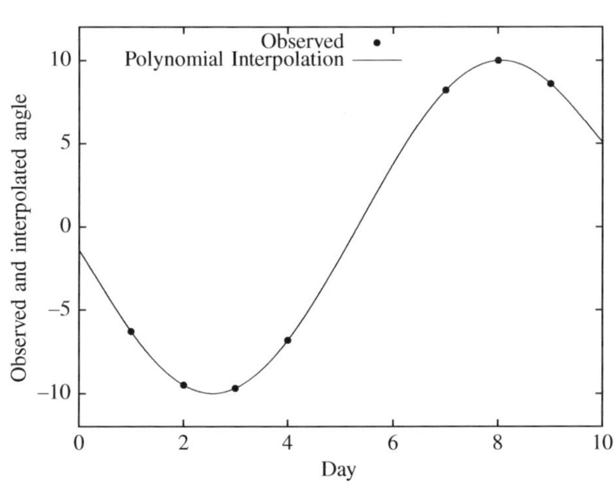

Figure 9.9 Polynomial interpolation of astronomical data points observing a moon's orbit.

and use the Linux text paging tool `less` to simply page through the intermediate interpolated coordinates. We would see two regions of interest, where the angular displacement reaches a maximum in both directions

```
  ...
2.52 -9.9998094
2.53 -10.000961
2.54 -10.001784
2.55 -10.002278
2.56 -10.002444
2.57 -10.002282
2.58 -10.001791
2.59 -10.000972
  ...
8.02 9.996753
8.03 9.9976342
8.04 9.9981853
8.05 9.9984066
8.06 9.9982981
8.07 9.9978599
8.08 9.9970923
8.09 9.9959955
  ...
```

We see a minimum angle at the extreme range to the left of −10.002444 seconds of arc at 2.56 days, and a maximum angle at the extreme right of 9.9984066 seconds of arc at 8.05 days. Subtracting the two times gives 5.49 days between the two extremes, which is half the orbital period. Therefore, the full orbital period is 10.98 days.

9.5 Application: Static Electric Potential

It is not only experimentally measured data points that often need to be filled in by interpolation, but also discrete points that have been obtained by some time-consuming computational process as well. In Chapter 8 we explored Monte Carlo simulation techniques, which are useful in many applications in which a multidimensional integration or stochastic simulation is required. However, statistically, we may need a very large number of pseudorandom trials to approach a reasonable accuracy in such simulations, and this may involve a significant expense of computation time. In this section we consider a Monte Carlo simulation of a surface integral to solve an electrostatics problem. But we will calculate only a small number of sparse data points, each obtained with a large pseudorandom sampling, and then use interpolation to fill in a physically reasonable dependence.

As an example, consider a standard long-playing (LP) $33\frac{1}{3}$ RPM vinyl audio disc that has dimensions shown in Figure 9.10. The annular disc has an outer radius R_2 of 30 cm, and the circular paper label in the center has a radius R_1 of 5 cm. As is commonly experienced, assume that by extracting the disc out of its paper sleeve to play it on a turntable, a total static electric charge of +50 nC (50×10^{-9} C) is deposited on the insulating vinyl surface by the triboelectric effect, but not on the paper label surface, and uniformly distributed around the annular vinyl area.

Suppose we need to know the electric potential that would be sensed above the vinyl record as a function of the radial coordinate off the axis of the annulus, as shown in Figure 9.11. Most textbooks for electromagnetic theory treat the problem of finding the electric potential above a uniform circular disc of charge, but only along the axis of the disc. This calculation involves integrating around a circular ring of charge on the disc, and then integrating over all the infinitesimally thin rings from the center of the disc out to its edge. Considering only potentials along the axis above the disc keeps these integrals symmetric and tractable. But for some knowledge of the potential at all radii spanning the space above the record when it is laid horizontally on an insulating turntable, let us approach this problem by performing a Monte Carlo integration over the charge on the annular vinyl surface.

Let's assume the potential should be calculated at a point P, with elevation coordinate z above the disc and a radius r from the central axis, as shown in Figure 9.11. We will choose two z elevation coordinate values 5 and 10 cm, and let the radius r range from 0 at the axis out to 30 cm, above the edge of the disc. The Monte Carlo simulation program will place 10^7 point charges randomly around the annular vinyl area that total to 50 nC, and sum their contributions to the electrical potential at the point P. The potential contributions can be calculated by a simple application of Coulomb's law from each point charge. The program will run the Monte Carlo simulations of the potential V over a coarse step of radius values and store the calculated points in an internal table. Then it will make use of the `interpolate.c` tool to interpolate a smooth

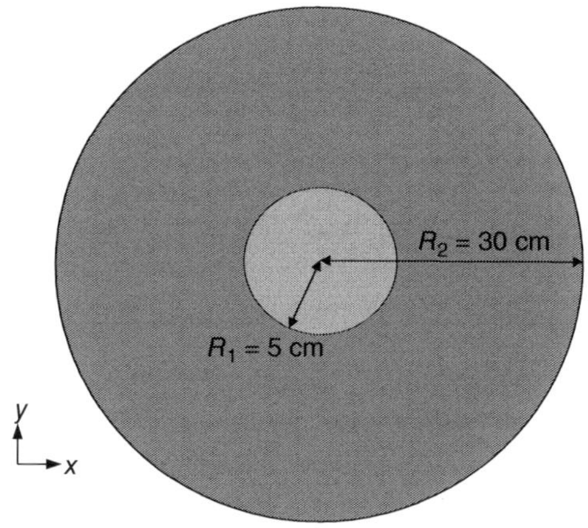

Figure 9.10 Dimensions of standard long-playing (LP) vinyl disc.

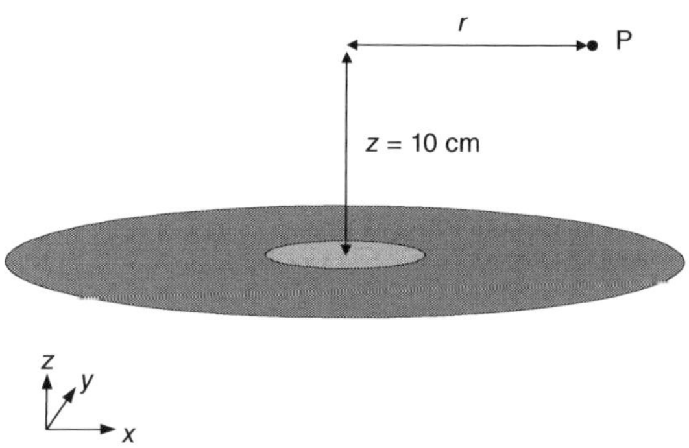

Figure 9.11 Simulating electric potential above a standard LP vinyl disc off the axis.

curve for the potential and generate an output stream from finely stepped values of the coordinate r along the polynomial curves. These curves will closely describe the actual dependence of potential versus radius.

See the main program `lprecord.c` available for download from the Student Companion website. The command line arguments that the program expects specify the physical dimensions of the annular disc, the total charge to be deposited, the z elevation, and the number of Monte Carlo point charges. There are also two grids of the radius coordinate to be listed on the command line. The first sets the small number of radius values at which the Monte Carlo simulation is to be run. The second, much finer, grid specifies the large number of radius values at which the interpolation polynomial is to be evaluated, once that polynomial is determined by the Monte Carlo simulations. Since these two grids cover the same range of radius, we have two command line arguments to set the common radius initial and final values, and then two more arguments to set the two different increment values.

The command line dialog shown below compiles the program and runs it for the geometry of the standard LP vinyl disc with $R_1 = 5$ cm and $R_2 = 30$ cm. The program is run twice, at two values of the z elevation coordinate of the potential measurement point – namely 5 and 10 cm above the disc surface. The output streams from the two runs are captured in two different data files. For each run we will compute the Monte Carlo simulation at a grid of five different r radius coordinates, with a course increment of 7.5 cm over the range of 0–30 cm. Then the interpolation polynomial is evaluated over the same range, but at a finer increment of 1 cm. Two different data sets are generated in the output stream to be plotted by `gnuplot`. The first is the dependence of the simulated electric potential over the five course steps of radius, and the second is the sampling of the interpolation polynomial over the fine radial steps.

```
$ gcc -Wall -Wshadow -O2 -o lprecord lprecord.c CPLibrary/interpolate.c CPLibrary/step.c
    ↪ CPLibrary/array.c CPLibrary/random.c -lm
$ ./lprecord
./lprecord <R1> <R2> <Q> <z> <N trials> <seed> <rinit> <rfinal> <rinc calc> <rinc interp>
$ ./lprecord 0.05 0.30 50e-9 0.05 10000000 1 0 0.3 0.075 0.01 >| lprecord_z0p05.dat
$ ./lprecord 0.05 0.30 50e-9 0.1 10000000 1 0 0.3 0.075 0.01 >| lprecord_z0p1.dat
```

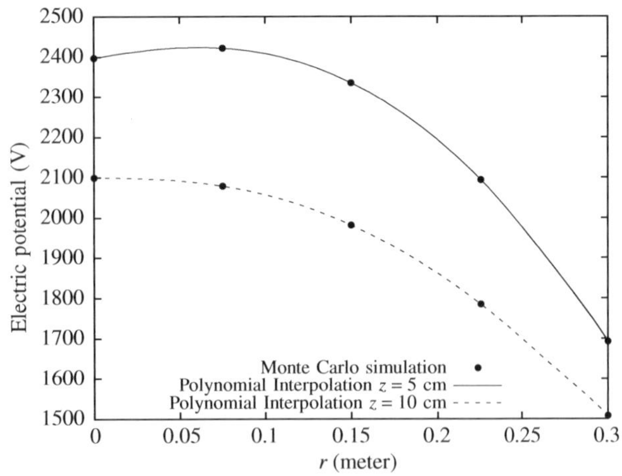

Figure 9.12 Interpolated electric potential above charged LP record off-axis.

We may plot the two simulation runs and their interpolations with the `gnuplot` dialog below. The resulting plot is shown in Figure 9.12. As the measurement point is moved farther off the axis toward the edge of the annulus, the electric potential drops off considerably. For an elevation of 5 cm, we can see that the potential falls off as we approach the axis of the disc as well, as the measurement point senses being over the uncharged paper hub rather than over the charged vinyl.

```
gnuplot> plot 'lprecord_z0p05.dat' i 0 w p lt -1 pt 7 t 'Monte Carlo simulation','' i 1 w l lt 1 t
    ↪ 'Polynomial Interpolation z=5 cm','lprecord_z0p1.dat' i 0 w p lt -1 pt 7 notitle,'' i 1 w l
    ↪ lt 2 t 'Polynomial Interpolation z=10 cm'
```

With numerical computations involving extensive coding, such as those in this program, it always boosts our confidence that our coding efforts have been correctly applied if we may reduce a physical problem to a simple case for which we have an analytic derivation to check the results against. For the case of a complete circular disc of radius R with no insulating center, and for measurement points limited to the axis of the disc, we may find the analytic prediction in introductory textbooks [17, Chap. 23; 18, Chap. 29; 19, Chap. 23; 20, Chap. 25] as

$$V = \frac{q}{2\pi R^2 \epsilon_0}\left(\sqrt{z^2 + R^2} - z\right) \tag{9.22}$$

Plugging in the example values of $R = 30$ cm, $z = 10$ cm, and $Q = 50$ nC that correspond to the simulations of the LP disc into this expression, the predicted potential is 2159.3327 V. Rerunning our program but specifying $R_1 = 0$ instead of 5 cm on the command line to eliminate the center insulator, and simply piping the standard output into the Linux pager tool `less`

```
$ ./lprecord 0 0.30 24.0e-9 0.1 10000000 1 0 0.3 0.075 0.01 | less
```

lets us inspect the first line of the output stream. The first line corresponds to an r radial value of 0, which gives 2159.1391 volts, in reasonably good agreement with the theoretical prediction.

What have we gained with this interpolation exercise? Consider the modified program `lprecord_mconly.c` also at the Student Companion website. This is a copy of `lprecord.c`, but with all the interpolation code removed. The electric potential above the disc is calculated completely by a Monte Carlo simulation performed at all 31 fine grid points of radius specified on the command line. Compiling and running with the Linux `time` command

```
$ gcc -Wall -Wshadow -O2 -o lprecord lprecord.c CPLibrary/interpolate.c CPLibrary/step.c
    ↪ CPLibrary/array.c CPLibrary/random.c -lm
$ ./lprecord
./lprecord <R1> <R2> <Q> <z> <N trials> <seed> <rinit> <rfinal> <rinc calc> <rinc interp>
$ time ./lprecord_mconly 0.05 0.30 50e-9 0.1 10000000 1 0 0.3 0.01 >| lprecord_mconly.dat

real   0m13.365s
user   0m13.311s
sys     0m0.003s
```

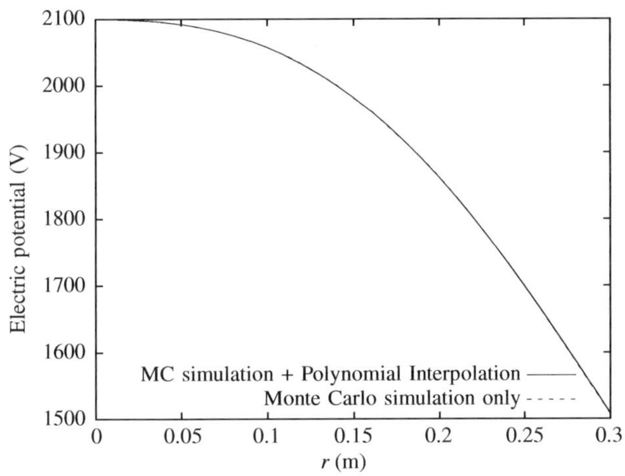

Figure 9.13 Comparison of interpolated electric potential with exclusively Monte Carlo simulations at z = 10 cm.

whereas the original program combining interpolation along only five Monte Carlo points shows

```
$ time ./lprecord 0.05 0.30 50e-9 0.1 10000000 1 0 0.3 0.075 0.01 >| lprecord.dat

real   0m2.159s
user   0m2.151s
sys     0m0.002s
```

Comparison of these two timing results shows a reduction of more than a factor of six in processing time by the use of interpolation. This reduction is accounted for by the factor of 6.2 fewer points, where the Monte Carlo simulation is run, as running the interpolation is virtually free compared with a Monte Carlo simulation with 10000000 pseudorandom points. Plotting the potential profiles computed by the two methods shows a very close correlation, to the resolution of the plot image seen in Figure 9.13.

9.6 Application: The Light Spectrum of a Prism

As another example of the need to interpolate values between some sparse collection of experimentally determined data points, Figure 9.14 shows a graphic diagram of a triangular prism with equal 60 degree angles at the corners of the triangular cross-section. The refraction angles of a beam of light traveling through this structure will vary with the wavelength of light. This is due to the chromatic dispersion of the prism glass material, as the index of refraction in the glass is dependent on wavelength. How may we predict the optical spectrum this prism will produce as light passes through it [17, Chap. 34; 19, Chap. 33; 20, Chap. 35]?

Using the labeling of the various angles in the diagram, and assuming the angle that an incoming ray of monochromatic light forms with the horizontal base of the prism is θ_{in}, we can calculate the outgoing light ray angle θ_{out} by following the procedure

$$\theta_1 = \theta_{in} + 30^\circ$$

From Snell's law we know

$$\sin(\theta_1) = n \sin(\theta_2)$$
$$\text{or } \theta_2 = \sin^{-1}\left(\frac{1}{n}\sin(\theta_1)\right)$$

where n is the index of refraction of the glass in the prism at the wavelength of light in the ray. Assume the index of refraction in air is 1.0. So

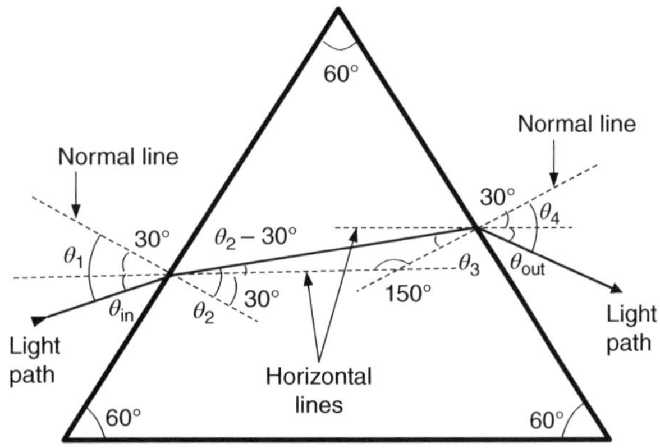

Figure 9.14 Light path through a prism.

From the triangle law

$$\theta_3 = 180° - (150° + \theta_2 - 30°) = 60° - \theta_2$$

Again from Snell's law

$$n\sin(\theta_3) = \sin(\theta_4)$$

$$\text{or } \theta_4 = \sin^{-1}(n\sin(\theta_3))$$

and finally

$$\theta_{out} = \theta_4 - 30°$$

Suppose we have experimental data available on the dependence of the index of refraction for a particular type of glass used for the prism, but only at certain discrete wavelengths where measurements were made. We would like to interpolate values of the index of refraction between the available data points in order to calculate a plot of a near-continuous function of the light output angle versus wavelength. This coding task is Exercise 9.4 at the end of this chapter.

9.7 Algorithms: Inverse Interpolation

In Sections 9.4–9.6, we formed interpolation polynomials from discrete data points, but the only application of those polynomials was simply evaluating intermediate points along its contour $y(x)$. In this section, we explore another application of the interpolation polynomial in which we locate the x value where the dependence is presumed to pass through some target y value. If that target y value is 0, then this is equivalent to finding the root of the underlying nonlinear dependence. Figure 9.15 shows the simple technique of fitting a polynomial through the sampled points with the roles of "x" and "y" interchanged, that is, with y now being considered the independent variable in the coordinate pairs and x now being the dependent variable. Finding the interpolated root is done by simply evaluating the inverse interpolation polynomial at an independent variable value of 0, or any other desired target value.

9.8 Application: Extraction of Local Gravitational Acceleration

Suppose now that an interplanetary probe is sent to the new dwarf planet of Section 9.4 with the intention of making measurements such as its gravitational acceleration. On the probe is a pendulum instrument with an image capture device capable of transmitting a sequence of periodic frame images of the pendulum's deflection angle, illustrated in Figure 9.16. Assuming band width limitations in the transmitted data stream back to Earth, we will have only the ability to measure the deflection angle at some sparse time points.

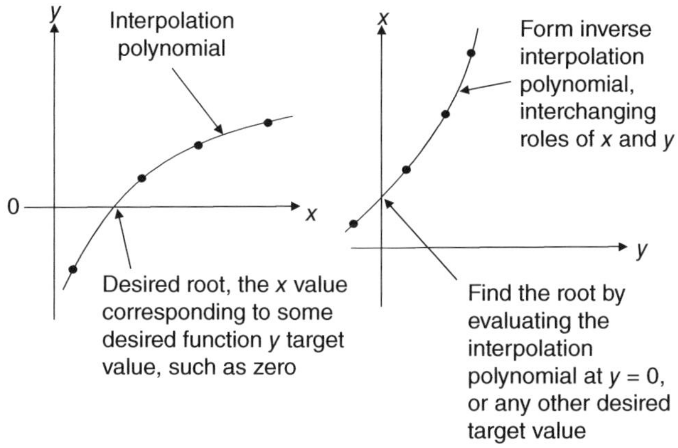

Figure 9.15 Finding a root with an inverse interpolating polynomial.

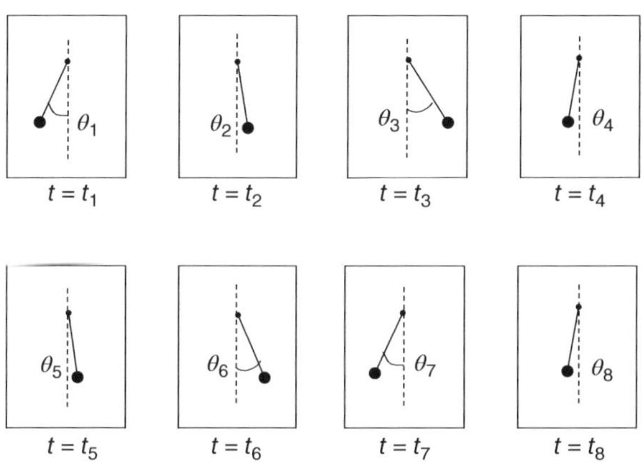

Figure 9.16 Sampled images showing the angle of a pendulum to measure gravitational acceleration.

If the deflection angle of the pendulum is kept small, then we know the period of the swinging oscillations is

$$T_{\text{pendulum}} = 2\pi\sqrt{\frac{L}{g}} \tag{9.23}$$

where L is the length of the pendulum from the rotational pivot to the center of rotational inertia and g is local acceleration due to gravity. The local gravitational acceleration may be inferred from a measured period as

$$g = L \cdot \left(\frac{2\pi}{T_{\text{pendulum}}}\right)^2 \tag{9.24}$$

In general, we may not assume that the periodic images we receive of the pendulum will be captured at time intervals exactly synchronized with its period. Therefore, in order to measure the period, we must use interpolation between the available time samples to solve for the time values where the oscillations pass through to some consistent deflection angle.

As an example scenario, some sampled angles, converted to degrees, versus time for a pendulum length of $L = 0.25$ m are in the file `interpolate_data_pend.dat` from the Student Companion website. The angles in these measurements are sampled at regular 0.1-second intervals.

The tool code for reading a two-column data file and evaluating an inverse interpolation polynomial fitting the data points from the file can be found in the file `interpolate_solve.c`, available for download from the Student Companion website. The inverse interpolation polynomial is evaluated at a point specified by a command line argument. If the argument is zero, this tool essentially finds the root of the interpolated function. So this general-purpose tool may be used to interpolate the multiple times that the swinging pendulum will pass through the zero angle condition. By subtracting successive times, we may measure the pendulum period and from it infer the dwarf planet's gravitational acceleration.

The tool first calls the `read_2col_data_file()` to fill the arrays `xdata` and `ydata` with the time and angle coordinates of the measured data points from the data file. Then the core of the coding method in this tool to scan through the coordinate arrays and interpolate roots of the angle dependence is shown here

```
ARRAY1D xdata,ydata,xtab,ytab;
xtab.n_elems = 4;
ytab.n_elems = 4;
for (i = 0; i < xdata.n_elems - 3; i++) {
  if ((ydata.elem[i + 1] - value) * (ydata.elem[i + 2] - value) <
      0.0) {
    xtab.elem = xdata.elem + i;
    ytab.elem = ydata.elem + i;
// X and Y coordinates exchanged for inverse interpolation
    printf("%.8g %.8g\n",lagrange(&ytab,&xtab,value),value);
  }
}
```

The iterative loop scans through the coordinate lists and searches for four consecutive data points that encompass a point where the interpolated polynomial passes through the target value. For each successive group of four data points, the loop tests for the condition that the sign of the difference of the y coordinate and the target value changes between the second and third point. This condition will be met if the y value passes through the specified target value somewhere in the center sampling interval among three consecutive intervals. Then two arrays `xtab` and `ytab` are constructed consisting of the four measured data points that encompass these three sampling intervals. These `xtab` and `ytab` are then passed to the `lagrange()` interpolation utility, but with the x and y coordinates reversed. The inverse interpolation result is then the x value of the bracketed y target value crossing. The scan then continues so that all y value crossings are reported to the standard output stream. We use the Lagrange interpolation formulation in this application since we need only one evaluation each time an interpolation polynomial is formed, and the computation efficiency of the two-step Newton method is not needed.

For the particular application of sensing the dwarf planet's gravitational field, we can have this tool process the `interpolate_data_pend.dat` data file, with the sample time column as the x variable, the deflection angle column as the y variable, and a y target value specified as zero. An example dialog at the command prompt is

```
$ gcc -Wall -Wshadow -O2 -o interpolate_solve interpolate_solve.c CPLibrary/interpolate.c
    ↪ CPLibrary/array.c CPLibrary/data_file.c -lm
$ ./interpolate_solve
./interpolate_solve <input file> <value>
$ ./interpolate_solve interpolate_data_pend.dat 0
Read 101 points from data file interpolate_data_pend.dat
0.78540286 0
2.3561962 0
3.9269859 0
5.4977882 0
7.0685879 0
8.639377 0
```

Taking, for example, the last complete period described by the two zero crossing times of 5.4977882 and 8.639377 seconds. Note that we would skip over the interpolated time of 7.0685879 seconds, as this represents a zero crossing in the opposite angular direction than for the two crossings we are using to define the period. Subtracting the time of 5.4977882 from 8.639377 seconds gives us an extracted period of 3.14158 seconds. With the pendulum length $L = 0.25$ m, this gives a value of 1.0000 m/s^2, or approximately one-tenth that on the surface of the Earth.

See the example of solving for the range of a ballistic projectile in Chapter 12. See the exercises in Chapter 12 for an example use of computing the oscillation period of a large-angle period with inverse interpolation.

9.9 Exercises

9.1 **(a)** Make a copy of the source code file `interpolate_step.c` and edit it to form a short program `interpolate_newtoncoeffs.c` that accepts a command line argument for a two-column data file name, reads the file, and finds the coefficients of the Newton interpolation polynomial fitting the data points. The program need not evaluate or output any points along the interpolation polynomial, but instead should simply list the Newton polynomial coefficients b_i from Equation (9.6) to the standard output stream.

(b) A colleague thinks the data points in Table 9.2 originate from an underlying cubic dependence of the variable y on the independent variable x. How can your colleague's hypothesis be tested very simply, using only the output from your program in part (a)?

Table 9.2 Set of experimental data points.

x	y
−2	1
−1	4
0	11
1	16
2	13
3	−4

9.2 **(a)** Download the data file `interpolate_data_cepheid30.dat` from the Student Companion website. This file has a table of 30 samples of representative observed data from a Cepheid variable star. Apply the `spline_step.c` tool from Section 9.1.3 to generate a cubic spline fit to these data coordinates. Generate a plot of the sampled data points together with the spline fit.

(b) Apply the `interpolate_solve.c` tool to extract an estimate of the period of the light curve.

9.3 Assume that the `stat_data_magnitude.dat` data file lists successive sampled measurements of the apparent magnitude of a star observed from a terrestrial telescope. The sampling time is listed in the first data column, and the observed magnitude in the second column. Due to atmospheric scintillation, or "twinkling," the observed magnitude exhibits a statistical variation. Make use of the `meanvar.c` and `meanvar.h` utility code and header file that was your solution to Exercise 3.6 to write a program `data_file_stats.c`. This program must accept a command line argument naming the data file to be analyzed. The program must first call the `read_2col_data_file()` function to read two columns of data from the specified file into arrays and then send a report of the mean, variance, and standard deviation (the square root of the variance) of the y coordinate in the second column of the data points to the standard output. Test your program with the `stat_data_magnitude.dat` data file.

9.4 Consider the prism problem discussed in Section 9.6.

Table 9.3 lists measurements of the index of refraction of a particular glass to be used in the prism taken at four light wavelengths.

Table 9.3 Index of refraction at measured wavelengths for prism glass.

Wavelength (nm)	Index of refraction
400	1.522
500	1.511
600	1.505
700	1.502

Write a program `prism_step.c` to use interpolation to generate the data for a plot of the output angle versus light wavelength predicted for an input angle of 30° below the horizontal. The plot should show output angle over the wavelength range of 400–700 nm, calculated for at least 100 wavelengths. Your program should first call the `read_2col_data_file()` function to read the experimental data into arrays from a data file. Then it should call `find_newton_coeffs()` to create an interpolation polynomial specified by an array of coefficients that will be available as a parameter to a C function. Then the `step()` routine should be called to predict the light output angle as a function of wavelength. The object function that you write to be sampled by the `step()` routine should call the `newton()` function to interpolate many different indices of refraction, and then trace through the angle calculations in Section 9.6 each index.

10

Numerical Integration

In Chapters 5 and 7, we encountered physical systems governed by equations of transcendental and other nonlinearities that required solution by numerical methods. In the same way, the evaluation of definite integrals, a task often required in physical modeling, can involve integrands that do not admit analytic methods of finding an "antiderivative" function. This chapter describes two straightforward methods of numerically evaluating definite integrals. These methods build on the principles of interpolation from Chapter 9.

10.1 Algorithms: Integration Methods

10.1.1 Trapezoidal Method

Figure 10.1 shows the plot of an arbitrary function $f(x)$ whose definite integral from x_0 to x_3 is needed.

The figure shows dividing the x range into three piecewise segments, and performing independent linear, or first-order, polynomial interpolations of the function over each segment. The area of the trapezoid thus defined is trivial to calculate, and we may then simply add the areas to arrive at an estimate of the definite integral. If the sampling on the x axis is on a uniform grid, then the trapezoidal area segments are

$$A = \frac{h}{2}(y_i + y_{i+1}) \tag{10.1}$$

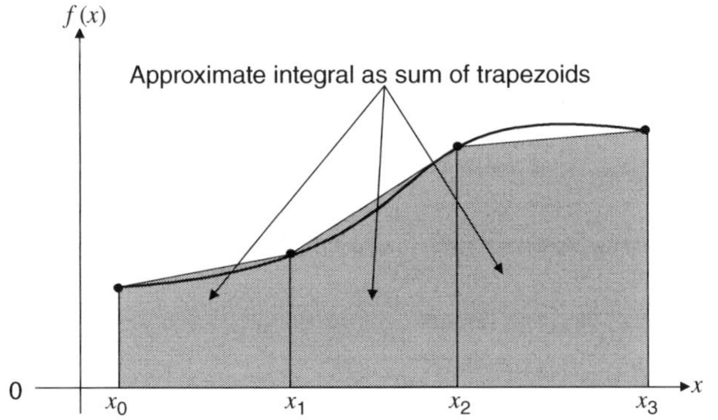

Figure 10.1 The trapezoidal method for numerical integration.

Computational Physics Using C: Efficient Programming with Ease, First Edition. John W. Fattaruso. © 2026 John Wiley & Sons, Inc.
All rights reserved, including rights for text and data mining and training of artificial intelligence technologies or similar technologies.
Published 2026 by John Wiley & Sons, Inc. Companion Website: http://www.wiley.com/go/CompPhysUsingC

where h is the step size for x, and the y_i coordinates are $y_i = f(x_i)$. Summing all the segment areas, where we align the terms containing a particular y_i coordinate in columns

$$
\begin{array}{l}
\frac{h}{2}y_0 + \frac{h}{2}y_1 \\
\qquad + \frac{h}{2}y_1 + \frac{h}{2}y_2 \\
\qquad\qquad + \frac{h}{2}y_2 + \frac{h}{2}y_3 \\
\qquad\qquad\qquad\qquad \vdots \\
\qquad\qquad\qquad\qquad\qquad + \frac{h}{2}y_{n-3} + \frac{h}{2}y_{n-2} \\
\qquad\qquad\qquad\qquad\qquad\qquad + \frac{h}{2}y_{n-2} + \frac{h}{2}y_{n-1} \\
\hline
= \frac{h}{2}y_0 + hy_1 + hy_2 + hy_3 + \cdots + hy_{n-3} + hy_{n-2} + \frac{h}{2}y_{n-1}
\end{array}
$$

By summing the terms in vertical columns, we arrive at the general formula for trapezoidal integration, valid for n uniformly spaced data points

$$\int_{x_0}^{x_{n-1}} f(x)dx \approx h \cdot \left[\frac{(y_0 + y_{n-1})}{2} + \sum_{i=1}^{n-2} y_i \right] \tag{10.2}$$

Assuming the integrand function has a reasonably well-behaved Taylor series representation, a careful analysis [10, Chap. 5] of the truncation error between the true integral and this trapezoidal approximation shows

$$E = -\left(\frac{1}{12}\right)(x_{n-1} - x_0)h^2 f''(\xi) \tag{10.3}$$

Note that the error term is of the second order, that is, every factor of two reduction in the step size h of the x sampling grid results in a factor of four less truncation error.

In general, an algorithm with only second-order error characteristics will require a fairly fine step size to reduce the error to a practical level. This fine step size then requires a large number of steps to cover the needed x range in Figure 10.1, and this requirement invites the accumulation of significant roundoff error. However, the trapezoidal method may easily be applied to a situation where the integrand function is defined only by a sequence of data points with monotonic, but not necessarily uniformly spaced, x coordinates. Although a uniform grid of x values was assumed for the straightforward derivation of Equation (10.2), there is nothing in the geometry of Figure 10.1 that requires this uniformity. A similar summation of adjoining trapezoidal areas can be made simply from lists of data point coordinates regardless of x coordinate spacing.

10.1.2 Simpson's Method

As the error term of the trapezoidal method is only of second order, we wish to find a method with a higher-order error characteristic so that the fine step size requirements may be relaxed. Consider the interpolation scheme pictured in Figure 10.2.

With this approach, we associate the sampled points along the integrand function in segments of three points, and we fit a second-order interpolation polynomial through each nonoverlapping group of three consecutive points. Although we consider the boundary points between each segment to be common to both adjoining segments, we consider each interpolation of a particular segment to be independent of the other segments. (This means that this method does not involve a similar interpolation process as the cubic spline fitting of Section 9.1.3 where continuity constraints had to be enforced simultaneously through the entire set of polynomials.) To start the derivation of a formula for Simpson's method

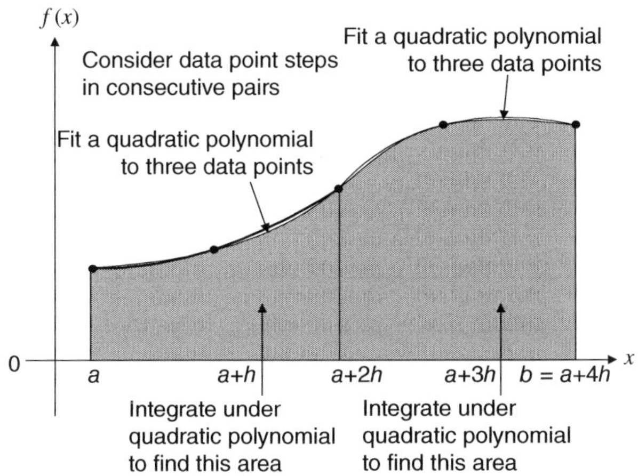

Figure 10.2 Simpson's method for numerical integration.

for uniformly spaced x coordinates, we fit a second-order Lagrange interpolation polynomial of Section 9.1.2 to the first segment of three data points

$$f(x) = \frac{(x - x_1)(x - x_2)}{(x_0 - x_1)(x_0 - x_2)} \cdot y_0 + \frac{(x - x_0)(x - x_2)}{(x_1 - x_0)(x_1 - x_2)} \cdot y_1 + \frac{(x - x_0)(x - x_1)}{(x_2 - x_0)(x_2 - x_1)} \cdot y_2 \tag{10.4}$$

We then change the x axis so that $x_0 = 0$ $x_1 = h$ $x_2 = 2h$, and rewrite the Lagrange interpolation polynomial for these uniformly spaced data points

$$f(x) = \frac{(x - h)(x - 2h)}{2h^2} \cdot y_0 - \frac{x(x - 2h)}{h^2} \cdot y_1 + \frac{x(x - h)}{2h^2} \cdot y_2 \tag{10.5}$$

Integration of the quadratic polynomial from 0 to $2h$ is straightforward, and gives an area for the first segment of

$$\int_0^{2h} f(x)dx \approx \left(\frac{h}{3}\right) \cdot [y_0 + 4y_1 + y_2] \tag{10.6}$$

Similar to the derivation for the trapezoidal method, we assemble a summation of all the segment areas in columns with common y coordinate values

$$\frac{h}{3}y_0 + \frac{4h}{3}y_1 + \frac{h}{3}y_2$$

$$+\frac{h}{3}y_2 + \frac{4h}{3}y_3 + \frac{h}{3}y_4$$

$$+\frac{h}{3}y_4 + \frac{4h}{3}y_5 + \frac{h}{3}y_6$$

$$\vdots$$

$$+\frac{h}{3}y_{n-5} + \frac{4h}{3}y_{n-4} + \frac{h}{3}y_{n-3}$$

$$+\frac{h}{3}y_{n-3} + \frac{4h}{3}y_{n-2} + \frac{h}{3}y_{n-1}$$

$$= \frac{h}{3}y_0 + \frac{4h}{3}y_1 + \frac{2h}{3}dy_2 + \frac{4h}{3}y_3 + \frac{2h}{3}y_4 + \frac{4h}{3}y_5 + \frac{2h}{3}y_6 + \cdots + \frac{2h}{3}y_{n-5} + \frac{4h}{3}y_{n-4} + \frac{2h}{3}y_{n-3} + \frac{4h}{3}y_{n-2} + \frac{h}{3}y_{n-1}$$

This leads to the following formula for Simpson's method integration for n uniformly spaced data points

$$\int_{x_0}^{x_{n-1}} f(x)dx \approx \frac{h}{3}\left[(y_0 + y_{n-1}) + 4\sum_{i=1,i\text{ odd}}^{n-2} y_i + 2\sum_{i=2,i\text{ even}}^{n-3} y_i\right] \tag{10.7}$$

Note that the formal Simpson's method is applicable only to the case of sampling the integrand with an even number of steps. Therefore, $n-1$ must be an even number, which implies the number of function points must be odd. However, we may compute these quadratic fits in segments of two x intervals through all but the last interval, and then append one trapezoidal step area at the right end to complete the numerical approximation to the integral.

The error term for Simpson's method may be derived [10, Chap. 5] as

$$E = -\left(\frac{1}{180}\right)(x_{n-1} - x_0)h^4 f^{(4)}(\xi) \tag{10.8}$$

We can see that by extending the order of the interpolation fit to the points along the integrand from first to only second order, we have achieved a substantial reduction of the error term from second to fourth order. Thus, for a modest increase in algorithm complexity, a relaxation of the step size to much coarser intervals is allowed to achieve a given error level, with a concomitant reduction in roundoff error and greater computation efficiency.

To make this advantage of Simpson's method more clear graphically, consider the following test program to calculate numerically the truncation errors of the trapezoidal and Simpson's methods. The function being integrated is the simple exponential function $f(x) = e^x$. This prototype function is chosen so that a polynomial fit of any finite order will not be exact, but its definite integral from $x = 0$ to $x = 1$ is known analytically as $e - 1$. This easily calculated value for the definite integral allows the error of a numerical integration relative to the analytic value to be characterized. The program makes use of integration functions for the trapezoidal and Simpson's methods that will be detailed in Section 10.2, and computes the integrals with double-precision floating point arithmetic.

```c
// File: precision_int.c
#include <stdio.h>
#include <stdlib.h>
#include <math.h>
#include "CPLibrary/grid.h"
#include "CPLibrary/integrate.h"

// Estimation of integral of f(x) = exp(x) from x=0 to 1

double f(double x,void *p) {
  return(exp(x));
}

int main(int argc,char *argv[]) {
  GRID xgrid;
  double exact,integral,decade;
  double error_trap,error_simp;
  double h[7] =
    {1.0,0.5,0.333333333333333,0.25,0.2,0.166666666666667,0.125};
  int i,j;

  xgrid.init = 0.0;
  xgrid.final = 1.0;
  exact = exp(1.0) - 1.0;
  decade = 0.1;
  for (i = 0; i < 8; i++) {
    for (j = 0; j < 7; j++) {
      xgrid.inc = decade * h[j];
      trapezoidal_func(f,&xgrid,NULL,&integral);
      error_trap = fabs(exact - integral);
      simpson_func(f,&xgrid,NULL,&integral);
      error_simp = fabs(exact - integral);
      printf("%.8e %.8e %.8e\n",xgrid.inc,error_trap,error_simp);
      if (i == 7) break;  // Only use h=1.0 of last decade
    }
    decade *= 0.1;
  }
  exit(0);
}
```

Listing 10.1 Test program for tradeoff between roundoff and truncation errors

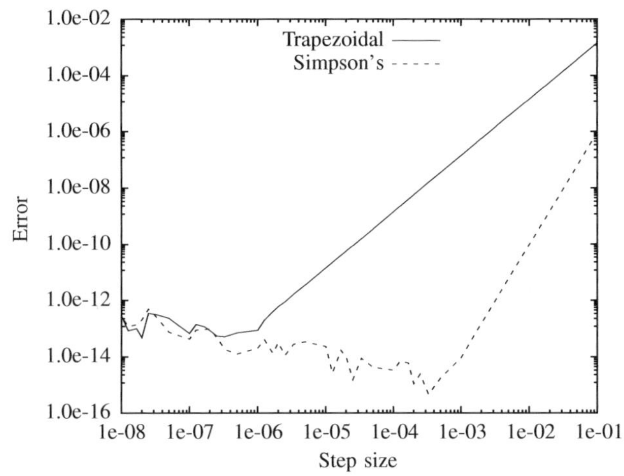

Figure 10.3 Tradeoff of roundoff and truncation error versus step size for different integration algorithms.

When we plot the truncation error for the two methods versus step size in Figure 10.3, we see a characteristic similar to that of Section 2.5 for numerical derivatives. We may reduce the step size on the right side to achieve integral approximations with lower truncation errors, and by looking up values on the plot axes, the student can verify the second- and fourth-order behavior of the two algorithms, respectively. But, as was the case with numerical differentiation, at some smaller step size the roundoff and loss of precision errors resulting from the binary bit length limitations of the double-precision standard will start to dominate.

10.2 Coding: The Integration Tool

The tool code for various integrating functions with either of the trapezoidal or Simpson's algorithms can be found in the files `integrate.c` and `integrate.h`, part of the `CPLibrary` package from the Student Companion website. Six routines are available in the `integrate.c` source code file, with arguments as shown here

```
int trapezoidal_func(double (*func)(double x,void *p),GRID *xgrid_ptr,
                     void *param,double *integral_ptr) {
  ...
}

int simpson_func(double (*func)(double x,void *p),GRID *xgrid_ptr,
                 void *param,double *integral_ptr) {
  ...
}

int simpson_vect_func(void (*func)(double x,void *p,ARRAY1D *f),
                      GRID *xgrid_ptr,void *param,
                      ARRAY1D *integral_ptr) {
  ...
}

int trapezoidal_array(ARRAY1D *y_ptr,double h,double *integral_ptr) {
  ...
}

int trapezoidal_array2(ARRAY1D *x_ptr,ARRAY1D *y_ptr,
                       double *integral_ptr) {
  ...
}

int simpson_array(ARRAY1D *y_ptr,double h,double *integral_ptr) {
  ...
}
```

The first two functions, `trapezoidal_func()` and `simpson_func()`, perform numerical integration with the trapezoidal and Simpson's method, respectively. The function that forms the integrand is assumed to be coded as a C function, and similar to functions analyzed by the methods of Chapters 4 and 5, the first argument to these integration functions must be a pointer to that C function with the prototype `double func(double x,void *p)`. This integrand function expects arguments of the independent variable `x` and a pointer to a parameter structure `p`. Then back to the argument list for the integration tools, the grid of uniform x values over which to perform the definite numerical integration is pointed to by the second argument, and a pointer to the structure of parameters to be passed directly on to the integrand function is the third argument. The final argument is a pointer to a `double` variable where the integration function is to store the value of the integral estimate.

We will see later in Section 10.6 a typical application of numerical integration for finding magnetic field strength with the Biot–Savart law. As with all applications of Maxwell's equations, the integration operation is performed over vector quantities. To make this operation more convenient, a third function is defined in the `integrate.c` source code file that applies Simpson's method to a multivalued function of a single variable. The `simpson_vect_func()` function is similar to `simpson_func()`, except that the first argument is expected to be a pointer to a function with a prototype `void func(double x,void *p,ARRAY1D *f)`. This class of functions accepts an argument value `x`, a pointer to structure of parameters `p`, and a pointer to an `ARRAY1D` structure where multiple function values are to be stored, all a function of the argument `x`. These function values are typically the components of a vector, and we would make use of this integration function when, for example, forming a line integral along a parameterized line coordinate of a three-dimensional magnetic field vector. The multiple vector components are integrated independently by the function over the `x` argument and returned in the argument array.

Finally, the last three integration routines in `integrate.c` are similar to the first two routines, except the y coordinates of the data points that define the integrand function are taken from the elements of an `ARRAY1D` table instead of computed by a C function. For the `trapezoidal_array()` and `simpson_array()` routines, the y values are taken in sequence from the `ARRAY1D` array pointed to by the first function argument, and assuming that the corresponding x values are separated by a uniform step size h that is passed as the second argument. However, for the `trapezoidal_array2()` routine, the first two arguments are both expected to be pointers to `ARRAY1D` tables that contain the sequence of x and y coordinates, respectively. This allows a trapezoidal estimate of the integral to be computed for a set of coordinates with x coordinates that may be nonuniformly spaced. For all three of these functions, the tabulated y and x values may come from external computations through a data file, or from other programmed calculations that may not be conveniently expressed as a C function.

In all these integration functions, the estimated value of the definite integral is first initialized with the contribution of the first point y_0 in the variable pointed to by dereferencing the formal argument `integral_ptr`, and then is updated as the subsequent sampled points along the integrand function are processed in sequence with an iterative loop. Note that some computation is saved by including the multiplicative factor of the step size h in Equations (10.2) and (10.7) only once, just before each function returns. As with other library functions, the integer return codes that characterize the integration process are defined in the corresponding header file, in this case, `integrate.h`.

In order to implement the computation of Simpson's method shown in Equation (10.7), the algorithms in the `simpson_func()` and `simpson_array()` functions must keep track of two possible states associated with each iteration. The state determines if the current point in the sequence is to be treated as having an odd index, and, therefore, included in the sum with a multiplicative factor of $\frac{4}{3}$, or as having an even index, and included with a factor of $\frac{2}{3}$. This alternation of state is implemented in the C code with an `int` variable named `state`. This state machine variable is initialized to 0 after the first point y_0 contributes to the sum, and then sequentially alternates between 1 and 0 in each iteration with a "state machine" that is implemented with the conditional statements

```
if (state == 0) {
  *integral_ptr += (4.0 / 3.0) * fx;
  state = 1;
}
else {
  *integral_ptr += (2.0 / 3.0) * fx;
  state = 0;
}
```

If the iterative loop that sequences through all but the last of the data points along the integrand terminates with the value of the `state` variable being 1, then we know that this routine has been called properly by specifying an odd number of samples. In this case, the integral sum may simply be updated with the weight factor of $\frac{1}{3}$ times the last point y_{n-1} per Equation (10.7), and the function returns with the integer return value `INTEGRATE_SUCCESS`. On the other hand, if the iterative loop terminates with the state machine variable at 0, then we know there are an even number of points, and the integral must be adjusted in a way that is equivalent to appending a final segment of the trapezoidal method. In this case, the routine sends a warning message to the standard error stream, and the return value is set to `INTEGRATE_EXTENDED`. This will inform the user and the calling program that a minimal fourth-order truncation error was likely not achieved in this function call.

Example programs making use of the `trapezoidal_func()` and `simpson_func()` integration routines are explored later in Sections 10.3 and 10.4. For a straightforward example program that makes use of the `trapezoidal_array()` and `simpson_array()` integration routines, download and study the source code for the `integrate_data_file.c` program. This example uses the `data_file.c` utility code described in Section 9.2 to read the data point coordinates contained in a file into `ARRAY1D` tables that sample some function to be integrated. The program then calls both the `trapezoidal_array()` and `simpson_array()` routines to generate two estimates of the integral that may be compared.

Let's test this program with a data file we used in Section 9.1.3 consisting of uniform samples of the function in Equation (9.16), as shown in Figure 9.3.

```
$ gcc -Wall -Wshadow -O2 -o integrate_data_file integrate_data_file.c CPLibrary/integrate.c
    ↪ CPLibrary/data_file.c CPLibrary/array.c -lm
$ ./integrate_data_file
./integrate_data_file <input file>
$ ./integrate_data_file interpolate_data_cosine9.dat
Read 9 points from data file interpolate_data_cosine9.dat:
Trapezoidal method: 0.31420872
Simpson's method: 0.31835272
```

The student may check these numerical results against the expected value known from a straightforward integration of Equation (9.16) over the domain of $-\frac{1}{4}$ to $\frac{1}{4}$. Note that with this quite coarse step size, the significantly larger truncation error of the second-order trapezoidal algorithm is readily apparent in comparison to that of the fourth-order Simpson's method.

10.3 Application: Gaussian Distribution

In this section, let's examine an example integration of the standard normal Gaussian distribution function using both the trapezoidal and Simpson's methods.

$$f(x) = \frac{1}{\sigma\sqrt{2\pi}} e^{-\frac{(x-\mu)^2}{2\sigma^2}} \tag{10.9}$$

Figure 10.4 shows an example of this function with the nominal parameter values of $\mu = 0$ and $\sigma = 1$.

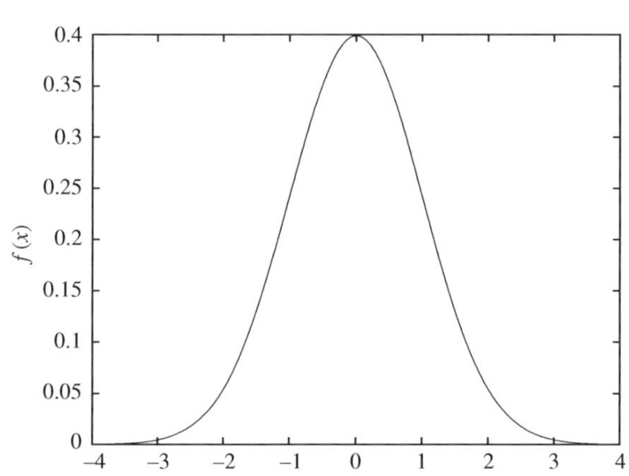

Figure 10.4 Gaussian distribution function with mean of 0 and standard deviation of 1.

Table 10.1 Definite integrals between integration limits of the Gaussian distribution function from published tables.

Lower x limit	Upper x limit	Integral
0	0.5	0.191 46
0	1	0.341 34
0	2	0.477 25
0	3	0.498 65

This function has been studied for centuries as one of the fundamental relationships of statistics. While no analytical integration is possible, the numerical integration of this function has been accurately tabulated for statistical work. See Table 10.1 for a sampling of several values from ubiquitous published tables.

Download the program `integrate_gaussian.c` from the Student Companion website. The C function `gaussian()` that computes the probability distribution of Equation (10.9) is defined in this source code. The program steps the function and sends a table of values to the standard output so that the plot in Figure 10.4 may be generated. Then both the `trapezoidal_func()` and `simpson_func()` routines are called to integrate the function over a specified grid of x values. Several integrations over different limits of the x values are performed in the following command shell dialog

```
$ gcc -Wall -Wshadow -O2 -o integrate_gaussian integrate_gaussian.c CPLibrary/integrate.c
    ↪ CPLibrary/step.c CPLibrary/array.c -lm
$ ./integrate_gaussian
./integrate_gaussian <mu> <sigma> <xinit> <xfinal> <xinc>
$ ./integrate_gaussian 0 1 0 0.5 0.05 >| integrate_gaussian.dat
Trapezoidal method: 0.19142578
Simpson's method: 0.19146248
$ ./integrate_gaussian 0 1 0 1 0.05 >| integrate_gaussian.dat
Trapezoidal method: 0.34129433
Simpson's method: 0.34134476
$ ./integrate_gaussian 0 1 0 2 0.05 >| integrate_gaussian.dat
Trapezoidal method: 0.47722737
Simpson's method: 0.47724986
$ ./integrate_gaussian 0 1 0 3 0.05 >| integrate_gaussian.dat
Trapezoidal method: 0.49864733
Simpson's method: 0.49865010
```

Even with the fairly coarse step size of 0.05, the fourth-order Simpson's method is capable of matching the published values of Table 10.1 to five significant figures. The truncation error from the second-order trapezoidal method is significantly larger.

10.4 Application: Orbital Circumference

Figure 10.5 depicts a comet in an elliptical orbit around the sun. The circumference of an elliptical orbit $C = 4aE(\epsilon)$, where a is the length of the semi-major axis, ϵ is the eccentricity, and the function E is the complete elliptic integral of the second kind

$$E(\epsilon) = \int_0^{\frac{\pi}{2}} \sqrt{1 - \epsilon^2 \sin^2(\theta)}\ \ d\theta \tag{10.10}$$

Finding the circumference of this orbit is Exercise 10.2 for the student.

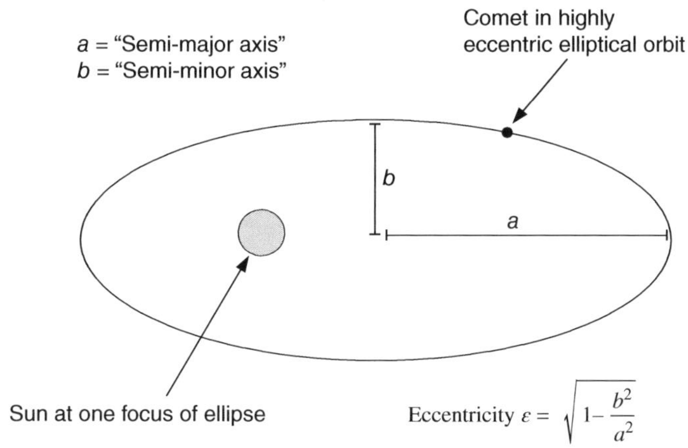

Figure 10.5 A comet in an elliptical orbit around the sun.

10.5 Application: Rotational Inertia

The rotational inertia of any body may be found by the definite integral over all elements of mass [17, Chap. 11; 18, Chap. 13; 19, Chap. 10;20, Chap. 10] as

$$I = \int r^2 dm \tag{10.11}$$

where r is the radial distance from the axis of rotation to the infinitesimal mass element dm. This integration can be straightforward for various homogeneous objects, where the mass density is constant throughout the body of the object. Let's see how we may evaluate this expression for the case of a nonhomogeneous but spherically symmetric object, such as a planet, and where using experimentally determined data points of mass density as a function of radius requires numerical integration.

First consider the geometry shown in Figure 10.6a where mass elements are distributed uniformly throughout a thin spherical shell of radius r_s and total mass m_s. We assume that the shell is thin enough so that surface areas on the inside and outside surfaces are essentially identical, and both equal to $A_s = 4\pi r_s^2$.

Now assume the spherical shell is rotated about any line through the center of the shell. We consider that the shell is composed of infinitesimal circular rings at an inclination angle θ as shown in the figure, and we may find the rotational inertia of the shell with an integration over all the rings. Since the infinitesimal surface area associated with each ring is

$$dA_r = 2\pi r_r \cdot r_s d\theta = 2\pi r_s \sin\theta \cdot r_s d\theta \tag{10.12}$$

the infinitesimal mass in each ring can be determined by an area ratio to be

$$dm_r = m_s \frac{dA_r}{A_s} \tag{10.13}$$

Then the rotational inertia integral of Equation (10.11) for the complete shell becomes

$$I_s = \int r_r^2 dm_r = \frac{2}{3} m_s r_s^2 \tag{10.14}$$

Consider now the solid sphere of radius R shown in Figure 10.6b. We may find the rotational inertia of the sphere by now considering the inertia of the shell I_s in Equation (10.14) to be an infinitesimal contribution to an integral due to an infinitesimal mass dm_s of each shell of infinitesimal thickness dr_s, and integrate over all the shells of radius r_s from 0 through R. If the sphere is assumed to be composed of a homogeneous material, then the mass of each infinitesimal shell is found from the volume ratio

$$dm_s = M \frac{4\pi r_s^2 dr_s}{\frac{4}{3}\pi R^3} \tag{10.15}$$

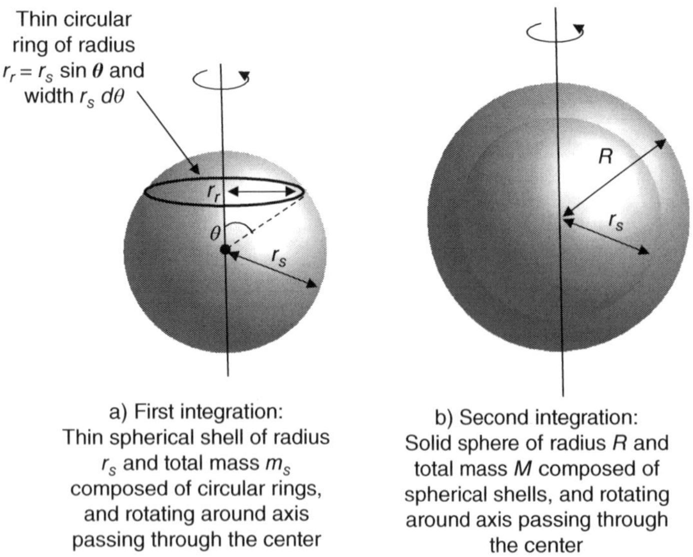

Figure 10.6 Integration of the rotational inertia of a nonhomogeneous sphere.

For this homogenous case, the rotational inertial of the solid shell may then be found with a simple integral, and is

$$I_{homogeneous} = \int dI_s = \frac{8\pi}{15}\mu R^5 = \frac{2}{5}MR^2 \tag{10.16}$$

where M is the total mass of the sphere. However, if the mass volume density of the material varies as a spherically symmetric function $\mu(r)$ of the radial coordinate r from the center, then the infinitesimal mass in each shell must be found from the volume of each shell as

$$dm_s = \mu(r_s) \cdot 4\pi r_s^2 \cdot dr_s \tag{10.17}$$

Then the rotational inertia elements may be summed with the integral

$$I_{nonhomogeneous} = \int dI_s = \frac{8\pi}{3}\int_0^R \mu(r_s) r_s^4 dr_s \tag{10.18}$$

If the mass density function $\mu(r)$ is known experimentally at discrete measured data points or by a piecewise interpolated function, then we can find the value of this definite integral with numerical integration methods. Examples of finding the rotational inertia of planets are Exercises 10.4 and 10.5 for the student.

10.6 Application: The Helmholtz Coil Revisited

We first analyzed the magnetic field generated by the Helmholtz coil in Section 4.4. However, in that section we were limited to the fields along the z axis, where the integrals evaluating the magnetic field following the Biot–Savart law are analytically tractable. In this section we expand the scope where we may calculate the field to the entire region around the coils by evaluating the required integrals numerically. The geometry of the two circular coils carrying electrical current is shown here as well in Figure 10.7.

We start the development of our programming tools by considering the field generated by the single circular loop of radius R carrying current I shown in Figure 10.8. The plane of the current loop is parallel to the x–y plane, and is located at some arbitrary z_c coordinate.

The incremental contribution to the magnetic field $d\vec{B}$ due to each current element around the loop measured at an arbitrary observation probe point located at cylindrical coordinates (r_p, θ_p, z_p) in the volume around the coil is given by the Biot–Savart law[6, Chap. 2; 7, Chap. 4; 8, Chap. 8; 15, Chap. 5]. The element of current path $d\vec{l}$ along the circular path is

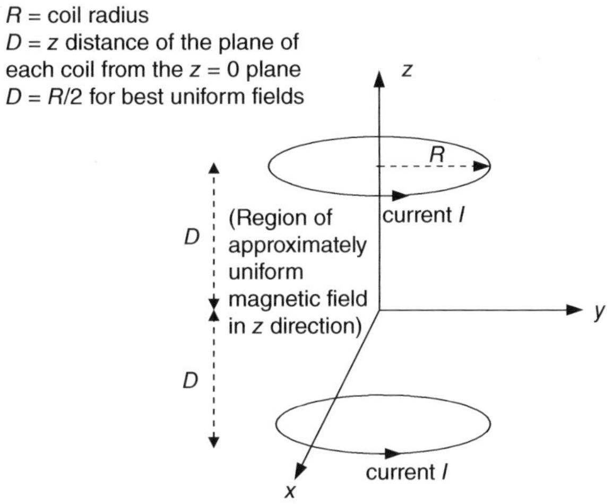

Figure 10.7 Geometry of a Helmholtz coil.

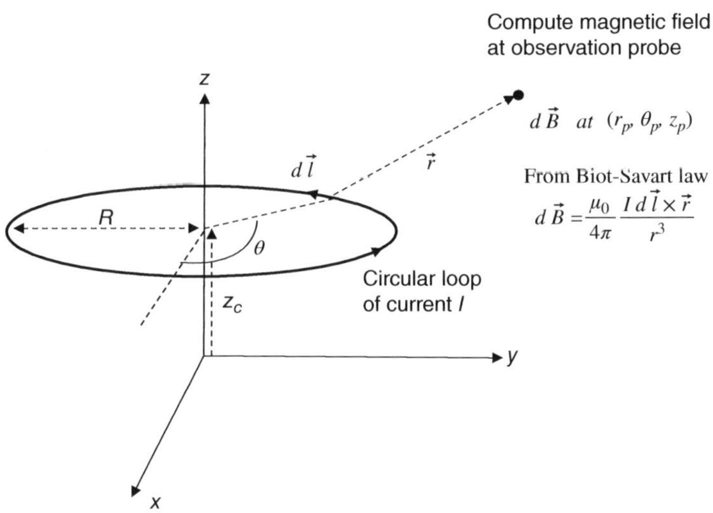

Figure 10.8 Geometry of a loop of electric current generating a magnetic field.

located at a cylindrical angle θ relative to the x axis, and the displacement vector from the current element to the field observation point is $\vec{r}$.

$$d\vec{B} = \frac{\mu_0}{4\pi}\frac{Id\vec{l}\times\vec{r}}{r^3} \tag{10.19}$$

We then find that the total field at the observation probe coordinates (r_p, θ_p, z_p) is the integration of all the incremental field contributions due to the elements around the circular loop

$$\vec{B} = \int d\vec{B} \tag{10.20}$$

Since we must form the displacement vector $\vec{r}$ in the Biot–Savart law, the required vector subtraction is most straightforward using Cartesian coordinates for the positions of $d\vec{l}$ and the observation probe. Then the vector cross-product $d\vec{l}\times\vec{r}$ operation in the expression for $d\vec{B}$ resolved into Cartesian coordinates is

$$\begin{aligned}
(d\vec{l}\times\vec{r})_x &= dl_y r_z - dl_z r_y \\
(d\vec{l}\times\vec{r})_y &= dl_z r_x - dl_x r_z \\
(d\vec{l}\times\vec{r})_z &= dl_x r_y - dl_y r_x
\end{aligned} \tag{10.21}$$

In forming this vector cross product, we may first simplify these coordinate expressions with the observation that the z component of $d\vec{l}$ is zero. Moreover, by the cylindrical symmetry of this geometry, we know the θ component of the integrated $\vec{B}$ field must be zero, so there is no need to compute and integrate any θ components of the $d\vec{B}$ contributions. This means we may simply locate the magnetic observation probe at only one convenient value of θ_p. So we may reduce the computation to only two cylindrical dimensions by assuming the location of the observation probe has, for example, an x_p component of zero, and assign the radial component of $B_r = B_y$ for any θ_p probe coordinate. Of course, reducing the computational problem from three to two dimensions will increase computational efficiency markedly. Using $dl_x = -R\sin(\theta)d\theta$ and $dl_y = R\cos(\theta)d\theta$, our cross product expressions may be reduced to

$$
\begin{aligned}
(d\vec{l} \times \vec{r})_r &= -dl_x r_z = R\sin(\theta) r_z d\theta \\
(d\vec{l} \times \vec{r})_z &= dl_x r_y - dl_y r_x = -(R\sin(\theta) r_y + R\cos(\theta) r_x) d\theta
\end{aligned}
\tag{10.22}
$$

To facilitate the computation of the magnetic fields, we first code a basic utility function to compute the magnetic field at any point around any number of circular loops of current stacked in a uniform vertical array. Refer to the source code file `current_loops.c` and its header file `current_loops.h` in the `CPLibrary` package. (We may think of this utility code as perhaps the magnetic counterpart to the electric `point_charges.c` utility in Chapter 4, except for the restricted geometry that the plane of the current loop is parallel to the *x-y* coordinate plane.) The utility function supplied by this source code file is named `magfield_loops()`, and this function in turn calls the `simpson_vect_func()` integration routine in `integrate.c` to perform the integration of Equation (10.20) numerically for each current loop and accumulates the sum of the integrals into the total magnetic field vector. The C function `loop()` that is passed to `simpson_vect_func()` that realizes the vector expressions in Equation (10.22) is

```
void loop(double theta,void *param_ptr,ARRAY1D *f) {
  double rx,ry,rz,rsq,r;
  CURLOOP *p;

  p = param_ptr;
  rx = -p->R * cos(theta);
  ry = p->rp - p->R * sin(theta);
  rz = p->zp - p->zc;
  rsq = rx*rx + ry*ry + rz*rz;
  r = sqrt(rsq);
  f->elem[0] = 1.0e-7 * p->cur * (p->R * sin(theta) * rz) / (r * rsq);
  f->elem[1] = 1.0e-7 * p->cur * -(p->R * sin(theta) * ry +
               p->R * cos(theta) * rx) / (r * rsq);
}
```

The `magfield_loops()` utility function is called from the main program, `helmholtz.c`, modeling the Helmholtz coil as a vertical array of two coils. The main program steps the probe point in both the r_p and z_p dimensions, and sends three columns of data to the output stream tabulating the step coordinates and the r and z components of the magnetic field. Two data sets are output, the first stepping in z_p and the second in r_p. The step size $d\theta$ in Equation (10.22) for the Simpson method integration is specified on the command line as the number of steps to take around each circular current loop.

These source code files are all available for download from the Student Companion website. We may compile and run the code at a Linux shell prompt window, together with the tool developed in Section 4.4 for calculating the field on the axis of a Helmholtz coil arrangement, with the dialog below. First we ask the general Helmholtz coil modeling program to calculate the magnetic field at a cylindrical r_p coordinate of zero, or along the axis of the Helmholtz coil structure. This should agree with the prediction of the `helmholtz_axis_step.c` program developed in Section 4.4, which simply stepped the evaluation of an expression derived with an analytical integration of the Biot–Savart law. That integral is analytically tractable along the axis.

```
$ gcc -Wall -Wshadow -O2 -o helmholtz helmholtz.c CPLibrary/current_loops.c CPLibrary/integrate.c
    ↪ CPLibrary/array.c -lm
$ ./helmholtz
./helmholtz <radius> <distance> <current> <rp> <zp> <num segments>
$ ./helmholtz_axis_step
./helmholtz_axis_step <radius> <distance from center> <current> <zinit> <zfinal> <zinc>
$ ./helmholtz 0.1 0.05 1 0 0 72 >| helmholtz_r0.dat
$ ./helmholtz_axis_step 0.1 0.05 1 -0.1 0.1 0.01 >| helmholtz_axis_step.dat'
```

To plot the z component of the magnetic field as a function of the z_p component along the axis, we may run an interactive `gnuplot` session and provide the commands

```
gnuplot> set xlabel 'Z coordinate (m)'
gnuplot> set xrange [-0.1:0.1]
gnuplot> set ylabel 'Z component of magnetic field (T)'
gnuplot> set format y "%0.1e"
gnuplot> set yrange [0:1.1e-5]
gnuplot> plot 'helmholtz_axis_step.dat' i 2 u 1:2 w l t 'Analytic','helmholtz_r0.dat' i 0 u 1:3 w
    ↪ l t 'Numerical'
```

Figure 10.9 shows good agreement, at least within the precision of the plot resolution, between these two modeling programs, as the angular interval of 72 steps around each circular loop ensures low truncation error with the fourth-order Simpson method.

With the confidence that the errors in our numerical model and integration routine are within reasonable limits, we can proceed to probe the field at locations with nonzero r_p components, or off the axis. We may, for example, run the program twice with

```
$ ./helmholtz 0.1 0.05 1 0.025 0 72 >| helmholtz_r0p025.dat
$ ./helmholtz 0.1 0.05 1 0.05 0 72 >| helmholtz_r0p05.dat
```

for $r_p = 2.5$ cm and $r_p = 5$ cm, respectively. The `gnuplot` dialog

```
gnuplot> set xlabel 'Z coordinate (m)'
gnuplot> set xrange [-0.1:0.1]
gnuplot> set ylabel 'Z component of magnetic field (T)'
gnuplot> set format y "%0.1e"
gnuplot> set yrange [0:1.1e-5]
gnuplot> plot 'helmholtz_r0.dat' i 0 u 1:3 w l t '$r_p=0$ m','helmholtz_r0p025.dat' i 0 u 1:3 w l
    ↪ t '$r_p=0.025$ m','helmholtz_r0p05.dat' i 0 u 1:3 w l t '$r_p=0.05$ m'
```

will again plot the z component of the magnetic field as a function of the z_p coordinate in the first data set from each program run, and produce the plot seen in Figure 10.10.

At an r_p coordinate of zero, along the axis of the coil arrangement, we again see the wide region of an approximately uniform magnetic field that the Helmholtz coil is intended to generate. However, as we stray off the axis at other r_p coordinates, we see regions of higher variability of the field as we approach the paths of the two current loops.

Another plot may be generated from the second output data set with the following `gnuplot` dialog to visualize the dependence as the r_p coordinate is varied at a z_p coordinate of zero. This plot is shown in Figure 10.11. We again may see the limits to the region of an approximately uniform magnetic field generated by the Helmholtz geometry.

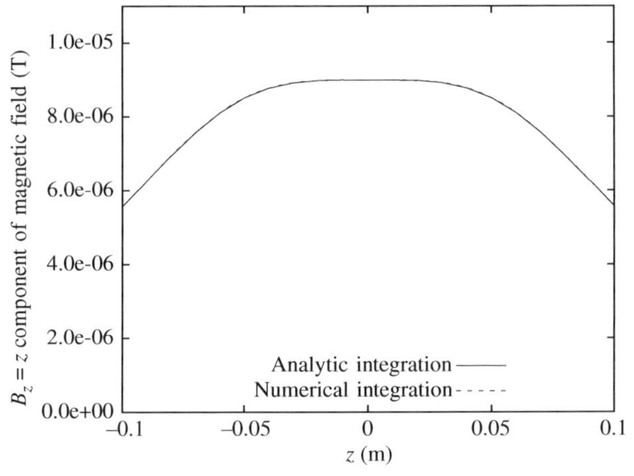

Figure 10.9 Helmholtz coil magnetic field on axis as a function of z_p; radius = 10 cm, distance = 5 cm.

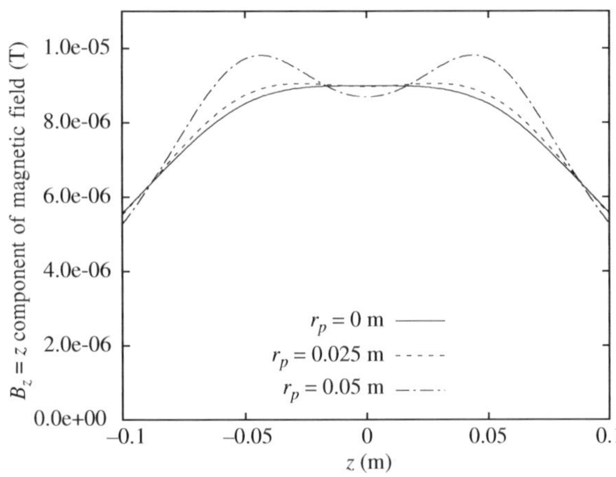

Figure 10.10 Helmholtz coil magnetic field off axis as a function of z_p; radius = 10 cm, distance = 5 cm.

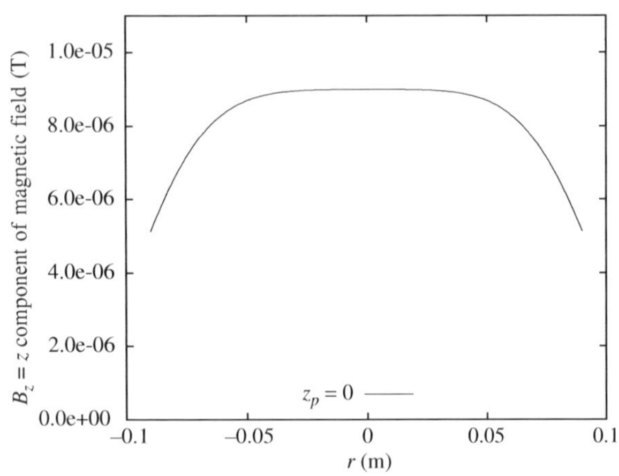

Figure 10.11 Helmholtz coil magnetic field off axis as a function of r; radius = 10 cm, distance = 5 cm.

```
gnuplot> set xlabel 'R coordinate (m)'
gnuplot> set xrange [-0.1:0.1]
gnuplot> set ylabel 'Z component of magnetic field (T)'
gnuplot> set yrange [0:1.1e-5]
gnuplot> plot 'helmholtz_r0.dat' i 1 u 1:3 w l t '$z_p=0$'
```

10.7 Application: Field Vectors Revisited

Most viewers find visualizing a two-dimensional vector quantity such as the magnetic field in this cylindrically symmetrical structure over a two-dimensional space to be difficult with the plotting techniques in Figures 10.10 and 10.11. In Section 4.11, we explored the vector plotting capability of `gnuplot` to help visualize the electric field vector over space. We return to that plotting technique here for the case of the magnetic field around the Helmholtz coil. The counterpart to the `efieldvectors.c` program from Section 4.11 is the program `magfieldvectors.c`, and the coding of this program and the rendering of representative graphical images is available as Exercise 10.6 for the student.

In order to render the two images of magnetic field vectors shown in Figures 10.12 and 10.13, we have run the `magfieldvectors.c` program twice, specifying a single circular coil in the first run, and then two coils in a Helmholtz arrangement in the second run. This lets us visualize the curving field vectors around a simple magnetic dipole, as well as the central region of approximately constant field vectors afforded by the Helmholtz geometry. For the two coil array, specifying a coil radius of 1 meter and an array height of 1 meter sets the distance from each coil to the x–y plane at half the radius, as prescribed by the Helmholtz geometry. Note that the grid of r_p and z_p values for the field probe, with initial, final, and increment values of −1.95, +1.95, and 0.1 meters, respectively, will skip over the values of −1.0 and +1.0 meters

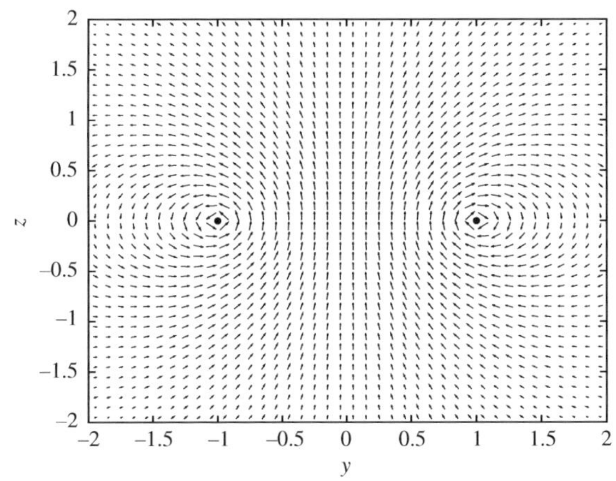

Figure 10.12 Magnetic field vectors around a single current loop, radius 1 meter, coded in length.

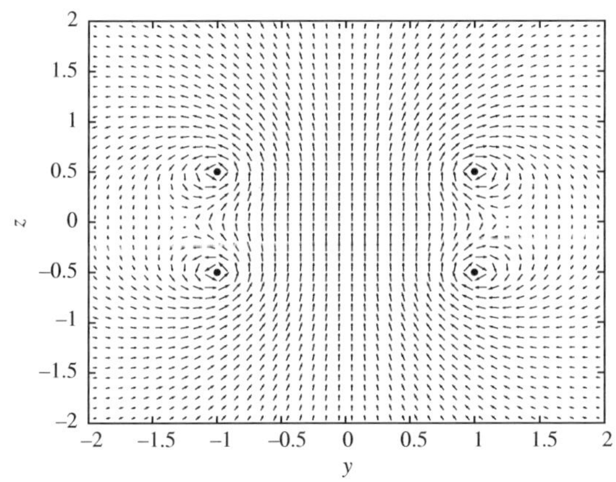

Figure 10.13 Magnetic field vectors around a Helmholtz coil, radius 1 m and distance 0.5 m, coded in length.

and avoid evaluating the magnetic field at the singular radius of the coil current paths. We may also make use of the plotting method with `gnuplot` demonstrated in Section 4.11 to represent the field magnitude coded into the color of drawn vectors. The resulting images `magfieldvectors_n1_color.png` and `magfieldvectors_n2_color.png` can be downloaded from the Student Companion website.

Following the `efieldvectors.c` program in Section 4.11, the `magfieldvectors.c` program should generate several data sets in its output stream to allow the plots to be generated. The first data set consists of a series of point coordinates that represent the cross-sections of the circular coils as they pierce through the y–z plane. These coordinates can be plotted with solid points in the graphical output, and give a visual reference around which field vectors can be seen to curl. The second output data set lists coordinates of vectors coding the magnitude as length, and produces the black color plots seen in Figures 10.12 and 10.13. The third data set is of vectors coding the magnitude in color, to give the downloadable color-coded images. As discussed in Section 4.11, we should bear in mind that these vectors are *not* the classical field lines, but rather are simple regular spatial samples of the magnitude and direction of the magnetic field vector. We will develop tools for plotting classical magnetic field lines, where the concentration of lines expresses flux density, in Section 13.4.

10.8 Exercises

10.1 The nth-order Bessel functions of the first kind $J_n(x)$ appear often in solutions to electrostatics problems with cylindrical symmetry. The function values must be calculated numerically from either a series or from the following integral:

$$J_n(x) = \frac{1}{\pi} \int_0^{\pi} \cos(nt - x\sin(t))dt \tag{10.23}$$

Write a program `bessel.c` that calculates the values of $J_0(1.0)$ and $J_1(1.0)$ by using the textbook routines `trapezoidal_func()` and `simpson_func()`. Bessel functions are needed so commonly that functions to compute them are included in the standard C library. Check your computation of the above integrals by also including in your main program the function calls $J_0(1.0)$ and $J_1(1.0)$ from the C library and display the return values in the standard output stream.

10.2 Write a program `ellipse.c` to integrate the elliptical function in Section 10.4 and calculate the circumference of an elliptical orbit of the comet. Assume a semi-major axis of 2.5×10^{12} m and an eccentricity of 0.9. Compare the elliptical circumference with the limiting circumference for two cases: a circular orbit with a radius equal to this elliptical semi-major axis, and the case where the eccentricity approaches 1 but with this same elliptical semi-major axis.

10.3 Write a program `integrate_maxwell_boltzmann.c` to integrate the Maxwell–Boltzmann speed distribution with Simpson's method.

$$\frac{n(v)dv}{N} = 4\pi\left(\frac{m}{2\pi kT}\right)^{\frac{3}{2}} v^2 \exp\left(\frac{-mv^2}{2kT}\right) \tag{10.24}$$

Use helium atoms with $m = 6.64647 \times 10^{-27}$ km at T =200, 250, and 300 °K, as shown in Figure 4.10. Integrate over velocity from 0 to 3000 m/sec with an increment of 20. Since this integrand is a probability distribution, the integral should be very close to 1.

10.4 The files `prem_density.c` and `prem_density.h` in the CPLibrary from the Student Companion website provide a function `density()` that computes the mass density per unit volume of the Earth in units of kg per m^3 as a function of radial distance from the center, according to the Preliminary Reference Earth Model (PREM) [22]. This model was also used in Exercise 4.2.

(a) Derive Equations (10.14), (10.16), and (10.18).

(b) Write a program `integrate_earth.c` to use the `density()` function to compute both the total mass and the rotational inertia of the earth. The program should also compute the rotational inertia factor C in the expression $I = \frac{C}{MR^2}$, for a total mass of the earth M and radius of the earth R. Compare with the ideal value of $C = \frac{2}{5}$ for a homogenous sphere.

10.5 Let's assume that the interplanetary probe of Section 9.8 that is launched to the dwarf planet includes seismic instruments capable of measuring the planet's mass density as a function of radial distance from its center. The experimentally determined mass density as a function of radius can be found in the file `planet_density.dat` on the Student Companion website.

(a) Derive Equations (10.14), (10.16), and (10.18), if not already derived for Exercise 10.4.

(b) Use the expression in Equation (10.18) to write a program `integrate_planet.c` to find both the total mass and the rotational inertia of the planet. The program should also compute the rotational inertia factor C in the expression $I = \frac{C}{MR^2}$, for a total mass M and radius R. Compare with the ideal value of $C = \frac{2}{5}$ for a homogenous sphere.

10.6 **(a)** Using the `efieldvectors.c` program developed in Section 4.11 as a model, and making use of the `current_loops.c` utility function developed in Section 10.6, write the program `magfieldvectors.c` discussed in Section 10.7 to plot the vectors of the magnetic field from any number of circular loops carrying the same electric current. The loops are all oriented parallel to the x–y plane, arrayed at uniform z intervals. Compile and run your `magfieldvectors.c` program to generate the plots shown in Figures 10.12 and 10.13 for single and double current loops.

(b) Use your `magfieldvectors.c` program to model the approximate magnetic field around a practical solenoid with 10 turns. A useful approximation to simplify the geometry of the current path is to assume each turn to be an isolated circular loop of current, with no actual current carrying connection between the turns.

11

Function Minimization and Fitting

In Chapter 5, we developed numerical tools for searching the values of some arbitrary function that is coded in C and converging on the root of the function. More precisely, the algorithms determine the value of the function argument that makes the function value to be in the vicinity of zero within some small tolerance. In this chapter, we develop similar tools for searching function values, but instead of roots, we aim at converging to a function minimum. The object of our algorithmic search is a function argument within some narrow vicinity of that argument that reduces the function's value to a local minimum. We will explore algorithms for the two cases of the function argument being either scalar- or vector-valued.

We will apply these minimization algorithms to finding two sorts of solutions: one finding a directly meaningful minimum for some physical system, and the other one finding the optimal fit of some nonlinear model to physical behavior.

11.1 Algorithms: Single Variable Function Minimization

First let's examine a reliable algorithm named the "Golden Section" method [10, Chap. 13; 11, Chap. 10] for converging on the minimum of a function $f(x)$ of a single scalar argument. The approach is similar to the Bisection method of Section 5.1.4 in that an initial bracketing region in the argument space is given to the algorithm, and each successive iteration narrows the width of the bracket in a prescribed sequence designed to converge on the minimum. We will specify two initial points $x = a$ and $x = c$, with $a < c$, that must "bracket" the minimum of the function, that is, the minimum to be found must be somewhere between these two x values. In each iteration, the values of either a or c will be adjusted to successively narrow the bracket width.

The iterations of this algorithm hold a state that governs the process of the search for the function minimum. The state of the algorithm can be one of two modes, coded as mode "0" and mode "1." The mode designates which side of the range from $x = a$ to $x = c$ that a third interior point $x = b$ lies. For mode 0, b lies in the left half, and for mode 1, b lies in the right half. This is illustrated in Figure 11.1.

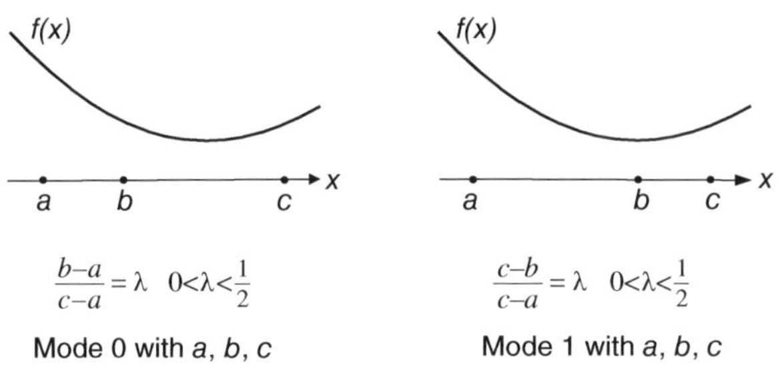

Figure 11.1 Two possible modes of the Golden Section minimization algorithm.

Computational Physics Using C: Efficient Programming with Ease, First Edition. John W. Fattaruso. © 2026 John Wiley & Sons, Inc.
All rights reserved, including rights for text and data mining and training of artificial intelligence technologies or similar technologies.
Published 2026 by John Wiley & Sons, Inc. Companion Website: http://www.wiley.com/go/CompPhysUsingC

The algorithm may be initialized arbitrarily in either mode, but the code in the textbook library chooses an initial b in mode 0. To start each iteration of the algorithm, a fourth test point $x = d$ is generated at the symmetrical point to b in the range $x = a$ to $x = c$. That is

$$b - a = c - d \tag{11.1}$$

If the current algorithm state is in mode 0, the new test point d will lie in the right half of the range $x = a$ to $x = c$, and if the current state is in mode 1, the new point d will lie in the left half. The function values $f(b)$ and $f(d)$ are compared, and a new interior point is chosen to be either b or d so that the function value is minimized. Then the two points surrounding the new interior points are chosen as the new bracketing points a and c that define the search range for the next iteration. In this way, the search range will contract around the function minimum as the iterations progress. The new mode state for the next iteration is chosen as shown in Figure 11.2 to match the new relationship of the interior point with the new range. While there are other minimization algorithms that will often converge with fewer iterations, the advantage of this method is the reliability of convergence.

For some numerical specifics, we set the distance from $x = a$ to $x = b$ to be some fraction λ of the distance from $x = a$ to $x = c$, with $\lambda < \frac{1}{2}$ for mode 0. This implies

$$\begin{aligned} \frac{b-a}{c-a} &= \lambda \\ \frac{c-b}{c-a} &= 1 - \lambda \end{aligned} \tag{11.2}$$

For a consistent algorithm where each step contracts the search range by the same factor in each iteration, in any sequence of modes, we also desire the new test point d to be placed in the larger portion b to c in the same proportion as b was placed in the range a to c or algebraically,

$$\begin{aligned} \frac{d-b}{c-b} &= \lambda \\ \frac{c-d}{c-b} &= 1 - \lambda \end{aligned} \tag{11.3}$$

Equations (11.1)–(11.3) may be collapsed into a single quadratic equation

$$\lambda^2 - 3\lambda + 1 = 0 \tag{11.4}$$

The solution to this equation defines the so-called classical Golden Section, with

$$\lambda = \frac{3 - \sqrt{5}}{2} \approx 0.38196601 \tag{11.5}$$

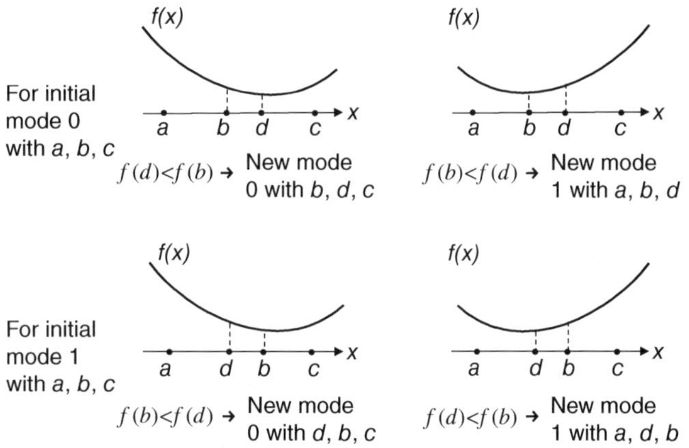

Figure 11.2 Iteration of the Golden Section minimization algorithm.

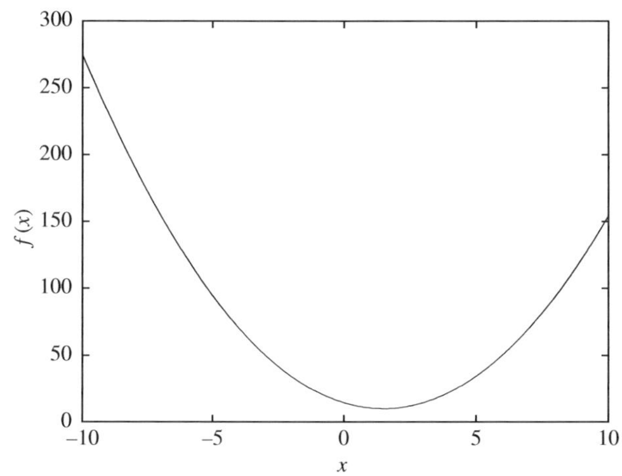

Figure 11.3 Test function for minimization with one-dimensional argument space.

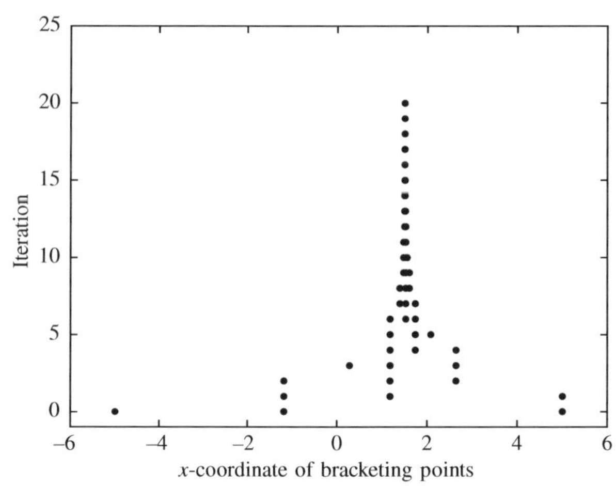

Figure 11.4 Minimization of the test function by successive selections of argument space bracketing points.

This fraction defines the method for creating the test points b and d in each iteration of the Golden Section search algorithm. Iterations continue until the distance from a to c has contracted to within a specified convergence tolerance.

As a demonstration of the convergence of the Golden Section algorithm, consider a simple example function $f(x) = 2(x - 1.5)^2 + 10$, plotted in Figure 11.3. By inspection, this quadratic function has a minimum at $x = 1.5$ where $f(x) = 10$. Figure 11.4 plots the process of the Golden Section method converging on this function minimum, after being initialized with the starting bracket of $a = -5$ to $c = +5$. The function argument is along the horizontal axis, and the three argument values a, b, and c at each iteration are plotted as points, with the iteration index along the vertical axis. This plot is reminiscent of Figure 5.10b in Chapter 5 that showed the methodical progression of two bracketing points converging on the root of a function with the Bisection algorithm. Here, with the Golden Section algorithm converging on a function minimum, the realignment of the three bracketing points following the steps of Figure 11.2 can be traced as the iterations proceed upward in Figure 11.4.

11.2 Algorithms: Multiple Variable Function Minimization

Next we consider a numerical method for converging on a local minimum of a function of more than one variable. We may conceive of such a function as being single-valued but expecting a vector $\vec{x}$ in its multidimensional argument space. We may imagine a numerical method that starts at an initial point $\vec{x}$ in the argument space, forms the function gradient vector $\nabla f(\vec{x})$, and takes iterative steps in the opposite vector direction, downhill to the minimum. However, evaluation of the gradient with numerical approximations to the component partial derivatives can involve considerable computation. We instead rely on a simple but practical algorithm that essentially senses the gradient by comparing the function value at a limited number of discrete sampling points through the argument space [10, Chap. 13; 11, Chap. 10; 31].

For an n-dimensional function argument space, we will form straight line simplex figures in the space with $n + 1$ vertices. As an example, for a function of two variables, the argument space is two-dimensional, and the simplex takes the form of a triangle with three corners. For a three-dimensional space, the simplex is a tetrahedron. Beyond this, visualization becomes difficult, but the algebra remains consistent. We will evaluate the function only at the $\vec{x}$ vectors that point to the corners, or vertices, of the simplex, and the algorithm navigates through the argument space toward the local function minimum by gradually mutating the simplex with a relatively small number of function evaluations. For a smooth function, this algorithm reliably converges on the minimum in a computationally efficient manner.

The Nelder–Mead method is sometimes called the "downhill simplex" algorithm, as we form simplices in the function argument space, and explore the argument space intending to head to the lowest elevation of the function value. For functions of a two-dimensional argument space, we may visualize the function value as the elevation of a physical terrain. Minimizing such a function is a process similar to walking downhill through a valley to the lowest elevation in the vicinity. The algorithm consists of several cleverly designed transformation rules, illustrated in Figure 11.5, to replace the vertices of the simplex in each iteration and, therefore, mutate the shape of the simplex so that it traverses the argument space in a "downhill" direction.

The $\vec{x}$ argument vector at each simplex vertex is associated with a corresponding function value $f(\vec{x})$, and these values may be ranked around the simplex. The vertices are ranked from the "worst" vertex with the highest function value to the "best" vertex with the lowest $f(\vec{x})$. To begin each iteration, with a given set of vertices forming the current simplex, the $n - 1$ dimensional reflection hyperplane is formed as the side of the simplex that excludes the worst vertex. For example, in a two-dimensional space with triangular simplices, the reflection hyperplane is simply a line segment connecting the two vertices excluding the worst vertex. In a three-dimensional space, the hyperplane is the triangular figure defined by the three simplex vertices excluding the worst. We can locate a center point in the hyperplane by simply computing the arithmetic mean of all vectors excluding the worst.

At the beginning of each iteration, a "reflection point" is located symmetrically opposite the worst vertex, reflected through the center point of the hyperplane. Then the value of the function is tested at this reflection point. This hopefully has the notion of heading "downhill" along the function contour, since we are heading in a direction away from the worst vertex of the current simplex. If the function value at the reflection point is lower, or "better," than the current best vertex, then an "expansion point" is located along the same line between the worst vertex and the reflection point, but at double the distance from the reflection hyperplane. The algorithm then tests the value of the function at the expansion point, intending to take advantage if the heading in the space through the reflection hyperplane is indeed heading directly down an extended slope. If the function value at the expansion point is still lower, or better, than the current best vertex, then the worst vertex in the simplex is replaced by the expansion point. This expands the simplex, and prepares future iterations to take larger steps downhill. However, if our expansion point gives a function value not better than the current best vertex, then we discard the expansion point, which evidently has climbed up the opposite side of a narrow valley, and we simply

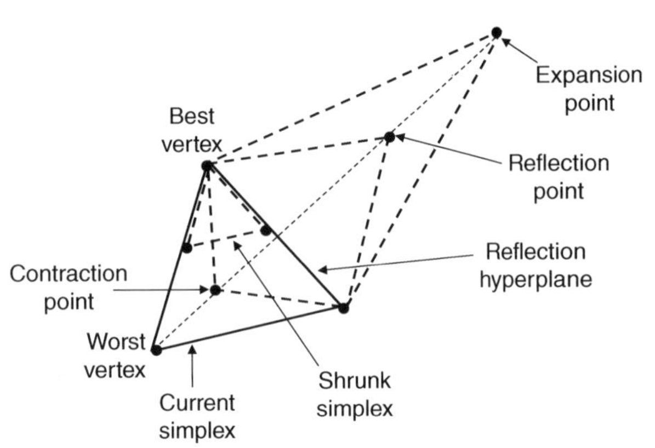

Figure 11.5 Simplex transformations in the Nelder–Mead algorithm in a two-dimensional parameter space.

replace the current worst vertex with the reflection point, which, after all, did improve the best simplex score. In either case we move on to the next iteration.

On the other hand, if the reflection point evaluates to a function value that is not better than the current best, but is at least better than the current worst vertex, then we still have an improvement over the worst vertex, so the worst vertex is replaced with the reflection point, but without any test at the expansion point, and the next iteration is started.

However, if the reflection point gives a function value that is worse than the current worst vertex, then we must conclude that the reflection heading is in an unfruitful direction, as perhaps we have bumped up against a valley wall. In that case, we may need to change direction to follow a curvature in the valley that leads downhill, and we should contract the simplex. So we then test the function at the "contraction point," which is along the reflection heading, but on the current side of the reflection hyperplane instead of the opposite side. If the contraction point gives a function value at least better than the worst vertex, then we replace the worst with the contraction point and move on to the next iteration.

Finally, if the contraction point gives a function value that is worse than the current worst vertex, then the best vertex is evidently very near the minimum, and any direction away from the best vertex is likely a step uphill. So we must start shrinking the simplex toward the best vertex by shortening all the simplex edges, shown as the shrunk simplex in Figure 11.5.

These are the algorithmic conditional tests in each iteration by which a small number of function evaluations at only the simplex vertices are interpreted to sense the gradient of the function and head in a downhill direction. But how do we start the algorithm with an initial simplex? The most straightforward way is to accept an initial guess value of the argument vector $\vec{x}$ assumed to be in the vicinity of the desired minimum, and also a set of displacement distances for each dimension in the argument space. Then an initial simplex is generated by including the nominal guess point, as well as the guess vector with each component displaced in turn by the specified distance. This will generate the $n+1$ points needed to define the vertices of a simplex in an n-dimensional argument space. Typically we choose a displacement increment for each vector component that would be considered significant enough to evaluate different function values so that a gradient may be sensed, but not so large that the function is taken into a wildly different regime. Once initialized, iteration steps continue until the simplex has shrunken to a width in each dimension that is less than a set of specified dimensional tolerances. So some general prior understanding of the terrain of the function to be minimized is important.

As a demonstration of the initialization and convergence of the Nelder–Mead algorithm, consider a simple example function $f(\vec{x}) = (x_1 - 10)^2 + 2(x_2 - 5)^2 - 20$, where x_1 and x_2 are the two components of the argument vector $\vec{x}$. This example function is plotted in Figure 11.6. By inspection, this two-dimensional quadratic function has a minimum where $x_1 = 10$ and $x_2 = 5$, at which point $f(\vec{x}) = -20$. Figure 11.7 plots the sequence of the argument space simplices in the Nelder–Mead method converging on this function minimum, after being initialized at the nominal guess point of $x_1 = 0$ and $x_2 = 0$ and with initial simplex displacement increments of 1 for both components. The initial simplex is the triangle at the lower-left end of the plot, defined by the three coordinates (0, 0), (1, 0), and (0, 1). The current simplex in the Nelder–Mead algorithm is transformed according to the algorithm rules above, and migrates generally to the top and right in the plot. We can see initial reflection and expansion steps that mutate the simplex rapidly to the vicinity of the minimum at (10, 5), and then contraction steps that shrink the simplex around the minimum point to a specified desired tolerance, in this case, 10^{-6} for each dimension.

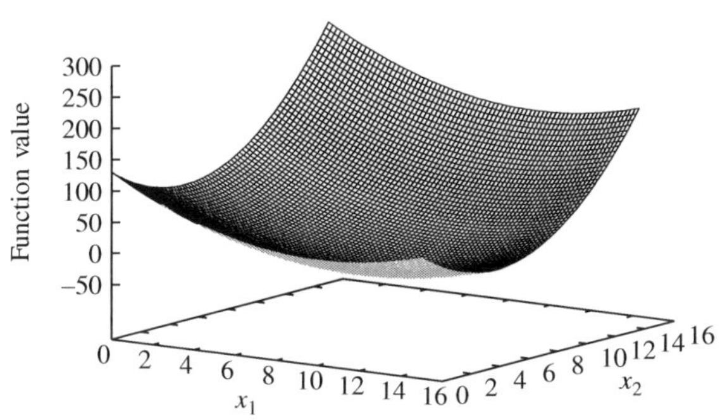

Figure 11.6 Test function for minimization with two-dimensional argument space.

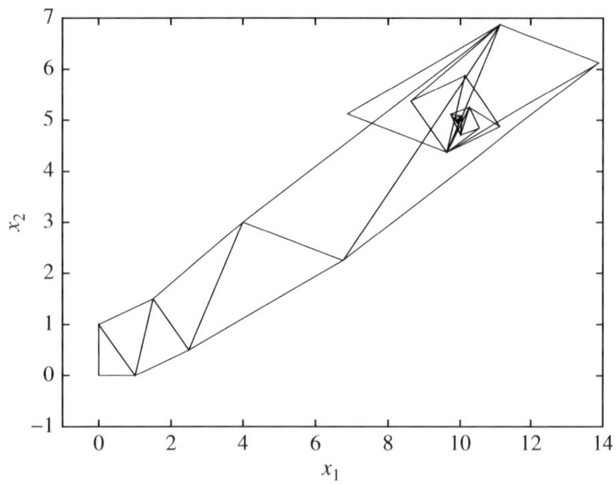

Figure 11.7 Minimization of the test function by successive mutation of argument space simplex.

11.3 Coding: The Function Minimization Tool

The tool code for converging on the minima of single-variable and multiple-variable functions with one of two different algorithms can be found in the files `minimum.c` and `minimum.h`, part of the `CPLibrary` package from the Student Companion website. Two C functions are available to realize the Golden Section and Nedler–Mead algorithms, with the function arguments shown here

```
int golden_section(double (*func)(double x,void *p),
                   double a,double c,void *p,
                   double tol,int n_iters,double *opt_ptr) {
  ...
}

int nelder_mead(double (*func)(ARRAY1D *x,void *p),
                ARRAY1D *init,ARRAY1D *inc,void *p,
                ARRAY1D *tol,int n_iters,ARRAY1D *opt_ptr) {
  ...
}
```

The pointer to the object function to be minimized is passed to the `golden_section()` tool as its first argument, and has an identical prototype as the object functions studied in Chapters 5 and 10. This must be a function that accepts a `double x` argument value and a `void *p` pointer to a structure with any physical parameters to pass to the function. The argument list continues with `double` values for the initial a and c bracketing values, the parameter set pointer, a `double` value for the convergence tolerance, an integer maximum iteration count, and finally a pointer to `double` where the function is to store the optimum argument if convergence is successful.

Arguments for the `nelder_mead()` tool are similar, except that the object function to be minimized is now a function of a vector argument. Accordingly, the object function expects its first argument to be an `ARRAY1D` array filled with the vector components. Likewise, the tool function arguments specifying the initial guess point in the argument space and a set of initial displacement increments are passed as `ARRAY1D` arrays as well. Since the convergence tolerance may be different in each argument vector component, the tolerances are also passed as an `ARRAY1D` array as well. The final argument is an array where the optimum values are to be stored. This array is assumed to be allocated by the calling code with a size equal to the number of components in the argument vector of the object function.

Both tool functions return integer codes that are defined in `minimum.h` to report to the calling code the success or failure of the convergence to the optimum minimum function point. To learn how to invoke these two minimization functions, the student may carefully study the main programs `test_golden_section.c` and `test_nelder_mead.c` from the Student Companion website. These are the simple programs demonstrated in the optimization tasks shown in Figures 11.4 and 11.7.

11.4 Application: Optimizing the Helmholtz Coil

Now we revisit the Helmholtz coil geometry from Section 4.4, but seek a somewhat different physical configuration requiring a numerical optimization approach. We return to the computation of the z component of the magnetic field only along the axis of the structure with two coils of electric current. In Section 4.4, we used the rule that positioning each coil at a symmetrical distance from the origin equal to half the coil radius gives an approximately constant magnetic field with respect to the z coordinate. This configuration is derived from setting to zero the derivative of the total B_z field component from two coils with respect to the z coordinate evaluated at *just one point* at the center of the structure. This analytic approach does indeed give us the coil separation for the optimum flatness of the field with respect to z, but only in the vicinity of this one center location. Suppose that for laboratory requirements, we are now interested in finding the coil separation that gives us a magnetic field with the least variation *over a wide range of z values*, perhaps even at the expense of some minimal fluctuation in B_z over that range.

Examine the source code file `helmholtz_axis_minvar.c` from the Student Companion website. This program contains the same function `helmholtz()` that was copied from the `helmholtz_axis_step.c` program of Chapter 4. This function computes the z component of the magnetic field along the axis of the Helmholtz coil structure from an analytically tractable integration, for some general distance of separation of each coil from the x–y plane. Then a second function `helmholtz_var()` steps the magnetic field evaluation over a z coordinate range from the bottom coil to the top coil, and computes the sum of the squared deviations of the field around its mean value for all the stepped samples. The return value of this variation function is the object function that is then minimized by the Golden Section method, with respect to a variable coil distance.

Compiling the program `helmholtz_axis_minvar.c` from the Student Companion website

```
$ gcc -Wall -Wshadow -O2 -o helmholtz_axis_minvar helmholtz_axis_minvar.c CPLibrary/minimum.c
    ↪ CPLibrary/step.c CPLibrary/array.c -lm
$ ./helmholtz_axis_minvar
./helmholtz_axis_minvar <radius>
$ ./helmholtz_axis_minvar 0.1 >| helmholtz_axis_minvar.dat
Minimum variance found at distance 0.057513822 m
```

Minimizing this metric of field variation, for a coil radius of 10 cm, the resulting optimum coil distance turns out to be 5.751 cm, which is close to the nominal value of 5 cm for the minimum slope at only the center point. At the optimum placement distance the program then uses the `step()` function stepper tool to visualize the optimized field variation.

To visualize the results, we start an interactive gnuplot session and give the commands

```
gnuplot> set key bottom center
gnuplot> set xlabel 'Z coordinate (m)'
gnuplot> set xrange [-0.06:0.06]
gnuplot> set ylabel 'Z component of magnetic field (T)'
gnuplot> set format y "%0.1e"
gnuplot> set yrange [0:1.1e-5]
gnuplot> plot 'helmholtz_axis_minvar.dat' u 1:2 w l t 'Optimum
    ↪ distance = 5.751 cm','helmholtz_axis.dat' i 2 u 1:2 w l t
    ↪ 'Nominal distance = 5 cm'
```

This optimized result is plotted in Figure 11.8, along with the prediction from the `helmholtz_axis_step.c` program at the nominal placement distance of 5 cm. We can see that, by design, the optimized result achieves less total variation over the wide range of z values, but does so by sacrificing a small amount of the ideal smoothness localized around the center at $z = 0$.

11.5 Application: Closest Approach of an Asteroid

Our planet has a long history of impacts by asteroids, and resources are devoted to tracking potentially dangerous orbital paths. In this section we examine a simple example of projecting the minimum separation distance between the Earth and an approaching asteroid in a highly elliptical orbit. For simplicity we only treat the problem in two dimensions, and assume the orbital planes of the asteroid and the earth are the same.

First, review Section 5.6 and the `orbit_step.c` example program. We assume now that the body in a highly elliptical orbit is an asteroid, and the orbit prediction code in the program will compute its time-sampled positions along its elliptical path. At the core of those orbital predictions is a numerical solution to Kepler's equation to translate between the angular

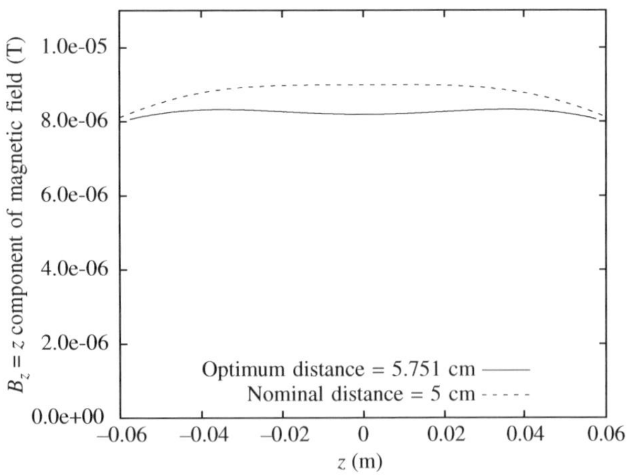

Figure 11.8 Magnetic field strength along axis of Helmholtz coil, radius = 10 cm, at optimized and nominal distances.

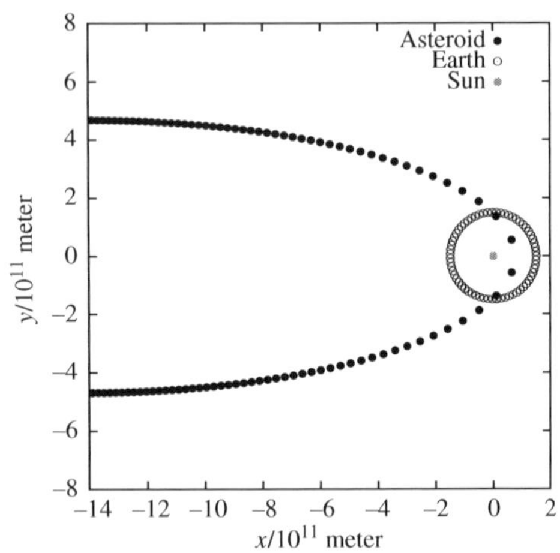

Figure 11.9 Elliptical orbit of an asteroid and circular orbit of the Earth, sampled at uniform time intervals.

representations of the orbital positions using the Newton–Raphson algorithm. If we assume the asteroid being predicted is orbiting the Sun in our solar system, then the mass of the primary body is known, and from the geometrical specifications of the orbit dimensions we can calculate the orbital period from Equation (5.15).

Now we assume that as the asteroid approaches the perihelion, the closest point along the orbital path to the Sun, it will also approach Earth's orbital path. The Earth orbits with its own period, unrelated to the asteroid's orbital period. How close does the asteroid pass by the Earth? Is there a danger that with slight gravitational perturbations a collision may occur?

Starting from the `orbit_step.c` program, it is straightforward to add a prediction of the Earth's position at the times corresponding to the asteroid's mean anomaly, since our own planet's orbital radius and period are known. For simplicity, we may assume with only small error that the Earth's orbit is circular. To allow the elliptical asteroid orbit to have any spatial relationship with the circular orbit of the Earth, we also introduce a counterclockwise phase angle offset parameter for the Earth's orbital time dependence. The code will be a solution to the first part of Exercise 11.1. Let's pick some example values for the asteroid's orbit that are numerically straightforward, such as a semi-major axis of 1.5×10^{12} m, or approximately 10 times the radius of the Earth's orbit, an eccentricity of 0.95, and a counterclockwise phase angle of 0.5 rad for the Earth's orbit at the time of perihelion for the asteroid. The time samples of the positions of both the asteroid and the Earth in a common plane are plotted in Figure 11.9.

As the program code computes the x and y coordinates in the orbital plane for both bodies, it is also straightforward to compute the Euclidean distance between the Earth and the asteroid, and send this measurement to another column in the output stream. This distance at each sampled time step is plotted with respect to the mean anomaly of the asteroid in

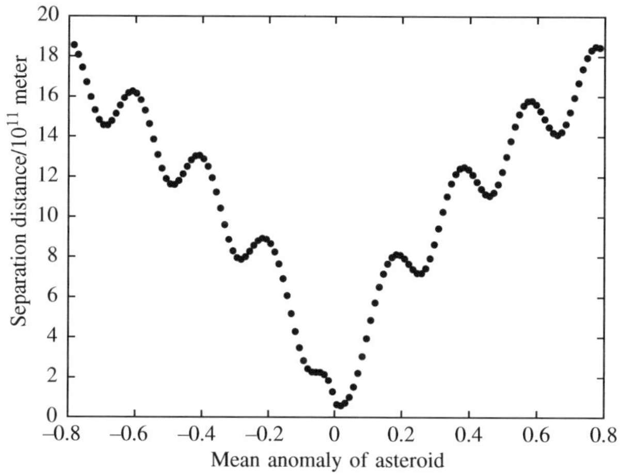

Figure 11.10 Distance between an asteroid and circular orbit of the Earth, sampled at uniform time intervals.

Figure 11.10, for an example Earth phase angle offset of 0.5 rad. As the values of the mean anomaly along the axis show, this plot represents only the portion of the highly elliptical asteroid orbit during its approach to the Sun.

Examining the plot in Figure 11.10, various local minima and one global minimum can be seen. During the relatively slow orbit of the asteroid, the Earth completes several complete orbits around the Sun during the asteroid's approach to perihelion, and each local reduction of the separation distance occurs when the Earth swings around in the direction of the current asteroid position. The global minimum occurs when the asteroid swings very close by, somewhere inside the Earth's orbit radius. We would like to use the Golden Section method to find the closest approach distance the asteroid makes to Earth, but the behavior of this separation distance shows it is often advisable to visualize the overall dependence first to let us give a narrowed bracketing range around a global minimum of interest to the minimization algorithm. Otherwise, if we specify too wide a bracketing range to initialize the search algorithm, we risk the search getting stuck in a local minimum that is likely not meaningful.

The C function that calculates the separation distance may be extracted from the code that generates the plot of Figure 11.10 with the `step()` tool and placed in a new main program. Instead of the `step()` tool, this new program calls the `golden_section()` minimization tool to find the asteroid's mean anomaly at the closest approach to the Earth, and to use this mean anomaly to evaluate the minimum separation distance. This program coding is left to the student as Exercise 11.1, and we may discover the minimum separation distance between us and the asteroid.

We have assumed at this point that we already know the Keplerian elements that describe the asteroid's orbit. Section 11.13 will address the important problem of extracting those orbital elements from astronomical observations of the asteroid. Moreover, for actually determining if an asteroid is on a collision course with Earth, we must also consider that the orbit of the asteroid will be perturbed from its Keplerian elliptical path when approaching the Earth, in a manner similar to the impact of the Shoemaker-Levy 9 comet with Jupiter in 1994. We will have to wait until Section 13.2 to be able to model properly the perturbations of orbits in n-body gravitational systems. For now we may consider only if the original Keplerian orbit prediction models the closest approach distance that is within some hazardous limit.

11.6 Coding: Linear Least Squares Fitting

In Section 11.4, we developed a C function to accumulate a sum of the squared errors between a characteristic function and an ideal target function sampled across some argument range. The value of that error function was then numerically minimized to set a physical parameter to optimize the fit to the ideal target. In this section and in Section 11.8, we make use of that same general technique to find the optimum fit of a model function to a set of data points describing a physical dependence. In this sense, we revisit the general idea of Chapter 9 of discovering some continuous function that plausibly describes some unknown functional dependence behind discrete data points. However, with the error minimization tools of this chapter, we now need not insist that the model function exactly pass through all the data points. But by minimizing the metric of the sum of squared sampled errors, we may achieve a "best fit" between a model and realistic data point values that may contain experimental errors.

To start with a simple first-order example, let's examine the very common case of presuming that the underlying dependence between a set of data points is some affine relationship $y = a_0 + a_1 x$. This is illustrated in Figure 11.11. We want to solve for the coefficients a_0 and a_1 that determine the line that best fits a collection of data points that are scattered around their intrinsic affine relationship with some experimental errors or unavoidable stochastic variations. For this simple first-order case, the best fit solution can be analytically found.

We define the error between the fit line and each data point as

$$e_i = |a_0 + a_1 x_i - y_i| \tag{11.6}$$

and then define the total accumulated fit error as the sum of the squared errors at each point

$$E = \sum_{i=0}^{n-1} e_i^2 = \sum_{i=0}^{n-1} (a_0 + a_1 x_i - y_i)^2 \tag{11.7}$$

Finding the "best" fit is defined to be finding the a_0 and a_1 coefficients that minimize E. Given a fixed set of data points (x_i, y_i), we note that E is a function of a_0 and a_1, so the minimum E will be where both conditions

$$\begin{aligned} \frac{\partial E}{\partial a_0} &= 0 \\ \frac{\partial E}{\partial a_1} &= 0 \end{aligned} \tag{11.8}$$

are simultaneously satisfied. The two equations in the simultaneous system of (11.8) expand to

$$\begin{aligned} a_0 &= \frac{\sum_i y_i - a_1 \sum_i x_i}{n} \\ a_0 \sum_i x_i &+ a_1 \sum_i x_i^2 - \sum_i x_i y_i = 0 \end{aligned} \tag{11.9}$$

Equation (11.9) forms a system of linear equations in the unknown coefficients a_0 and a_1, and with some algebraic substitution we can solve this system to find

$$\begin{aligned} a_1 &= \frac{n \sum_i (x_i y_i) - \left(\sum_i x_i\right)\left(\sum_i y_i\right)}{n \sum_i (x_i^2) - \left(\sum_i x_i\right)^2} \\ a_0 &= \frac{\left(\sum_i x_i^2\right)\left(\sum_i y_i\right) - \left(\sum_i x_i\right)\left(\sum_i x_i y_i\right)}{n \sum_i (x_i^2) - \left(\sum_i x_i\right)^2} \end{aligned} \tag{11.10}$$

For Exercise 11.3, the student can write a simple program tool to solve for the best linear fit to a collection of data points from a file. This application is further explored in Section 11.7.

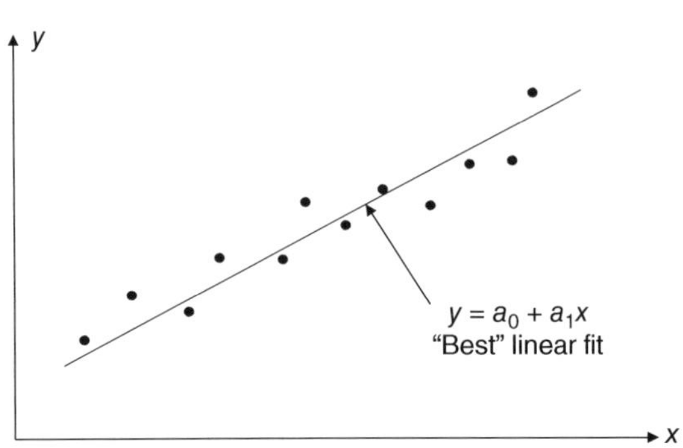

Figure 11.11 Least squares fitting to a linear function.

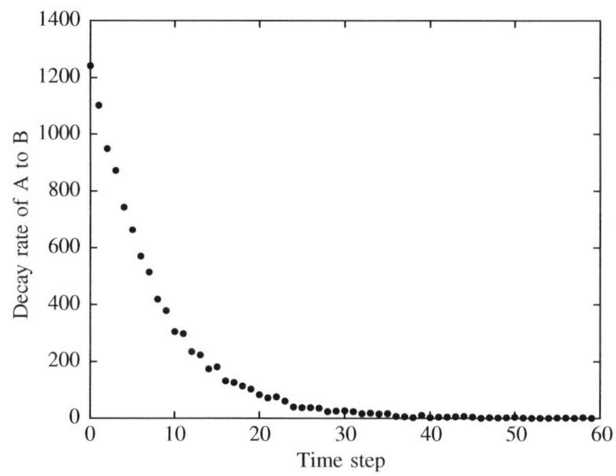

Figure 11.12 Example experimental measurements of the rate of nuclear decay events.

11.7 Application: Fitting Radioactive Decay

The basic expressions governing nuclear decay between two nuclides were developed in Section 5.9. As was mentioned in Section 8.7 covering Monte Carlo simulations of decay processes, when such processes are characterized experimentally, generally the measurable event is the emission of particles with a signature of the decay event itself, rather than the remaining population of a particular nuclide. Therefore, differentiating Equation (8.11) gives us

$$R_{AB}(t) = R_{AB0}e^{-\lambda_{AB}t} \tag{11.11}$$

as the rate of measured decay events from nuclide A to B per unit time as a function of time, where R_{AB0} is the initial rate when measurements begin. As an example, suppose Figure 11.12 shows the plot of measured decay events per unit time as a function of time step for some nuclear process with an unknown half life.

Taking the natural logarithm of both sides of Equation (11.11) gives

$$\log(R_{AB}(t)) = \log(R_{AB0}) - \lambda_{AB}t \tag{11.12}$$

The linear model of Equation (11.12) may be fit to the *natural logarithm* of experimentally measured decay rates as a function of time with the unknown coefficients $\log(R_{AB0})$ and $-\lambda_{AB}$. This will allow us to determine the best estimate of the value of the probability parameter λ_{AB} for the decay process from data points with inherent random deviations from the ideal model. From the λ parameter we can determine the corresponding half life from Equation (5.36). The coding work of a program for this fitting application is available as Exercise 11.4.

11.8 Coding: Nonlinear Least Squares Fitting

Many useful nonlinear fitting problems, illustrated in Figure 11.13, will not admit a convenient solution with a set of linear equations. Similar to other transcendental problems encountered in Chapters 5 and 7, the optimum fit of the unknown function parameters $a_0, a_1, a_2, \ldots$ of some nonlinear function $f(x)$ in the figure to the known data points must be found by numerical minimization methods that comprehend the nonlinear relationships between the fit parameters and the least squared fit error.

Conceptually, it can be challenging to keep track of the hierarchy of functions involved in the task of nonlinear model fitting. The *model function* being shaped to closely fit the data is a function that has an argument and value that mimics the independent and dependent variables associated with the experimental data points. However, the *function being minimized* in the fitting process is one whose argument or arguments play the role of general parameters in the model function, and its value is the sum of the squared fit errors. This hierarchy of mathematical functions will be reflected in a hierarchy of C function calls as we code these problems. For example, suppose that we have experimentally arrived at a set of n data points expressed as pairs of coordinates (x_0, y_0), (x_1, y_1), ..., (x_{n-1}, y_{n-1}) in some space. The model function to be fit can be coded in the form

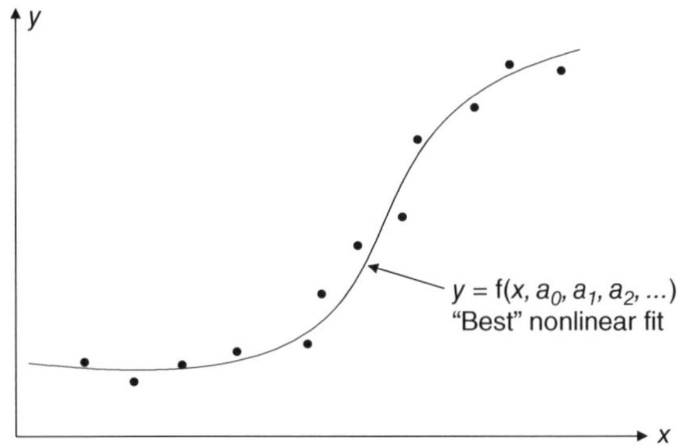

Figure 11.13 Least squares fitting to a nonlinear function.

```
double model(double x,void *param_ptr) {
    ...
}
```

where the argument x models the x coordinate of the space and the return value of `model()` models the y coordinate. A pointer to a structure of parameters that control the shape of the model function is passed as a second argument. Suppose that among those needed parameters are a set of m variable parameters $a_0, a_1, ..., a_{m-1}$ that we seek to optimize to discover the best estimate of the intrinsic relationship between x and y.

We then code one level up in the function hierarchy, a function that calculates the sum of squared errors between the data points and the model. The coded form may be either

```
double fit_error(double a,void *param_ptr) {
    ...
}
```

in the case of $m = 1$, or only one model function parameter to be optimized, or

```
double fit_error(ARRAY1D *a,void *param_ptr) {
    ...
}
```

in the case of a multiple variable optimization with $m > 1$. The parameter structure pointed to by the second argument of this `fit_error()` function is a different set of parameters than those passed to the `model()` function above, and in particular must include pointers to the experimental (x_i, y_i) data points so that the `fit_error()` function may compute the fit error relative to the values returned by calls to the `model()` function. It is, of course, the `fit_error()` function that is submitted to an appropriate minimization tool, either `golden_section()` or `nelder_mead()`, for single or multiple variable minimization.

The application of these coding practices can likely be best understood by example, starting in the sections that follow. We will see that the parameter C structures in the coding system developed for the numerical library for this textbook play an important part of communicating values through this hierarchy.

11.9 Application: Fitting the Maxwell–Boltzmann Distribution

As a simple example of a single variable nonlinear model fitting, we revisit the Maxwell–Boltzmann distribution from Sections 4.5 and 8.11. This distribution describes the probability of the particles of an ideal gas being measured with some speed. The distribution function will have particle speed as a function argument, but will be shaped and spread through the argument space according to the parameter of gas temperature, as shown in Figure 4.10. We presume that some experimental data has been acquired describing the distribution of velocities in an ideal gas at an unknown temperature, and that there is some unavoidable experimental error present in the samples. We wish to find the most likely fit of the ideal Maxwell–Boltzmann distribution to this set of experimental data points in order to infer the temperature of the gas.

The main program for this task is in the file `fit_maxwell_boltzmann.c` from the Student Companion website. Examining the main program in the source code, the data file named as the first command line argument is first read with the statements

```
if (read_2col_data_file(argv[1],&(fparam.xsample),
    &(fparam.ysample),10) != DATA_FILE_SUCCESS) {
  exit(1);
}
```

The data file samples are read into two arrays in the `fparam` structure. For these measurements, the `fparam.xsample` array contains a particle speed and the `fparam.ysample` array contains the corresponding probability. The `fparam` structure holds parameters for the fitting function, but the model function describing the Maxwell–Boltzmann distribution will require a different C structure holding its parameters, such as the gas temperature and particle mass. It is only the gas temperature parameter that will be adjusted to minimize the fit error for this problem. Since there is only one parameter being optimized, we may use the one-dimensional Golden Section method. This tool is called with the following statements

```
if (golden_section(fit_error,Ta,Tc,&fparam,1.0e-3,100,&Topt) ==
    MINIMUM_SUCCESS) {
  fprintf(stderr,"Minimum found at %.8g\n",Topt);
}
```

The function to be minimized is passed to `golden_section()` as its first argument, and is named `fit_error()`. This function is defined earlier in the source code as

```
double fit_error(double T,void *param_ptr) {
  int i;
  double err,sumsq;
  MB mbparam;
  FIT_DATA *p;

  p = param_ptr;
  mbparam.m = p->m;
  mbparam.T = T;
  sumsq = 0.0;
  for (i = 0; i < p->xsample.n_elems; i++) {
    err = p->ysample.elem[i] - maxwell(p->xsample.elem[i],&mbparam);
    sumsq += err * err;
  }
  return(sumsq);
}
```

The argument of this function is the trial gas temperature that will be adjusted by the Golden Section algorithm to converge gradually on the optimum fit. This C function defines a parameter block for the Maxwell–Boltzmann distribution function `maxwell()`, and fills in the gas particle mass and the trial temperature parameters that are needed for the distribution calculation from either known fixed parameter values or the `fit_error()` function argument variable `T`. The function subsequently accumulates the sum of squared errors between the trial distribution function and the measured data points. The fit errors are defined relative to the same `maxwell()` model function we developed for our solution to Exercise 4.1, and the sum of the squared fit errors is then returned to be minimized by the Golden Section tool.

Before the `golden_section()` tool is called to perform this fit, however, the `step()` tool from Chapter 4 is first called to generate a plotting data set in the output stream. This allows us to plot the fit error value as a function of trial temperature and visualize the shape of the function that we are expecting the Golden Section algorithm to minimize. Then after the `golden_section()` tool is called, and the temperature parameter that minimizes the fit error is found, the final resulting Maxwell–Boltzmann distribution is stepped as a function of particle speed so that we can assess the fidelity of the optimized fit.

Compiling and running the program with an example set of gas measurements that have been synthesized with an ideal temperature of 270 K, but also including a small representative amount of stochastic experimental error

```
$ gcc -Wall -Wshadow -O2 -o fit_maxwell_boltzmann fit_maxwell_boltzmann.c CPLibrary/minimum.c
    ↪ CPLibrary/step.c CPLibrary/data_file.c CPLibrary/array.c -lm
$ ./fit_maxwell_boltzmann
./fit_maxwell_boltzmann <input file> <mass> <temp a> <temp c>
$ ./fit_maxwell_boltzmann fit_data_maxwell_boltzmann.dat 6.64647e-27 100 400 >|
    ↪ fit_maxwell_boltzmann.dat
Read 11 points from data file fit_data_maxwell_boltzmann.dat
Minimum found at 269.98511
```

shows that the minimization process has arrived at a minimum fit error with a trial gas temperature that has been skewed very slightly away from 270 K by the included measurement errors. Starting an interactive gnuplot session and giving the commands

```
gnuplot> unset key
gnuplot> set xlabel 'Temperature degrees K'
gnuplot> set ylabel 'Fit error'
gnuplot> plot 'fit_maxwell_boltzmann.dat' i 0 w l
gnuplot> set key default
gnuplot> set xlabel 'Speed m/sec'
gnuplot> set ylabel 'Probability per unit speed'
gnuplot> plot 'fit_maxwell_boltzmann.dat' i 1 w l t 'Fit','fit_data_maxwell_boltzmann.dat' w p pt
    ↪ 7 t 'Samples'
```

produces the plots seen in Figures 11.14 and 11.15. The first plot is from the first data set output from the program describing the fit error as a function of trial temperature to be minimized. This fit error characteristic has a prominent minimum at the actual gas temperature, and the Golden Section method will easily converge to this temperature. Although on the scale of this plot it may appear that the minimum error will be zero, with some experimental error sources there will be some residual fit error at the minimum. The second plot is the final fitting result, which is the probability versus speed along with the original data points that included some example experimental error. We can see that this nonlinear fitting process, though somewhat complicated, results in a very good fit and a meaningful value for the measured gas temperature.

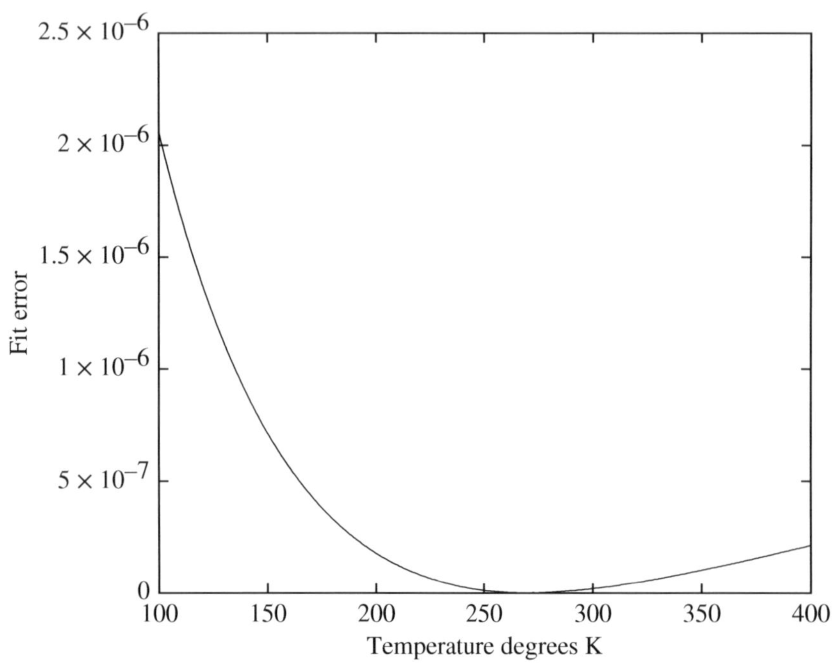

Figure 11.14 One-dimensional parameter space with respect to temperature with mass = 6.64647×10^{-27} kg.

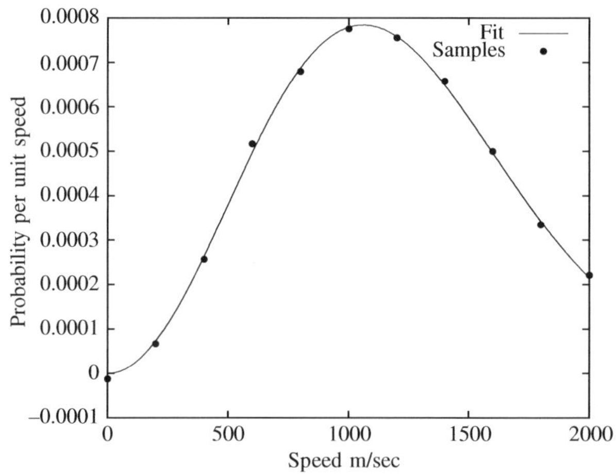

Figure 11.15 Gas particle probability versus speed with mass = 6.64647×10^{-27} kg and optimum fit temperature of 269.99 K.

11.10 Application: Fitting Planck's Radiation Law

In Section 5.8, we examined the problem of finding the wavelength at which the radiation energy density of Planck's model of black body radiation reaches its maximum.

$$u(\lambda) = \frac{8\pi hc}{\lambda^5} \frac{1}{e^{\frac{hc}{\lambda kT}} - 1} \tag{11.13}$$

In a manner very similar to that used for the Maxwell–Boltzmann distribution in Section 11.9, experimental data can be fit to this relationship. Historically, Planck fit the measured spectrum from a black body radiator at a known temperature to the statistical law of Equation (11.13) to determine the value of the constant h that bears his name.

With the value of Planck's constant known, we may then use Equation (11.13) to measure the effective temperature of any other radiating black body by fitting an observed spectrum with respect to the temperature parameter T. If we are fitting the radiation spectrum to microwave radio measurements, which is the part of the electromagnetic spectrum explored for characterizing the cosmic microwave background, it is more convenient to use the form of Planck's law in terms of frequency rather than wavelength.

$$u(f) = \frac{8\pi hf^3}{c^3} \frac{1}{e^{\frac{hf}{kT}} - 1} \tag{11.14}$$

Writing the code for this solution is Exercise 11.5 for the student. The code provided in the Student Companion website for fitting the Maxwell–Boltzmann distribution `fit_maxwell_boltzmann.c` is a good starting point.

11.11 Application: Fitting the Damped Harmonic Oscillator

Now we revisit the damped harmonic oscillator from Section 4.2. We suppose that measurements have been made of an oscillatory behavior with unknown parameters of some physical system, and consider the problem of finding the best fit of both the natural frequency and damping factor of the oscillation to a set of periodically sampled oscillator displacements versus time. We assume that the displacement will follow the time dependence of

$$x(t) = x_0 e^{-\zeta \omega_0 t} \cos(\omega_d t) \tag{11.15}$$

with the damping factor $\zeta < 1$ and the damped oscillation frequency $\omega_d = \omega_0 \sqrt{1 - \zeta^2}$. The initial displacement parameter x_0 may be found as the first time sampled data point, so we can consider that parameter to be a known constant. The two unknown parameters ω_0 and ζ are embedded in various transcendental functions, and this leads us to a numerical optimization approach. See the source code in the program `fit_damped_osc.c` from the Student Companion website.

Since there are two variable parameters to be adjusted in the search for the minimum fit error, we must now employ the Nelder–Mead multidimensional minimization algorithm for this search. This code in the main program block supplies the tool with the needed arguments

```
  if (nelder_mead(fit_error,&init,&inc,&fparam,&tol,100,&opt) ==
        MINIMUM_SUCCESS) {
    fprintf(stderr,"Minimum found at %.8g %.8g\n",
            opt.elem[0],opt.elem[1]);
  }
```

As in the case of a single variable minimization, we define a fit error function as

```
double fit_error(ARRAY1D *a,void *param_ptr) {
  int i;
  double err,sumsq;
  OSC oparam;
  FIT_DATA *p;

  p = param_ptr;
// Argument a[0] is frequency in Hertz
// Argument a[1] is damping factor
  oparam.w0 = TWO_PI * a->elem[0];
  oparam.zeta = a->elem[1];
  oparam.wd = sqrt(oparam.w0 * oparam.w0 *
              (1.0 - (oparam.zeta * oparam.zeta)));
  oparam.x0 = p->x0;
  sumsq = 0.0;
  for (i = 0; i < p->xsample.n_elems; i++) {
    err = p->ysample.elem[i] - osc(p->xsample.elem[i],&oparam);
    sumsq += err * err;
  }
  return(sumsq);
}
```

This error function is designed to be called by the Nelder–Mead routine as discussed in Section 11.3. The first C function argument is a pointer to an `ARRAY1D` structure that stores the components of the multidimensional argument vector `a` that is adjusted in search of a minimum error. As in the case of the single-dimensional search, this intermediate function maps components of the argument vector to the model parameters in any desired manner. In this example, we map the first argument vector component to the natural frequency, but with an angular scale factor that lets us treat the first component to be a real frequency in hertz. The second vector component is the damping factor. The two arrays from reading the data file are assumed to be the time and corresponding displacement. The data point errors are defined relative to the `osc()` model function copied from Section 4.2.

To test this program, we can make use of the `damped_oscillator.c` program developed in Section 4.2. Let us fill an example data file `fit_data_osc.dat` with coarsely spaced samples of a damped harmonic oscillator with

```
$ ./damped_oscillator 0.8 0.075 0.25 10 0.1 >| fit_data_osc.dat
```

This data file mimics experimental measurements of an oscillator with a natural frequency of 0.8 inverse time units and a damping factor of 0.075, except that the points in an actual data file would also reflect some level of measurement error. We use an initial displacement of 0.25 units. Now compiling and running the demonstration fitting program, we run it assigning an initial displacement of 0.25. This amplitude value may be ascertained directly from the first sample in the data file. We also provide two initial guesses for the frequency and damping factor on the command line.

```
$ gcc -Wall -Wshadow -O2 -o fit_damped_osc fit_damped_osc.c CPLibrary/minimum.c CPLibrary/step.c
    ↪ CPLibrary/data_file.c CPLibrary/array.c -lm
$ ./fit_damped_osc
./fit_damped_osc <input file> <x0> <init f0> <init zeta>
$ ./fit_damped_osc fit_data_osc.dat 0.25 1 0.1 >| fit_damped_osc.dat
Read 101 points from data file fit_data_osc.dat
Minimum found at 1.0000002 0.04999994
```

We suppose that the user inspected the "measured" data and estimated the frequency and damping factor as 1 and 0.1, respectively. We can see that from this starting vector the Nelder–Mead algorithm correctly converged on the values of 0.8 and 0.075 for these parameters.

Inspection of the `fit_damped_osc.c` will show that two data sets are generated in the output stream. The first data set describes the fit error function over a two-dimensional vector argument, and the second will allow plotting the time dependence of the oscillator displacement with the final optimized parameter values together with the original "measured" data points. Starting an interactive gnuplot session and giving the commands

```
gnuplot> unset key
gnuplot> set hidden3d
gnuplot> set xlabel 'Natural frequency (Hz)'
gnuplot> set ylabel 'Damping factor'
gnuplot> set zlabel 'Fit error' rotate by 90
gnuplot> set ticslevel 0.25
gnuplot> set view 67,100
gnuplot> splot 'fit_damped_osc.dat' i 0 w l
gnuplot> set view 72,350
gnuplot> splot 'fit_damped_osc.dat' i 0 w l
gnuplot> set key default
gnuplot> set xlabel 'Time (seconds)'
gnuplot> set ylabel 'Displacement (centimeters)'
gnuplot> plot 'fit_damped_osc.dat' i 1 w l t 'Fit','damped_oscillator_z0p05.dat' w p pt 7 t
    ↪ 'Samples'
```

produces the plots seen in Figures 11.16 and 11.17. There are two views of the fit error function shown in Figure 11.16, and we can see that although the steepness of the slopes in the two parameter dimensions is markedly different, the Nelder–Mead algorithm has successfully converged on an excellent fit.

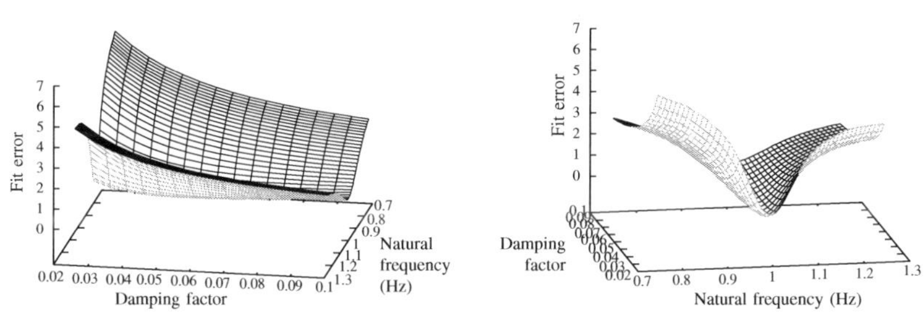

(a) First view (b) Second view

Figure 11.16 Two-dimensional parameter space with respect to natural frequency and damping factor.

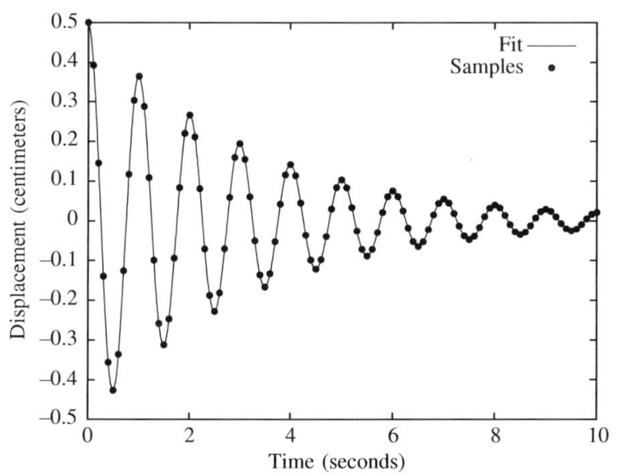

Figure 11.17 Oscillator displacement versus time with optimum fit natural frequency and damping factor.

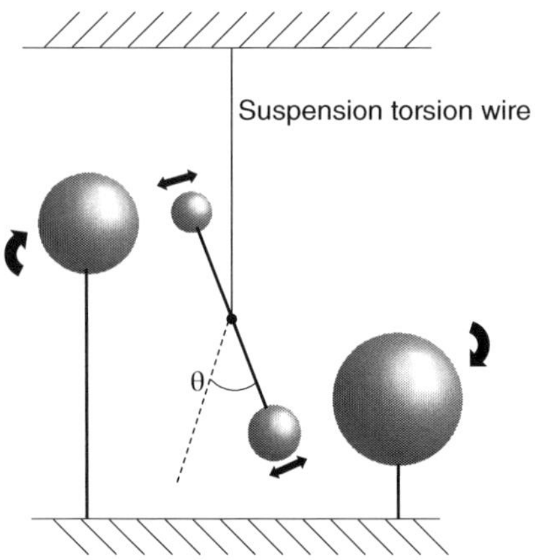

Figure 11.18 Sketch of the Cavendish experiment to measure the gravitational constant.

11.12 Application: The Cavendish Experiment

Figure 11.18 shows a schematic diagram of the torsion balance used in the famous Cavendish experiment, the historical procedure used to make the first measurement of G, the universal constant in the Newtonian gravitational law. Two smaller spheres are at either end of an axis rod, and the center of the rod is suspended down from a fixed support with a torsion wire. The rod and small spheres are free to rotate about in a horizontal plane as the torsion support wire twists back and forth, but small frictional losses in the torsion wire will slowly dampen the oscillations. When undisturbed, the orientation of the axis rod lies along a reference direction relative to the surroundings. Then two massive spherical objects are brought into the vicinity of two smaller spheres, and, in general, this disturbance will initiate some small oscillation of this torsional pendulum. If we trace the angular displacement of the rod θ relative to the reference direction as a function of time, in the case of small angles and a linear torsion characteristic, it will trace out a damped harmonic oscillation. However, there will be a small offset angle relative to the reference direction due to the gravitational forces between the spheres. To measure the universal gravitational constant, we must extract the parameters of the harmonic oscillation, including the offset angle.

We add a small offset angle θ_g due to gravitation to the expected damped harmonic oscillation of the rod angle

$$\theta(t) = \theta_g + \theta_0 e^{-\zeta\omega_0 t} \cos(\omega_d t) \tag{11.16}$$

Note that now the parameter θ_0 represents the displacement amplitude of the sinusoidal component, but not the total angular displacement at time $t = 0$. In order to measure the gravitational forces at work, we must know independently the masses of the spheres. By measuring the displacement angle $\theta(t)$ versus time at a sequence of sample times and find the best fit of the measurements to the damped harmonic oscillation model of Equation (11.16), we may extract the parameters θ_g, θ_0, ω_0, and ζ. From the oscillation natural frequency, and knowing the masses, we may calculate the torsion characteristics of the wire, since for a torsional pendulum we know the natural frequency is

$$\omega_0 = \sqrt{\frac{\kappa}{I}} \tag{11.17}$$

where κ is the torsional spring constant and I is the rotational inertia [17, Chap. 16; 19, Chap. 14; 20, Chap. 13]. From the spring constant and the offset angle θ_g we may then infer the universal gravitation constant. The program to fit the measured oscillation to the model and extract the four needed parameters is an extension of the program `fit_damped_osc.c` from Section 11.11. Writing the code for this solution is Exercise 11.7 for the student.

11.13 Application: Fitting an Asteroid Orbit

In this section, we suppose that an asteroid in a highly elliptical orbit has been spotted approaching its perihelion, and a series of measurements of position radius r and position angle θ_r are obtained over a span of time with parallax and ranging technology. We seek the best fit to the nonlinear Keplerian model of the asteroid's orbit in order to project the likelihood of a collision with the Earth using the analysis of Section 11.5. Continuing, for the sake of simplicity, to treat this challenge

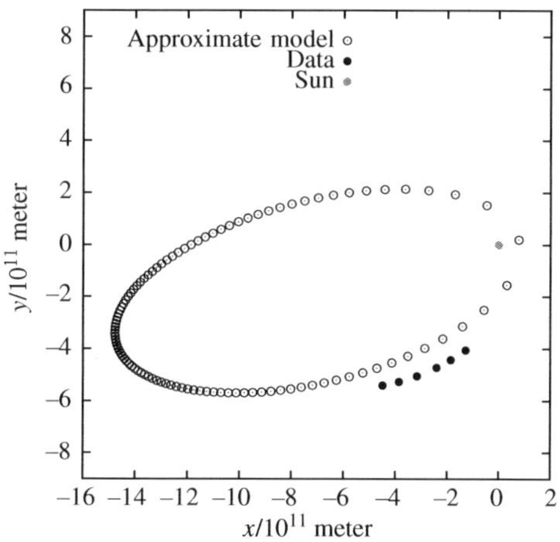

Figure 11.19 Measured astronomical locations of approaching asteroid and an initial trial and error orbital guess.

only in two dimensions, we will make use of the numerical optimization of the four Keplerian elements of eccentricity ϵ, semi-major axis a, argument of periapsis ω, and time of periapsis T. In Figure 11.19, we see an example of six position measurements plotted with solid black circles.

In Section 5.6, we appreciated the complexity of the nonlinear Keplerian model governing the motion of a single body orbiting a dominant primary mass. Although this model allows a considerable savings of computation for a single orbiting body relative to the full solution of the Newtonian equations of motion we will treat in Section 13.2, there is a sequence of several nonlinear equations and a numerical root solution involved in applying it to predict the motion of the body. Therefore, it is important to obtain a set of initial values of the four fitting parameters within a reasonable neighborhood of a proper fit before seeking convergence with numerical optimization of the full fitting problem.

Suppose we run the `orbit_step.c` program from Section 5.6 and plot the progression of orbital positions together with the measured data, and by manually adjusting the parameters of eccentricity, semi-major axis, and argument of periapsis, ascertain an approximate orbit that visually lies in the neighborhood of fitting the data points. An example result of such a manual coarse fitting step is also plotted in Figure 11.19 in the empty circular points. The grey circle marks the origin at essentially the center of the Sun.

The task of coding the program to perform a least squares fit of the measured asteroid position to an orbit specified by its Keplerian elements is available to the student in Exercise 11.8.

11.14 Exercises

11.1 **(a)** Modify the program `orbit_step.c` from Section 5.6 to write a new program `asteroid_step.c`. The new program must compute the distance between an asteroid in a highly elliptical orbit and the Earth as a function of time. Assume a simplified two-dimensional arrangement of the Sun, Earth, and asteroid all in a single plane. For the asteroid's orbit, assume a semi-major axis of 1.5×10^{12} m, an eccentricity of 0.95, and an argument of periapsis of zero. A simplified version of the `kepler_orbit()` function will provide the spatial coordinates of the asteroid (x_a, y_a), and the velocity computations may be removed. For the Earth's orbit at spatial coordinates (x_e, y_e), assume a simple circular path, with $x_e = R_e \cos(\frac{2\pi t}{P_e} + \phi_e)$ and $y_e = R_e \sin(\frac{2\pi t}{P_e} + \phi_e)$, where R_e is the orbital radius, P_e is the orbital period, and ϕ_e is a counterclockwise phase angle of 0.5 rad for the Earth at the time of asteroid perihelion. Generate plots resembling those in Figures 11.9 and 11.10.

(b) Use the solution code for part (a) to write a function returning the separation distance between the Earth and an approaching asteroid. Write a second program `asteroid_min.c` to use the Golden Section algorithm to find the time and separation distance of the closest approach. The program should accept two additional command line arguments to set the bracketing values of the mean anomaly of the asteroid orbit for the desired minimum, found approximately from the plot in Figure 11.10.

11.2 In Exercise 5.8, we used a root-finding method to converge on the series of times of the extrema of the displacement of a damped harmonic oscillator described in Section 5.1. Write a new program `damped_oscillator_1min.c` using the Golden Section method to find the time of the first displacement minimum, that is, the time of the first displacement with maximum negative displacement. Use the same oscillator parameters as in Exercise 5.8, namely, an initial displacement of 1 unit, a natural period of 1 time unit, and a damping factor of 0.5. Compare your result to the first time found in Exercise 5.8.

11.3 Write a general purpose tool program to read a set of data points from a two-column data file and compute the two coefficients a_0 and a_1 of the best fit line according to the solution of Equation (11.10). The two data point coordinates represent time and position of some particle, but the position measurements incur random experimental errors. Run the program to read the example data points in the file `fit_data_velocity.dat` from the Student Companion website and produce a plot similar to that in Figure 11.20.

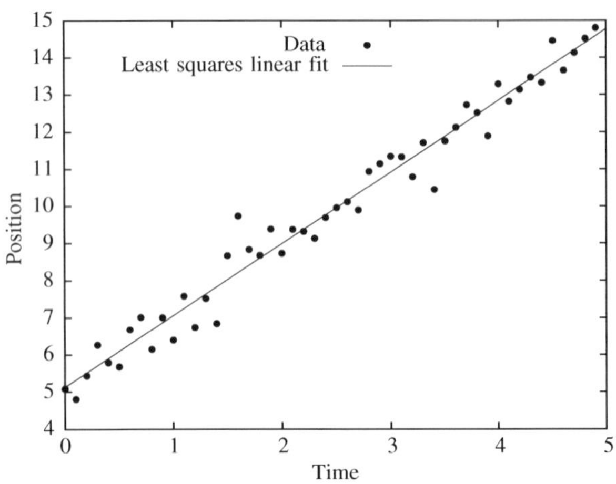

Figure 11.20 Least squares linear fit to example data measuring position and velocity of a particle.

Similar to the code structure used in Exercise 3.6 for the `meanvar.c` utility code, structure your programming solution into two code files, one that contains a main program `fitlinear_step.c` that reads a file of data points and then calls a utility function `fitlinear()` to perform the computations of Equation (11.10). The `fitlinear()` function should reside in a separate code file `fitlinear.c`, along with its attendant header file `fitlinear.h`, so that this utility module may be plugged into other programs. The form of the `fitlinear()` utility function should be

```
void fitlinear(ARRAY1D *x_ptr,ARRAY1D *y_ptr,double *a0ptr,
               double *a1ptr) {
}
```

The first two arguments are pointers to two `ARRAY1D` structures that hold the x and y coordinates of the data points. The last two arguments are pointers to where the `fitlinear()` should store the solved fitting coefficients a_0 and a_1.

When compiled linked to the `fitlinear.c` utility module, the main program module `fitlinear_step.c` should generate a data file output stream that will allow the optimum fit line to be plotted against the original data points, as in Figure 11.20. (Note that for a straight line, only two coordinate points need to be output over a "grid" of two independent variable values, x_{init} and x_{final}, specified on the command line.) The main program must also report to the standard error stream the values of the two fitting coefficients a_0 and a_1, representing the best estimates of the initial position at time $t = 0$ and the particle velocity. If you wish, you may add your `fitlinear.c` and `fitlinear.h` utility files to your `CPLibrary` directory as a new library tool.

11.4 Make use of the `fitlinear.c` and `fitlinear.h` utility module of the program solution of Exercise 11.3 together with a new main program `fit_decay.c` to read a data file `fit_data_decay.dat` from the Student Companion website describing a nuclear decay rate as a function of time and perform a least squares linear fit of the *natural logarithm* of the rate. Care must be taken to consider rate measurements only until the random event count happens to reach zero for some time step so that the logarithm is defined. The linear model fit should resemble the plot shown in Figure 11.21. From the linear fit coefficients, the program should calculate the best estimate of the half life of the decay process.

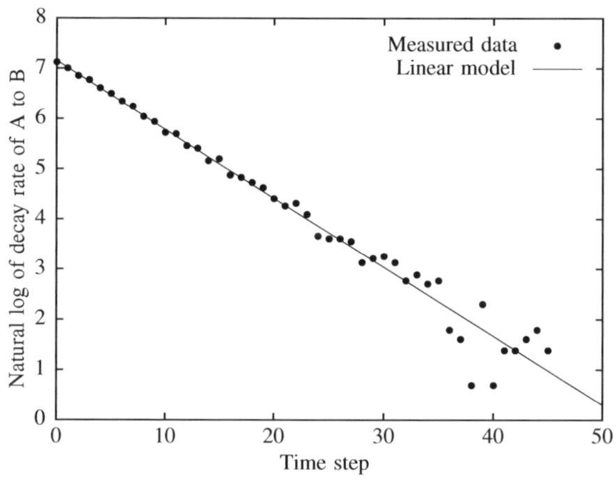

Figure 11.21 Linear fit to the logarithm of nuclear decay rate.

11.5 Write a program `fit_planck.c` to fit the model of Planck's law to experimental data to find the value of the temperature of the cosmic microwave background radiation T. Edit the program `fit_maxwell_boltzmann.c` discussed in Section 11.9 as a starting point. Use the experimental data in the file `fit_data_planck.dat` listing microwave power density per unit frequency versus frequency in hertz, which has been synthesized to include a representative amount of stochastic measurement error. Generate a plot showing the original data points and the fit curve.

11.6 Modify the `fit_damped_osc.c` program of Section 11.11 to write a new program `fit_damped_osc_gpanim.c`. This new program need not perform any actual fitting, but must include a `gen_output_gpanim()` function so that a rotating animation of the surface plot in Figure 11.16 can be generated using the techniques developed in Section 4.13. Your animation should resemble that shown in the downloadable file `fit_damped_osc_gpanim.gif`.

11.7 Extend the code of Section 11.11 used to fit a damped harmonic oscillator to write a new program `fit_cavendish.c` that includes an amplitude and bias displacement. Then use this new program to extract the oscillation natural frequency, damping coefficient, displacement amplitude, and offset displacement of the mirror of a Cavendish experiment to measure the gravitational constant, as described in Section 11.12. Download a sample data file `fit_data_cavendish.dat` for an example series of data points that include a representative level of measurement error.

11.8 Six astronomical measurements of an approaching asteroid taken over a period of time are tabulated in Table 11.1.

Table 11.1 Measured polar coordinates of an approaching asteroid.

Time t (seconds)	**Radius r (m)**	**Polar angle θ_r (radians)**
23 095 901	7.03×10^{11}	−2.265
24 016 448	6.53×10^{11}	−2.204
24 936 994	5.96e $\times 10^{11}$	−2.131
25 857 541	5.30×10^{11}	−2.042
26 471 238	4.80×10^{11}	−1.968
27 084 936	4.25×10^{11}	−1.876

These measurements are also listed in the downloadable file `fit_data_orbit.dat`.

(a) Using the `orbit_step.c` and `fit_damped_osc.c` programs as starting points, write a program `fit_orbit.c` to read the measured coordinates and fit to an orbit specified by the Keplerian elements of eccentricity ϵ, semi-major axis a, argument of periapsis ω, and time of periapsis T. Since we will be using the Nelder–Mead algorithm from the textbook library to optimize the multivariable nonlinear fit, the program must accept command line arguments that set the initial, increment, and tolerance values of all four fitting parameters. Use the library tool `read_3col_data_file()` to read the three coordinates of each measurement into arrays. The program should report the four optimized parameters to the terminal window, and send a stepped progression of the orbit determined by the optimized fit to the standard output stream for plotting.

(b) As described in Section 11.13, plot the data points against the predictions of the `orbit_step.c` program from Section 5.6. Ascertain an approximately plausible set of orbit elements of eccentricity ϵ, semi-major axis a, and argument of periapsis ω by a manual adjustment of the command line arguments over a few runs. Also propose a reasonable guess of the time of periapsis parameter T by making use of the time coordinates listed in Table 11.1.

(c) Run your `fit_orbit.c` program with your initial guess, report the optimized parameter values, and generate a plot of the measured data along with the fit orbit.

12

Explicit Methods for Ordinary Differential Equations

In this chapter, and Chapters 13 and 14, we will explore numerical methods for solving ordinary differential equations. In some ways, we will see some similarities to the numerical tools developed in Chapter 4 for visualizing functions in that we will develop coding tools that fill arrays with sampled function values over some grid of argument points. However, in Chapter 4, the function being sampled was described directly with a coded C function, whereas when we "solve" differential equations numerically in this chapter, we will be developing an array of sampled values of a function that is described instead by a relationship between its derivatives. This sort of differential relationship is at the heart of many physical laws, and just as with other types of equations, it is common that numerical methods are essential for solving equations burdened by nonlinearities or high orders.

12.1 Algorithms: Vector Fields

Let's start our algorithm development examining the most basic first-order ordinary differential equation in the canonical form of Equation (12.1). We will seek some function $y(x)$ whose first derivative follows a given nonlinear function that may, in general, depend on both the solution function value $y(x)$ and its argument x.

$$\frac{dy}{dx} = f(x, y) \tag{12.1}$$

If the function $f(x, y)$ in Equation (12.1) were only a function of the argument x, then this equation could be solved by "simply" integrating the function $f(x)$. However, if the function has some implicit dependence on the solution function $y(x)$ being sought, then it falls under the class of a differential equation. Consider an example case of the classical first-order differential equation known as the "logistics equation," which finds application in a wide variety of fields of study, including Fermi–Dirac statistics.

$$\frac{dy}{dx} = ky(1 - y) \tag{12.2}$$

This differential equation may be solved analytically by the method of separation of variables, giving

$$y(x) = \frac{e^{kx}}{e^{kx} + \frac{1-y_0}{y_0}} \tag{12.3}$$

where $y_0 = y(0)$ is the initial value of y at $x = 0$. We can gain a graphical insight into how differential equations describe the solution function by plotting the "vector field" in Figure 12.1. The value of the function $f(x, y)$ on the right-hand side of the differential equation (12.2) is plotted as the *slope* of the short vectors in Figure 12.1, in general as a function of x and y, although for this particular differential equation there is no dependence on x.

Consider the process of initializing the solution function $y(x)$ at four different initial conditions $y(0) = y_0$, and then visualizing the evolution of the four different subsequent solutions $y(x)$ for increasing x values corresponding to each initial condition. Since the slope of each of the solution functions $\frac{dy}{dx}$ is equated to the right-hand side function in Equation (12.2),

Computational Physics Using C: Efficient Programming with Ease, First Edition. John W. Fattaruso. © 2026 John Wiley & Sons, Inc.
All rights reserved, including rights for text and data mining and training of artificial intelligence technologies or similar technologies.
Published 2026 by John Wiley & Sons, Inc. Companion Website: http://www.wiley.com/go/CompPhysUsingC

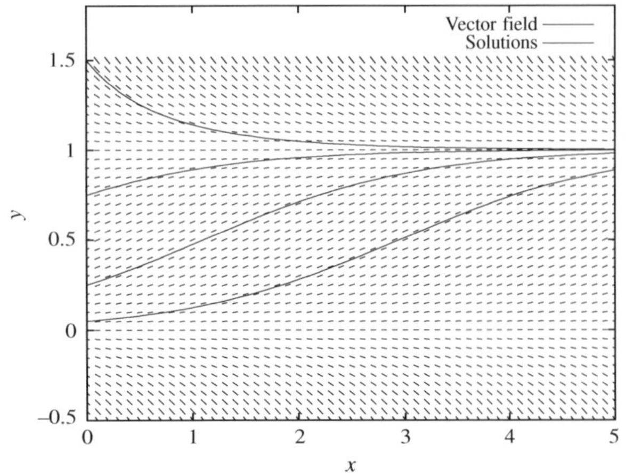

Figure 12.1 Vector field for the logistics differential equation.

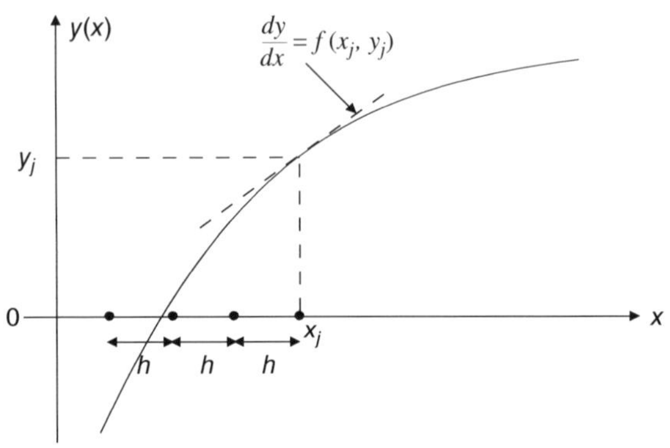

Figure 12.2 Solving differential equations with stepped algorithms.

and the two-dimensional sampled function values are plotted as the *slope* of the plot vectors, we may visualize the trajectories of the four solutions in the plot as they start their journeys at their initial conditions and then uniquely follow the "channels" formed by the plotted vectors.

How do we develop practical algorithms that will numerically accomplish the same task as what we may graphically imagine as "following the channels" of Figure 12.1? We will develop two classes of such algorithms in both this chapter and in Chapter 14. All these approaches will generate a series of uniformly sampled values along the solution function $y(x)$, under the assumption that $y(x)$ is represented with acceptable accuracy by a Taylor series of Section 2.4 truncated to some finite number of terms. The general approach is illustrated in Figure 12.2. If we assume that the curve plotted in the figure is the true solution to the differential equation and an initial condition, and that we have found points on that curve for the x values on some grid of uniform stepping up through the current state of (x_j, y_j), then the only certain information we have about the solution trajectory for the following samples is the slope of the solution at the current point, which by definition is $f(x_j, y_j)$. The slopes of the solution at x values starting at x_{j+1} are unknown, since that is where $f(x, y)$ is dependent on unknown values of y. The challenge of algorithms to solve differential equations is to project the location of points "in the channel" without a priori knowledge of the slope at the following sample points. Generally, we need a solution to the differential equation over some known range of the x values. The more terms from the Taylor series an algorithm can essentially include, the more accurate each projection to the next point will be, and the greater the allowed step size will be to achieve a given solution error level. As with all numerical algorithms based on Taylor series approximations, the overall computation requirements are reduced if we are allowed coarser steps to cover a given range.

We call algorithms "explicit" methods that make use of only the values of the right-hand side function $f(x, y)$ that are already known at prior sampled points in the x–y space to project the location of the next sample point. Another class of algorithms are the "implicit" methods that use the values of $f(x, y)$ at the prior sampled points as well as at the next point that is sought in the x–y space at each step. Since we involve the value of the $f(x, y)$ function at a projected point that is still unknown in order to discern where that projected point will be located, an additional solution task must be included for each step along the x axis in implicit methods. Depending on whether or not the $f(x, y)$ function is linear or nonlinear, that solution task is accomplished by the methods of Chapter 6 or 7. This additional numerical task in each step reduces the computational efficiency of implicit methods relative to explicit methods.

The most straightforward and efficient explicit solution methods developed in this chapter are applicable to many useful physical scenarios, and can be used to develop solutions with numerical errors limited by truncation and roundoff errors as we have seen in Sections 2.5 and 10.1.2. These methods are named by the assumption that we may solve for the next projected point in the sampled solution space within the desired error explicitly by some appropriate numerical computations that explicitly involve only the prior points. However, we will see in Section 14.1 how, with a certain class of differential equations, these methods can become unstable and catastrophically experience errors unbounded by any consideration of truncation or roundoff. For these scenarios, we will have to move on to the more elaborate and less efficient implicit methods of Chapter 14.

12.2 Algorithms: Explicit Methods for Differential Equations

12.2.1 Euler's Method

A simple first-order numerical method to develop the next sample of the solution function $y(x)$ is found by truncating the Taylor series of Equation (2.7) of the solution $y(x)$ to the term with the first derivative. Suppose that we have started from the initial condition and have accumulated the solution samples up through some point (x_j, y_j) in the solution space. Then taking the next step along the x axis with the uniform step size h, we may find the next sample y_{j+1} with the simple relation

$$y_{j+1} = y_j + h \cdot f(x_j, y_j) \tag{12.4}$$

Note that we evaluate only the right-hand side function $f(x, y)$ at the current, known point to form this explicit formula for y_{j+1}. A solution step for this method is illustrated in Figure 12.3. Graphically we can see that with only a first-order term included in the truncated Taylor series, we are estimating the next point as lying along a straight line with a slope of $f(x_j, y_j)$.

To trace algebraically the resulting sequence of y_j samples for some simple prototype case, let us take the example of a very simple right-hand side function $f(x, y) = -ay$ for some constant a. In that case the differential equation is

$$\frac{dy}{dx} = -ay \tag{12.5}$$

We already know that the straightforward solution to this equation is the decaying exponential $y(x) = y_0 e^{-ax}$, where the initial condition $y(0) = y_0$. The Euler method attempts to trace this true solution at discrete steps at an interval h in the independent variable, $x_j = jh$. From Equation (12.4) the relationship from the current step to the next step is

$$y_j = y(x_j) = (1 - ah)y_j \tag{12.6}$$

Therefore, beginning from the initial condition, the solution that is developed by this algorithm is

$$y_j = y_0(1 - ah)^j \tag{12.7}$$

If the step interval h is small enough, this solution will approximate the true solution $y_j = y_0 e^{-ax_j}$. In the next subsection, we will examine how closely this solution approximates the true solution in comparison to second order algorithms.

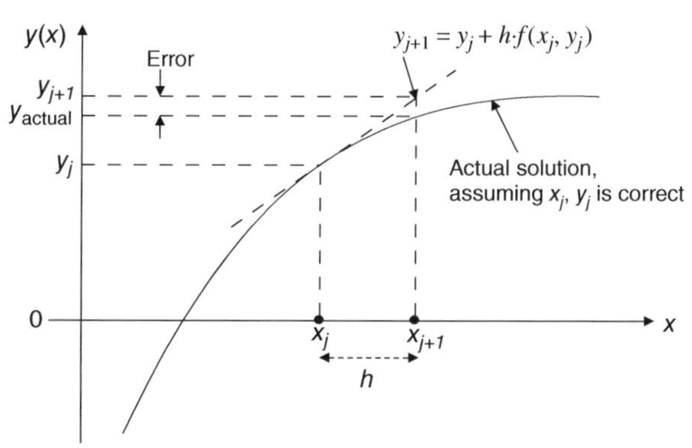

Figure 12.3 Solution step of Euler's method.

12.2.2 Heun's Method

Next we look at methods of including terms of higher order from the Taylor series of the solution so that future points may follow the curvature of the solution with smaller errors. Heun's method for tracing out solutions to differential equations can be seen as two parts to each step: first the prediction step in Figure 12.4 and then the correction step in Figure 12.5. The prediction step uses the first-order Euler estimate with a first slope $k_1 = f(x_j, y_j)$ to locate the predicted point $(x_{j+1}, \hat{y}_{j+1})$, where $\hat{y}_{j+1}$ denotes a first predicted point for the sample point that is indexed $j + 1$.

Then the predicted point is corrected by evaluating a second slope given by the right-hand side function $k_2 = f(x_{j+1}, \hat{y}_{j+1})$. The second slope k_2 is greater or less than the first slope k_1 if the vector field is curving upward or downward, respectively, for increasing x. We can imagine correcting the trajectory toward the actual solution by averaging the two slopes as $\frac{k_1+k_2}{2}$. This average slope is used finally to take a step to the next point with

$$y_{j+1} = y_j + h \cdot \frac{k_1 + k_2}{2} \tag{12.8}$$

By computing an effective slope with this two-step prediction and correction method, the second-order term of the Taylor series is essentially included in the projection to the next point (x_{j+1}, y_{j+1}).

Returning to the reference differential equation (12.5), with Heun's algorithm applied to this particular equation now gives a relationship from one sample to the next of

$$y_j = y(x_j) = \left(1 - ah + \frac{a^2h^2}{2}\right) y_j \tag{12.9}$$

This implies that the solution developed by this algorithm is

$$y_j = y_0 \left(1 - ah + \frac{a^2h^2}{2}\right)^j \tag{12.10}$$

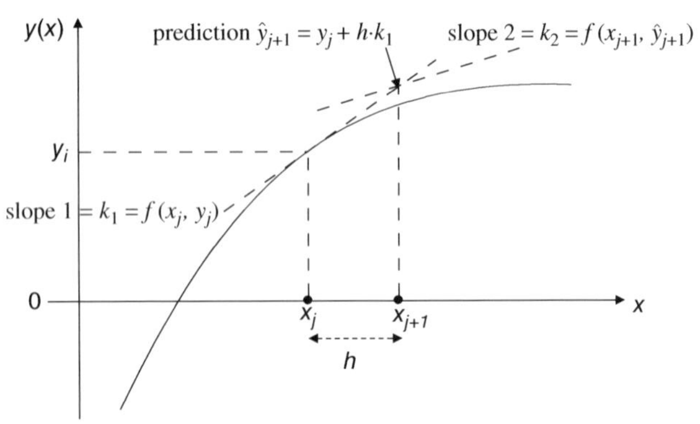

Figure 12.4 Prediction step of Heun's method.

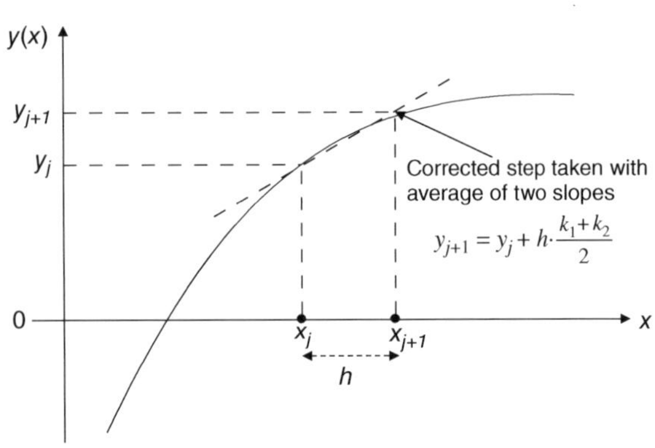

Figure 12.5 Correction step of Heun's method.

Since this solution comprehends an additional term in the Taylor series, the step size can be relaxed to larger values, saving computation time to cover a needed range in the x variable, and the approximation to the true solution $y_j = y_0 e^{-ax_j}$ will be improved.

12.2.3 Improved Polygon Method

Similar to Heun's method, the Improved Polygon (or Modified Euler's) method can also be seen as involving two parts to each step: first the prediction in Figure 12.6 and then the correction in Figure 12.7. The prediction step attempts only a projection covering half of the eventual step size, using a first slope $k_1 = f(x_j, y_j)$ from Euler's method. This results in a midpoint estimate of

$$y_{j+\frac{1}{2}} = y_j + \frac{h}{2} \cdot f(x_j, y_j) \tag{12.11}$$

We then evaluate the right-hand side function of the differential equation at this midpoint prediction to obtain a second slope $k_2 = f\left(x_{j+\frac{1}{2}}, y_{j+\frac{1}{2}}\right)$. As the vector field curves upward or downward, the second slope k_2 may be greater or less than the first slope k_1.

The corrected next point in the solution sequence is then found with a full step with the midpoint slope k_2, as

$$y_{j+1} = y_j + h \cdot k_2 \tag{12.12}$$

This two-step method also essentially takes account of the second-order term in the Taylor series approximation for the solution by adapting the projection with some measurement of the vector field curvature. Equation (12.12) can also be seen as applying the two-sided, second-order estimate of the first derivative we first saw as Equation (2.10) for a half step size of $\frac{h}{2}$ at the midpoint of $x_{j+\frac{1}{2}} = x_j + \frac{h}{2}$, assuming the slope k_2 above is a close approximation to the midpoint slope.

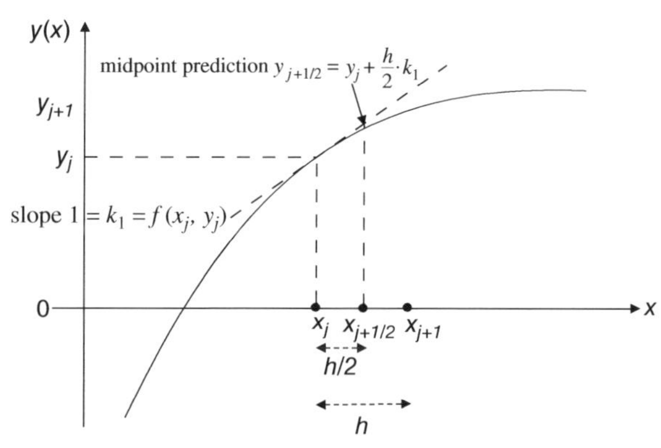

Figure 12.6 Prediction step of Improved Polygon method.

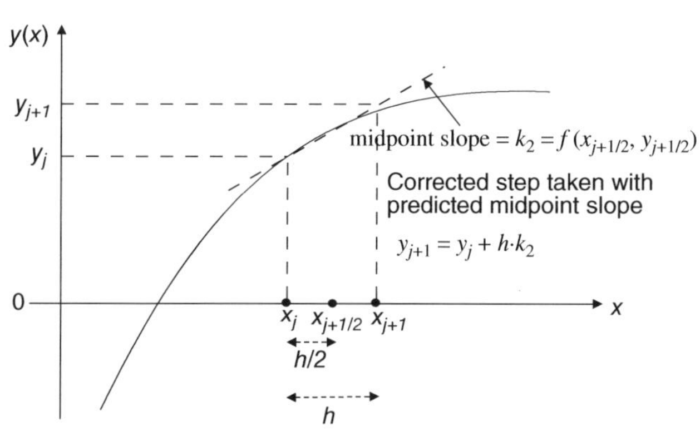

Figure 12.7 Correction step of Improved Polygon method.

In fact, returning to the reference differential equation (12.5), an analysis gives exactly the same relationship between successive steps as in Equation (12.9), so the solution developed will be exactly the same as that with Heun's algorithm as in Equation (12.10).

Since we know the solution to this simple reference equation, we can easily plot the error between the first- and second-order explicit solution methods with the known solution. The plot of Figure 12.8 shows three solutions for the reference equation (12.5) with an initial condition of $y(0) = 1$ over the range of $x = 0$ to $x = 1$. The solid line is the exact known solution, and the two dashed lines are the numerical solutions with Euler's and Heun's method using a fairly coarse step size of 0.1 in order to accentuate truncation error. We can see in the small window showing a zoomed-in portion of the plot that the second-order Heun's method (and, therefore, also the Improved Polygon method) gives a solution with much smaller truncation error relative to the first-order Euler method.

We can exercise Equations (12.7) and (12.10) using double-precision floating point arithmetic over a range of step sizes to visualize the tradeoff between truncation and roundoff errors that influences the choice of step size for solving differential equations. Figure 12.9 plots the error between these two numerical solutions and the exact known solution of the reference differential equation (12.5). Not unexpectedly, it shows the same sort of dependence as we have seen earlier in Figures 2.16 and 2.18 for numerical differentiation and in Figure 10.3 for numerical integration. With large step sizes on the right side of the plot, the error is dominated by truncation error, and we can see the error being reduced with two different slopes with respect to step size according to the order of the Taylor series truncation. As before with the other numerical examples, at

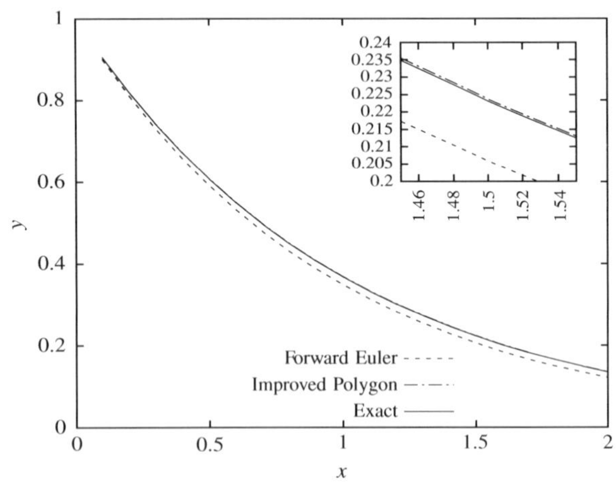

Figure 12.8 Example solution of reference differential equation with explicit methods, step size 0.1.

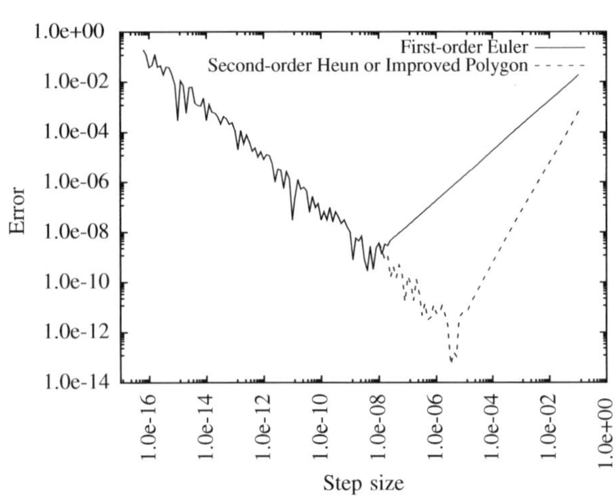

Figure 12.9 Tradeoff of roundoff and truncation error versus step size for first- and second-order differential equation explicit solution estimates.

some point the error encounters a lower limit and becomes dominated by the roundoff errors inevitable with floating point arithmetic.

Although the truncation error shown in Figure 12.9 for a second-order algorithm can be reduced to practical levels for some problems with careful tuning of the step size, in general we want to explore methods of higher order that are capable of a small error over a wider range of step sizes. We will next examine the general class of Runge–Kutta methods that will reduce truncation error even further.

12.2.4 Runge–Kutta Methods

In the discussion of the Heun's and Improved Polygon methods above, we examined two approaches for improving the value of the slope of the trajectory from the current point to the next actual point on the solution curve. Assuming a continuous solution function, there is indeed one exact slope that, when multiplied by the step size, will increment the current y_i to an exactly correct y_{i+1}. How do we obtain the best value of that slope, making use of higher-order terms in the Taylor series for the solution? A generalized technique was developed by Runge and Kutta, and Figure 12.10 illustrates the general scheme of the Runge–Kutta solution methods. In a manner similar to the Improved Polygon method above, the right-hand side function $f(x, y)$ may be evaluated a number of times within the intermediate range between the current x_i and the next $x_{i+1} = x_i + h$. Each evaluation gives more information about the curvature of the vector field ahead, and contributes to the value of the exact slope that incorporates corrections from higher order Taylor series terms. In fact, the Improved Polygon method of Section 12.2.3 conforms to the general formulation of Runge–Kutta methods of the second order.

In search of a practical algorithm for the solution of differential equations that exhibits fourth-order truncation error with respect to the step size, Equation (12.13) shows a Taylor series representation to the solution function $y(x)$ truncated at four terms.

$$y(x+h) = y(x) + y'(x)h + \frac{y''(x)}{2}h^2 + \frac{y'''(x)}{6}h^3 + \frac{y''''(x)}{24}h^4 + E_5 \tag{12.13}$$

Suppose we evaluate the vector field function $f(x, y)$ four times at intermediate points and arrive at four different values for the trajectory slope, k_1, k_2, k_3, and k_4. The derivation of any of the class of fourth-order Runge–Kutta methods starts by assuming that the truncated series of Equation (12.13) can be made equal to the current y_i value incremented by the step size multiplied by some linear combination of the four contributing slopes, as shown in Equation (12.14)

$$y(x+h) = y(x) + h \cdot (a_1k_1 + a_2k_2 + a_3k_3 + a_4k_4) \tag{12.14}$$

We must solve for some suitable selection of the unknown coefficients a_i and slope values k_i. These relations do not specify a unique solution, but one feasible and popular solution is

$$y_{j+1} = y_j + \frac{h}{6} \cdot [k_1 + 2k_2 + 2k_3 + k_4] \tag{12.15}$$

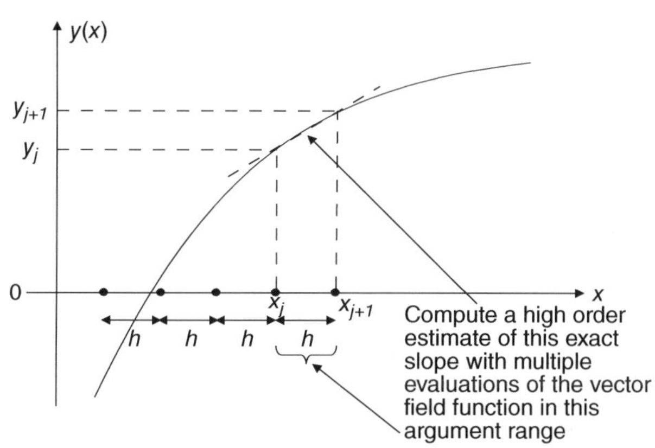

Figure 12.10 General scheme of the class of Runge–Kutta methods.

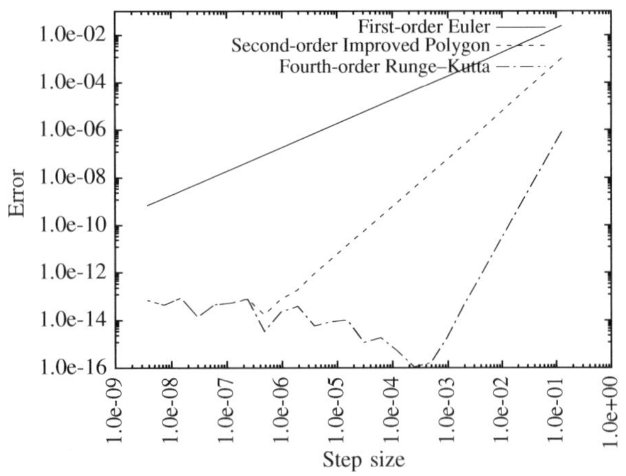

Figure 12.11 Tradeoff of roundoff and truncation error versus step size for first-, second-, and fourth-order differential equation explicit solution estimates.

where the four k_i slope values are given by

$$\begin{aligned} k_1 &= f(x_j, y_j) \\ k_2 &= f(x_j + \frac{h}{2}, y_j + \frac{h}{2}k_1) \\ k_3 &= f(x_j + \frac{h}{2}, y_j + \frac{h}{2}k_2) \\ k_4 &= f(x_j + h, y_j + hk_3) \end{aligned} \tag{12.16}$$

Figure 12.11 plots the error characteristic versus step size for the three complete algorithms examined so far, the first-order Euler, the second-order Improved Polygon, and the fourth-order Runge–Kutta. The differential equation being solved is the reference equation of Figure 12.5 with $a = 1$, with an initial condition of $y(0) = 1$. The error recorded is the deviation of the numerical solution from the known value of e^{-1} at $x = 1$. The step size ranges from 2^{-28} to 2^{-3}. Unless we were very careful to optimize the step size, the truncation error accumulated with Euler's method is typically at a level that would limit that algorithm to only the most approximate applications. Indeed, we can see from this plot that the truncation error is so large that the floor of roundoff error is never reached for step sizes even as small as 2^{-28}. The second-order Heun and Improved Polygon methods require only modest computation and exhibit error levels that make them useful in many graphical applications. We will see this method applied in Section 13.4 when we draw classical field lines for static electric and magnetic fields. However, for most applications where numerical accuracy is important, the additional computational requirements of a fourth-order algorithm such as Runge–Kutta are worthwhile.

12.3 Algorithms: Solving Higher-order Equations and Systems of Differential Equations

In the previous sections, we have seen methods by which we can develop a sampled version of the solution of a simple first-order differential equation that can be cast in the form

$$\frac{dy}{dx} = f(x, y) \tag{12.17}$$

Although we could identify some physical processes that are governed by this simple equation, we need to expand the scope of equations that may be solved to model the great variety of possible physical systems. To do this, we only need to treat the first-order equation (12.17) as a fundamental prototype, and cast more elaborate systems into the form of a simultaneous system of equations following this form.

We can write a system of n first-order differential equations as

$$
\begin{aligned}
\frac{dy_0}{dx} &= f_0(x, y_0, y_1, y_2, \dots, y_{n-1}) \\
\frac{dy_1}{dx} &= f_1(x, y_0, y_1, y_2, \dots, y_{n-1}) \\
\frac{dy_2}{dx} &= f_2(x, y_0, y_1, y_2, \dots, y_{n-1}) \\
&\vdots \\
\frac{dy_{n-1}}{dx} &= f_{n-1}(x, y_0, y_1, y_2, \dots, y_{n-1})
\end{aligned}
\tag{12.18}
$$

Customarily, the subscript notation for a system of n equations range from 1 to n, but to be consistent with the array subscripts of the C programming language, we will write subscripts ranging from 0 to $n-1$. In this formulation, we are using the subscript y_i to denote the solution function $y(x)$ to the differential equation indexed by i in the system. Note that now the "right-hand side" functions f_i may be, in the most general case, functions of the independent variable x as well as all the solution functions $y_i(x)$. We will be applying the algorithms developed in the prior sections to follow the interrelated trajectories of all the $y_i(x)$ solutions sampled at a uniform grid of x_j values, so we will generate a two-dimensional array of $y_{i,j}(x)$ values, where i is the equation index and j is the sample index.

Solving the system of Equation (12.18) can be accomplished in a straightforward manner. We may simply sequentially work down the list of n equations in the system from index $i=0$ to $i=n-1$ taking a single step in each equation from the current step index j to step $j+1$ with an appropriate algorithm above. It is important that during the step j to $j+1$ we evaluate all the right-hand side functions $f_i(x, y_0, y_1, y_2, \dots, y_{n-1})$ at their values for the current step j. That is, during step j we evaluate all the f_i as $f_i(x, y_{0,j}, y_{1,j}, y_{2,j}, \dots, y_{n-1,j})$ in order to compute the slopes that project to the next $y_{i,j+1}$ values. Then once the current step has been completed for all n equations, all the $y_{i,j}$ values can be updated to their new $y_{i,j+1}$ values and a new step initiated. Careful use of C array structures will help keep these values in their proper storage locations.

The differential equations in the system (12.18) are only of the first order. How do we approach higher-order equations that describe common behavior in the physical universe? We may always transform a higher-order differential equation of the form

$$
a_n \frac{d^n y}{dx^n} + a_{n-1} \frac{d^{n-1} y}{dx^{n-1}} + \cdots + a_2 \frac{d^2 y}{dx^2} + a_1 \frac{dy}{dx} = g(x) \tag{12.19}
$$

into a system of Equation (12.18) by use of the systematic substitutions

$$
\begin{aligned}
y_0 &= y \\
y_1 &= \frac{dy_0}{dx} \\
y_2 &= \frac{dy_1}{dx} \\
y_3 &= \frac{dy_2}{dx} \\
&\vdots
\end{aligned}
\tag{12.20}
$$

The functions $y_i(x)$ are known as the state variables of the system, and the solution $y(x)$ to the original high-order equation (12.19) is simply $y_0(x)$. This transformation process will become more clear with the examples and exercises in the following sections.

For problems in physics described by vectors, such as classical mechanics or electromagnetic fields, a vector differential equation may also be decomposed into component equations corresponding to each vector component. Therefore, a second- or higher-order vector differential equation can be applied to numerical solvers by defining a state variable both for each equation order and for each vector component in the system. Again, the following sections show examples of this decomposition.

12.4 Coding: The Differential Equation Solver Tool

The tool code for solving systems of differential equations with explicit algorithms can be found in the files `diffeq.c` and `diffeq.h`, part of the `CPLibrary` package from the Student Companion website. Three routines can be found in the source code, as follows.

```
int rk2(void (*func)(double x,ARRAY1D *y,void *p,ARRAY1D *yp),
        GRID *xgrid_ptr,void *param,ARRAY1D *init,ARRAY2D *ytab_ptr) {
  ...
}

int rk2step(void (*func)(double x,ARRAY1D *y,void *p,ARRAY1D *yp),
            double x,double xinc,void *param,ARRAY1D *y) {
  ...
}

int rk4(void (*func)(double x,ARRAY1D *y,void *p,ARRAY1D *yp),
        GRID *xgrid_ptr,void *param,ARRAY1D *init,ARRAY2D *ytab_ptr) {
  ...
}
```

The `rk2()` and `rk4()` functions are called to solve systems of first-order differential equations with second- and fourth-order Runge–Kutta methods, respectively, over a specified grid of x values. The second-order method is described in Section 12.2.3, and the fourth-order algorithm is described in Section 12.2.4. Both the `rk2()` and `rk4()` functions expect the same sequence of arguments. The first argument is a pointer to a C function that computes the set of n right-hand side functions in Equation (12.18), and, therefore, can be seen as the C function that defines the vector field of the entire system of differential equations. The prototype of that function is `void (*func)(double x,ARRAY1D*y,void *p,ARRAY1D*yp)`. This vector field function expects the current value of x_j at each step, a pointer to an array of the current values of $y_{i,j}$ at each step, and a pointer to a set of physical parameters that may be needed for the function calculations. Its final argument is a pointer to an array where the function should store all the right-hand side values, which are defined to be the derivatives $y'_{i,j}$ by the differential equation form.

Returning to the argument list for `rk2()` and `rk4()`, those functions next expect a pointer to a grid of the x independent variable over which the solution is to be found, the parameter structure pointer that will be passed on to the vector field function, a pointer to an array of the initial conditions for the state variables y_i, and finally a pointer to a two-dimensional array where the solver is to store the solution. The memory for the array pointed to by the `ARRAY2D` structure must be allocated by the code that calls these solver routines. The row index i of the array indexes the equations in the system and the state variables $y_i(x)$, so the number of rows in the allocated array must match the count of state variables in the system being solved. The column index j of the array indexes the samples in each state variable $y_{i,j}$ over the grid of x_j values. The sample index j corresponds to the independent variable value $x = x_{init} + jx_{inc}$, and the `ARRAY2D` must be allocated by the calling code with sufficient size to accommodate $y_{i,j}$ samples for x_j values up to and including x_{final}, where x_{init}, x_{final}, and x_{inc} are specified in the `xgrid` argument. As with the `step()` tool introduced in Section 4.1, invoking the `NUM_GRID_PTS` macro with the argument of `xgrid` will be a convenient way to set the number of columns. As an example, suppose that a C function `f` provides the right-hand side slopes for a system of differential equations with four state variables. The typical sequence to develop a sampled solution for this system is the code

```
ARRAY2D ytable;
GRID xgrid;
...
alloc_array2d(&ytable,4,NUM_GRID_PTS(xgrid));
rk4(f,&xgrid,&param,&init,&ytable);
...
free_array2d(&ytable);
```

The following sections provide examples and exercises demonstrating the use of these solvers.

The remaining routine in the `diffeq.c` source code file is `rk2step()`, which implements a single step of the Improved Polygon solver method in `rk2()`. In most cases, we are solving a system of differential equations over a fixed desired range of the independent variable x. However, there are some cases in which we will initialize a solution at some value of x with the initial conditions on the $y_i(x)$ state variables, and then project a sequence of steps in the solution space indefinitely until some termination condition is met among the state variables. One example of this case is drawing the classical field lines demonstrated in Section 13.4. For these limited cases the `rk2step()` tool is available that realizes only one step of the Improved Polygon method at a particular value of the independent variable x. This tool requires a somewhat different argument list, without a full grid specification, but instead a simple value for the x increment h. The final argument is a one-dimensional `ARRAY1D` array to hold the next set of $y_{i,j+1}$ state variables that result from the single solution step.

12.5 Application: The Large-angle Pendulum

Figure 12.12 shows a single pendulum, a deceptively simple nonlinear physical system capable of interesting and rich behavior. For simplicity, we'll assume an ideal structure with the mass m concentrated at a single point at the end of a massless arm of length l, but this may be made more realistic with the usual application of the rotational center of inertia. Most introductory textbooks analyze this structure, but generally only within the limit of small angular deviations θ from the vertical. This limit leads to an approximately linear differential equation describing simple harmonic motion [17, Chap. 16; 18, Chap. 14; 19, Chap. 14; 20, Chap. 13]. However, by bringing numerical tools to bear on the analysis, we are able to solve the governing transcendental differential equation over any angular range. Although the equation of motion for this system can be derived easily from the simple Newtonian relation $\vec{F} = m\vec{a}$, we'll be using the Lagrangian formulation here, as this will provide a basis for also using the Lagrangian approach for applying mechanical forcing to the single pendulum and for the more complex double pendulum in subsequent sections. For the Lagrangian method, we assume a coordinate system with its origin at the pivot point of the pendulum, and write the coordinates of the mass $x = l\sin\theta$ and $y = -l\cos\theta$. The Lagrangian $L = T - V$, where the kinetic energy term is

$$T = \frac{m|\vec{v}|^2}{2} \tag{12.21}$$

Using the usual "dot" notation for the time derivative, the square of the velocity $|\vec{v}|^2$ is

$$\begin{aligned} |\vec{v}|^2 &= \dot{x}^2 + \dot{y}^2 \\ &= l^2(\cos^2\theta)\dot{\theta}^2 + l^2(\sin^2\theta)\dot{\theta}^2 \\ &= l^2\dot{\theta}^2 \end{aligned} \tag{12.22}$$

So the kinetic energy term is

$$T = \frac{ml^2}{2}\dot{\theta}^2 \tag{12.23}$$

The potential energy term of the Lagrangian is

$$\begin{aligned} V &= mgy \\ &= -mgl\cos\theta \end{aligned} \tag{12.24}$$

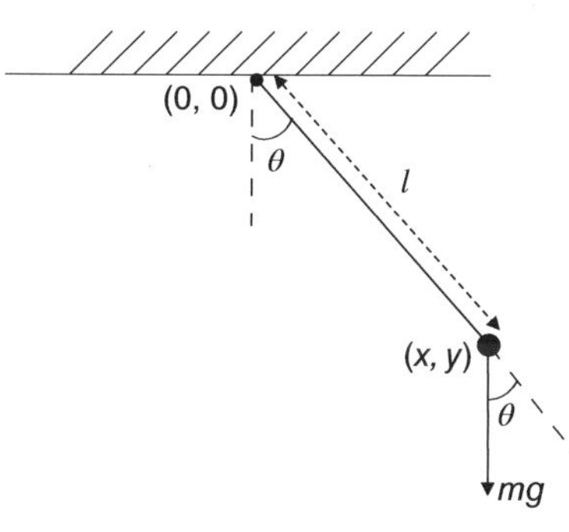

Figure 12.12 Single pendulum.

This makes the total Lagrangian

$$L = \frac{ml^2}{2}\dot{\theta}^2 + mgl\cos\theta \tag{12.25}$$

We find the equation of motion by applying the Euler–Lagrange equation for the single degree of freedom θ

$$\frac{d}{dt}\left[\frac{\partial L}{\partial \dot{\theta}}\right] = \frac{\partial L}{\partial \theta} \tag{12.26}$$

$$\frac{d}{dt}\left[ml^2\dot{\theta}\right] = -mgl\sin\theta \tag{12.27}$$

$$\ddot{\theta} = -\frac{g}{l}\sin\theta \tag{12.28}$$

In introductory physics textbooks, Equation (12.28) is usually solved only for small angles with the linearizing approximation that $\sin\theta \approx \theta$. This simplifies the equation of motion to a second-order linear homogeneous differential equation with constant coefficients, with a solution of simple harmonic motion with an angular frequency $\omega_0 \approx \sqrt{\frac{g}{l}}$, and therefore a period $T \approx 2\pi\sqrt{\frac{l}{g}}$. We can identify the left-hand side of Equation (12.28) as the angular acceleration, and therefore the right-hand side must be the ratio of the torque acting on the pendulum to the rotational inertia. To make our solution even more complete, we can now assume that there will also be some damping torque acting in the system, perhaps from friction in the pivot mechanism or from air drag on the mass. We will simply model the damping torque as being proportional to the angular velocity, and, therefore, add a term to Equation (12.28) proportional to $\dot{\theta}$. For convenience, let's choose a constant of proportionality so that in the limiting case of small oscillation angles the damping parameter will be aligned with the harmonic oscillator damping coefficient ζ in Equation (4.1) of Section 4.2. This choice makes the effective equation of motion

$$\ddot{\theta} = -\frac{g}{l}\sin\theta - 2\zeta\sqrt{\frac{g}{l}}\dot{\theta} \tag{12.29}$$

Beyond a linear approximation, to get a more complete solution of this nonlinear system that demonstrates the period of non-sinusoidal oscillation varying with the amplitude of the oscillation, we decompose this second-order differential equation into a system of two first-order equations that we then may easily submit to a numerical solver. Following the scheme of Equation (12.20), we define the two state variables

$$\begin{aligned} y_0 &= \theta \\ y_1 &= \frac{dy_0}{dt} = \dot{\theta} \end{aligned} \tag{12.30}$$

and thereby transform the differential equation (12.29) into the system

$$\begin{aligned} \frac{dy_0}{dt} &= y_1 \\ \frac{dy_1}{dt} &= -\frac{g}{l}\sin(y_0) - 2\zeta\sqrt{\frac{g}{l}}y_1 \end{aligned} \tag{12.31}$$

An example main program for solving for the deflection angle $\theta(t)$ and the deflection angular velocity $\dot{\theta}(t)$ versus time is in the file `pendulum.c`, available for download from the Student Companion website. The following excerpt shows the structure storing two physical parameters of the pendulum, its length and damping coefficient, and the C function to be passed to the `rk4()` differential equation solver tool

```
typedef struct pend_param {
  double length;  // pendulum length in meters
  double zeta;  // damping coefficient in 1/sec
} PEND;

// State variable 0 is the displacement angle
// State variable 1 is the angular velocity
void pendulum(double x,ARRAY1D *y,void *param_ptr,ARRAY1D *yp) {
  double omegasq,omega;
```

```
    PEND *p;

    p = param_ptr;
    omegasq = GRAVITATIONAL_ACCEL/p->length;
    omega = sqrt(omegasq);
    yp->elem[0] = y->elem[1];
    yp->elem[1] = -omegasq * sin(y->elem[0]) -
                  2.0 * p->zeta * omega * y->elem[1];
}
```

In the case of equations of motion, the independent variable named x in the general form of Equation (12.19), and represented by the `double` variable `x` in these solver routines, is time. The `ARRAY1D` elements `y->elem[0]` and `y->elem[1]` are the state variables θ and $\dot{\theta}$. The function `pendulum()` calculates the derivatives in Equation (12.31). The rest of the main program interprets arguments from the command line, fills instances of a `GRID` and `PEND` structure with values, and calls the `rk4()` solver with

```
rk4(pendulum,&tgrid,&param,&init,&ytable);
```

After the solver routine fills the `ytable` two-dimensional array with the sampled values of the solution state variable functions, the program calls a `gen_output()` function to generate an output stream describing the solution function. This output function works in a manner similar to the corresponding output functions in the programs of Chapter 4, except that multiple state variables that are stored in the rows of a two-dimensional `ARRAY2D` structure are output in multiple columns. We'll start with an example pendulum with a length of 1 m under the influence of the gravitational acceleration at the Earth's surface. We would expect an oscillation period of $2\pi\sqrt{\frac{l}{g}} = 2.0064093$ s for small deflection angles. Choosing a simulation time step of 0.01 s would provide about 200 steps in the Runge–Kutta algorithm per cycle, and we can expect the fourth-order algorithm to follow the oscillation motion with minimal truncation error. Compiling and running the program

```
$ gcc -Wall -Wshadow -O2 -o pendulum pendulum.c CPLibrary/diffeq.c CPLibrary/CPLibrary/array.c -lm
$ ./pendulum
./pendulum <length> <zeta> <init y0> <init y1> <tfinal> <tinc>
$ ./pendulum 1 0 3 0 10 0.01 >| pendulum.dat
```

We can think of this program as an experimental numerical laboratory with which we can observe the pendulum's behavior with various parameter choices. First we initialize the pendulum with a straightforward scenario of holding an initial deflection angle with no angular velocity and releasing the mass, and with no significant damping. The plots in Figure 12.13 show the deflection angle versus time for four different initial angles. For the smallest deflection, the numerical solution approximates the small angle limit of $\sin\theta \approx \theta$ and shows a near sinusoidal dependence of angle versus time. The expected oscillation period of approximately 2 s is apparent from the time axis labels. As the initial angle is made larger, the angular dependence with time is distorted more and more from a sinusoid, with the period slowing considerably at maximum deflection angles. This behavior is due to the nonlinear $\sin(\theta)$ factor in the equation of motion (12.29). Physically, this angle makes the vertically oriented gravitational force acting on the pendulum mass less and less effective at applying rotational torque as the deflection approaches vertical. We can see the oscillation period lengthened considerably for an angle amplitude of 3 rad, which is just short of straight vertically upward.

When the `pendulum.c` program was run at the command line above, we supplied a time step of 10 ms for the Runge–Kutta algorithm. Since we expected an oscillation period of about 2 s in the small angle regime, this time step would cover one period with at least 200 steps. How do we have confidence that this selected time step places us in an area along the fourth-order curve of Figure 12.11 that reduces the error level to acceptable levels for this particular differential equation? Recall that for the root-finding algorithms of Chapters 5 and 7 and the minimization algorithms of Chapter 11, we were able to specify a tolerance parameter that directly governs the accuracy with which the algorithm would converge to the correct solution. It should be mentioned that there are more sophisticated algorithms for solving differential equations that measure their own local truncation error at each step and automatically adjust the local step size to maintain the errors within some specification. These variable time step methods are beyond the scope of this text. However, we may use several methods to measure the truncation error resulting from a fixed time step selection, and one method for this pendulum system appeals to the fundamental physics principle of the conservation of energy. By adding Equations (12.23) and (12.24), we may find the total energy of the pendulum. With no damping, the total energy must remain conserved as the kinetic and potential terms trade energy back and forth with each swing. We have the time sampled values of θ and $\dot{\theta}$ available

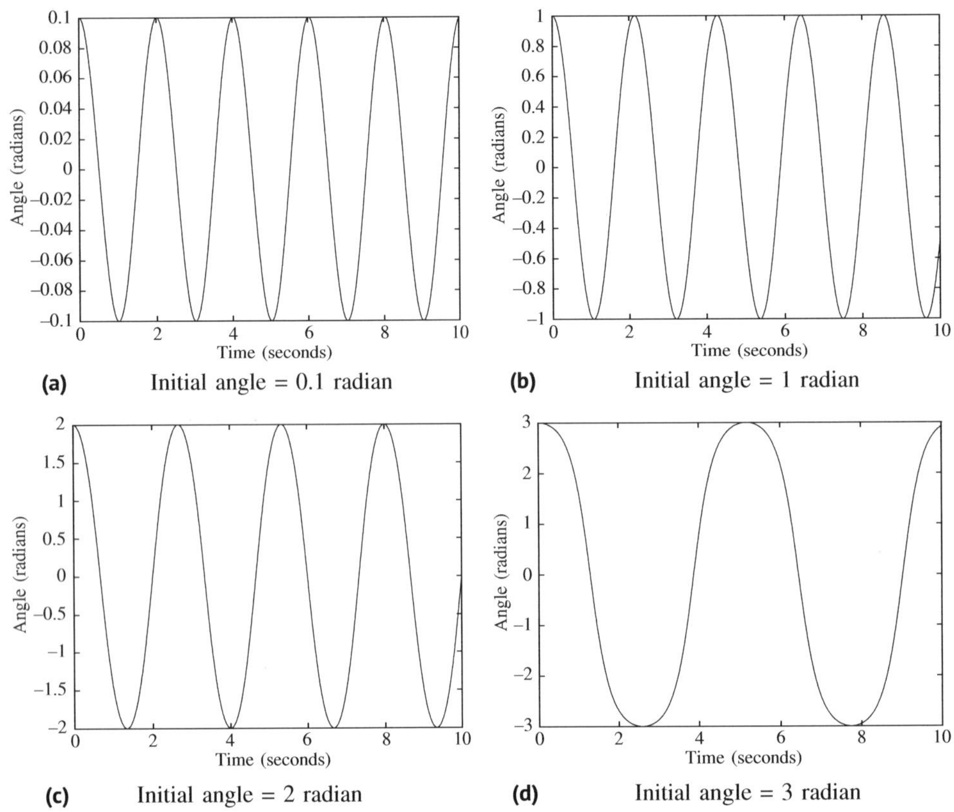

Figure 12.13 Pendulum with large angles of initial deflection and no damping.

throughout the simulation as the two state variables, so we may make use of the principle of conservation of energy as a check of our numerical precision. The ratio of total energy to the pendulum mass is given by Equation (12.32).

$$\frac{E}{m} = \frac{l^2}{2}\dot{\theta}^2 - gl\cos\theta \tag{12.32}$$

Note that the `gen_output()` function in the `pendulum.c` program outputs this computation of the total energy per unit mass, including all 16 significant digits, in a fourth column after the two state variables. Figure 12.14 plots this quantity over a 30-s simulation using a trial time step of 20 ms after an initial angle of 3 rad.

For an ideal simulation with no numerical errors, we would expect a perfectly horizontal line of constant energy in Figure 12.14. Although we do see some accumulation of the truncation error in the simulated energy that follows each pendulum swing, note that the span of the vertical axis shows that the total conservation error is on the order of only 1 ppm. Then rerunning the simulation with the step size reduced by half, to 10 ms, produces the plot shown in Figure 12.15. In this case, we can measure the accumulated conservation error to be on the order of 50 ppb, less than that with a 20-ms step size by a factor of approximately 16. This reduction is consistent with the expected truncation error of a fourth-order algorithm, as shown by the slope in Figure 12.11. This gives us the confidence that we may select an appropriate step size to satisfy the accuracy requirements we have for the simulation, down to the limit of the double-precision floating point roundoff error.

Another way to initialize the motion of the pendulum is to start with an initial deflection angle of 0 with the mass dangling vertically downward, and imparting an impulse of angular momentum to the mass. We simulate this scenario with the initial state variable θ at 0 but giving a nonzero initial value to the state variable $\dot{\theta}$. Let us plot the pendulum behavior with the initial angular velocity of 6.25 radians per second in Figure 12.16a. This is just enough velocity to swing the mass upward just short of vertical, which would be with $\theta = \pm\pi$, at the height of its motion. Besides the deflection angle, we also see the angular velocity in the plot, and can observe the high speed of the mass as it passes through the bottom of its excursion, and the very low speed during the slow reversal of direction at the top.

What happens if we impart an initial angular velocity that exceeds that shown in Figure 12.16a by a small amount? Figure 12.16b shows the simulation result with an initial angular velocity of 6.3 radians per second. This slightly higher initial momentum markedly changes the behavior. How do we interpret this plot? This plotted curve is the result of the

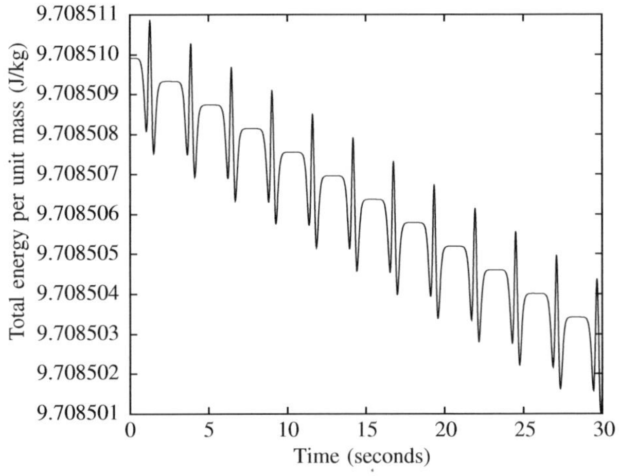

Figure 12.14 Total energy versus time of pendulum with initial angle of 3 rad using simulation step size 0.02 s showing accumulating truncation error.

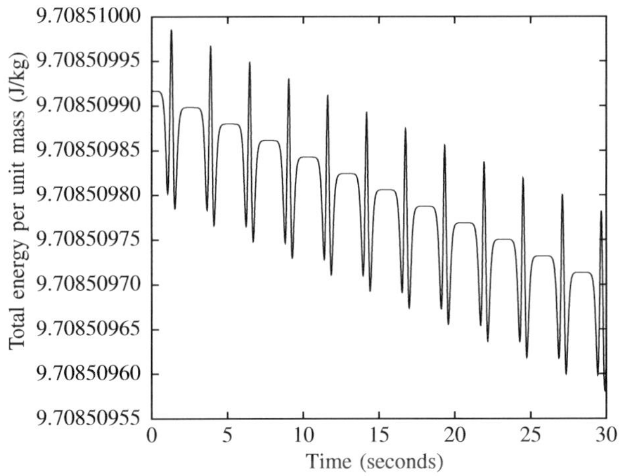

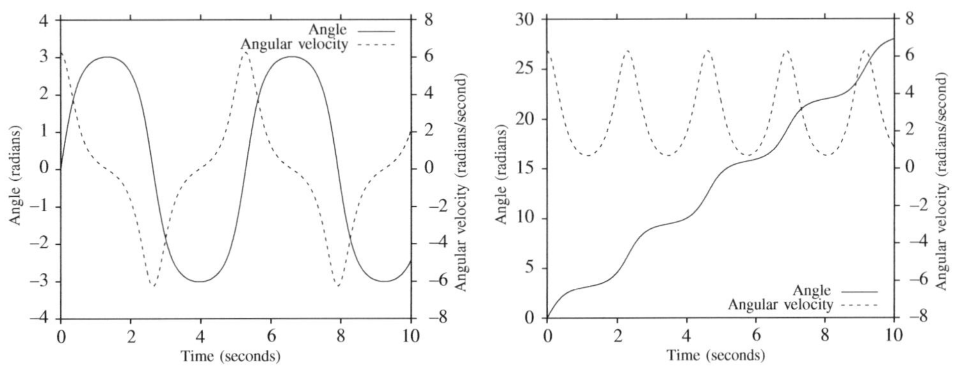

Figure 12.15 Total energy versus time of pendulum with initial angle of 3 rad using simulation step size 0.01 s showing accumulating truncation error.

(a) Initial angular velocity = 6.25 radian/sec **(b)** Initial angular velocity = 6.3 radian/sec

Figure 12.16 Pendulum with large initial angular velocity and no damping.

mass having enough momentum climbing the right-hand side of the pendulum to continue counterclockwise just barely through the completely vertical position. After passing through the vertical position, it will continue in counterclockwise motion and fall down the left-hand side. With the assumption of no damping, energy will be conserved, and each swing of the pendulum similarly will just clear the straight vertical orientation and continue revolving around the pivot counterclockwise. The mathematical relationships of the equation of motion continue to be valid as the angle grows past π. We can see the angle growing monotonically in the plot as the mass continues to rotate around the pivot. The angular velocity varies between slow and fast in a cyclical manner, but remains positive as the rotation never reverses.

We can now run an experiment with a small damping factor of 0.05 introduced. The plot of Figure 12.17 shows the deflection amplitude slowly reducing from its initial value of 3 rad as the pendulum's energy is lost to heat due to the damping. Moreover, we can see the time dependence of both the angle and angular velocity approaching sinusoidal as the amplitude reduces and the small angle approximation becomes valid.

An interesting experiment to run is providing enough initial angular momentum to the mass to start the pendulum in its rotation mode, but then observing the effect of the gradual dissipation of energy due to damping losses. Plotted in Figure 12.18 is the case of an initial angular velocity of 7 rad/s with a small damping coefficient of 0.01. We can count a few initial rotations around the pivot, but then with the loss of enough energy, we see the pendulum motion settles into its swinging mode with large angles.

While plotting the deflection angle and velocity versus time is a natural way to visualize the motion of the pendulum, other interesting features can be gleaned through another representation. If we plot the trajectory of the motion with the two plotting axes being the two state variables, we are able to visualize the behavior in *state-space*. While there is no axis for time in this plotting mode, the parameter of time migrates the state of the dynamic system around the loci of points

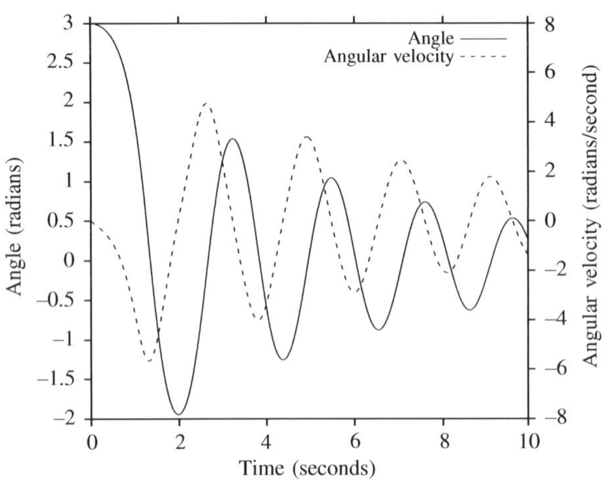

Figure 12.17 Pendulum with initial angle of 3 rad, damping coefficient 0.05.

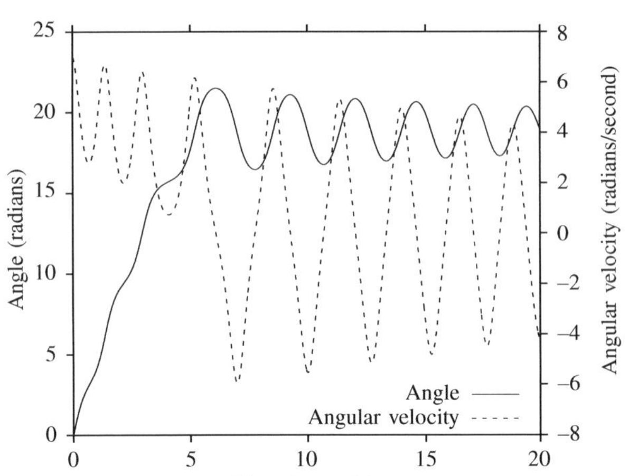

Figure 12.18 Pendulum with initial angular velocity of 7 rad/s, damping coefficient 0.01.

in the plot in the prescribed trajectory. We are plotting a locus of points in a space spanned by all the state variables of this dynamic system. Therefore, each point in this space corresponds to a complete initial condition for the system of differential equations that govern future behavior. As an example, Figure 12.19 plots two of the scenarios in Figure 12.13, with initial deflection angles of 1 and 3 rad. The shape of the trajectory figure traced out in time changes markedly with the two initial conditions. When the speed of the oscillations is slowed at large deflection angles as in Figure 12.19b, the ends of the trajectory figure grow more pointed. Figure 12.20 provides a state-space view of the experiments of Figure 12.16. The differences between the oscillatory and rotational modes are graphically apparent.

Let us now see the effect in state space of the level of damping shown in Figure 12.17. Figure 12.21a is a state-space representation with a large initial angle and an example damping coefficient of 0.05. As the system energy is lost to friction and drag, the trajectory figure reshapes from having pointed ends down to an approximate ellipse. Figure 12.21b is the state-space representation of the scenario of Figure 12.18 where both the rotational and swinging modes are seen.

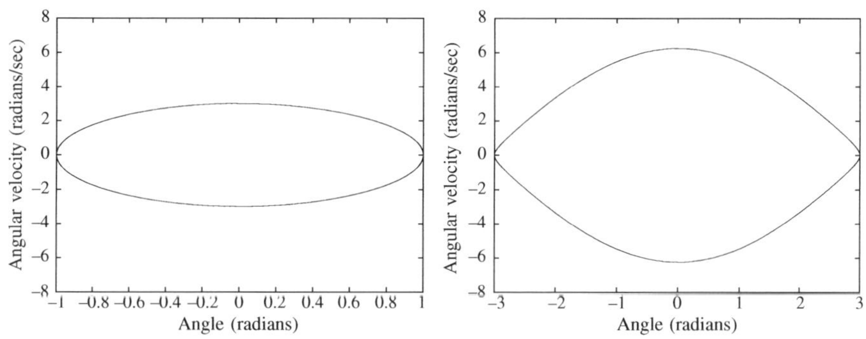

(a) Initial angle = 1 radian **(b)** Initial angle = 3 radian

Figure 12.19 Pendulum in state space with large initial angles and no damping.

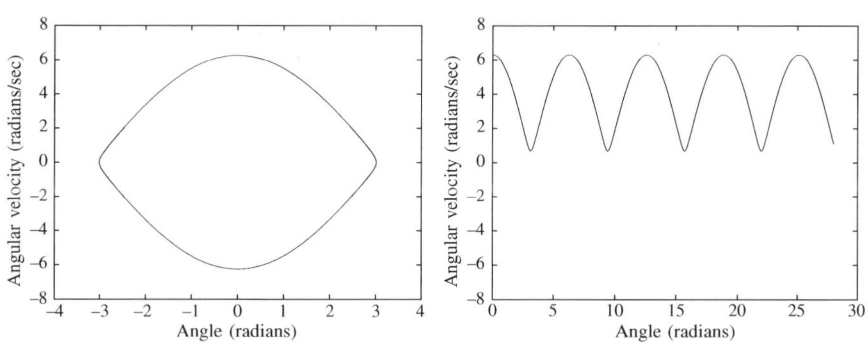

(a) Initial angular velocity = 6.25 radian/sec **(b)** Initial angular velocity = 6.3 radian/sec

Figure 12.20 Pendulum in state space with large initial angular velocity and no damping.

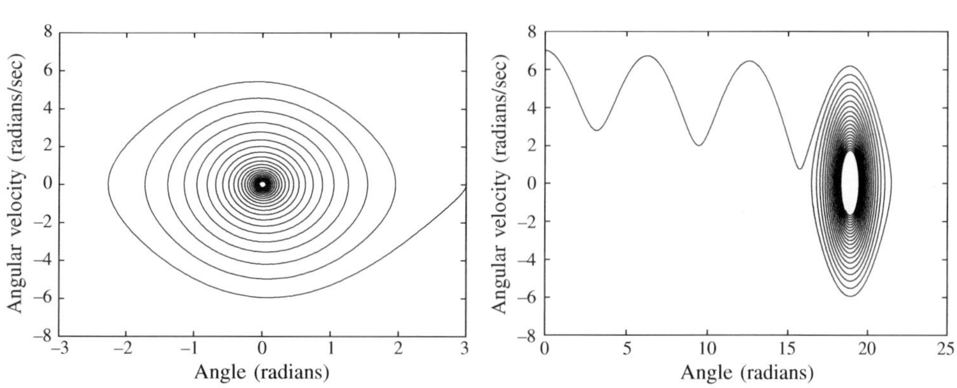

(a) Initial $\theta = 3$ radian, $\zeta = 0.025$ **(b)** Initial $\dot{\theta} = 7$ radians/sec, $\zeta = 0.01$

Figure 12.21 Pendulum in state space with damping.

12.6 Application: Forced Pendulum

Let us now add an additional mechanical feature to our pendulum laboratory. Instead of a stationary rotation pivot located at the origin of the (x, y) coordinate system, we arrange for the pivot to oscillate vertically up and down in a reciprocating motion, at a fixed frequency independent of the oscillation of the pendulum. The pivot is assumed now to be located at coordinates $(0, y_0(t))$, where $y_0(t)$ is some function of time that describes the pivot reciprocation, as shown in Figure 12.22. We may add some additional terms to the Lagrangian derivation to arrive at the equation of motion for this scheme

The square of the velocity $|\vec{v}|^2$ is now

$$\begin{aligned} |\vec{v}|^2 &= \dot{x}^2 + \dot{y}^2 \\ &= l^2\dot{\theta}^2 + \dot{y}_0^2 + 2\dot{y}_0 l\dot{\theta} \sin\theta \end{aligned} \tag{12.33}$$

So the kinetic energy term is

$$T = \frac{m}{2}\left[l^2\dot{\theta}^2 + \dot{y}_0^2 + 2\dot{y}_0 l\dot{\theta} \sin\theta\right] \tag{12.34}$$

The potential energy term of the Lagrangian is

$$\begin{aligned} V &= mgy \\ &= mg\left[\, y_0 - l\cos\theta\right] \end{aligned} \tag{12.35}$$

This makes the Lagrangian

$$L = \frac{m}{2}\left[l^2\dot{\theta}^2 + \dot{y}_0^2 + 2\dot{y}_0 l\dot{\theta} \sin\theta\right] - mg\left[\, y_0 - l\cos\theta\right] \tag{12.36}$$

We find the equation of motion by applying the Euler–Lagrange equation for the single degree of freedom θ

$$\frac{d}{dt}\left[\frac{\partial L}{\partial \dot{\theta}}\right] = \frac{\partial L}{\partial \theta} \tag{12.37}$$

$$\ddot{\theta} = \frac{\sin\theta}{l}\left[-g - \ddot{y}_0\right] \tag{12.38}$$

Comparing Equation (12.38) with Equation (12.28), we may interpret the equation of motion for this forced pendulum as the local gravitational acceleration being modified by the acceleration of the reciprocation motion. Let us consider the case where the reciprocation of the pivot is arranged as a piston driven by a variable-speed rotary motor, such as with a mechanism called the "Scotch yoke," and approximated by the function $y_0(t) = A_f \sin(\omega_f t)$, where A_f and ω_f are the amplitude and angular frequency of the sinusoidal forcing, respectively. This makes the equation of motion

$$\ddot{\theta} = \frac{\sin\theta}{l}\left[-g + \omega_f^2 A_f \sin(\omega_f t)\right] \tag{12.39}$$

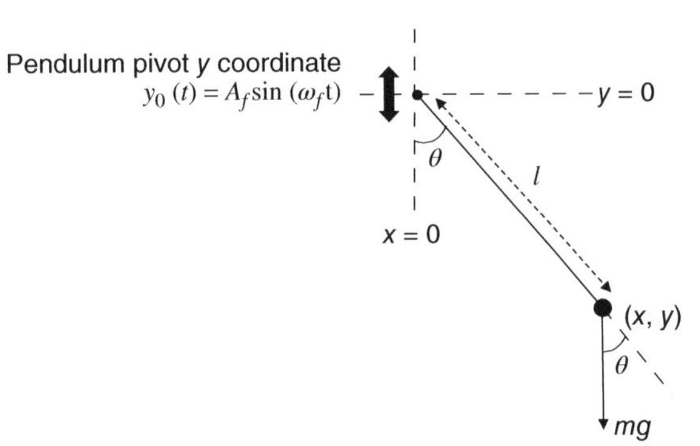

Figure 12.22 Single forced pendulum.

The system of first-order differential equations we must solve numerically is now

$$\frac{dy_0}{dt} = y_1$$
$$\frac{dy_1}{dt} = \frac{\sin\theta}{l}\left[-g + \omega_f^2 A_f \sin(\omega_f t)\right] \tag{12.40}$$

This system more fully reflects the general form of Equation (12.18), with a right-hand side function that is dependent on both the independent variable, in this case, time, as well as the current values of the state variables. To solve this system, see the source code file `forcedpendulum.c` from the Student Companion website, where the modified object function passed to the `rk4()` solver tool is now

```
typedef struct pend_param {
  double length;  // pendulum length in meters
  double Af;  // Forcing amplitude in meters
  double wf;  // Forcing angular frequency in 1/sec
} PEND;

// State variable 0 is the displacement angle
// State variable 1 is the angular velocity
void pendulum_forced(double x,ARRAY1D *y,void *param_ptr,
                     ARRAY1D *yp) {
  PEND *p;

  p = param_ptr;
  yp->elem[0]  = y->elem[1];
  yp->elem[1]  = (p->wf * p->wf * p->Af *
               sin(p->wf * x) - GRAVITATIONAL_ACCEL) *
               sin(y->elem[0]) / p->length;
}
```

A dialog to compile and run this new program to simulate the complex dynamics plotted in Figure 12.23a is

```
$ gcc -Wall -Wshadow -O2 -o forcedpendulum forcedpendulum.c CPLibrary/diffeq.c CPLibrary/array.c
    ↪ -lm
$ ./forcedpendulum
./forcedpendulum <length> <Aforced> <fforced> <init y0> <init y1> <tfinal> <tinc>
$./forcedpendulum 1 0.15 1.5 0.1 0 100 0.01 >| forcedpendulum.dat'
```

The amplitude of the reciprocating motion is set to be 0.15 m in these simulations, a small fraction of the 1-m length of the pendulum arm. Figure 12.23b shows the same sort of simulation with the forcing frequency changed to 3 Hz. Unlike the periodic behavior shown in Figure 12.19, the dynamic behavior for both sets of parameters in Figure 12.23 is now classified as *quasiperiodic*. There is more than one period of oscillation apparent, circulating around the phase diagram at a faster rate, and then modulating the shape of the trajectory at a slower rate. These rates depend on the complex nonlinear interaction between the natural oscillation frequency of the pendulum and the forcing frequency.

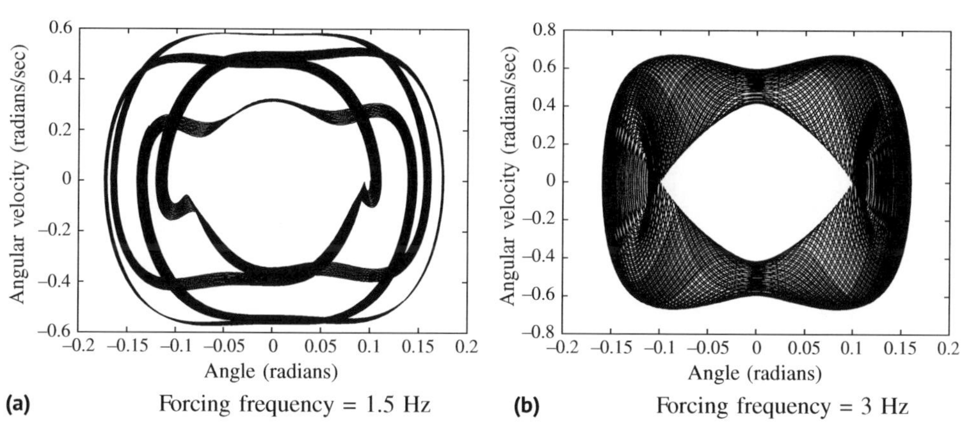

Figure 12.23 Forced pendulum in state space with small initial angle = 0.1 rad, forcing amplitude = 0.15 m.

12.7 Application: Chaotic Dynamics

With another change of parameters of the forced pendulum, we are able to simulate a solution to Equation (12.39) that represents another class of dynamic behavior. We are able to get a glimpse here of the rich field of chaotic behavior with this simple system. Consider initializing the forced pendulum with an initial angle somewhat above horizontal, with an initial $\theta = 2$, by running a simulation with

```
$./forcedpendulum 1 0.15 2 2 0 100 0.01 >| forcedpendulum.dat'
```

This initializes more swinging energy in the pendulum to interact with the forcing motion. We see the complex result in Figure 12.24a. With the increased interaction through the nonlinear dependencies of Equation (12.39), we can see from this state-space plot alternating episodes of swinging and rotational motion. We can test if this regime represents chaotic motion by running another simulation with some tiny modifications in the initial conditions with

```
$./forcedpendulum 1 0.15 2 2.000001 0 100 0.01 >| forcedpendulum.dat'
```

Figure 12.24b shows the motion resulting from an initial $\theta = 2.000001$, which differs from the first simulation by less than 1 ppm. We see in this second plot an evolution of completely different dynamics. In fact, the pendulum is being sent rotating in the opposite direction to that in Figure 12.24a. This hypersensitivity to the initial conditions is a hallmark of chaotic dynamics.

12.8 Application: Inverted Pendulum

Let us now consider an interesting special case of the forced pendulum, when the initial deflection angle is extended to being close to the value of π. This means that the pendulum arm is near its vertical position, as illustrated in Figure 12.25. This configuration is referred to as an inverted pendulum. The behavior in this case reveals a surprising property of the nonlinear dynamics of the forced pendulum.

Consider initializing this pendulum with an initial angle of 3 rad and no initial velocity. With no forcing, the behavior of this system would follow the simulation shown in Figure 12.13d. The pendulum arm would swing down and start an oscillation covering almost the entire angular range. Let's now see the result of adding a forcing motion, initially with a forcing frequency increased to 5 Hz, by running the commands

```
$ ./forcedpendulum 1 0.15 5 3 0 100 0.01 >| forcedpendulum.dat
```

the result is pictured in Figure 12.26a. This plot shows the arm of the pendulum in a rotational mode and the deflection angle steadily climbing with time, due to the forcing adding energy into the pendulum system. However, when the forcing frequency is increased to 6 Hz with

```
$ ./forcedpendulum 1 0.15 6 3 0 100 0.01 >| forcedpendulum.dat
```

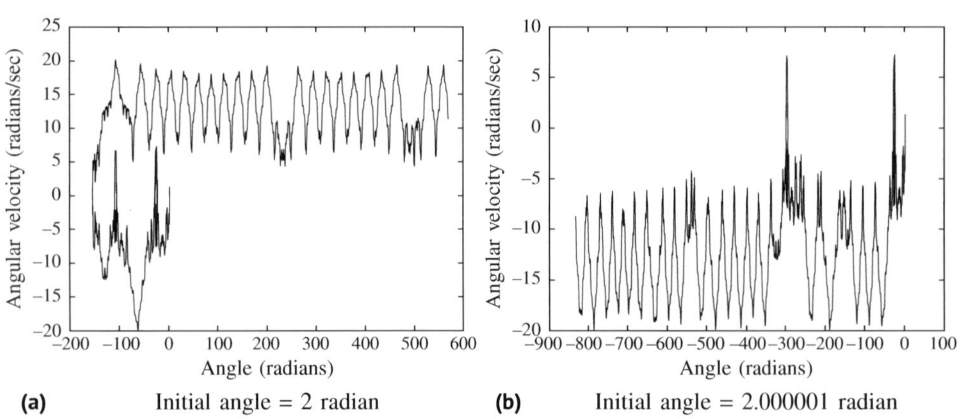

Figure 12.24 Chaotic forced pendulum in state space with large initial angle, forcing frequency = 2 Hz, and forcing amplitude = 0.15 m.

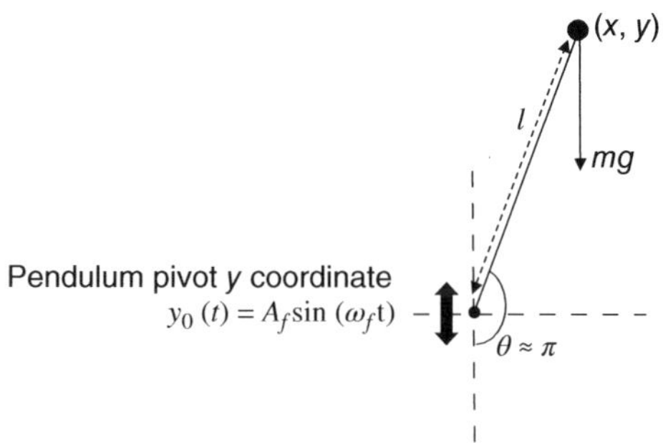

Figure 12.25 Inverted pendulum.

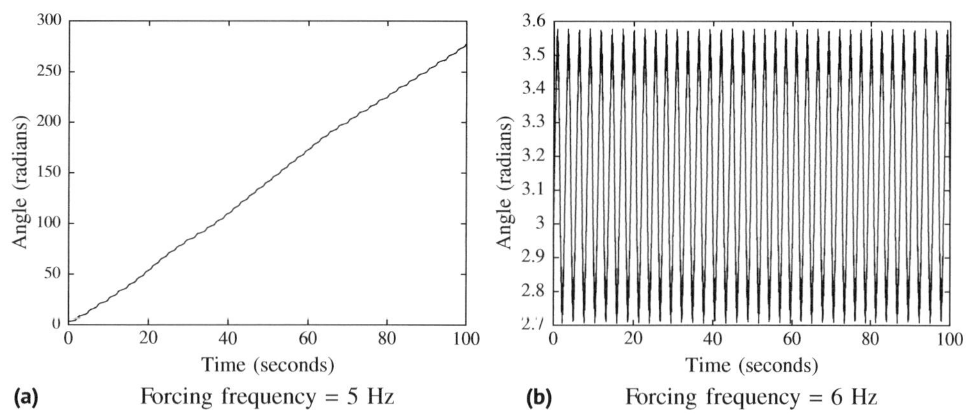

Figure 12.26 Forced pendulum with large initial angle = 3 rad, forcing amplitude = 0.15 m.

we can see in Figure 12.26b that we are into a curious regime where the pendulum remains in a nearly vertical orientation, with θ oscillating in the vicinity of the value of π. At this forcing frequency, the nonlinear interaction of the forcing motion is able to bring the inverted pendulum into a condition of being nearly stable and balanced in a vertical position, despite this angle being a position of dynamic instability. Many entertaining acts of acrobatics are based on this nonlinear phenomenon.

12.9 Application: Double Pendulum

The double pendulum is a physical system that adds a second degree of freedom to the conventional single pendulum, and thereby provides rich opportunities to study different regimes of operation of a nonlinear dynamic system numerically. Unlike the single pendulum, we can observe interesting nonlinear interactions between the system states solely with the natural motion of the pendulum, and not requiring any external time-dependent forcing. Figure 12.27 shows a simple physical diagram. Again we will assume the mass of the system concentrated into two ideal point masses m_1 and m_2, but the treatment can be modified for more physical reality with centers of mass distribution. The outer mass m_2 is attached to the inner mass m_1 with a second massless rigid arm, but one that may rotate freely with the location of mass m_1 as its pivot. The two arms make independent angles θ_1 and θ_2 with respect to vertical, as shown.

For a system of this complexity, the Lagrangian formulation provides the most straightforward framework for deriving the equations of motion of the two masses, even if somewhat messy algebraically. With the origin of our coordinate system at the pivot of the inner arm, the coordinates of the masses are

$$x_1 = l_1 \sin\theta_1 \qquad y_1 = -l_1 \cos\theta_1$$

$$x_2 = l_1 \sin\theta_1 + l_2 \sin\theta_2 \qquad y_2 = -l_1 \cos\theta_1 - l_2 \cos\theta_2$$

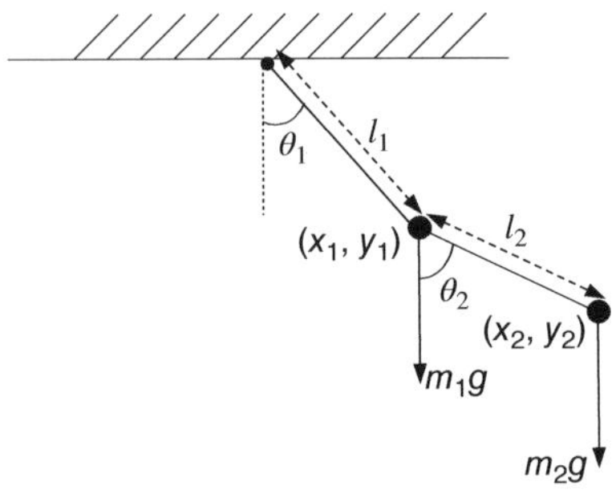

Figure 12.27 Double pendulum.

We form the Lagrangian as $L = T - V$, where the kinetic energy term T is

$$T = \frac{m_1|\vec{v}_1|^2}{2} + \frac{m_2|\vec{v}_2|^2}{2} \tag{12.41}$$

The velocity magnitudes are

$$\begin{aligned} |\vec{v}_1|^2 &= \left(\frac{dx_1}{dt}\right)^2 + \left(\frac{dy_1}{dt}\right)^2 = l_1^2\dot{\theta}_1^2 \\ |\vec{v}_2|^2 &= \left(\frac{dx_2}{dt}\right)^2 + \left(\frac{dy_2}{dt}\right)^2 = l_1^2\dot{\theta}_1^2 + l_2^2\dot{\theta}_2^2 + 2l_1l_2\dot{\theta}_1\dot{\theta}_2\cos(\theta_1 - \theta_2) \end{aligned} \tag{12.42}$$

Therefore the kinetic energy term T in the Lagrangian is

$$T = \frac{m_1l_1^2}{2}\dot{\theta}_1^2 + \frac{m_2l_1^2}{2}\dot{\theta}_1^2 + \frac{m_2l_2^2}{2}\dot{\theta}_2^2 + m_2l_1l_2\dot{\theta}_1\dot{\theta}_2\cos(\theta_1 - \theta_2) \tag{12.43}$$

The potential energy term V in the Lagrangian is

$$\begin{aligned} V &= m_1gy_1 + m_2gy_2 \\ &= -m_1gl_1\cos\theta_1 - m_2gl_1\cos\theta_1 - m_2gl_2\cos\theta_2 \end{aligned} \tag{12.44}$$

So the total Lagrangian is

$$\begin{aligned} L = &\frac{m_1l_1^2}{2}\dot{\theta}_1^2 + \frac{m_2l_1^2}{2}\dot{\theta}_1^2 + \frac{m_2l_2^2}{2}\dot{\theta}_2^2 + m_2l_1l_2\dot{\theta}_1\dot{\theta}_2\cos(\theta_1 - \theta_2) \\ &+ m_1gl_1\cos\theta_1 + m_2gl_1\cos\theta_1 + m_2gl_2\cos\theta_2 \end{aligned} \tag{12.45}$$

We apply the Euler–Lagrange equation twice, with respect to each of the two degrees of freedom θ_1 and θ_2

$$\begin{aligned} \frac{d}{dt}\left[\frac{\partial L}{\partial \dot{\theta}_1}\right] &= \frac{\partial L}{\partial \theta_1} \\ \frac{d}{dt}\left[\frac{\partial L}{\partial \dot{\theta}_2}\right] &= \frac{\partial L}{\partial \theta_2} \end{aligned} \tag{12.46}$$

which reduces to a system of two second-order differential equations

$$\begin{aligned} &(m_1 + m_2)l_1^2\ddot{\theta}_1 + m_2l_1l_2\ddot{\theta}_2\cos(\theta_1 - \theta_2) + m_2l_1l_2\dot{\theta}_2^2\sin(\theta_1 - \theta_2) + (m_1 + m_2)gl_1\sin(\theta_1) = 0 \\ &m_2l_2^2\ddot{\theta}_2 + m_2l_1l_2\ddot{\theta}_1\cos(\theta_1 - \theta_2) - m_2l_1l_2\dot{\theta}_1^2\sin(\theta_1 - \theta_2) + m_2gl_2\sin(\theta_2) = 0 \end{aligned} \tag{12.47}$$

We also define the convenient constant

$$\alpha = \frac{m_2}{(m_1 + m_2)} \tag{12.48}$$

as the fraction of the total mass in the outer mass. That is, $\alpha = \frac{1}{2}$ corresponds to a pendulum with equal masses, and α values greater than or less than $\frac{1}{2}$ describe systems with heavier or lighter outer masses than the inner mass, respectively.

Unfortunately, even as elaborate as it is, Equation (12.47) does not conform to the standard form for systems of first-order differential equations to be solved by the Runge–Kutta algorithm. To apply our system of substitutions in Equation (12.20) in order to define a system of first-order equations compatible with our numerical algorithms, there must be only one second derivative of each angle variable appearing in each equation of motion. However, we may render this system in the proper form by eliminating one second derivative between each of Equations (12.47). We may solve the first equation for $\ddot{\theta}_1$ and substitute it into the second equation, and then solve the second equation for $\ddot{\theta}_2$ and substitute it back into the first equation. Finally, the system that is in a form suitable for numerical solution then emerges as

$$\begin{aligned}\ddot{\theta}_1 &= \frac{\alpha l_1 \dot{\theta}_1^2 \sin(\theta_1 - \theta_2)\cos(\theta_1 - \theta_2) + \alpha g \sin(\theta_2)\cos(\theta_1 - \theta_2) - \alpha l_2 \dot{\theta}_2^2 \sin(\theta_1 - \theta_2) - g\sin(\theta_1)}{l_1(1 - \alpha\cos^2(\theta_1 - \theta_2))} \\ \ddot{\theta}_2 &= \frac{\alpha l_2 \dot{\theta}_2^2 \sin(\theta_1 - \theta_2)\cos(\theta_1 - \theta_2) + g\sin(\theta_1)\cos(\theta_1 - \theta_2) + l_1\dot{\theta}_1^2 \sin(\theta_1 - \theta_2) - g\sin(\theta_2)}{l_2(1 - \alpha\cos^2(\theta_1 - \theta_2))}\end{aligned} \tag{12.49}$$

As we have seen, we may transform these two higher-order equations into four first-order equations with the definition of four state variables

$$\begin{aligned} y_0 &= \theta_1 \\ y_1 &= \theta_2 \\ y_2 &= \frac{dy_0}{dt} = \dot{\theta}_1 \\ y_3 &= \frac{dy_1}{dt} = \dot{\theta}_2 \end{aligned} \tag{12.50}$$

The system that may be coded in C and submitted to the Runge–Kutta algorithm becomes

$$\begin{aligned} \dot{y}_0 &= y_2 \\ \dot{y}_1 &= y_3 \\ \dot{y}_2 &= \frac{\alpha l_1 y_2^2 \sin(y_0 - y_1)\cos(y_0 - y_1) + \alpha g\sin(y_1)\cos(y_0 - y_1) - \alpha l_2 y_3^2 \sin(y_0 - y_1) - g\sin(y_0)}{l_1(1 - \alpha\cos^2(y_0 - y_1))} \\ \dot{y}_3 &= \frac{\alpha l_2 y_3^2 \sin(y_0 - y_1)\cos(y_0 - y_1) + g\sin(y_0)\cos(y_0 - y_1) + l_1 y_2^2 \sin(y_0 - y_1) - g\sin(y_1)}{l_2(1 - \alpha\cos^2(y_0 - y_1))} \end{aligned} \tag{12.51}$$

As we turn our attention to planning out a series of simulations that will lend understanding to this physical system, following the usual treatment of the single pendulum, we first examine this derivation with the assumption of small deflection angles, but with the expansion into two degrees of freedom. Under this linearizing approximation, we will see oscillations that may be resolved into two normal modes, very similar to the treatment of the coupled spring-mass oscillator in Section 6.5 [27, Chap. 11; 28, Chap. 4; 29, Chap. 11]. If we apply the approximations that $\theta_1 \ll 1$, $\theta_2 \ll 1$, $\dot{\theta}_1 \ll 1$, $\dot{\theta}_2 \ll 1$ to Equation (12.49), we may discard any quadratic term in the angles, and approximate $\cos(\theta_1 - \theta_2) \approx 1$, $\sin(\theta_1) \approx \theta_1$, and $\sin(\theta_2) \approx \theta_2$. With these small angle limits, the system of equations of motion become

$$\begin{aligned} \ddot{\theta}_1 &\approx -\frac{g}{l_1}\left[\frac{1}{1-\alpha}\theta_1 - \frac{\alpha}{1-\alpha}\theta_2\right] \\ \ddot{\theta}_2 &\approx -\frac{g}{l_2}\left[\frac{1}{1-\alpha}\right](\theta_2 - \theta_1) \end{aligned} \tag{12.52}$$

Following the treatment of coupled oscillators, we now assume that solutions for the two deflection angles may be written in the form

$$\begin{aligned} \theta_1 &= a_1 e^{i\omega t} \\ \theta_2 &= a_2 e^{i\omega t} \end{aligned} \tag{12.53}$$

with ω being the common physical frequency with which both angles oscillate, and a_1 and a_2 are two allowed oscillation amplitudes. Then substituting back into the linearized equations of motion (12.52) gives the conditions that the parameters must meet for all values of time t as

$$\begin{aligned} -\omega^2 a_1 e^{i\omega t} &= -\frac{g}{l_1}\frac{1}{1-\alpha}a_1 e^{i\omega t} + \frac{g}{l_1}\frac{\alpha}{1-\alpha}a_2 e^{i\omega t} \\ -\omega^2 a_2 e^{i\omega t} &= \frac{g}{l_2}\frac{1}{1-\alpha}a_1 e^{i\omega t} - \frac{g}{l_2}\frac{1}{1-\alpha}a_2 e^{i\omega t} \end{aligned} \tag{12.54}$$

Define another convenient constant

$$\beta = \frac{l_2}{(l_1+l_2)} \tag{12.55}$$

as the ratio of the total pendulum length that the outer arm length l_2 occupies, and with the definition

$$\lambda = \frac{\omega^2(l_1+l_2)}{g} \tag{12.56}$$

we find that the parameter conditions for these linearized sinusoidal solutions reduce to another eigenvalue problem

$$\begin{bmatrix} \frac{1}{1-\beta}\frac{1}{1-\alpha} & -\frac{1}{1-\beta}\frac{\alpha}{1-\alpha} \\ -\frac{1}{\beta}\frac{1}{1-\alpha} & \frac{1}{\beta}\frac{1}{1-\alpha} \end{bmatrix}\begin{bmatrix} a_1 \\ a_2 \end{bmatrix} = \lambda \begin{bmatrix} a_1 \\ a_2 \end{bmatrix} \tag{12.57}$$

This eigenvalue problem may be solved by constructing the quadratic characteristic polynomial and solving with the quadratic equation. The allowed eigenvalues of the parameter λ are then

$$\lambda = \frac{1 \pm \sqrt{1-4\beta(1-\beta)(1-\alpha)}}{2\beta(1-\beta)(1-\alpha)} \tag{12.58}$$

These values of λ correspond to two allowed values of the physical frequency ω by Equation (12.56). By substituting the allowed λ values into the top row of Equation (12.57), we can find the ratio of amplitudes of the two angular oscillations corresponding to each allowed frequency as

$$\frac{a_2}{a_1} = \frac{1-(1-\beta)(1-\alpha)\lambda}{\alpha} \tag{12.59}$$

This ratio gives a requirement of the initial conditions for the two pendulum angles that will excite the two normal modes of oscillation, again under the approximation that the angular displacements remain small. For a nominal case of equal masses and equal arm length segments, the constants $\alpha = \frac{1}{2}$ and $\beta = \frac{1}{2}$, and the eigenvalues evaluate to $\lambda = 4 \pm 2\sqrt{2}$. The corresponding physical frequencies are $\omega = \frac{\sqrt{(4\pm2\sqrt{2})g}}{l_1+l_2}$, and the amplitude ratios are $\frac{a_2}{a_1} = \pm\sqrt{2}$.

The plots of Figures 12.28 and 12.29 illustrate simulations of these two approximate normal modes for an example total length of $l_1 + l_2 = 1$ m and our terrestrial gravitational acceleration. Figure 12.28 shows the mode excited with $a_1 = 0.1$ rad and $a_2 = -0.1414$ rad, and Figure 12.29 shows the mode excited with $a_1 = 0.1$ rad and $a_2 = 0.1414$ rad. These simulations solve the full nonlinear system of Equation (12.51) with the fourth-order Runge–Kutta solver tool, but with the initial conditions above that match the normal modes we have seen emerge from the small-angle linearized analysis. A step size of 0.01 s is sufficiently fine for simulating these smoothly oscillating modes.

Then picking some other arbitrary small-angle initial conditions not conforming to these two normal modes, for example, $a_1 = 0.1$ rad and $a_2 = -0.05$ rad, we may excite a mixed mode oscillation, as shown in Figure 12.30.

Next, with an interest in exploring larger angles for initial conditions, the linearizing approximations that led to Equation (12.52) will, in general, no longer apply. With initial angles only a fraction of their total range of $\pm\pi$, we can observe quasiperiodic behavior arise from the nonlinearities in the double pendulum system. As in the case of the forced single pendulum, a state-space representation is useful here, but a full visualization with the double pendulum would be challenging indeed with four state variables. Figure 12.31 plots the motion with only the inner and outer angles θ_1 and θ_2 as a subset of the state variables for two example initial conditions. The step size has been reduced to 0.001 s for these more complex and erratic oscillation modes.

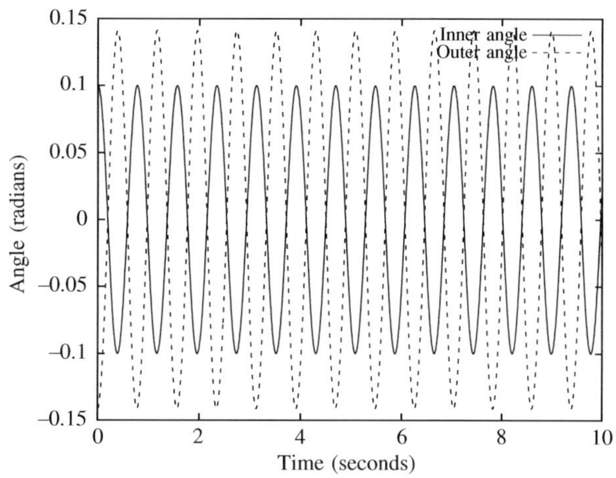

Figure 12.28 Double pendulum in small-angle complementary oscillation, initial angles θ_1 = 0.1 rad, θ_2 = −0.1414 rad.

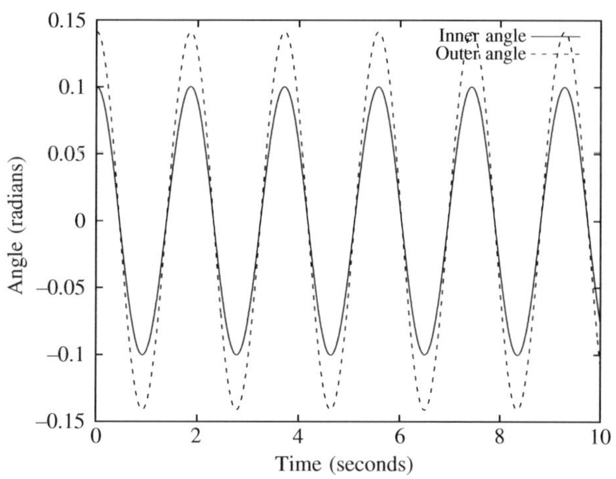

Figure 12.29 Double pendulum in small-angle similar oscillation, initial angles θ_1 = 0.1 rad, θ_2 = 0.1414 rad.

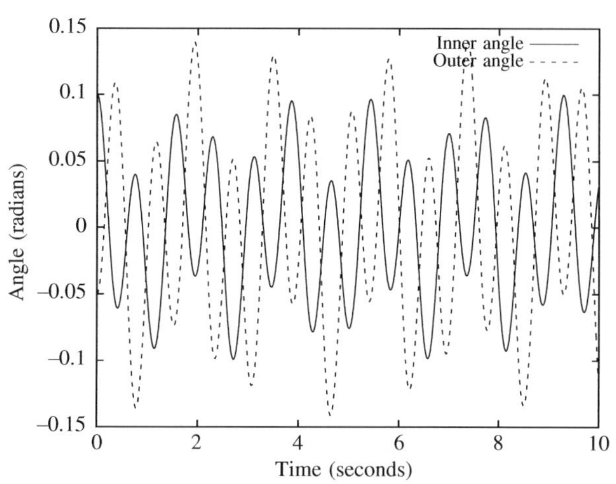

Figure 12.30 Double pendulum in small-angle mixed mode oscillation, initial angles θ_1 = 0.1 rad, θ_2 = −0.05 rad.

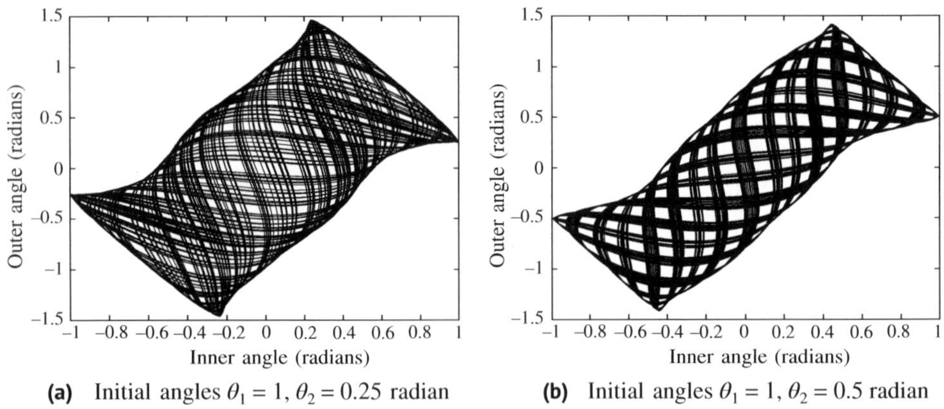

Figure 12.31 Two examples of the double pendulum with large-angle initial conditions in quasiperiodic motion.

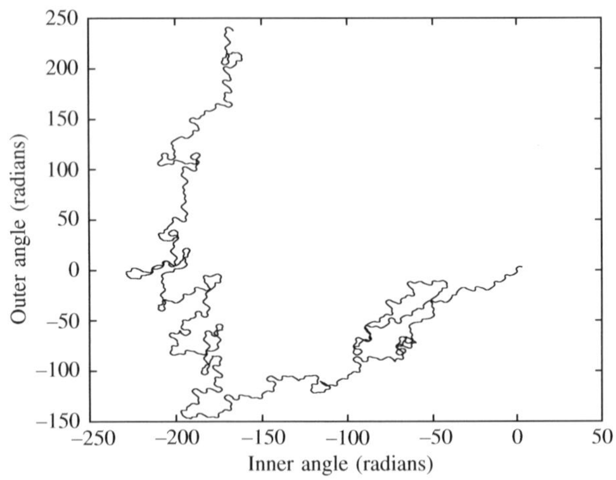

Figure 12.32 Double pendulum in chaotic oscillation, initial angles, $\theta_1 = 3$ rad, $\theta_2 = 3$ rad.

As in the case of the forced single pendulum, by applying initial conditions of the double pendulum with a higher energy configuration, we may enter a widespread regime of chaotic behavior. An example case is shown in Figure 12.32, simulated again with a step size of 0.001 s, with both pendulum arms extended nearly vertical with θ_1 and θ_2 both initially set to 3 rad. As the masses fall down from their starting positions, chaotic oscillations ensue with no external forcing.

We may confirm that we are into a chaotic regime by making some tiny alteration of the initial angles of the pendulum arms. For example, in Figure 12.33 we see a totally different evolution of the arm angles from the initial conditions $\theta_1 = 3$ rad and $\theta_2 = 3.000001$ rad.

In order to be confident of the accuracy of the results of these simulations with these step sizes, let us again invoke the principle of the conservation of energy for the double pendulum, as in Section 12.5 for the single pendulum. The ratio of the total energy of the double pendulum to its total mass may be found by summing Equations (12.43) and (12.44), giving

$$\begin{aligned}\frac{E}{m_1 + m_2} &= \frac{l_1^2}{2}\dot{\theta}_1^2 + \frac{\alpha l_2^2}{2}\dot{\theta}_2^2 + \alpha l_1 l_2 \dot{\theta}_1 \dot{\theta}_2 \cos(\theta_1 - \theta_2) \\ &\quad - g l_1 \cos\theta_1 - \alpha g l_2 \cos\theta_2\end{aligned} \tag{12.60}$$

Plotting this measurement of the total energy in the chaotic mode in Figure 12.34, which is expected to appear ideally as a horizontal constant, shows a small level of truncation error being accumulated through this simulation. This is attributable to the very erratic motion with high frequency components in this dynamic mode. However, the error in the energy plot is still less than 1 ppm.

The student will find exercises at the end of the chapter to explore the rich dynamics of the double pendulum, including the coding exercise for two main programs, one for the linearized approximation and another for the full nonlinear analysis. This latter coding problem is most straightforward by extending the downloadable example program for the single

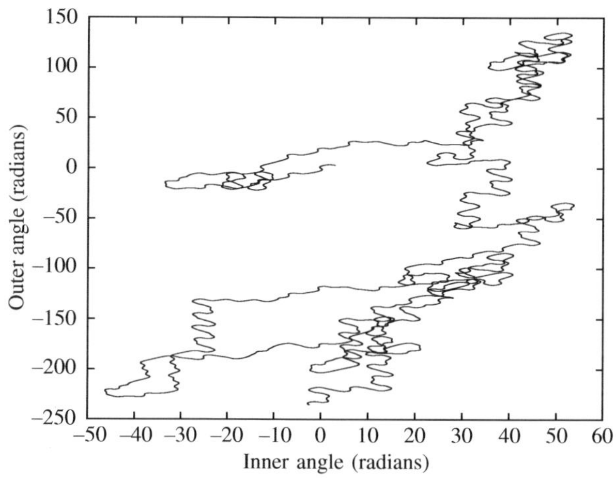

Figure 12.33 Double pendulum in chaotic oscillation, initial angles θ_1 = 3 rad, θ_2 = 3.000001 rad.

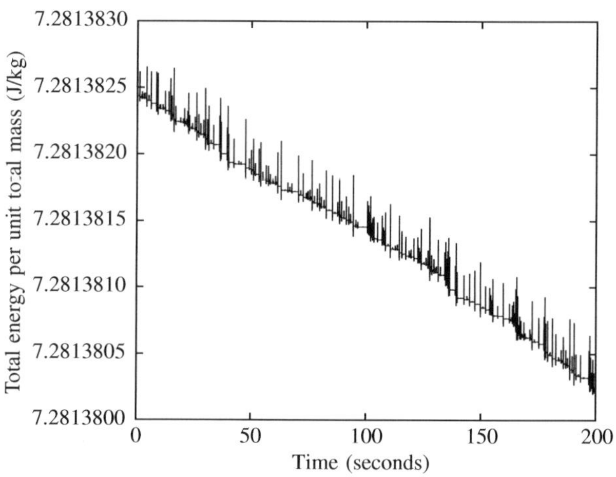

Figure 12.34 Total energy versus time of double pendulum in chaotic oscillation, initial angles θ_1 = 3 rad, θ_2 = 3 rad showing accumulating truncation error.

pendulum. In particular, applying the animation techniques of Sections 4.13 and 6.9 will provide a much fuller appreciation for the regime of chaotic behavior than can the static plots of Figures 12.32 and 12.33.

12.10 Application: The Electronic Oscillator

In Sections 6.7 and 7.4, we explored Kirchoff's current law (KCL) and solved for the electrical potentials in a network with electric currents that were constant with respect to time. In this section, we introduce electrical components with current and voltage characteristics that are time dependent. The capacitor in Figure 12.35a is an electrical element where energy is stored in an internal electric field. The circuit behavior that relates the current i_C through it and the electric potential drop v_C across it is governed by the differential relationship

$$i_C(t) = C\frac{dv_C}{dt} \tag{12.61}$$

where the constant C is the capacitance of the element. The inductor in Figure 12.35b is an electrical element where energy is stored in an internal magnetic field. The circuit behavior that relates the current i_L through it and the potential drop v_L across it is

$$v_L(t) = L\frac{di_L}{dt} \tag{12.62}$$

where the constant L is the self-inductance of the element.

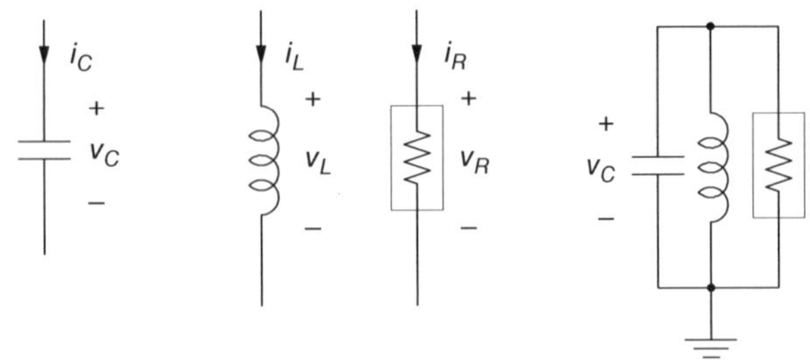

(a) Capacitor element **(b)** Inductor element **(c)** Nonlinear resistor element **(d)** Oscillator circuit

Figure 12.35 Circuit elements and network for an inductor-capacitor-resistor oscillator.

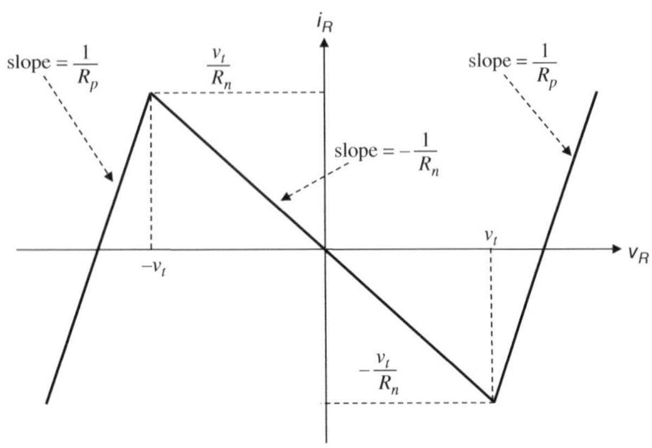

Figure 12.36 Current–voltage characteristic of the nonlinear resistor in an oscillator network.

In Section 6.7, we analyzed a circuit network containing only linear resistors. For those circuit elements, the current and voltage were governed by the linear time-independent relationship

$$i_R = \frac{1}{R} v_R \tag{12.63}$$

with resistance R. These components have no capability of storing energy, but will dissipate electrical energy into heat. In this section, we introduce a conceptual nonlinear resistance, which can be formulated as a representative model for the behavior of a more complex actual circuit network interconnected with two terminals. We assume the connected circuit contains nonlinear elements but no energy storage elements. In Figure 12.35c, we assume that the nonlinear resistor is modeled by a characteristic plotted in Figure 12.36. A potential difference v_R across this element of either transition voltage $\pm v_t$ forms the boundary between an incremental negative resistance at the center of the characteristic and an incremental positive resistance.at either side. Algebraically, we can express this model with the time-independent relationships

$$i_R = \begin{cases} \dfrac{v_t}{R_n} + \dfrac{v_R + v_t}{R_p} & v_R < -v_t \\ -\dfrac{v_R}{R_n} & -v_t < v_R < v_t \\ -\dfrac{v_t}{R_n} + \dfrac{v_R - v_t}{R_p} & v_t < v_R \end{cases} \tag{12.64}$$

where the constants R_n and R_p determine the slopes of the characteristic in the regimes with negative and positive incremental resistance, respectively.

The oscillator shown in Figure 12.35d may be analyzed by applying KCL to the single network node at the top of the diagram. We will see that the symmetry of Equations (12.61) and (12.62) gives rise to a second-order differential equation describing the circuit behavior. We start by asserting that if by KCL

$$i_C + i_L + i_R = 0 \tag{12.65}$$

then so must

$$\frac{di_C}{dt} + \frac{di_L}{dt} + \frac{di_R}{dt} = 0 \tag{12.66}$$

Observing that $v_L = v_R = v_C$ allows us to express the potential drop across all three of the circuit elements as the capacitor voltage drop v_C. Using the relationships for the elements in Equations (12.61) and (12.62), we rewrite Equation (12.66) as

$$C\ddot{v}_C + \frac{1}{L}v_C + \frac{di_R}{dv_C}\dot{v}_C = 0 \tag{12.67}$$

where we have now adopted the conventional "dot" notation for the time derivative and have applied the chain rule of differentiation for the third term. Note that we may obtain $\frac{di_R}{dv_C}$ easily from the characteristic in Equation (12.64), but that this derivative will be a piecewise and strongly nonlinear function of v_C that renders this equation very difficult to solve without numerical tools. Rearranging the second-order differential equation (12.66) into a form readily applicable to the numerical tools of this chapter,

$$\ddot{v}_C = -\frac{1}{LC}v_C - \frac{1}{C}\frac{di_R}{dv_C}\dot{v}_C \tag{12.68}$$

With the resistance values R_n and R_p designed to be sufficiently high, this second-order differential equation will exhibit an oscillatory solution. For example, choosing an inductance of 200 μH, a capacitance of 125 pF, and resistance slopes of $R_n = 20$ kΩ in the negative incremental resistance regime and $R_p = 5$ kΩ in the positive incremental resistance regime will give an oscillation frequency very close to 1 MHz. The solution of this example case with an initial capacitor voltage v_C of 10 mV and a transition voltage v_t of 1 V is shown in Figure 12.37. Note that the amplitude of the oscillations grows exponentially in the negative incremental resistance regime and then stabilizes when the peaks of the oscillations are high enough to range into the regime of positive incremental resistance. At this point, the oscillation amplitude is stabilized for future times and may be used for practical instrumentation or wireless communication applications.

An actual physical resistor component commonly used in electrical circuits governed by Equation (12.63) can have only a positive resistance R. These components are generally formed with some poor electrical conductor such as carbon and can only dissipate electrical energy into heat. From an energy point of view, the rate of energy dissipation is given by the product of the current i_R and potential drop v_R

$$P_R = i_R \cdot v_R \tag{12.69}$$

Since this expression is derived from the fundamental process of bringing a quantity of electrical charge through a potential change at a given rate, this relation also holds for the case of a current flowing through any electric network interconnected with two terminals where the potential may be raised by the application of electric energy. In that case, we may model the connected network with a conceptual negative resistance so that the power dissipation expression of Equation (12.69) is negative. This sign indicates a transfer of electric power into the network connected to the negative

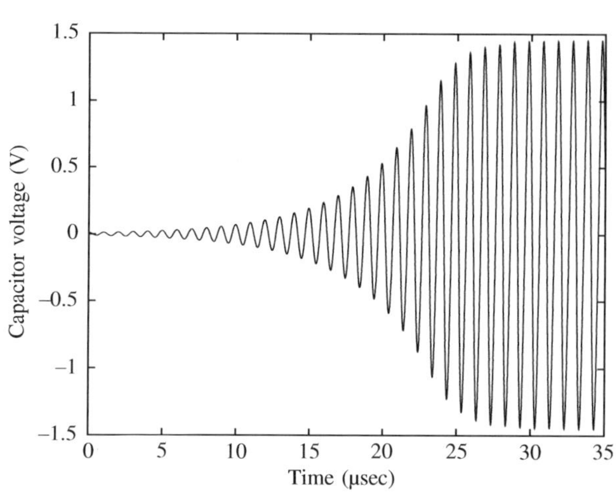

Figure 12.37 Simulation of electronic oscillator with nonlinear resistive model.

Figure 12.38 The historic 201-A vacuum tube from the early twentieth century.

resistance, which, in the case of Figure 12.35d, is the inductor–capacitor resonant oscillator network. The piecewise nonlinear circuit networks that are modeled by the nonlinear resistance of Figure 12.36 will provide regenerative energy when the oscillatory potential levels are of low amplitude, covering a regime where the incremental resistance over the oscillatory cycle is negative, but will incrementally dissipate electrical energy when the potentials are high. With this arrangement, the oscillation amplitude grows initially when power is first applied to the network and then stabilizes at some target amplitude determined by the transition voltage v_t. The instantaneous time-dependent power either dissipated or supplied by the nonlinear resistance model is the product of the resistance current and voltage drop, given by Equation (12.70).

$$P_R = i_R \cdot v_R = i_R \cdot v_C(t) \tag{12.70}$$

A plot of this power will show the power dissipation as initially negative, and then as the oscillation amplitude grows, positive power dissipation excursions appear. As the oscillation amplitude grows larger, the energy dissipation during each cycle grows until it balances the energy supplied during each cycle.

Early inventors such as Armstrong, Meissner, Colpitts, and Hartley developed different forms of the electronic network equivalent to a piecewise negative resistance. All these realizations require the inclusion of one or more "active" signal amplifying devices, which were vacuum tubes in the first half of the twentieth century, and then various forms of transistors following the invention of solid state electronics in 1948. The historic vacuum tube shown in Figure 12.38 was in common use when these circuits were first being developed. These inventions allowed the generation of essentially spectrally pure sinusoidal signals at radio frequencies in the electromagnetic spectrum and enabled the era of wireless communication that is an integral part of life in the present day. Other oscillators make use of resonators composed of piezoelectric materials such as crystalline quartz with electric energy storage properties that are equivalent to the inductor and capacitor of Figure 12.35d. These oscillators have very stable oscillation frequencies and are used to generate clock signals for computation and other digital applications.

The coding work for a time simulation showing the generation of a radio frequency sinusoid with stabilized amplitude and circuit power balance through each cycle is left as Exercise 12.8 for the student. The source code for the simulation of the pendulum in Section 12.5 may be adapted for this exercise.

12.11 Application: Synchronized Oscillators

With the development of a model for the electronic oscillator in Section 12.10, we are in a position to model another remarkable property of nonlinear dynamic systems. The phenomenon of lightly coupled oscillatory systems locking in synchronized phases has been observed for many years. A common attribution is Huygens' observations of a pair of pendulum clocks coupled through some mechanical means synchronizing their tempos. The mechanical spring and nonlinear

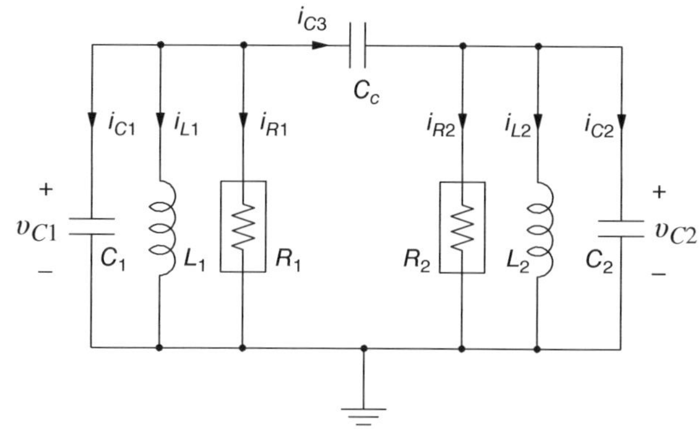

Figure 12.39 Two coupled inductor-capacitor-resistor oscillators that demonstrate synchronization.

escapement mechanism in a pendulum clock can be characterized by a similar relationship as that for the nonlinear resistor plotted in Figure 12.36. The tendency of a pair of electronic oscillators with some small mutual electrical coupling to synchronize their frequencies and phases is well known, and synchronizing a single oscillator to an electrically injected signal is used in many designs. Simulating this synchronizing phenomenon also gives us insight into other systems with nonlinear dynamics, such as tidal locking in orbiting gravitational systems.

Let us now assemble two oscillators of the type analyzed in Section 12.10 in the network pictured in Figure 12.39. In addition to two instances of the oscillator with the nonlinear resistance characterized by the plot in Figure 12.36, a third capacitance, C_c, has been added to couple a small amount of energy between the oscillators.

Extending the analysis that led to Equation (12.66), we take the time derivatives of the two equations reflecting KCL at the two nodes of the circuit. This leads to a system of equations that govern the potential drop across the resonant capacitors of the two oscillators as

$$\begin{aligned} \dot{i}_{C1} + \dot{i}_{L1} + \dot{i}_{R1} + \dot{i}_{C3} &= 0 \\ \dot{i}_{C2} + \dot{i}_{L2} + \dot{i}_{R2} - \dot{i}_{C3} &= 0 \end{aligned} \tag{12.71}$$

We note that the potential drop across the coupling capacitance C_c is $v_{C1} - v_{C2}$, the difference in the potentials between the two oscillator capacitances. This allows us to write the system of equations as

$$\begin{aligned} (C_1 + C_c)\ddot{v}_{C1} - C_c\ddot{v}_{C2} + \frac{1}{L_1}v_{C1} + \frac{di_{R1}}{dv_{C1}}\dot{v}_{C1} &= 0 \\ (C_2 + C_c)\ddot{v}_{C2} - C_c\ddot{v}_{C1} + \frac{1}{L_2}v_{C2} + \frac{di_{R2}}{dv_{C2}}\dot{v}_{C2} &= 0 \end{aligned} \tag{12.72}$$

Note that if the coupling capacitance C_c is very small relative to the resonant capacitances C_1 and C_2, the physical system governed by Equation (12.72) reduces to two independent oscillators, each governed by Equation (12.67). Unfortunately, Equation (12.72) cannot be directly submitted to our Runge–Kutta solver tool, as two second derivatives of both state variables v_{C1} and v_{C2} appear in each equation. However, after some algebraic manipulation, for convenience we define

$$C_x^2 = (C_1 + C_c)(C_2 + C_c) - C_c^2 \tag{12.73}$$

and with this definition we can isolate the two second derivatives into separate equations as

$$\begin{aligned} \ddot{v}_{C1} &= -\frac{C_2 + C_c}{L_1 C_x^2}v_{C1} - \frac{C_c}{L_2 C_x^2}v_{C2} - \frac{C_2 + C_c}{C_x^2}\frac{di_{R1}}{dv_{C1}}\dot{v}_{C1} - \frac{C_c}{C_x^2}\frac{di_{R2}}{dv_{C2}}\dot{v}_{C2} \\ \ddot{v}_{C2} &= -\frac{C_c}{L_1 C_x^2}v_{C1} - \frac{C_1 + C_c}{L_2 C_x^2}v_{C2} - \frac{C_c}{C_x^2}\frac{di_{R1}}{dv_{C1}}\dot{v}_{C1} - \frac{C_1 + C_c}{C_x^2}\frac{di_{R2}}{dv_{C2}}\dot{v}_{C2} \end{aligned} \tag{12.74}$$

These two equations will lead to the definition of four state variables, for the two capacitor voltages and their derivatives, and four state equations that may be readily solved with the Runge–Kutta tool in our library. Again, in the limit of a very small coupling capacitance $C_c \to 0$, $C_x^2 \approx C_1 C_2$, and the two equations above approach two decoupled equations, each in the form of (12.68). We look now at the example case of the two oscillators being almost identical to the single example in Section 12.10. We'll assume that both capacitors, C_1 and C_2, have equal capacitances of 125 pF, but that the two inductances are slightly different. Instead of identical inductances of 200 μH, we assume $L_1 = 201$ μH and $L_2 = 199$ μH. The two oscillation frequencies will be quite close to 1 MHz, but will differ by 0.5 percent. First we look at the case of the coupling capacitance $C_c = 0$ and essentially two independent, uncoupled oscillators. Figure 12.40 shows two plots of the solution after the amplitudes have stabilized, each plot covering a time span of 10 μs, with one time segment early in the simulation and the other at a later time. We note that the relative phases between the oscillations of the two capacitor voltages v_{C1} and v_{C2} have slowly drifted between the two time segments by a portion of the oscillation period, reflecting the slightly different oscillation frequencies.

Finally the case with a coupling capacitor is simulated. We let $C_c = 10$ pF, a small fraction of the two capacitances used in the resonant circuits. Even with this coupling path where a small amount energy may be exchanged between the oscillators, the two time segments plotted in Figure 12.41 show that the two oscillators have synchronized, and their phases will remain locked together as time progresses. With different initial conditions it is also possible for the phases to lock with an opposite relationship.

The capability of two oscillators to lock into synchronized phases as demonstrated in this simulation is due to the nonlinear dynamics of the amplitude stabilization process driven by the nonlinear resistor characterized by Figure 12.36. The linear systems analyzed in Section 6.5, which will be further simulated directly with the differential equations of motion later in Section 14.1, consist of multiple tightly coupled mechanical oscillators. In that case, the coupling leads to dynamic behavior that is characterized by linearly independent normal modes of oscillation, with the energy in each mode determined indefinitely solely by the initial conditions, and with no tendency for the phases of each mode to lock. In the coupled

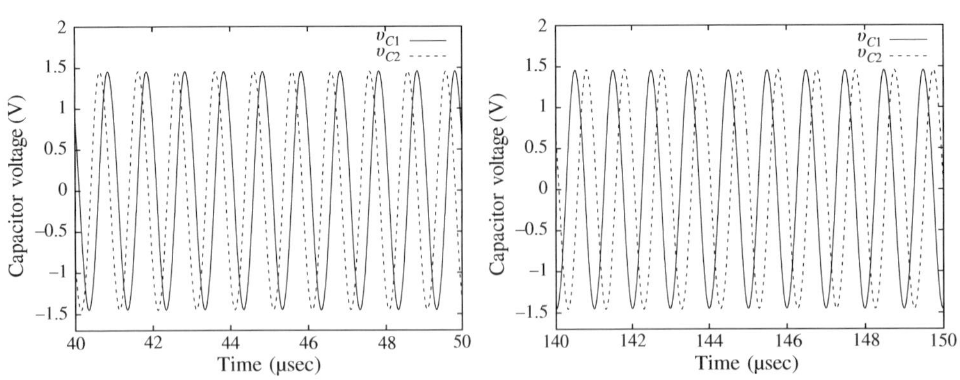

(a) Time samples between 40 and 50 μsec **(b)** Time samples between 140 and 150 μsec

Figure 12.40 Two inductor-capacitor-resistor oscillators with zero coupling capacitance showing phase drift from two different frequencies.

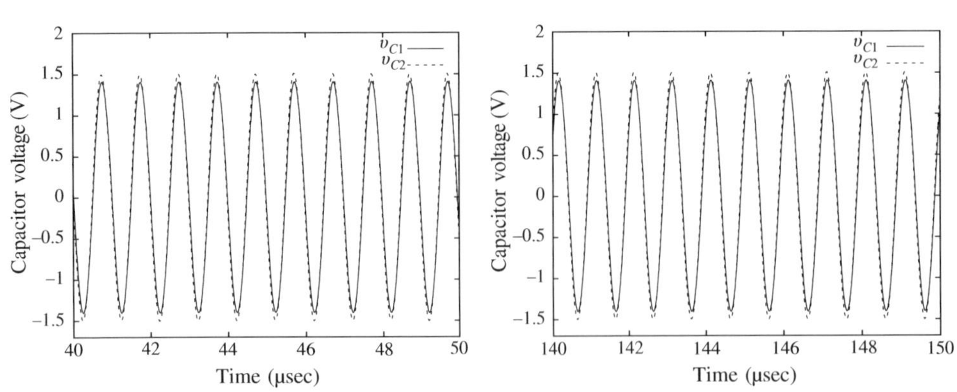

(a) Time samples between 40 and 50 μsec **(b)** Time samples between 140 and 150 μsec

Figure 12.41 Two inductor-capacitor-resistor oscillators with 10-pF coupling capacitance showing phase synchronization of two different frequencies.

oscillators studied in this section, the phase locking is due to the interaction between the lightly coupled energy and the nonlinear resistance, as the latter acts to stabilize the amplitude of each oscillator. The student may code this simulation and further explore synchronization in Exercise 12.9.

12.12 Application: Deflecting Charges in Magnetic Fields

We know from electromagnetic theory that the force $\vec{F}$ experienced by a particle with electric charge q moving with velocity $\vec{v}$ through a magnetic field of strength $\vec{B}$ is given by the vector cross product [17, Chap. 29; 18, Chap. 32; 19, Chap. 27; 20, Chap. 29]

$$\vec{F} = q\vec{v} \times \vec{B} \tag{12.75}$$

For a particle mass of m, this gives us the vector equation of motion

$$\frac{d\vec{v}}{dt} = \frac{q}{m}\vec{v} \times \vec{B} \tag{12.76}$$

We'll assume that a uniform magnetic field is created in a three-dimensional volume of space defined by a rectangular prism, by some means similar to the Helmholtz coil analyzed in Sections 4.4, 10.6, and 11.4. For simplicity, we may assume at first that the magnetic field is essentially zero outside the field region.

Given an initial position outside the field region and a velocity with a trajectory toward the region, the evolving dynamics of the path of the charged particle may be simulated by numerically solving the equation of motion (12.76) in three dimensions. This will require six state variables: the x, y, and z components of the position vector of the particle, and the x, y, and z components of the velocity vector. The coding of this simulation is available as Exercise 12.10 for the student.

Figure 12.42 shows the result of an example simulation where a positively charged particle starts at position coordinates (−0.5, −5, 1) in three-dimensional space with some system of units with an initial velocity vector (0.05, 0.7, 0). After 10 time units, the charge will reach the point (0, 2, 1) and will enter a cubical region defined by the diagonal from (0, 0, 0) to (4, 4, 4). Inside this region, an idealized magnetic field is assumed to be uniform and directed toward the $-x$ direction. With a velocity vector primarily in the y direction, once the charged particle enters the uniform field region, we can see from the simulation that its trajectory changes from a straight line to a helical path that circulates around the magnetic field lines. Eventually, with the relatively small component of the velocity in the x direction, the particle emerges from the field region and resumes a linear trajectory. These sorts of paths are measured in particle accelerators and similar experiments. If the field strength and particle velocity are such that only a small angular portion of the helix is traversed before the particle emerges out of the field region, this arrangement is also used for mass spectroscopy and the magnetic scanning required by electron microscopy. We will return to this topic in Section 14.8 and simulate a charged particle confined in more realistic magnetic field regions.

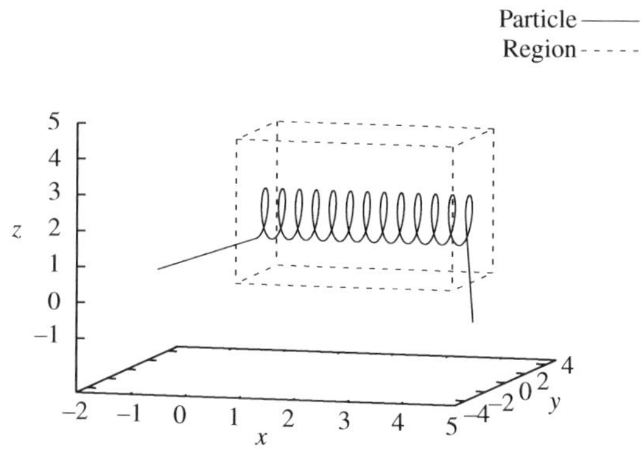

Figure 12.42 Simulation of charged particle trajectory in a uniform magnetic field in the $-x$ direction.

12.13 Coding: Generating Audio Simulations

Let's now return to the simple classical mechanics problem of analyzing the large-angle pendulum from Section 12.5, but explore another way that we may experience the results of physical simulations. In Sections 4.13 and 6.9, we developed an alternative technique of coding a `gen_output()` procedure to directly generate commands for the `gnuplot` tool in a program output stream. In this section, we consider another method for coding a `gen_output()` function to generate lines of data suitable as input to the `sox` utility. This useful program, available on Linux and MSYS2, transforms files representing audio streams from one format to another. We can easily adapt the `gen_output()` function included in the `pendulum.c` simulation program of Section 12.5 into a function that generates an output stream of lines in the format expected by the `sox` tool, and use that tool to transform the audio description to any common file format, such as a RIFF wave file, that can be played by universal audio applications or included in multimedia presentations.

The pendulums simulated in Section 12.5 had a length of 1 m, and, therefore, with small oscillation amplitudes would have oscillation period of $\frac{1}{2\pi}\sqrt{\frac{g}{l}}$, or approximately 1/2 Hz. If we were to form an audio signal from the variation of the pendulum angle with respect to time, the resulting frequency would be well below the range of human hearing. However, we are free to scale the time coordinate to anything convenient in our `gen_output()` function, and as an example we may substitute the following function in a pendulum simulator `pendulum_audio.c`

```
void gen_output_audio(ARRAY2D *ytab_ptr,GRID *tgrid_ptr) {
  int j;
  double t,min,max,scale;

  minmax_array2d_row(ytab_ptr,0,&min,&max);
  scale = 1.0 / (max - min);
  printf("; Sample Rate %.8g\n",1.0e3 / tgrid_ptr->inc);
  printf("; Channels 1\n");
  for (j = 0; j < ytab_ptr->n_cols; j++) {
    t = tgrid_ptr->init + ((double) j) * tgrid_ptr->inc;
    printf("%.8g %.8g\n",1.0e-3 * t,ytab_ptr->elem[0][j] * scale);
  }
}
```

In this example, the time coordinate is scaled down by a factor of one thousand, which places the tone of the pendulum swings in a comfortable musical range. The angle coordinate is also scaled to be well within the range of -1 to 1, as required by `sox`. After substituting this output function in the source code, we may compile and run simulations with the following commands

```
$ gcc -Wall -Wshadow -O2 -o pendulum_audio pendulum_audio.c CPLibrary/diffeq.c CPLibrary/array.c
    ↪ -lm
$ ./pendulum_audio
./pendulum_audio <length> <alpha> <Aforced> <fforced> <init y0> <init y1> <tfinal> <tinc>
$ ./pendulum_audio 1 0 0 0 0.1 0 1000 0.01 >| pendulum_audio.dat
$ sox pendulum_audio.dat -c 1 -r 44100 -b 16 -e signed-integer pendulum_audio_th0p1.wav
```

The last command runs the `sox` utility and transforms the data output stream written by our new `gen_output_audio()` function into a standard RIFF audio wave file. The file format is selected to be a 16-bit signed integer encoding, a shorter version of the bit mapping shown in Figure 2.4. This standard format may be played by any media application available on the student's operating system. Note that the pendulum simulation was first run with an initial deflection angle θ of only 0.1 rad, an initial condition that is coded in the name of the audio file. We know from the plot in Figure 12.13a that the oscillation wave will be at a relatively high frequency and close to a pure sinusoidal shape. When playing the audio file generated above, we will hear a high pitch tone with a pure, flute-like timbre. If we rerun the simulation with the sequences

```
$ ./pendulum_audio 1 0 0 0 1 0 1000 0.01 >| pendulum_audio.dat
$ sox pendulum_audio.dat -c 1 -r 44100 -b 16 -e signed-integer pendulum_audio_th1.wav
```

and

```
$ ./pendulum_audio 1 0 0 0 3 0 1000 0.01 >| pendulum_audio.dat
$ sox pendulum_audio.dat -c 1 -r 44100 -b 16 -e signed-integer pendulum_audio_th3.wav
```

we will run simulations corresponding to the plots of Figures 12.13b and 12.13d with initial deflection angles of 1 and 3 rad, and we will generate musical tones with successively lower pitches and with timbres of larger harmonic content, such as those from string instruments. This lets us experience the consequences of different physical simulation parameters with a different human sense. See Exercise 12.5 for comparing the character of the generated audio for the case of quasiperiodic and chaotic dynamics of the forced pendulum.

12.14 Exercises

12.1 Write a new program `pendulum_disponly.c` by copying the modifying the `pendulum.c` program of Section 12.5. Edit the `gen_output()` function to send only two columns of simulation data to the standard output stream containing the time coordinate and the pendulum angular displacement. A two-column data file may then be read by the `interpolate_solve.c` program from Sections 9.7 and 9.8. Then use the inverse interpolation method of Sections 9.7 and 9.8 to find the zero crossing times of the large-angle pendulum of length 1 m and no damping. Compute the periods of oscillation for the three cases of initial angles of 0.1, 1, and 3 rad with no initial angular velocity. Compare the case of 0.1 rad to the small angle harmonic oscillator prediction from Section 12.5.

12.2 Refer to the `forced_pendulum.c` program modeling the motion of the forced pendulum developed in Section 12.6 and also applied to the inverted pendulum in Section 12.8. The two plots in Figure 12.26 show an unstable and vertically stabilized inverted pendulum system resulting from two different forcing frequencies. With all the other physical parameters remaining the same as in the figure, discover an approximate forcing frequency between these two limits that appears to be the boundary between instability and stability.

12.3 Review the problem of nuclear decay of two successive nuclides from Section 5.9.

(a) Write a program `decay2_rk4.c` that solves the system of differential equations (5.39) numerically with the RK4 algorithm. This can be most easily accomplished by starting with the program `pendulum.c` and editing. The three state variables will be the three nuclide populations N_A, N_B, and N_C.

(b) Plot the solution of the three state variables over time and compare with the analytic results plotted in Figure 5.24.

12.4 **(a)** Apply the coding techniques developed in Sections 4.13 and 6.9 for a `gen_output_gpanim()` function to substitute into the `pendulum.c` program. Write a new program `pendulum_gpanim.c` to generate animated graphics of the motion of a large-angle single pendulum of length 1 m with an initial angle of 3 rad and with damping coefficient $\zeta = 0.01$, as described in Section 12.5. Include a text label in the plotting area with the running time coordinate. Your animation should resemble that shown in the downloadable file `pendulum_gpanim_z0p01th3.gif`.

(b) Expand the `gen_output_gpanim()` function in `pendulum_gpanim.c` to substitute into the `forcedpendulum.c` program from Section 12.6 to write a new program `forcedpendulum_gpanim.c`. The graphical representation should include the positions of both the pendulum mass and the pivot point, with the massless connecting member. Generate animated graphics of the motion of the forced single pendulum of length 1 m. Animate the quasiperiodic and chaotic motion described in Sections 12.6 and 12.7 to start with, and then discover other sets of parameters leading to these modes of oscillation.

(c) Rerun the animated graphics program for the forced single pendulum, but with the parameters described in Section 12.8 for the inverted pendulum to start with. Animate the motion of both unstable and stabilized inverted motion, and discover other parameter regimes for these two modes. Also observe the behavior for the frequency you discovered in Exercise 12.2.

12.5 Incorporate the `gen_output_audio()` function developed in Section 12.13 into the `forcedpendulum.c` program of Section 12.6 to write a new program `forcedpendulum_audio.c`. Use parameters described in Sections 12.6 and 12.7 to generate audio files for the two quasiperiodic modes and the chaotic mode, and characterize the audio you generate.

12.6 Write a program `doublependulum.c` that solves the system of Equation (12.51).

(a) Simulate the two approximate normal modes plotted in Figures 12.28 and 12.29 under the condition of small displacement angles, and the mixed mode motion shown in Figure 12.30.

(b) Simulate the large angle quasiperiodic motion shown in Figure 12.31.

(c) Simulate the chaotic motion shown in Figures 12.32 and 12.33.

12.7 Write a program `doublependulum_gpanim.c` that applies the coding techniques developed in Sections 4.13 and 6.9 for a `gen_output_gpanim()` function to substitute into the your `doublependulum.c` program solution to Exercise 12.6. Include a text label in the plotting area with the running time coordinate. Generate animated graphics of the motion of the double pendulum for the same cases in Exercise 12.6. Hint: Write the `gen_output_gpanim()` function to only output every tenth time sample. This will accommodate the finer time step size necessary for the erratic motion of the double pendulum in some motional modes, yet keep the animated motion at an interesting rate for human observation, as well as maintaining compact video file sizes.

12.8 Write a program `LCnegRosc.c` to simulate the equations of an electronic oscillator with a linear inductance and capacitance, but a nonlinear piecewise resistance that stabilizes the oscillation amplitude, as described in Section 12.10. The physical parameters should be set with command line arguments, and the solution should be sent to the standard output stream. Run the program with an inductance of 200 μH, a capacitance of 125 pF, resistance slopes of $R_n = 20$ kΩ in the negative incremental resistance regime and $R_p = 5$ kΩ in the positive incremental resistance regime, and a transition voltage v_t of 1 V. With these inductance and capacitance values, the oscillation frequency will be very close to 1 MHz. Plan your simulation for a duration of about 35 cycles and about 100 time steps per cycle. Generate a plot of the capacitance voltage versus time.

12.9 Following the solution to Exercise 12.8, write a program `LCnegR2osc.c` to simulate the equations of two coupled electronic oscillators as described in Section 12.11. Use the values of 125 picofarads for the resonating capacitances C_1 and C_2, a 10-pF coupling capacitance, and inductances of μH and $L_2 = 199$ μH. As in Exercise 12.8, assume resistance slopes of $R_n = 20$ kΩ in the negative incremental resistance regime and $R_p = 5$ kΩ in the positive incremental resistance regime, and a transition voltage v_t of 1 V.

(a) First run the program to replicate the simulation results in Figure 12.40 where the two oscillators have locked in synchronous and aligned phases.

(b) Find the program parameters that will also synchronously lock the phases of the two oscillators, but with opposing phases. Generate plots of this behavior similar to Figure 12.40.

12.10 Write a program `deflection.c` to simulate the deflection of a charged particle in a region of uniform magnetic field as discussed in Section 12.12. The program should accept command line arguments to specify the physical parameters of the system in some consistent system of units. Start the simulation with a particle of +1 charge unit and 1 mass unit at the position vector $(-0.5, -5, 1)$ with the velocity vector $(0.05, 0.7, 0)$. Run the simulation for 100 time units and with a time step of 0.01 units. The program should send two data sets to the standard output, the first defining a set of lines that outline the boundary of the field region for plotting, and the second the solution of the equations of motion describing the particle trajectory. Generate a three-dimensional plot of the particle trajectory, along with an indication of the region in space with the magnetic field, to resemble Figure 12.42. Rerun with various different particle properties to observe different trajectories.

13

More Extensive Systems with Ordinary Differential Equations

In this chapter, we continue the development of the explicit numerical methods for solving ordinary differential equations detailed in Chapter 12. The systems and analyses we will develop in this chapter are somewhat more elaborate than those in Chapter 12, either as the consequence of surrounding the differential equation solutions with other numerical algorithms, or by the number of state variables to keep track of in our code. Studying these systems can represent significant coding challenges, but provide some valuable insight into physical behavior.

13.1 Application: Ballistic Trajectories

Many popular sports such as baseball, cricket, tennis, and golf are highly dependent on the ballistic trajectories of spheroidal balls. There are complex forces acting on spinning balls that can be significant, such as the so-called "Magnus force" associated with the stitches on a baseball or the dimples on a golf ball surface. But beyond the first-order gravitational forces, the most straightforward second-order effect we can incorporate into a simulation of a ballistic trajectory is a simplified model of air drag.

Figure 13.1 shows a simple ballistics problem, predicting the trajectory of a projectile through a fluid exhibiting drag forces such as air. A ball of mass m is impacted by a bat or club and launched with an initial velocity $\vec{v}_0$ at an angle θ with respect to the horizontal. Along its trajectory, it is subject to the gravitational force $m\vec{g}$ and drag force $\vec{F}_d$. Although there are more comprehensive models for the complexity of drag forces, we'll assume here a simple quadratic model where the magnitude of the drag force $|\vec{F}_d|$ is proportional to the square of the velocity $|\vec{v}|^2$, and the direction of $\vec{F}_d$ is aligned opposite that of the instantaneous velocity vector $\vec{v}$

$$\vec{F}_d = -\frac{C_d \rho A}{2}|\vec{v}|\vec{v} \tag{13.1}$$

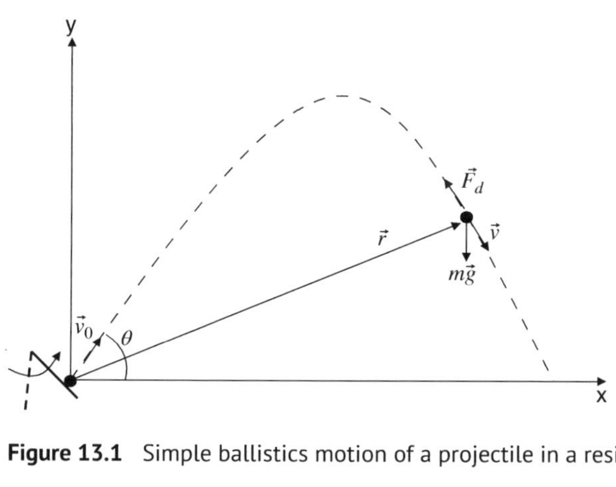

Figure 13.1 Simple ballistics motion of a projectile in a resistant fluid.

Computational Physics Using C: Efficient Programming with Ease, First Edition. John W. Fattaruso. © 2026 John Wiley & Sons, Inc.
All rights reserved, including rights for text and data mining and training of artificial intelligence technologies or similar technologies.
Published 2026 by John Wiley & Sons, Inc. Companion Website: http://www.wiley.com/go/CompPhysUsingC

where C_d is the so-called drag coefficient, ρ is the density of the air, and A is the cross-sectional area of the ball. For the purpose of the simulation, we will bundle all these constants into one factor α

$$\vec{F}_d = -\alpha|\vec{v}|\vec{v} \tag{13.2}$$

With the ball located at the instantaneous position vector $\vec{r}$, we can easily apply the Newtonian law of motion $\vec{a} = \frac{\vec{F}}{m}$ and find the vector equation of motion

$$\frac{d^2\vec{r}}{d^2t} = -\frac{\alpha|\vec{v}|\vec{v}}{m} - \vec{g} \tag{13.3}$$

For simplicity, we can assume a coplanar trajectory solution $\vec{r}(t)$ in two dimensions, and then both decompose the vector Equation (13.3) into its x and y components and also include the expression for the velocity magnitude

$$\begin{aligned}\frac{d^2r_x}{d^2t} &= -\frac{\alpha}{m}\sqrt{v_x^2+v_y^2}\cdot v_x \\ \frac{d^2r_y}{d^2t} &= -\frac{\alpha}{m}\sqrt{v_x^2+v_y^2}\cdot v_y - g\end{aligned} \tag{13.4}$$

We then follow the scheme of Equation (12.20) and define the four state variables

$$\begin{aligned}y_0 &= r_x \qquad y_2 = \frac{dy_0}{dt} = v_x \\ y_1 &= r_y \qquad y_3 = \frac{dy_1}{dt} = v_y\end{aligned} \tag{13.5}$$

Note that with this composition of the state variables we have the velocity vector components needed for the drag force computation available in two of the state variables. The system of four first-order differential equations in the form suitable for the numerical solver is then

$$\begin{aligned}\frac{dy_0}{dt} &= y_2 \qquad \frac{dy_2}{dt} = -\frac{\alpha}{m}\sqrt{y_2^2+y_3^2}\cdot y_2 \\ \frac{dy_1}{dt} &= y_3 \qquad \frac{dy_3}{dt} = -\frac{\alpha}{m}\sqrt{y_2^2+y_3^2}\cdot y_3 - g\end{aligned} \tag{13.6}$$

The main program that predicts the trajectory governed by this differential equation is in the file `ballistic.c`, available for download from the Student Companion website. The function passed to the fourth-order Runge–Kutta differential equation solver is named `ballistic()`, and computes the four derivatives in Equation (13.6). This function, together with its physical parameter structure, is excerpted here

```
typedef struct bal_param {
  double mass;  // Projectile mass
  double v0;    // Launch speed
  double theta; // Launch angle
  double alpha; // Drag coefficient, for drag force = alpha * speed^2
} BALLISTIC;

// State variable 0 is x component of position
// State variable 1 is y component of position
// State variable 2 is x component of velocity
// State variable 3 is y component of velocity
void ballistic(double x,ARRAY1D *y,void *param_ptr,ARRAY1D *yp) {
  double v;
  BALLISTIC *p;

  p = param_ptr;
  v = sqrt(y->elem[2]*y->elem[2] + y->elem[3]*y->elem[3]);
  yp->elem[0] = y->elem[2];
  yp->elem[1] = y->elem[3];
  yp->elem[2] = -p->alpha * v * y->elem[2] / p->mass;
  yp->elem[3] = -p->alpha * v * y->elem[3] / p->mass -
                GRAVITATIONAL_ACCEL;
}
```

After the time sample grid has been set in the `GRID` structure `tgrid`, the `BALLISTIC` parameter structure `param` has been filled from the command line arguments, and the initial condition array set, the solver routine is called with

```
rk4(ballistic,&tgrid,&param,&init,&xytable);
```

We may compile and run this program for an example case with the command shell dialog

```
$ gcc -Wall -Wshadow -O2 -o ballistic ballistic.c CPLibrary/diffeq.c CPLibrary/array.c -lm
$ ./ballistic
./ballistic <mass> <launch velocity> <launch angle> <alpha> <tfinal> <tinc>
$ ./ballistic 0.1 40 45 0 7 0.01 >| ballistic.dat
```

This initial simulation is for launching a ball with mass 0.1 kg with a velocity of 40 m/s at an angle of 45° with an ideal case of no drag force. We run the simulation for 7 s with a time step of 10 ms. Since we assume initially that the dynamics are ideal with no air drag, the trajectory should follow an ideal parabolic path. Inspection of the `gen_output()` function in the program will show that three columns are sent to the output stream, the first being the simulation time, and the second and third being the x and y components of the position vector $\vec{r}(t)$. The simple `gnuplot` command

```
gnuplot> plot 'ballistic.dat' u 2:3 w l
```

plots the parabolic trajectory shown in Figure 13.2.

Next we simulate a more complete case that requires numerical treatment, including a significant influence from air drag. We set the drag coefficient $\alpha = 10^{-3}\frac{\mathrm{N\,s^2}}{\mathrm{m^2}}$, and to compensate for the loss of energy through the trajectory, we increase the launch velocity to 65 m/s. With the command line

```
$ ./ballistic 0.1 65 45 1e-3 7 0.01 >| ballistic.dat
```

and the same plotting command as for Figure 13.2, we generate the trajectory shown in Figure 13.3. Note that we now experience a considerable distortion from the ideal parabolic profile due to the drag losses, and that even covering a shorter horizontal range seen in Figure 13.2, the faster initial launch velocity lifts the ball significantly higher before the accumulated air drag shortens the trajectory.

Although tracing the trajectory curve of the ball required the numerical differential equation tools of this chapter, we may extend our analysis of this dynamic problem by bringing to bear tools from previous chapters of this text. For example, how do we find the exact horizontal range of the ball, which is the x coordinate where the y coordinate crosses zero, assuming a level playing ground? This is an analysis task that the inverse interpolation functions in Section 9.7 are well suited for. A practical strategy for solving for the range is to step through the time samples of the differential equation trajectory solution to find a sequence of four time samples over which the sign of the y coordinate changes between the second and third samples. Then we may construct an interpolating cubic polynomial fitting the four samples of the y versus x trajectory and find the intermediate x value where y is zero. We can then bundle this computation into a function that computes this effective ball range with an argument of the launch angle. An excerpt from the range calculation function in the

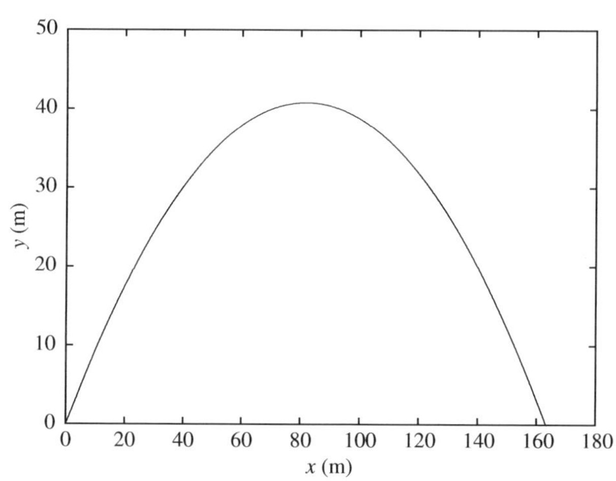

Figure 13.2 Ballistic trajectory with no air drag, $v_0 = 40$ m/s, $\theta = 45°$.

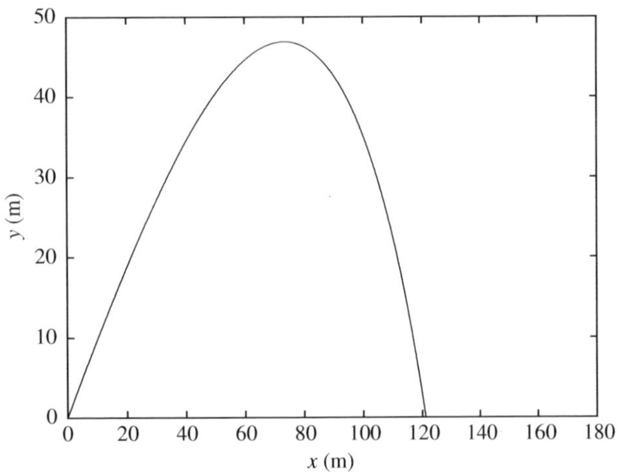

Figure 13.3 Ballistic trajectory with air drag $\alpha = 10^{-3}$, $v_0 = 65$ m/s, $\theta = 45°$.

example program `ballistic_range_step.c` is shown below. After the parameters for the ballistic launch are set, the Runge–Kutta solver tool is called to fill the x and y coordinates of the ball at each time sample in the two-dimensional array `xytable` as before. Then the `for` loop steps through all the time samples, searching for four consecutive samples that bracket a time when the y coordinate crosses the zero elevation line. If this condition is found, the `lagrange()` tool from Section 9.3 is called with the order of arguments chosen to implement inverse interpolation among the four bracketing samples. Note yet another flexible and efficient use of pointers, with the elements of the two `ARRAY1D` types that the `lagrange()` tool expects are filled in with pointer arithmetic so that they point to a sequence of four data coordinates that are actually stored in the memory allocated for the RK4 ballistic solution. The interpolated x value is returned as the range of the trajectory.

```
double range(double angle,void *p) {
  .....
  alloc_array2d(&xytable,4,NUM_GRID_PTS(tgrid));
  rk4(ballistic,&tgrid,&param,&init,&xytable);
  for (i_init = 0; i_init < xytable.n_cols - 3; i_init++) {
    if (xytable.elem[1][i_init + 1] * xytable.elem[1][i_init + 2]
        < 0.0) {
      xtab.n_elems = 4;
      xtab.elem = xytable.elem[0] + i_init;
      ytab.n_elems = 4;
      ytab.elem = xytable.elem[1] + i_init;
// X and Y coordinates exchanged for inverse interpolation of range
      range = lagrange(&ytab,&xtab,0.0);
      free_array2d(&xytable);
      return(range);
    }
  }
  .....
}
```

We may step this range prediction function with the `step()` tool of Chapter 4 over the practical range allowed by the player with a range of bats or clubs, say 10° to 80°. The resulting ballistic prediction curve is plotted in Figure 13.4.

Also appearing on this plot is a horizontal line for a range of 100 m, as an example of the range where a particular target lies. Note that there are two launch angles that will end the trajectory at this particular target. The next step in our analysis is to solve numerically for these two required launch angles that will land the ball at the target range of 100 m under the influence of air drag. For this solution, we can make use of the `step_roots.c` tool from Chapter 5 to solve for these two launch angles. This implies that we need to form a function whose roots are the required launch angle to land the trajectory at the target range. In the program `ballistic_range_solve.c` is a function that contains essentially the same code as the `range()` function above, but with a minor modification of returning an error measurement between the trajectory range and the desired target range. That desired range is one of the physical parameters now included in the

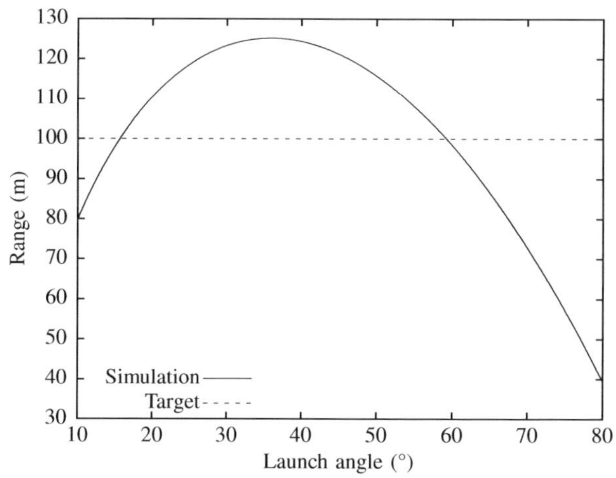

Figure 13.4 Range of ballistic trajectory with air drag $\alpha = 10^{-3}$, $v_0 = 65\,\text{m/s}$.

RANGE structure. Physically, we can think of the return value as being the distance that we would measure to the actual landing spot of the ball if we stood at the target landing spot. A code excerpt is shown here

```
double range_error(double angle,void *p) {
   .....
   alloc_array2d(&xytable,4,NUM_GRID_PTS(tgrid));
   rk4(ballistic,&tgrid,&param,&init,&xytable);
   for (i_init = 0; i_init < xytable.n_cols - 3; i_init++) {
      if (xytable.elem[1][i_init + 1] * xytable.elem[1][i_init + 2]
           < 0.0) {
         xtab.n_elems = 4;
         xtab.elem = xytable.elem[0] + i_init;
         ytab.n_elems = 4;
         ytab.elem = xytable.elem[1] + i_init;
// X and Y coordinates exchanged for inverse interpolation of range
         error = lagrange(&ytab,&xtab,0.0) - p->des_range;
         free_array2d(&xytable);
         return(error);
      }
   }
   .....
}
```

Since there may be more than one launch angle that lands the ball at the desired range, we call the `step_secant()` tool in the main program to converge on both angles. Compiling and running this program with some example parameters is shown here

```
$ gcc -Wall -Wshadow -O2 -o ballistic_range_solve ballistic_range_solve.c CPLibrary/diffeq.c
    ↪ CPLibrary/interpolate.c CPLibrary/step_roots.c CPLibrary/root.c CPLibrary/array.c -lm
$ ./ballistic_range_solve
./ballistic_range_solve <mass> <launch velocity> <alpha> <tfinal> <tinc> <desired range>
$ ./ballistic_range_solve 0.1 65 1e-3 10 0.01 100
2 roots found (in degrees):
15.61 59.24
```

Or perhaps we are interested in the launch angle that maximizes the range of the projectile. We know from introductory mechanics that with no air drag, the maximum range of the ideal parabolic flight path is obtained at a launch angle of 45° from the horizontal. But we can apply the minimization tool developed in Chapter 11 to find numerically the different optimum angle in the case of drag forces.

The main program is in the file `ballistic_range_max.c`, and calls the function `golden_section()` to find the minimum of an object function. That object function is `double range_recip()` in the program file, and is basically identical to the `double range()` called to step through ranges to draw Figure 13.4, but the `double range_recip()` function returns the reciprocal of the range that it calculates. This will let the `golden_section()` find the angle that maximizes the range instead of minimizing it. That optimum range is reported to the standard error stream if the minimization is successful. The function minimization tool is called with the statement

```
retval = golden_section(range_recip,agrid.init,agrid.final,&param,
                        1.0e-6,100,&opt_angle);
```

to search for an optimum angle within a tolerance of $10^{-6\circ}$. We may compile and run the program with some example values with

```
$ gcc -Wall -Wshadow -O2 -o ballistic_range_max ballistic_range_max.c CPLibrary/diffeq.c
    ↪ CPLibrary/interpolate.c CPLibrary/minimum.c CPLibrary/array.c -lm
$ ./ballistic_range_max
./ballistic_range_max <mass> <launch velocity> <alpha> <tfinal> <tinc>
$ ./ballistic_range_max 0.1 40 0 10 0.01
Optimum angle = 44.999996 degrees
$ ./ballistic_range_max 0.1 65 1e-3 10 0.01
Optimum angle = 35.906896 degrees
```

Note that with no air drag forces, the program confirms the analytic result that the maximum range is obtained by launching at a 45° angle. But with some moderate drag on the projectile, we see that the launch angle must be reduced significantly to maximize the range. A lower trajectory angle will reduce the flight time of the ball and reduce the energy eventually lost to air drag.

13.2 Application: *n*-body Gravitational Systems

Classical analytical solutions of the equations of motion of bodies interacting gravitationally are generally limited to two bodies, or three bodies in some restricted cases. Applying our numerical tools greatly extends the scope of such systems where solutions may be sought. We will see in Section 14.7 that some gravitational systems in which the bodies experience velocity variations of orders of magnitude are beyond treatment with the explicit algorithms of this chapter and must be submitted to the implicit algorithms that will be developed in Chapter 14. However, there are many gravitational systems in which the velocities of the interacting bodies remain within bounds and may be readily simulated with the explicit algorithms of this chapter.

In Figure 13.5 several massive bodies are interacting gravitationally. Body i has a position vector $\vec{r}_i$, a velocity vector $\vec{v}_i$ and mass m_i. The force on body i is related to its acceleration with Newton's law of motion

$$\vec{F}_i = m_i\vec{a}_i = m_i\frac{d\vec{v}_i}{dt} = m_i\frac{d^2\vec{r}_i}{dt^2} \tag{13.7}$$

We assume that the force on body i is given by the sum of the Newtonian gravitation forces due to the mass of each of the other bodies $j \neq i$. The force is directed along the vector pointing from body i to body j, with a magnitude that varies as the inverse square of the separation distance. So we may formulate the equation of motion for body i as

$$\frac{d^2\vec{r}_i}{dt^2} = G\sum_{j\neq i}\frac{m_j(\vec{r}_j - \vec{r}_i)}{|\vec{r}_j - \vec{r}_i|^3} \tag{13.8}$$

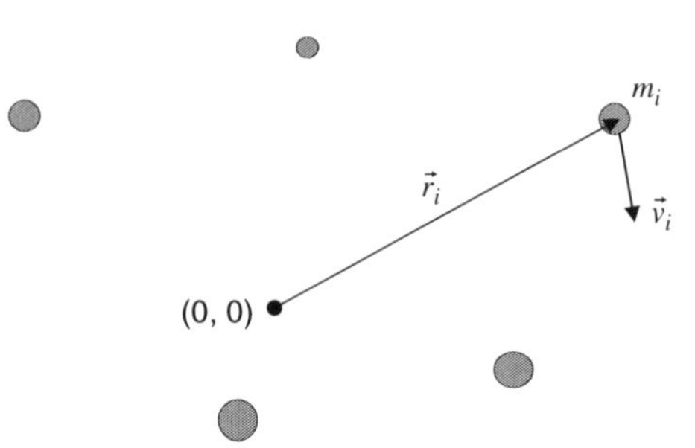

Figure 13.5 An *n*-body gravitational system.

If, for simplicity, we limit our treatment to systems where the orbits are coplanar, we may solve the system by resolving Equation (13.8) into two vector components

$$\begin{aligned} \frac{d^2 r_{ix}}{dt^2} &= G \sum_{j \neq i} \frac{m_j (r_{jx} - r_{ix})}{|\vec{r}_j - \vec{r}_i|^3} \\ \frac{d^2 r_{iy}}{dt^2} &= G \sum_{j \neq i} \frac{m_j (r_{jy} - r_{iy})}{|\vec{r}_j - \vec{r}_i|^3} \end{aligned} \tag{13.9}$$

Since these two equations are of the second order, each body in the physical system will contribute four first-order differential equations to the mathematical system. We follow the procedure outlined in Equation (12.20) to define the state variables in the system. Similar to the case of solving ballistics problems in Section 13.1, there will be two state variables for each component of the vector differential equation (13.9), totaling to four state variables *for each body*. For a system involving N bodies, we may index the bodies as 0 to $N-1$, and there will be a total of $4N$ state variables and first-order differential equations in the system to solve. The state equations are indexed 0 to $4N-1$ for the solver, and body i will contribute the state equations with indices $4i+0$, $4i+1$, $4i+2$, and $4i+3$. We then follow the scheme of Equation (12.20) and define for each body in the system, the four state variables

$$\begin{aligned} y_{4i} &= r_{ix} & y_{4i+2} &= \frac{dy_{4i}}{dt} = v_{ix} \\ y_{4i+1} &= r_{iy} & y_{4i+3} &= \frac{dy_{4i+1}}{dt} = v_{iy} \end{aligned} \tag{13.10}$$

Note that the cube of the separation distance $|\vec{r}_j - \vec{r}_i|^3$ appears in the denominator of both expressions in Equation (13.9). This can be most efficiently computed by first calculating and storing the square of the distance

$$d_{ij}^2 = (r_{jx} - r_{ix})^2 + (r_{jy} - r_{iy})^2 \tag{13.11}$$

and then multiplying by the square root of the distance squared

$$d_{ij}^3 = d_{ij}^2 \sqrt{d_{ij}^2} \tag{13.12}$$

So for each body i, four out of the total $4N$ equations we will submit to the Runge–Kutta solver are

$$\begin{aligned} \frac{dy_{4i}}{dt} &= y_{4i+2} & \frac{dy_{4i+2}}{dt} &= G \sum_{j \neq i} \frac{m_j (y_{4j} - y_{4i})}{d_{ij}^2 \sqrt{d_{ij}^2}} \\ \frac{dy_{4i+1}}{dt} &= y_{4i+3} & \frac{dy_{4i+3}}{dt} &= G \sum_{j \neq i} \frac{m_j (y_{4j+1} - y_{4i+1})}{d_{ij}^2 \sqrt{d_{ij}^2}} \end{aligned} \tag{13.13}$$

where $d_{ij}^2 = (y_{4j} - y_{4i})^2 + (y_{4j+1} - y_{4i+1})^2$.

The C code to provide the derivatives following Equation (13.13) for all N bodies to the Runge–Kutta solver tool is shown below. The structure GRAV holds the necessary physical parameters of the gravitational system, which in this case is the set of masses of each of the bodies. In this C structure, the masses are stored as elements of the `mass` array, which is allocated by the main program once it ascertains from the command line how many bodies are to be simulated.

```
typedef struct grav_param {
  ARRAY1D mass;
} GRAV;

void gravitation(double x,ARRAY1D *y,void *param_ptr,ARRAY1D *yp) {
  int i_body,j_body,n_bodies,i,j;
  double dsq;
  GRAV *p;

  p = param_ptr;
  n_bodies = y->n_elems / 4;
  for (i_body = 0; i_body < n_bodies; i_body++) {
    i = 4 * i_body;
```

```
      yp->elem[i]   = y->elem[i+2];
      yp->elem[i+1] = y->elem[i+3];
      yp->elem[i+2] = 0.0;
      yp->elem[i+3] = 0.0;
      for (j_body = 0; j_body < n_bodies; j_body++) {
        if (j_body != i_body) {
          j = 4 * j_body;
          dsq = (y->elem[j]-y->elem[i])*(y->elem[j]-y->elem[i]) +
                (y->elem[j+1]-y->elem[i+1])*(y->elem[j+1]-y->elem[i+1]);
          yp->elem[i+2] += p->mass.elem[j_body] *
                           (y->elem[j]-y->elem[i]) / (dsq * sqrt(dsq));
          yp->elem[i+3] += p->mass.elem[j_body] *
                           (y->elem[j+1]-y->elem[i+1]) /
                           (dsq * sqrt(dsq));
        }
      }
      yp->elem[i+2] *= GRAVITY;
      yp->elem[i+3] *= GRAVITY;
    }
    return;
}
```

The Runge–Kutta solver is called with the statement

```
rk4(gravitation,&tgrid,&param,&init,&ytable);
```

in the main program, and then a suitable `gen_output()` function is called to send to the output stream the simulation time and the x, y coordinates of all the bodies in multiple columns for plotting. The full program code is in the file `gravitation.c` from the Student Companion website.

Consider first a simple example of a solar system centered around a star of the same size as our Sun, and with only two planets in orbit. These planets, however, are two gas giants the size of our Jupiter, the first orbiting at a radius and with a period the same as our Jupiter, and the second orbiting at a radius 1.25 times farther from the Sun. This system may be simulated with the compilation and execution dialog

```
$ gcc -Wall -Wshadow -O2 -o gravitation gravitation.c CPLibrary/diffeq.c CPLibrary/array.c -lm
$ ./gravitation
./gravitation <tfinal> <tinc> <num bodies> <x1,y1,vx1,vy1,m1> <x2,y2,vx2,vy2,m2> ...
$ ./gravitation 1e10 1e6 3 0,0,0,0,1.99e30 7.78e11,0,0,1.3066e4,1.90e27
    ↪ -9.725e11,0,0,-1.1686e4,1.90e27 >| gravitation.dat
```

The simulation has been run for 10^{10} s, or over 300 years, with a step size of 10^6 s. The plot of the simulation in Figure 13.6 shows the coplanar trajectories of the two planets. These trajectories exhibit significant interaction between the two planets in their orbits. As the planets approach each other, the two orbits are perturbed to paths that do not conform to the pure elliptical paths of independent Keplerian orbits.

To verify that our time step size is sufficiently fine to provide a low enough truncation error for our simulation interest, we return to the method from Section 12.5 of plotting the simulated total energy of the system. The total energy of an n-body gravitational system is the sum of the kinetic energies of the bodies and the pair-wise sums of the gravitational potential energies between the bodies.

$$E = \sum_i \left[\frac{m_i v_i^2}{2} - G \sum_{j<i} \frac{m_i m_j}{|\vec{r}_j - \vec{r}_i|} \right] \tag{13.14}$$

In Figure 13.7, we see the simulated total energy versus time, which must obey the law of conservation of energy. With a step size of 10^6 s, the span of the vertical axis shows that the deviation from numerical conservation due to accumulated truncation error is about 20 ppb.

For the solution of gravitational systems described by the Newtonian laws treated in this section, we must find initial values for the state variables of the position and velocity vectors in Equation (13.10) for the orbital paths that interest us. In Section 5.6, we found the roots of the basic nonlinear equation describing Keplerian orbital elements. In Section 5.6, the position coordinates at a given point in the elliptical orbit can be found from the true anomaly and the radius in Equations (5.17) and (5.18), and the velocity components along the orbital path by Equations (5.20) and (5.22). These

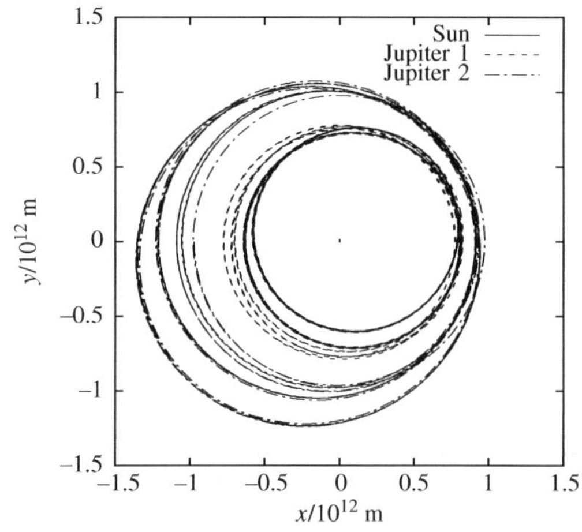

Figure 13.6 Simulation of a solar system with interaction between two massive planets.

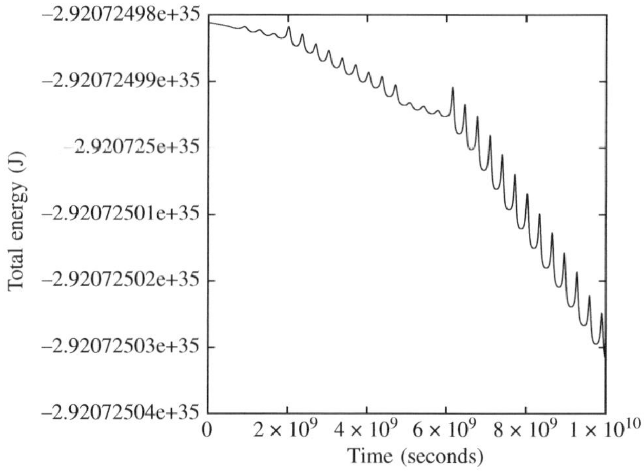

Figure 13.7 Total energy of a simulated solar system showing accumulating truncation error.

expressions may be used to relate the Keplerian elements to the initial conditions for the vector orbital elements. The student may explore this relation in Exercise 13.1.

13.3 Application: *n*-body Collisions

In Section 4.8, we created an idealized model for the collisions of spherical or circular objects and treated various problems dependent on tracing the trajectories of the objects after the collisions. That model required the assumption that the impulse imparted during a collision acted within an infinitesimal length of time, so that we could use a simple geometric calculation to ascertain the direction of the impulse vector. In this section, we will relax the idealized assumption of the dynamics of the collision and predict trajectories with a more realistic model.

This more realistic model requires a full solution of the differential equations of motion during the collision as a surface spring constant model of the objects is compressed and released. The model is detailed in Figure 13.8. We still assume that the only forces acting on the balls involved in collisions will be directed along the lines drawn between their centers, but we allow some compression of the surfaces of the balls so that the impulse is not assumed to be instantaneous. Note that this force vector direction is similar to the condition assumed in the analysis of central forces in classical mechanics, but with the inclusion of a conditional constraint that beyond a certain distance separating the centers of the balls, the surfaces are fully released and no further interaction force applies.

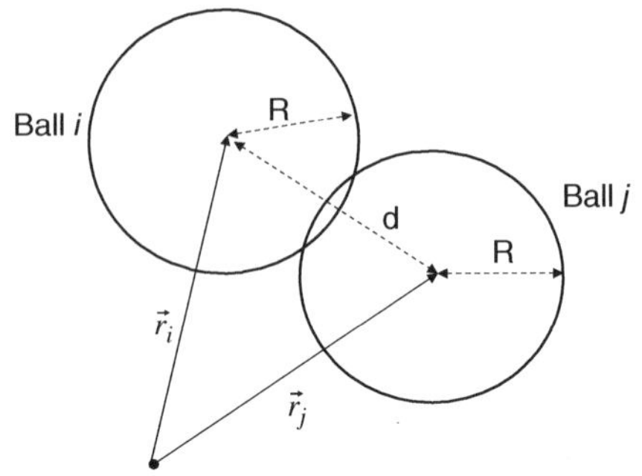

Figure 13.8 Collision model with spring constant.

The position vectors $\vec{r}_i$ and $\vec{r}_j$ locate the centers of the two colliding objects. Again we assume a spherical or circular object so that each uncompressed surface is isotropically at a radius R from each center. The distance between the centers is $d = \left|\vec{r}_i - \vec{r}_j\right|$. We will assume that the compression behavior of the object surfaces acts with a linear spring constant, so we can algebraically model the force on object i due to object j as

$$\begin{aligned} \vec{F}_{ij} &= k(2R-d)\cdot\frac{\vec{r}_i-\vec{r}_j}{d} \text{ when } d<2R \\ \vec{F}_{ij} &= 0 \text{ otherwise} \end{aligned} \tag{13.15}$$

where k is a spring constant.

The file `collision2.c` from the Student Companion website is an example program written to solve the equations of motion for two balls possibly engaged in a collision using the impact model of Equation (13.15). There are eight state variables needed, storing the two-dimensional position and velocity of the two balls. The initial positions and velocities of the balls are collected from the command line in some arbitrary system of units, and the following code collects the needed physical parameters and computes the motion derivatives for the Runge–Kutta algorithm.

```
typedef struct col_param {
  double R;  // ball radius
  double k;  // ball spring constant
  double m;  // ball mass
} COLL;

void collisions2D(double x,ARRAY1D *y,void *param_ptr,ARRAY1D *yp) {
  double d,overlap;
  COLL *p;

  p = param_ptr;
  yp->elem[0]  = y->elem[2];
  yp->elem[1]  = y->elem[3];
  yp->elem[4]  = y->elem[6];
  yp->elem[5]  = y->elem[7];
  d = sqrt((y->elem[4]-y->elem[0])*(y->elem[4]-y->elem[0]) +
           (y->elem[5]-y->elem[1])*(y->elem[5]-y->elem[1]));
// Test for overlap between surfaces of balls
  overlap = 2.0 * p->R - d;
  if (overlap > 0.0) {
    yp->elem[2]  =
      (p->k * overlap * (y->elem[0]-y->elem[4])) / (d * p->m);
    yp->elem[3]  =
      (p->k * overlap * (y->elem[1]-y->elem[5])) / (d * p->m);
    yp->elem[6]  =
      (p->k * overlap * (y->elem[4]-y->elem[0])) / (d * p->m);
    yp->elem[7]  =
      (p->k * overlap * (y->elem[5]-y->elem[1])) / (d * p->m);
  }
  else {
    yp->elem[2]  = 0.0;
```

```
      yp->elem[3] = 0.0;
      yp->elem[6] = 0.0;
      yp->elem[7] = 0.0;
    }
}
```

We may compile and run this program for the somewhat "springy" example case of the spring constant of 10 force units per distance unit with the dialog

```
$ gcc -Wall -Wshadow -O2 -o collision2 collision2.c CPLibrary/diffeq.c CPLibrary/array.c -lm
$ ./collision2
./collision2 <radius> <spring constant> <rx1,ry1,vx1,vy1> <rx2,ry2,vx2,vy2> <tfinal> <tinc>
$ ./collision2 1 10 -10,-9.75,2,2 10,-10.25,-2,2 10 0.001 >| collision2.dat
```

If we use this program to model the dynamics of two objects on a collision course, we may summarize the predicted trajectories in the plot of Figure 13.9. The solid line in the plot is the ideal model of Section 4.8, and the dashed lines with various patterns are the trajectories resulting from various values of the surface spring constant k. We can see from the plot that as the spring constant is increased on the command line to very stiff values, up to 100 units, the collision trajectory approaches the model of the instantaneous impulse analysis of Section 4.8 that assumed the objects touch at only one contact point around their surfaces. It should be understood that the solid lines in the plot of Figure 13.9 and the dashed lines are from two completely different models.

We can visualize the elastic process of the collision with a different sort of plot, with a similar consideration of the conservation of energy as in Figure 13.7, but in this case we examine only the total kinetic energy of the two balls. The calculation of the total kinetic energy at each time step has been added to the `gen_output()` function of the `collision2.c` program, and it is displayed with full 16-digit precision as the last output column. In Figure 13.10, we see the kinetic energy plotted versus time, and zoomed into the collision event. During the initial time before the surfaces of the balls overlap, the kinetic energy is constant. Then while the surfaces overlap, the kinetic energy is converted to the potential energy of the compression of the spring surfaces. Finally, the total kinetic energy is restored with the new trajectories after the collision. As the spring constant takes on higher values, the time duration of the collision process shortens and approaches the instantaneous ideal collision. Capturing the sharp corners where the surface overlap just starts and stops, as well as the short collision impulse duration, is the motivation for choosing the short time step of 0.001 units in the simulations above.

The surface spring model and the differential equation solver tool can be used effectively for modeling collisions of more than two spherical balls, when at any point in time there may be more than one collision impulse acting on various balls. The solution to the multiple equations of motion is very similar to the solution of the attractive gravitational forces among several bodies analyzed in Section 13.2. In the simple two-dimensional coplanar case, the number of state variables needed is four times the number of colliding balls.

We may also include the effects of a confining wall with a spring constant, as illustrated in Figure 13.11. We assume that a boundary wall will exert only a confining force on an incident ball with a component normal to the surface of the wall and that the surface of the ball will allow it to maintain the component of velocity directed along the wall unchanged.

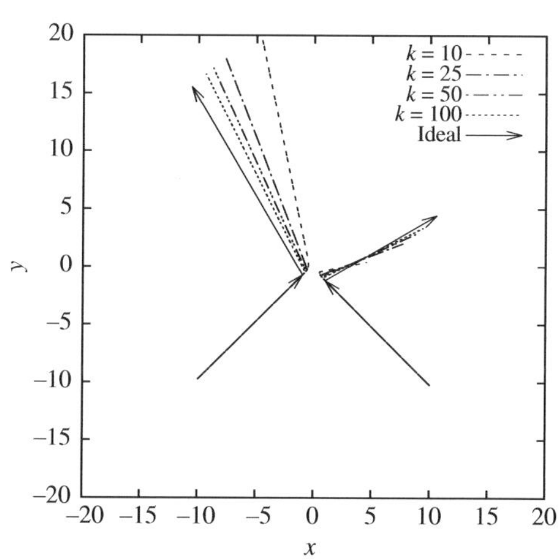

Figure 13.9 Two-ball collision simulation with varying spring constant.

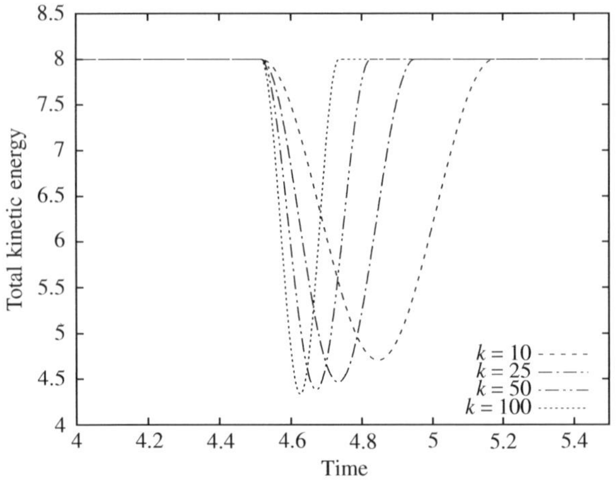

Figure 13.10 Total kinetic energy of two-ball collision simulation with varying spring constant.

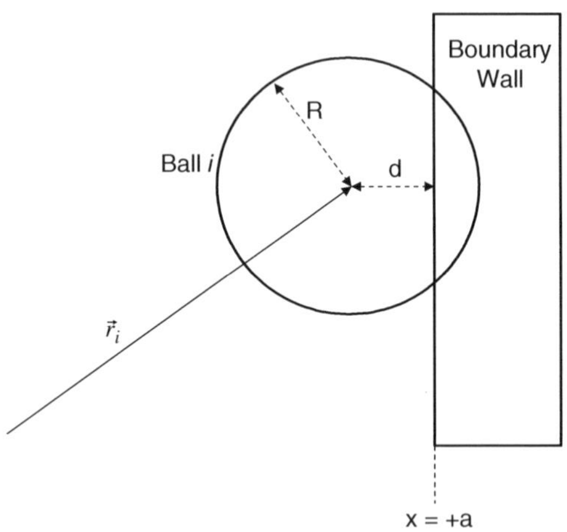

Figure 13.11 Confining wall model with spring constant.

Figure 13.11 shows an example collision of the ball with a wall that establishes the boundary in the positive x dimension at coordinate $x = a$. The only force component on the ball is in the negative x direction, and will depend linearly with a spring constant on the compression of the ball into the wall surface. The distance from the center of ball i to the wall is $d = a - r_{ix}$ and the confining force component is given by

$$
\begin{aligned}
F_x &= k(d - R) \text{ when } d < R \\
F_x &= 0 \text{ otherwise}
\end{aligned}
\tag{13.16}
$$

where k is a spring constant.. Similar models may be derived for boundary walls in the other directions, and allow for the possibility that a ball may collide with a corner and compress two different walls in two components at the same time. When the total confining vector force is found, its contribution to the vector acceleration of each ball is added to the forces due to the ball collisions.

To illustrate the capability of this model, Exercise 13.2 simulates what is called the break shot of a rack of pool balls, confined with walls to the surface of a pool table, as shown in Figure 13.12.

Another interesting graphical view of the behavior of this physical system is to calculate and plot the position of the center of mass at each time step. This visualization gives us some insight into the fundamental principle that the motion of the center of mass of any system is unchanged unless external forces are applied. Figure 13.13 plots the x and y coordinates of the center of mass over all the pool balls as a function of time. Initially the motion of the center of mass has a uniform velocity vector in the positive x direction, after the impulse of the cue stick has exerted its external force on the cue ball. We can see in Figure 13.12 that the first impact of the cue ball with the formation of stationary balls occurs at time 2.4, but

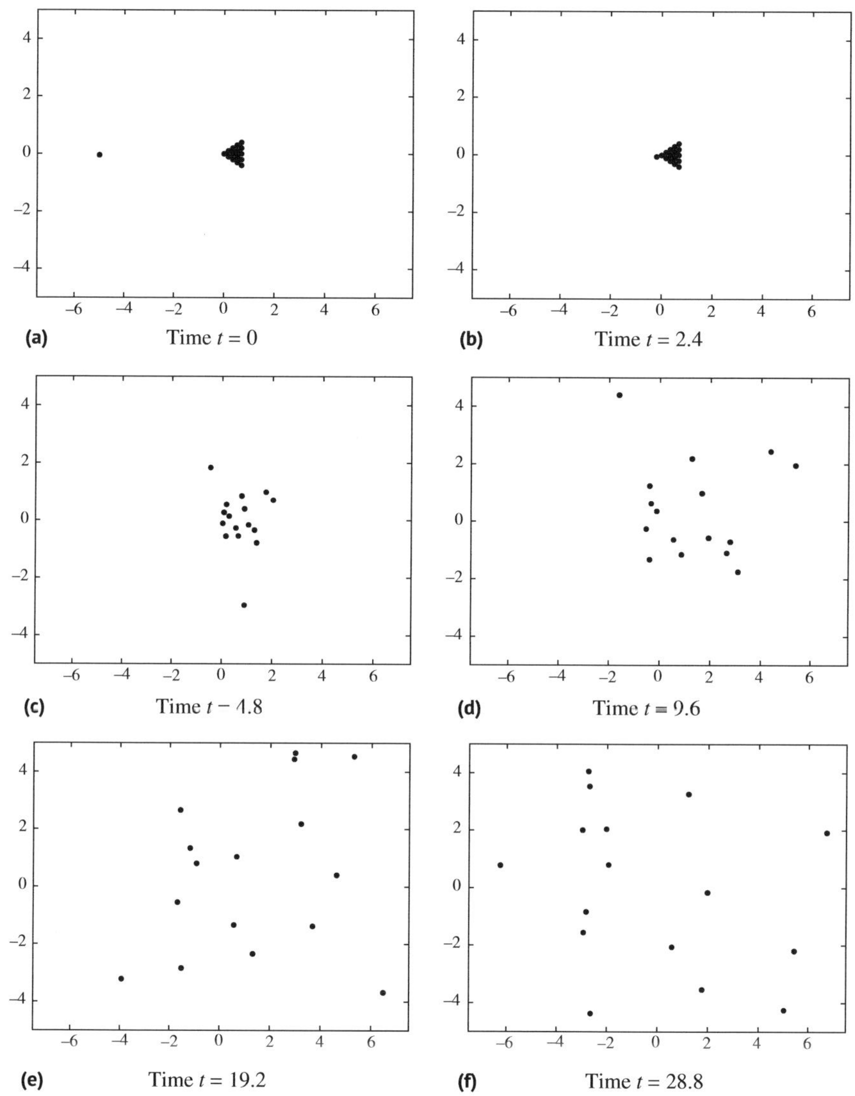

Figure 13.12 Simulated progression of a pool break shot.

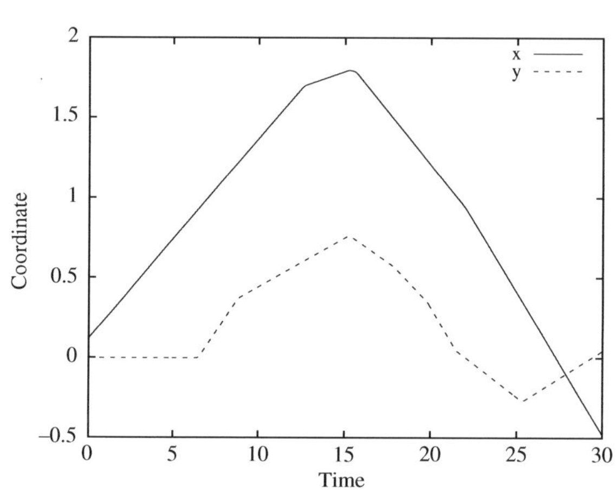

Figure 13.13 The x and y coordinates of center of mass of pool ball system as a function of time.

from then until time 6.4, through all the ball collisions involving 16 balls in the breaking process when only internal forces are applied, the motion of the center of mass remains uniform. It is only after time 6.4, when balls start to rebound from the confining walls, that external forces are again exerted. At each collision, the velocity of the center of mass is altered in one coordinate direction or the other.

13.4 Application: Classical Field Lines

Our goal in this section is to develop the numerical techniques for drawing classical "field lines" that are typically used for visualizing electric and magnetic flux. This graphics problem reduces to solving a spatial differential equation that we may submit to the Runge–Kutta algorithm.

We start with an example exercise of drawing curving contours through a two-dimensional space that start from coordinates near a point electrical charge and proceed through the space with the constraint that at each point along the contour, the tangent line of our constructed curve follows the unit vector that points in the direction of the local electric field. Mathematically, we solve for a curve that is parameterized by a scalar distance s traveled along it. The points on the field line are located on the vector $\vec{F}(s)$, and the vector differential equation to solve to maintain the tangent of the field line alighted with the local field direction is

$$\frac{d\vec{F}(s)}{ds} = \frac{\vec{E}}{|\vec{E}|} \tag{13.17}$$

where $\vec{E}$ is the local electric field vector at the location of the current value of $\vec{F}$, and $|\vec{E}|$ is the electric field magnitude. The ratio on the right-hand side of Equation (13.17) is, therefore, the unit vector pointing in the direction of the field. This vector differential equation may be split into scalar equations for each component as

$$\frac{dF_x(s)}{ds} = \frac{E_x}{|\vec{E}|} \qquad \frac{dF_y(s)}{ds} = \frac{E_y}{|\vec{E}|} \qquad \frac{dF_z(s)}{ds} = \frac{E_z}{|\vec{E}|} \tag{13.18}$$

In most of the differential equations we encounter in physics, the derivatives are with respect to a spatial coordinate or time. But in this case, the derivative is with respect to a distance along the contour of a field line, no matter how the tangent to that line may change its orientation in the coordinate system.

We seek to draw a set of lines in this manner through the space around charges that are useful for visualizing the local strength of the resulting electric field. For this purpose, we should arrange the field lines so that the "channels" between all the lines will carry the same quantity of electric flux. To achieve this graphical goal, the initial points from which we launch the evolution of each field line should be a set of uniformly spaced points over a path where we are confident the flux density is uniform. The best locations for this condition are in the close vicinity of each positive point charge, near enough to the charge so that the local field will be almost completely dominated by the local charge, and circling around the charge with uniform angular intervals. The total number of initial points should be scaled by the strength of the originating positive charge. This will ensure that the channel formed between the lines we draw will all carry the same flux outward from the positive charge.

Inspect the source code in the program file `efieldlines.c` from the Student Companion website. The program accepts a specification of an arbitrary collection of electrical charges and their locations in the y–z plane on the command line. Note that the first two data sets sent to the output stream are simply a listing of the coordinates of the positive and negative charges in the collection. These may be plotted as points by `gnuplot` to aid the visualization of the sources and sinks of the electric flux, positive charges with black points and negative charges in gray. The C function that calculates the derivatives of Equation (13.18) and that is passed to the Runge–Kutta solver is as follows

```
void efieldlines(double x,ARRAY1D *y,void *param_ptr,ARRAY1D *yp) {
  double r_comp[3];
  ARRAY1D r = {.elem = r_comp,.n_elems = 3};
  double E_comp[3];
  ARRAY1D E = {.elem = E_comp,.n_elems = 3};
  double emag;
  CHGS *p;

  p = param_ptr;
  r.elem[0] = 0.0;
```

```
  r.elem[1] = y->elem[0];
  r.elem[2] = y->elem[1];
  efield_charges(&r,p,&E);
  emag = sqrt(E.elem[1]*E.elem[1] + E.elem[2]*E.elem[2]);
  yp->elem[0] = E.elem[1] / emag;
  yp->elem[1] = E.elem[2] / emag;
}
```

This function calls the utility routine `efield_charges()` from Chapter 4 to calculate the electric field due to some collection of point charges. The locations and values of the charges are specified in the parameter structure passed to the function as the pointer p. Note that the `efield_charges()` utility function was written generally to calculate the electric field in three dimensions, but we are only plotting the trajectories of the field lines in the two-dimensional y-z plane here.Therefore this function also discards the x component of the field and calculates only two-dimensional contour derivatives.

For the other simulation problems approached in this chapter, the total stepping range of the independent variable over which the differential equations vary is known in the statement of the problem. For example, in simulating the equations of motion of the pendulum, the user specifies to the program the total grid of time steps the Runge–Kutta algorithm is to sample the motion. However, for classic field lines, the independent variable is the distance along the line, and though we may specify an initial point and step size, the point at which the field line ends is dependent on the curved path it takes through the region of interest with its constraint to follow the electric field orientation. The total count of steps that are required to terminate the curve in the close vicinity of a negative charge is not known a priori. Therefore, this is an application for the `rk2_step()` tool in the `diffeq.c` utility. This tool merely takes a single step and relies on the calling program to test whether or not some graphical condition for terminating the sequence of steps has been reached. Since our goal is the generation of graphic images for visualization, which requires only modest numerical accuracy, the second-order Improved Polygon algorithm should be appropriate with reasonable distance step sizes. The code in the `efieldlines.c` program that sends the sequence of x, y coordinates along the field line as the third data set in the output stream for plotting is

```
done = 0;
s = 0.0;
while (!done && s < sfinal) {
  rk2step(efieldlines,s,sinc,&param,&y);
  printf("%.8e %.8e\n",y.elem[0],y.elem[1]);
  k = 0;
  while (k < param.n_chgs) {
    rsq = (y.elem[0]-param.chg[k].y) *
          (y.elem[0]-param.chg[k].y) +
          (y.elem[1]-param.chg[k].z) *
          (y.elem[1]-param.chg[k].z);
    r = sqrt(rsq);
    if (param.chg[k].q < 0.0) {
      if (r < smin) done = 1;
    }
    k++;
  }
  s += sinc;
}
```

The floating point variable s accumulates the distance traversed along the field line. An integer flag variable `done` is used to test for the termination conditions of the Runge–Kutta solution iterations when there is a negative charge in the collection that is within some minimum distance from the current drawn point. The dialog for compiling and running this program for the simple example of an electric dipole is

```
$ gcc -Wall -Wshadow -O2 -o efieldlines efieldlines.c CPLibrary/diffeq.c CPLibrary/point_charges.c
    ↪ CPLibrary/array.c -lm
$ ./efieldlines
./efieldlines <num charges> <s min> <s final> <s inc> <num lines per unit charge> <x1,y1,q1>
    ↪ <x2,y2,q2> ...
$ ./efieldlines 2 0.1 10000 0.01 48 0,1,1 0,-1,-1 >| efieldlines.dat
```

We have specified initializing the field line solutions at 48 points in a circular arrangement around each unit positive charge. Figure 13.14 shows the plot resulting from the example electric dipole drawn with the `gnuplot` command

```
plot 'efieldlines.dat' i 0 w p pt 7 lc 'black','' i 1 w p pt 7 lc 'gray','' i 2 u 1:2 w l lt -1
```

Figure 13.15 shows similar field lines around an example electric quadrupole generated by running

```
$ ./efieldlines 4 0.1 10000 0.01 48 -1,1,1 1,1,-1 -1,-1,-1 1,-1,1 >| efieldlines.dat
```

This same technique may be used to draw two-dimensional classic lines for magnetic fields as well. Instead of point charges, we can form lines around closed loops of electric current. This problem is presented as Exercise 13.4 for the student. We will again form a system of first-order differential equations to solve with a numerical solver that will evolve the contour of each field line from an initial point, this time with the tangent of the line guided by the magnetic field vector

$$\frac{dF_x(s)}{ds} = \frac{B_x}{|\vec{B}|} \qquad \frac{dF_y(s)}{ds} = \frac{B_y}{|\vec{B}|} \tag{13.19}$$

We would still wish to draw a set of field lines in the case of a magnetic field that defines channels through which the same magnetic flux is flowing. However, in general, the geometry of loops of electric current does not provide a region in which we know the field flux is close to uniform. We can approximate this goal by initiating our magnetic field lines with a uniform spacing through the center area of the current loop. In the case of the electric field, we drew our lines originating in the close vicinity of positive charges and terminated them in the close vicinity of negative charges in the

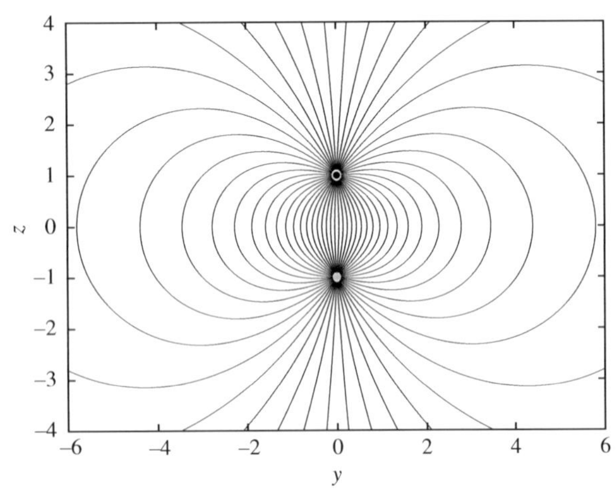

Figure 13.14 Classic electric field lines around an electric dipole.

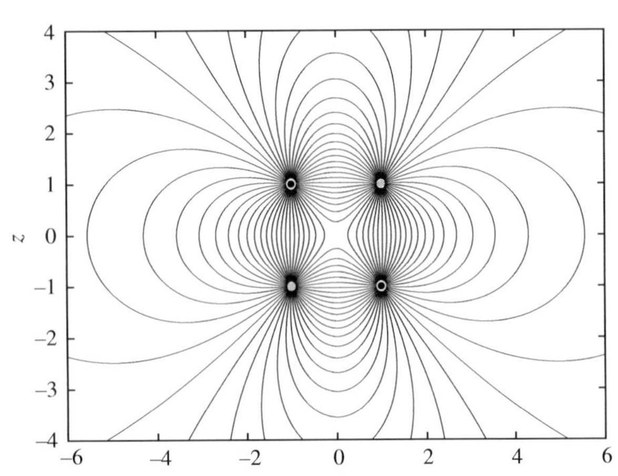

Figure 13.15 Classic electric field lines around an electric quadrupole.

collection. In the case of the magnetic fields, however, the field lines will originate at locations between the current loops and terminate when their contours form complete loops that return to the close vicinity of their origin points. The program for Exercise 13.4 should be used to draw field lines through a single current loop forming the magnetic dipole in Figure 13.16 and a practical solenoid with 10 turns in Figure 13.17. A useful approximation to the solenoid to simplify the geometry of the current path is to assume each turn to be an isolated circular loop of current, all in parallel planes, and with no actual current carrying connection between the turns.

In Section 1.3, we saw the Linux `time` command used to measure the execution time of two programs for modeling two-dimensional diffraction patterns and a Monte Carlo simulation. The solution program to Exercise 13.4 that generates the field line contours in Figures 13.16 and 13.17 involves a significant amount of computation. The contours of several field lines must be algorithmically traced, with each line requiring a large number of spatial steps in the differential equation solution. At each spatial step, a numerical integration of the Biot–Savart law must be performed to calculate the local magnetic field and to prepare the differential equation vector field. Running the solution code for the 10-turn solenoid of Exercise 13.4 through the Linux `time` command reports that the simulation code written efficiently in C and optimized by the compiler consumes about 18 s of processor time on an example of typical contemporary hardware. From the simple benchmarking exercises in Section 1.3 showing the substantial execution time penalties involved in coding intense numerical tasks in an interpreted language such as Python relative to optimized C code, we might expect to have to wait on the order of 15 min for our computation results with an interpreted programming language.

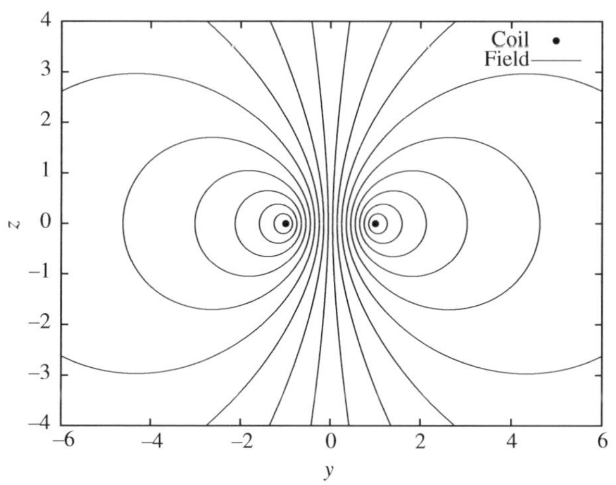

Figure 13.16 Field lines through a magnetic dipole.

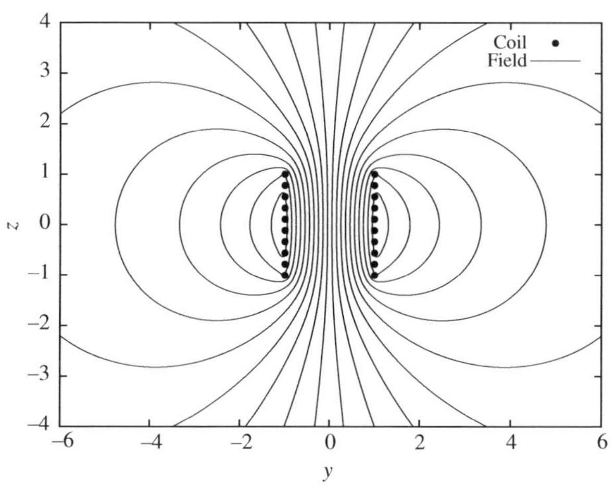

Figure 13.17 Field lines through a solenoid with finite height.

13.5 Application: Quantum Scattering Revisited

Let us now revisit the topic of the one-dimensional scattering of an incident quantum particle from a region of a potential barrier, which we first treated in Section 4.10. In that treatment, we assumed the commonly analyzed ideal profiles for the potential well, with an abrupt transition of the potential between neighboring regions of constant potential. This was done to be able to write down known normalizable wave function solutions of the time-independent Schrödinger equation

$$\frac{d^2\psi(x)}{dx^2} = -\frac{2m}{\hbar^2}\left[E - V(x)\right]\psi(x) \tag{13.20}$$

within the two regions, and solve for the conditions of continuity of the stationary wave function solutions across the potential transition. But abrupt potential transitions are not very physically realistic, as they require infinitesimally thin spots where infinite force would be exerted on a particle. With the tools for solving differential equations in this chapter, we can now find numerical solutions of the Schrödinger equation directly, and thereby allow more realistic potential profiles to be analyzed. Initially we will develop our solution strategies again with the ideal problem of the abrupt potential step to be able to compare these solutions to those in Section 4.10, but then extend our solutions to those for more realistic potential profiles.

First we examine the one-dimensional quantum mechanical problem of a particle represented by a wave packet approaching a positive step in potential from the left, illustrated in Figure 13.18. To start with, as in Section 4.10, we will assume that the step potential change is greater than the particle's energy, so the particle will eventually rebound from the step.

Here we will approach the problem of solving for a wave packet representing the incident and reflected particle in a manner different from that in Section 4.10. There we assumed that the regions of two different but constant potentials necessarily covered the entire range of the x coordinate, and meeting at a discontinuous step in potential at $x = 0$. As we prepare for treating a more realistic problem of a graded change in potential, we assume that there will still be a field region of zero potential, constant in x, somewhere on the left side through which we know the wave function solutions must follow

$$\psi(x) = A_1 e^{+ik_1x} + B_1 e^{-ik_1x} \tag{13.21}$$

where $k_1 = \frac{\sqrt{2mE}}{\hbar}$, but that this constant potential region does not necessarily extend all the way to $x = 0$. The coefficient A_1 determines the flux of particles incident from the left, and the coefficient B_1 scales the flux of particles scattered by the collision with the potential barrier and reflected back to the left. We similarly assume that there will still be a field region somewhere on the right side with a constant potential at the height of the barrier, but that this region does not necessarily start abruptly at $x = 0$. Since we are considering only wave functions that are normalizable, we set the coefficient $A_2 = 0$, and the solution in this constant potential region must follow the decaying exponential form

$$\psi(x) = B_2 e^{-k_2x} \tag{13.22}$$

where $k_2 = \frac{\sqrt{2m(V_0-E)}}{\hbar}$.

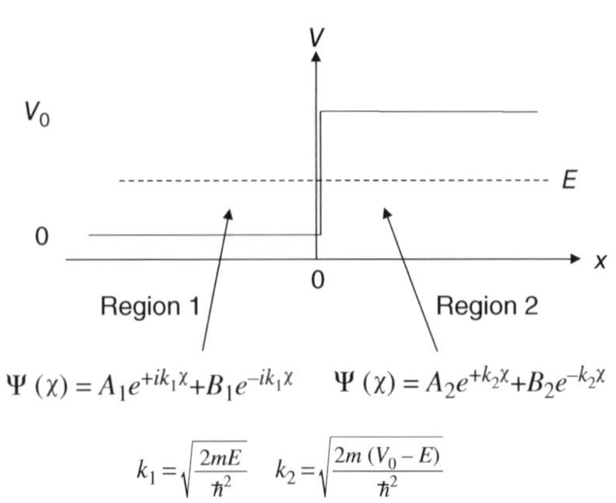

Figure 13.18 Profile of an abrupt potential step.

As we have seen in this chapter solving other second-order differential equations, the solution process requires the definition of two state variables and the assignment of an appropriate initial condition to both of them. As before, we follow the scheme of Equation (12.20) and define the two state variables

$$y_0 = \psi(x) \qquad y_1 = \frac{dy_0}{dx} = \frac{d\psi}{dx} \tag{13.23}$$

The system of two first-order differential equations in the form suitable for the numerical solver is then

$$\frac{dy_0}{dt} = y_1 \qquad \frac{dy_1}{dt} = \frac{2m}{\hbar^2}(V(x) - E)y_0 \tag{13.24}$$

for some potential profile $V(x)$.

Instead of assuming that these wave function solution forms meet at $x = 0$ across an infinitesimally thin transition and solving for the continuity conditions at that interface, we will now make use of a numerical solution to the Schrödinger equation (13.20) through the entire range of x. As in the examples in the previous sections, to initiate a numerical solution of the governing differential equation, we will make use of the assumed form of Equation (13.21) in the initiation region only for the purpose of defining the initial values of the two state variables in the differential system, in this case the stationary wave function $\psi(x)$ and its first derivative $\frac{d\psi}{dx}$. Similarly, in the final solution region on the right side, we will make use of the assumed form of Equation (13.22) only for the purpose of defining two state variable values that will take part in a numerical solution of a consistency constraint between incident and reflected waves. Otherwise, we rely solely on the Runge–Kutta algorithm to trace out the required trajectory of the stationary wave function over x. This solution will indeed follow Equations (13.21) and (13.22) where x passes through regions of constant potential (within the allowed truncation error, of course), but will also dutifully follow other trajectories with challenging analytic solutions through regions of nonuniform potential, and will maintain the required continuity of the wave function and its derivative over all x.

Since wave function solutions of the linear and homogenous Equation (13.20) may be determined only within an arbitrary multiplicative constant, how do we then establish the consistency constraints that may be solved numerically for the unknown coefficients A_1, B_1, and B_2 that characterize the particle scattering process? We know from the properties of linear homogeneous differential equations that any linear combination of solutions is also a solution. That allows us to write a system of two linear (and complex-valued) equations that relates the two initial state variable values and the two final values where the Runge–Kutta iterations terminate. Therefore, there will be some complex-valued matrix T such that

$$\begin{bmatrix} t_{11} & t_{12} \\ t_{21} & t_{22} \end{bmatrix} \begin{bmatrix} \psi_{\text{init}} \\ \psi'_{\text{init}} \end{bmatrix} = \begin{bmatrix} \psi_{\text{final}} \\ \psi'_{\text{final}} \end{bmatrix} \tag{13.25}$$

We plan on initializing a solution that will follow Equation (13.21) within a left-side region of constant potential, so we will set the initial values of the two state variables as

$$\begin{aligned} \psi(x_{\text{init}}) &= A_1 e^{+ik_1 x_{\text{init}}} + B_1 e^{-ik_1 x_{\text{init}}} \\ \psi'(x_{\text{init}}) &= ik_1 A_1 e^{+ik_1 x_{\text{init}}} - ik_1 B_1 e^{-ik_1 x_{\text{init}}} \end{aligned} \tag{13.26}$$

with the understanding that the complex coefficients A_1 and B_1 are yet to be solved for. Note that the complex exponential functions are evaluated at $x = x_{\text{init}}$. We assume that we are working toward analyzing the case of a potential barrier with some gradual profile centered at $x = 0$, so the initial value of x_{init} on the left side will be less than zero. As we intend to assemble solutions into wave packets to represent a particle, initializing the state variables in the manner of Equation (13.26) will follow the similar treatment in Section 4.10, where all the constituent solutions of the wave packet will follow a trajectory heading toward a phase alignment in the vicinity of $x = 0$. We may absorb the values of the complex exponentials into the coefficients of the T matrix and rewrite the constraint system with a different matrix U as

$$\begin{bmatrix} u_{11} & u_{12} \\ u_{21} & u_{22} \end{bmatrix} \begin{bmatrix} A_1 \\ B_1 \end{bmatrix} = \begin{bmatrix} \psi_{\text{final}} \\ \psi'_{\text{final}} \end{bmatrix} \tag{13.27}$$

The elements of the matrix U are dependent on the trajectories over x that the wave function solutions must follow to conform to the potential profiles of a particular physical system. Knowing the $u_{i,j}$ elements will allow us to solve for the unknown coefficients A_1 and B_1. How may we find these four matrix elements? One simple method is to initialize and solve two wave function solutions with the two fixed combinations of $A_1 = 1, B_1 = 0$ and $A_1 = 0, B_1 = 1$, respectively, and fill in the columns of the U matrix with the values of the state variables found at the termination of the Runge–Kutta iterations. This method is described by the two relations

$$\begin{bmatrix} u_{11} & u_{12} \\ u_{21} & u_{22} \end{bmatrix}\begin{bmatrix} 1 \\ 0 \end{bmatrix} = \begin{bmatrix} \psi_{\text{final1}} \\ \psi'_{\text{final1}} \end{bmatrix} = \begin{bmatrix} u_{11} \\ u_{21} \end{bmatrix} \tag{13.28}$$

$$\begin{bmatrix} u_{11} & u_{12} \\ u_{21} & u_{22} \end{bmatrix}\begin{bmatrix} 0 \\ 1 \end{bmatrix} = \begin{bmatrix} \psi_{\text{final2}} \\ \psi'_{\text{final2}} \end{bmatrix} = \begin{bmatrix} u_{12} \\ u_{22} \end{bmatrix} \tag{13.29}$$

Now the two scenarios of $A_1 = 1, B_1 = 0$ and $A_1 = 0, B_1 = 1$ are not particularly physically meaningful. We know that eventually a particle will be reflected from a potential greater in height than its energy, so a scenario where an incident particle is not reflected, or where a particle is reflected when none was incident, is not physically feasible. However, the actual physical solution will be in the space spanned by these two vectors, and so these two test conditions are mathematically useful to find the elements of the matrix U.

All that remains to find a solution for A_1 and B_1 is to set the right-hand vector of the linear system of Equation (13.27) to that vector required to yield a physically meaningful normalizable wave function. We know that Equation (13.22) must, therefore, describe the solution in the termination region of constant potential on the right side. We are only able to solve any linear, homogeneous differential equation within an arbitrary multiplicative constant, so we may set $B_2 = 1$ and evaluate the required values of the final state variables as

$$\begin{aligned} \psi_{\text{final}} &= e^{-k_2 x_{\text{final}}} \\ \psi'_{\text{final}} &= -k_2 e^{-k_2 x_{\text{final}}} \end{aligned} \tag{13.30}$$

We now have all the pieces in place to solve the linear system of Equation (13.27). Since this is only a system of two equations in two unknowns, we can calculate the A_1 and B_1 values simply by applying Cramer's rule with complex arithmetic.

$$A_1 = \frac{\det\begin{bmatrix} \psi_{\text{final}} & u_{12} \\ \psi'_{\text{final}} & u_{22} \end{bmatrix}}{\det\begin{bmatrix} u_{11} & u_{12} \\ u_{21} & u_{22} \end{bmatrix}} \qquad B_1 = \frac{\det\begin{bmatrix} u_{11} & \psi_{\text{final}} \\ u_{21} & \psi'_{\text{final}} \end{bmatrix}}{\det\begin{bmatrix} u_{11} & u_{12} \\ u_{21} & u_{22} \end{bmatrix}} \tag{13.31}$$

As in Section 4.10, we construct a wave packet centered at some nominal energy level E_0 that is less than the height of the potential step, find the wave number $k_0 = \frac{\sqrt{2mE_0}}{\hbar}$ that corresponds to this single wave solution, and then construct a weighted sum of wave solutions. The wave numbers k in the packet are uniformly and symmetrically surrounding k_0 with the weighting coefficient following Equation (4.26). Then the initializing values for each constituent wave are calculated with Equation (13.26) at the corresponding energy level $E = \frac{\hbar^2 k^2}{2m}$. Figure 13.19 plots 50 of the 101 stationary wave function solutions for an abrupt potential step that will be combined into a wave packet that represents the scattered particle.

When the wave packet is constructed with the Gaussian weighting coefficients and the time-dependent factor included in a manner identical to the model developed in Section 4.10, we can visualize the time sequence of the particle approaching and rebounding from the potential step in Figure 13.20.

Another interesting case is when the incident particle has an energy level higher than the height of the potential barrier. In that case, we know that the solution on the right side will follow

$$\psi(x) = A_2 e^{ik_2 x} \tag{13.32}$$

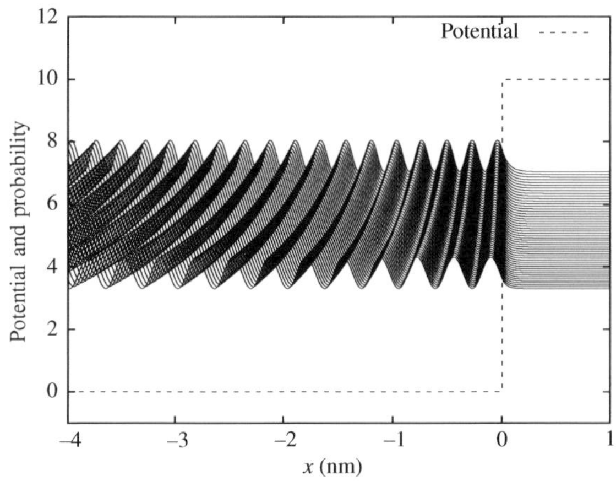

Figure 13.19 The family of wave functions composing a wave packet for a particle with energy below an abrupt potential step.

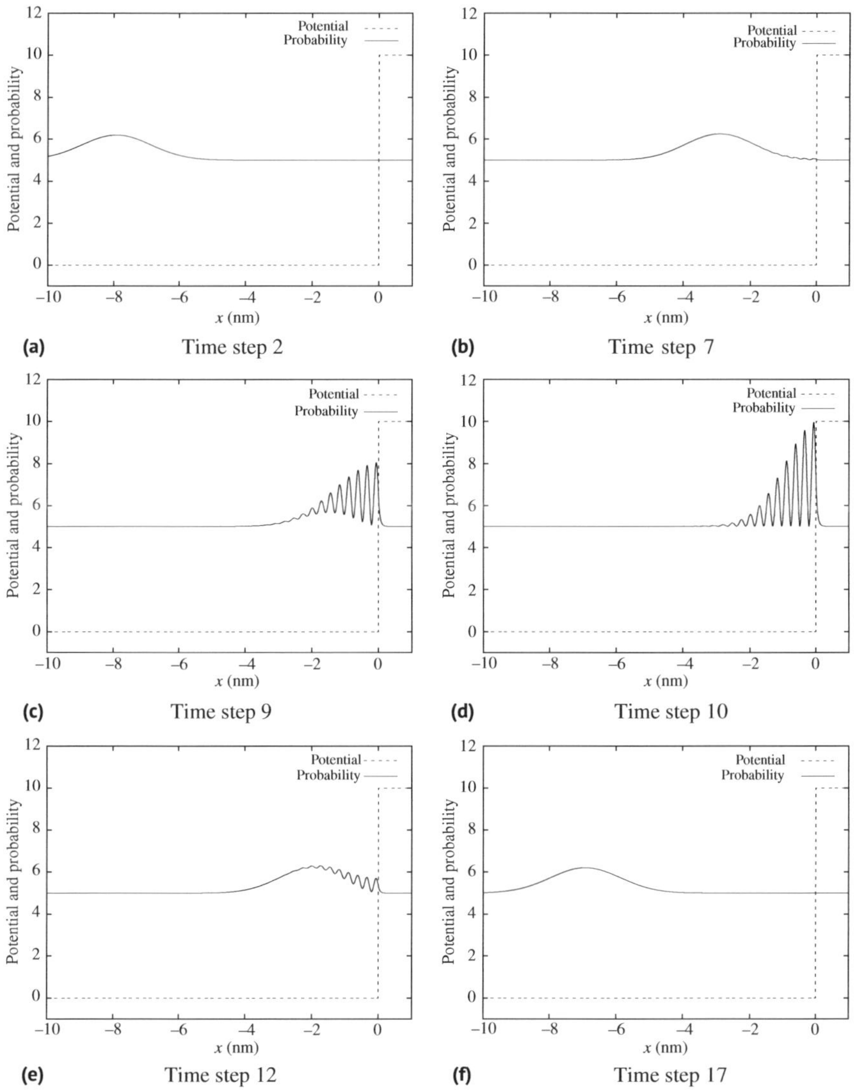

Figure 13.20 Time solution of an electron rebounding from an abrupt quantum potential step.

describing a particle continuing to propagate to the right after passing the barrier. If a conditional test in the program reveals that the particle energy is greater than the potential in the final termination region, we change the assignment of the final state variables in the linear system with the arbitrary choice of $A_2 = 1$ to

$$\begin{aligned} \psi(x_{\text{final}}) &= e^{ik_2 x_{\text{final}}} \\ \psi'(x_{\text{final}}) &= ik_2 e^{ik_2 x_{\text{final}}} \end{aligned} \tag{13.33}$$

Figure 13.21 plots 50 of the 101 stationary wave function solutions for the case of an energy level of $E = 11$ eV, lifting it above the 10-eV barrier height. These are the solutions that will be combined into a wave packet that represents the overall probability distribution of the particle. For the constituent waves corresponding to energies in the lower portion of the range in the packet, the stationary wave functions terminate on the right side in the form of Equation (13.22), but for higher energies, the waves terminate in the form of Equation (13.32), and the probabilities given by the square of the wave function will be constant with respect to x on the right end.

Combining this collection of stationary waves with Gaussian weighting and multiplying by the time dependence factor for an electron energy level above the barrier height shows the preponderance of the wave packet probability passing the barrier to the right as expected, but also shows a small probability of the particle rebounding from the barrier. This result, plotted in Figure 13.22, is impossible for a physical system in the realm of classical mechanics.

These numerical solutions for an abrupt potential step are really duplicating the solution we found analytically in Section 4.10. Of course, the unique value of approaching this quantum scattering problem numerically is to make available solutions with much more realistic potential profiles that do not demand a localized unbounded force. So we now consider a linearly sloped potential profile of Figure 13.23 as an example of an arbitrary profile that may be coded. Figure 13.24 plots 50 of the 101 stationary wave function solutions that will be combined into a wave packet that represents a particle approaching this barrier.

When the wave packet is constructed with the Gaussian weighting coefficients and the time-dependent factor included, we can visualize the time sequence of the particle approaching and rebounding from the potential step in Figure 13.25.

Comparing the progress of the wave packet in Figure 13.25 with the plots of Figure 13.20, we can see that the probability that the electron spends many more simulated time steps in the vicinity of the potential barrier. We may also compare the progress of the rebounding particle shown in Figures 13.20f and 13.25f we see that scattering off of the abrupt step involves a significantly shorter time delay than for the linearly graded step. We can interpret this behavior as comparable to a classical particle "rolling up" a graded potential, and then "falling" back down on the rebound. This analysis shows that the longer rebounding delay is a complex combination of the penetration into the classically forbidden region and "rolling up" the potential gradient. The problem of modeling an electron scattering against a potential that has the quadratic profile in Figure 13.26, comparable to compressing a quantum "spring," is detailed in Exercise 13.5.

A required enhancement of the numerical routines for differential equations that must be made for the problems of this section, however, becomes clear when we consider that the state variables and coefficients in Equation (13.24) are,

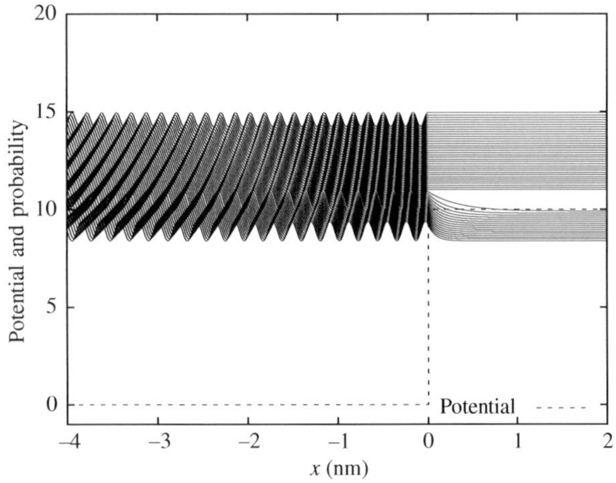

Figure 13.21 The family of wave functions composing a wave packet for a particle with energy above an abrupt potential step.

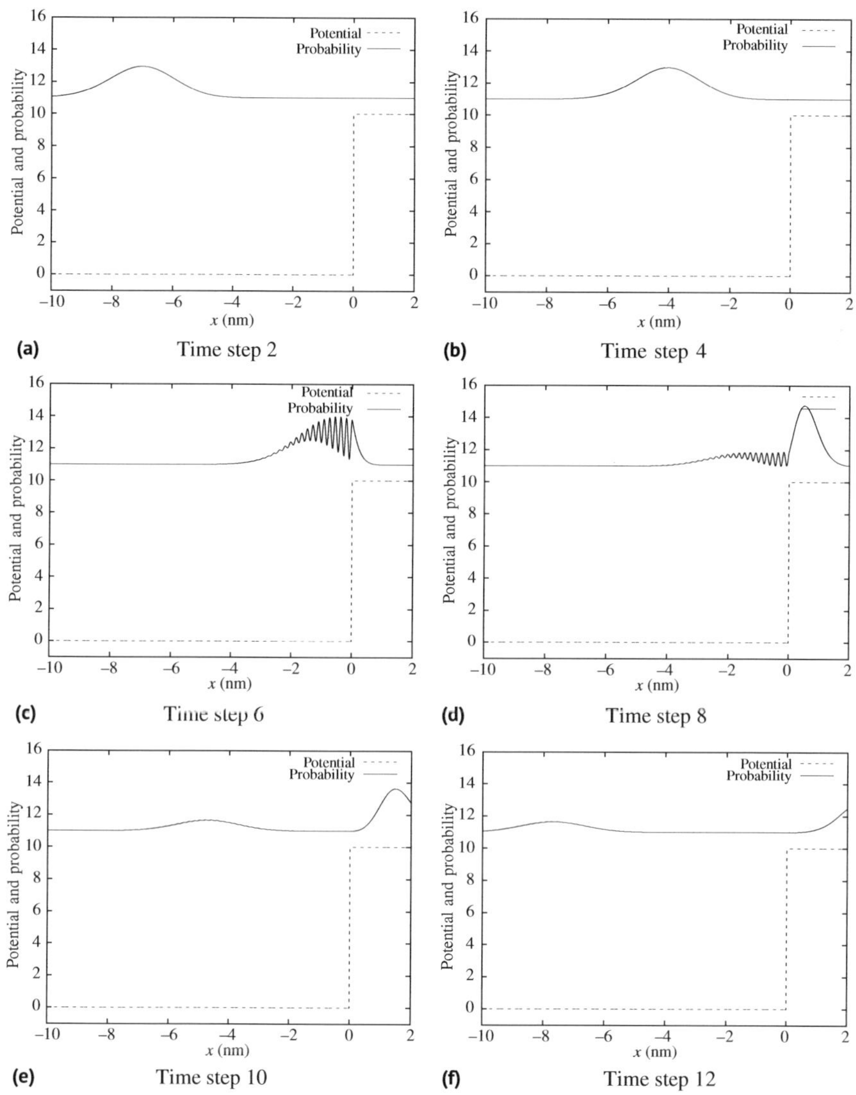

Figure 13.22 Time solution of an electron passing over an abrupt quantum potential step.

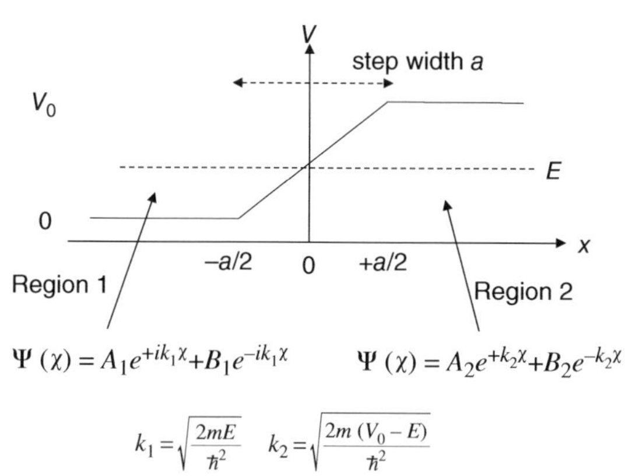

Figure 13.23 Profile of a linearly sloped potential step.

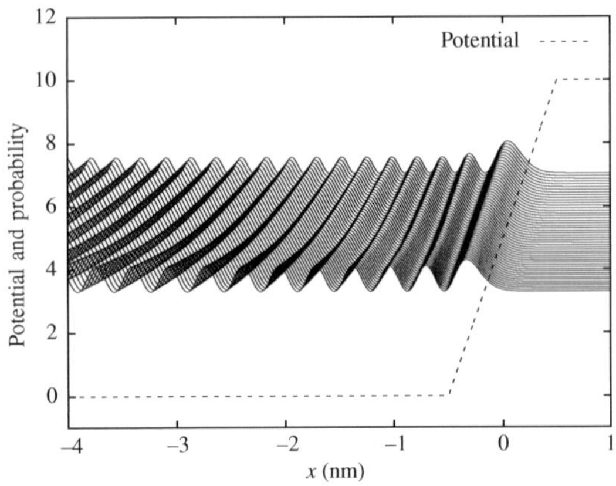

Figure 13.24 The family of wave functions composing a wave packet for a linearly sloped potential step.

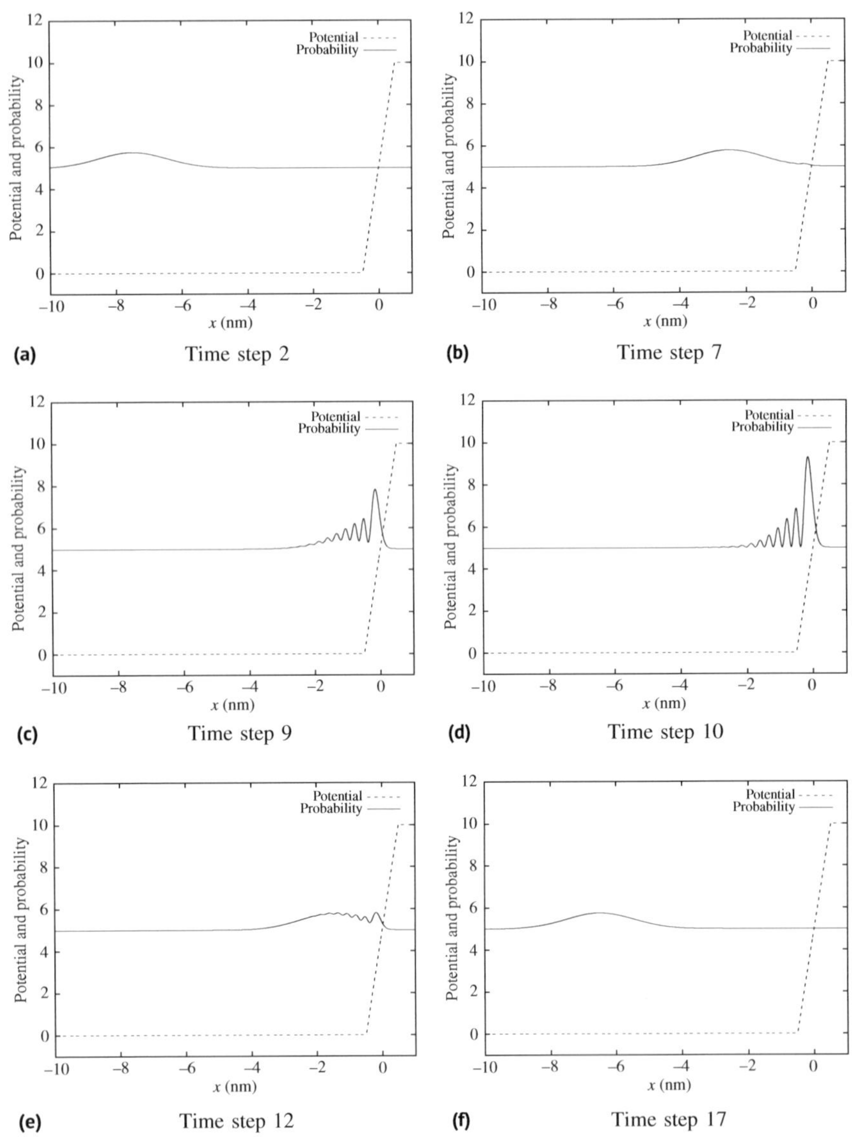

Figure 13.25 Time solution of an electron rebounding from a linearly sloped quantum potential step.

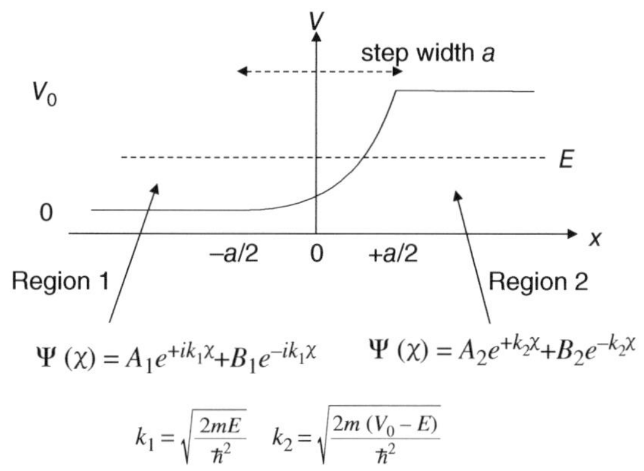

Figure 13.26 Profile of a quadratic potential step.

in general, complex valued. A set of routines that correspond to the routines described in Section 12.4 are provided in the additional tool code files `diffeq_complex.c` and `diffeq_complex.h` in the Student Companion download site. These functions are basically copies of the real-valued algorithms in `diffeq.c` but with the `double` type for the state variables replaced by the C type `double complex`, similar to what we saw with the `step_complex.c` tool in Section 4.9.

See the source code in the file `quantum_linbarrier_packet.c` from the Student Companion website for the complete solution to the simulation problem of a wave packet scattering from a linearly graded barrier. The parameter structure and the function computing the derivatives of Equation (13.24) are

```
typedef struct qbar_wave_param {
  double mass;  // particle mass in kg
  double energy_J;  // particle energy in J
  GRID xgrid_nm;  // xgrid for evaluation of wave function in nm
  GRID xgrid_m;  // xgrid for evaluation of wave function in m
  QBARPOT *qbpp;  // pointer to quantum barrier potential parameters
} QBARWAVE;

void quantum_barrier_wave(double x,ARRAY1D_COMPLEX *y,void *param_ptr,
                          ARRAY1D_COMPLEX *yp) {
  QBARWAVE *p;

  p = param_ptr;
  yp->elem[0]  = y->elem[1];
  yp->elem[1]  = (2.0 / (PLANCK_REDUCED * PLANCK_REDUCED)) * p->mass *
                 (quantum_barrier_potential(x,p->qbpp) -
                 p->energy_J) * y->elem[0];
}
```

This code expects that a function `quantum_barrier_potential()` has been defined in the source code that calculates the potential function $V(x)$ in Equation (13.24).

Compiling and running the program

```
$ gcc -Wall -Wshadow -O2 -o quantum_linbarrier_packet quantum_linbarrier_packet.c
   ↪ CPLibrary/diffeq_complex.c CPLibrary/step.c CPLibrary/array_complex.c CPLibrary/array.c -lm
$ ./quantum_linbarrier_packet
./quantum_linbarrier_packet <width in nm> <height in eV> <mass * c^2 in Mev> <E0> <a in nm> <xinit
   ↪ in nm> <xfinal in nm> <xinc in nm>
$ ./quantum_linbarrier_packet 1 10 0.5110 5 0.5 -4 1 0.0025 >| quantum_linbarrier_packet.dat
```

Now we consider a particle approaching a barrier with elevated potential over only some restricted range of x, as shown in Figure 13.27. This exercise will let us simulate the possibility of the incident particle tunneling through the barrier. The probability of a tunneling event gives us a visualization of the nuclear decay events that we simulated with Monte Carlo events in Section 8.7.

The wave function on the left side now is a complex exponential function, giving an oscillatory spatial dependence corresponding to a flux of particles. These particles may be traveling to the right or left. As with the case of a potential step,

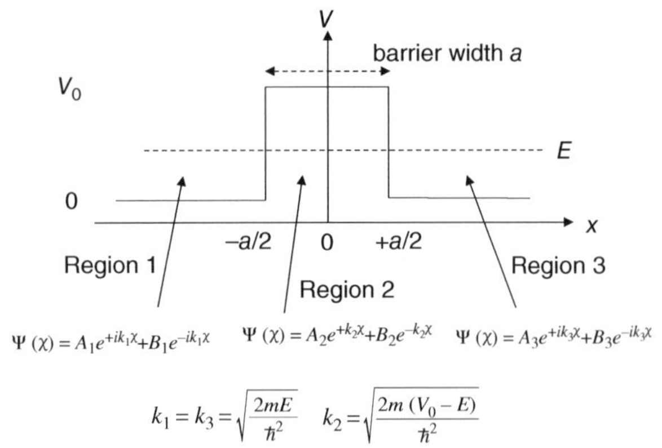

Figure 13.27 Profile of a rectangular potential barrier.

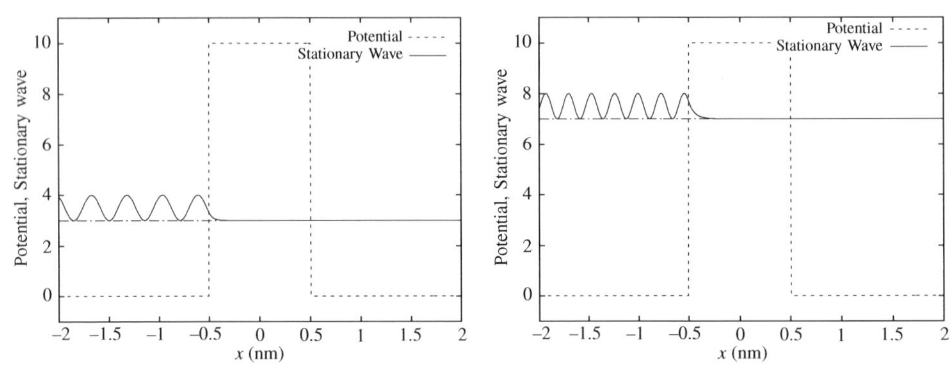

(a) Energy level 3 eV **(b)** Energy level 7 eV

Figure 13.28 Stationary wave function for an electron approaching a thick rectangular barrier potential, width 1 nm.

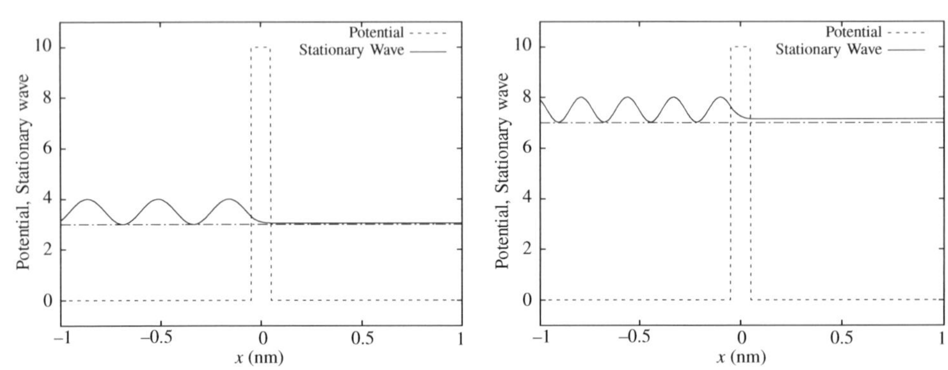

(a) Energy level 3 eV **(b)** Energy level 7 eV

Figure 13.29 Stationary wave function for an electron approaching a thin rectangular barrier potential, width 0.1 mm.

the electron is not bound to a particular region in space, so the electron energies are not limited to certain quantized states, but rather solutions are available at any arbitrary energy level and wave number.

Similarly the wave function solution on the right side of the barrier will also be oscillatory. We are now targeting solutions in this regime with waves propagating only to the right, and with no incident component propagating to the left. Just as in the case of scattering from a potential step, the appropriate final state variables to assign in the consistency solution are those of Equation (13.33).

The wave functions plotted in Figures 13.28 and 13.29 correspond to a simple rectangular potential profile. This case can actually be analyzed analytically.

With the numerical approach, we are free to analyze systems with any arbitrary potential that may be coded. The potential barriers with a trapezoidal profile in Figures 13.30 and 13.31 represent any coded potential profile.

Figure 13.32 plots a wave packet representing a particle impacting a thick trapezoidal barrier of 1-nm width.

Then the case of a thin trapezoidal barrier of only 0.2-nm thickness is plotted in Figure 13.33. In this case, we can see a much larger wave probability tunneling through the barrier and propagating off to the right compared with the case of the 1-nm thick barrier of Figure 13.32.

To provide more effective visualization of the various cases of quantum scattering analyzed above, example animated plots in the style of those generated by the coding techniques of Section 4.13 are available on the Student Companion website in the files `quantum_barrier_packet_gpanim_pa.gif`, `quantum_barrier_packet_gpanim_pl1.gif`, `quantum_barrier_packet_gpanim_pq1.gif`, `quantum_barrier_packet_gpanim_pt1.gif`, and `quantum_barrier_packet_gpanim_pt0p2.gif`.

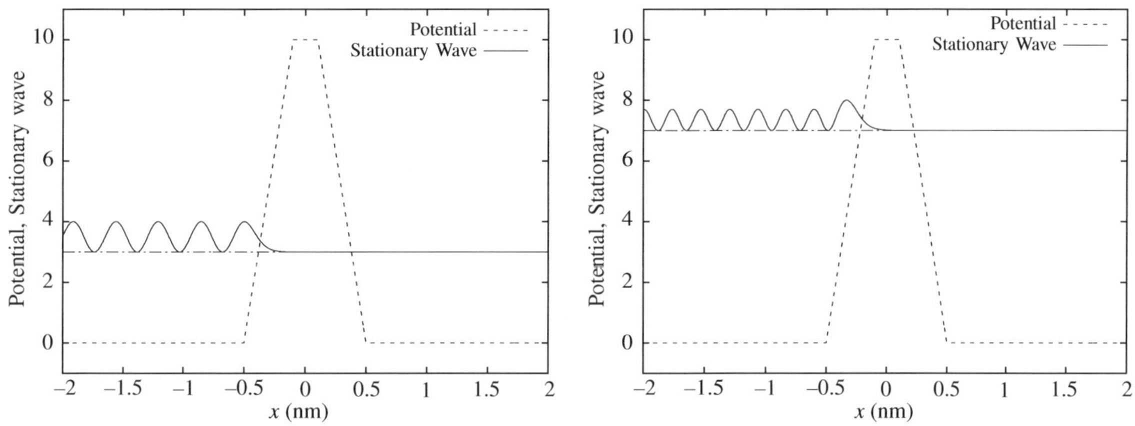

(a) Energy level 3 eV (b) Energy level 7 eV

Figure 13.30: Stationary wave function for an electron approaching a thick trapezoidal barrier potential, width 1 nm.

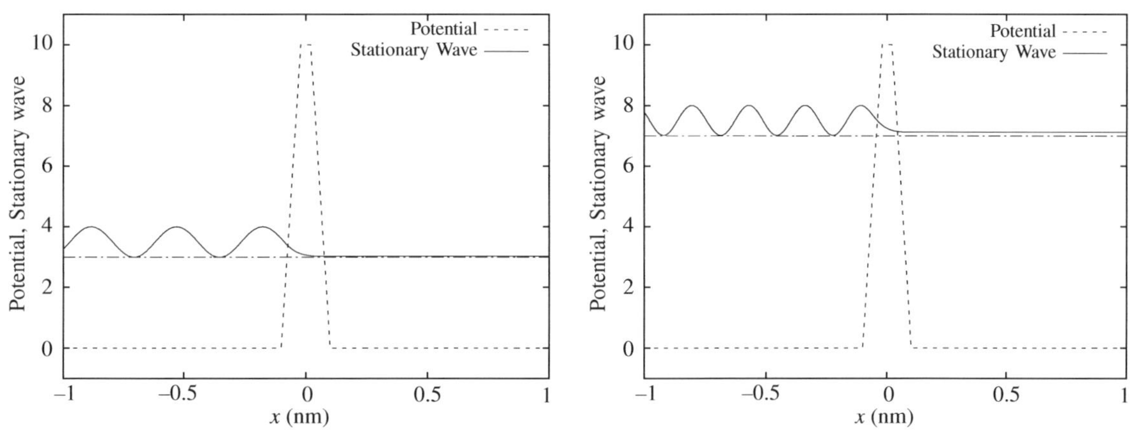

(a) Energy level 3 eV (b) Energy level 7 eV

Figure 13.31 Stationary wave function for an electron approaching a thin trapezoidal barrier potential, width 0.2 mm.

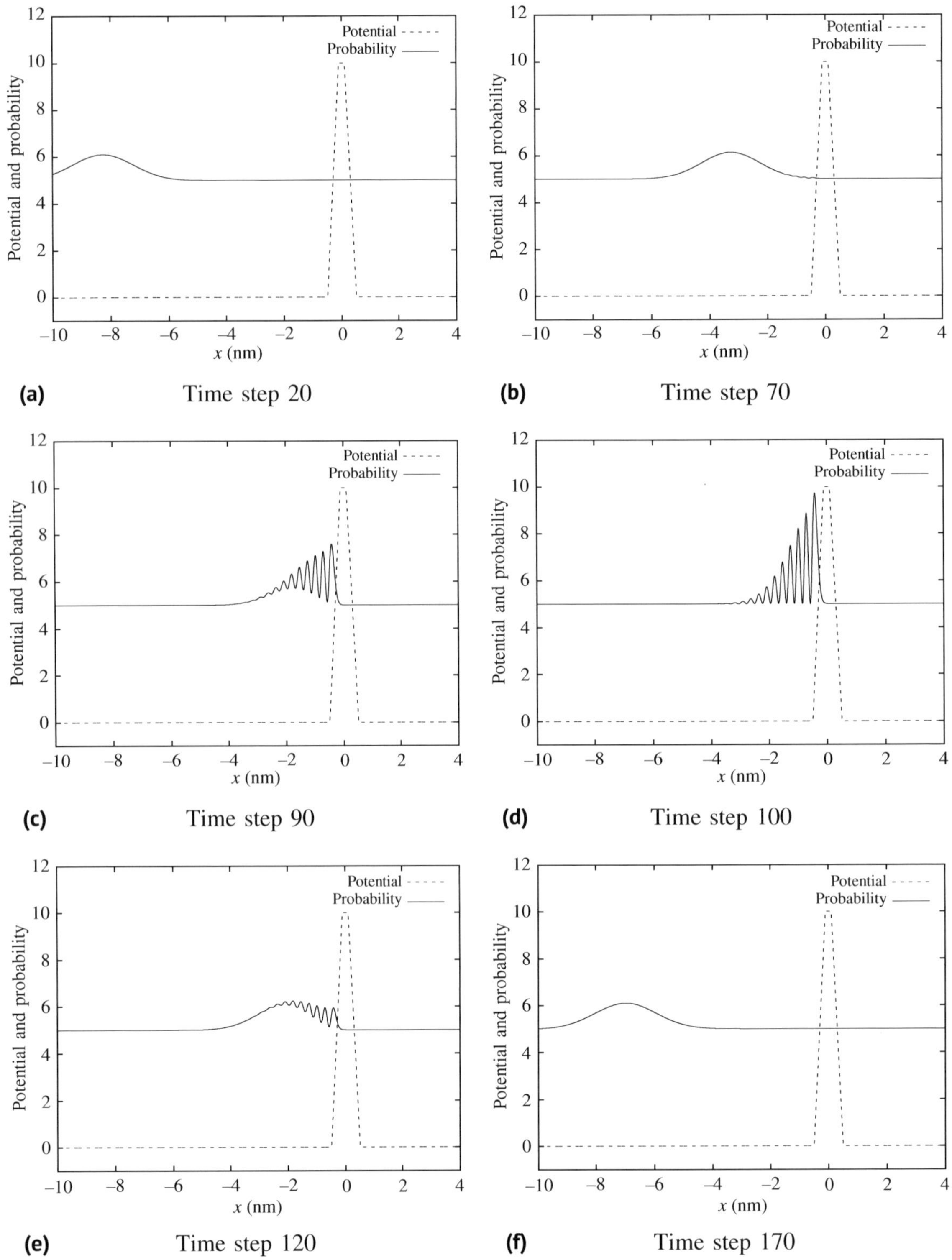

Figure 13.32 Time solution of an electron rebounding from a thick quantum potential barrier.

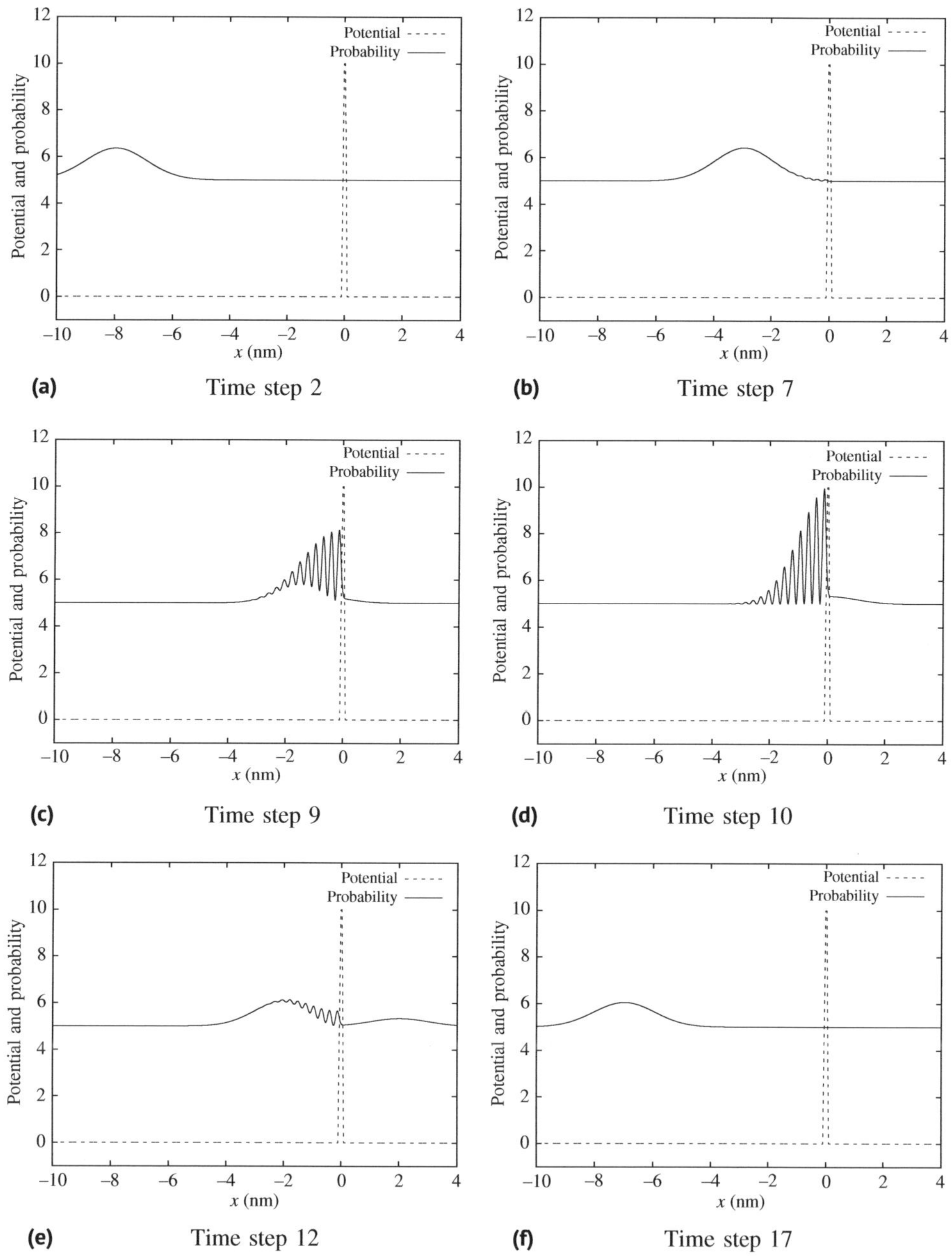

Figure 13.33 Time solution of an electron both rebounding from and tunneling through a thin quantum potential barrier.

13.6 Application: Solid State Physics

In Sections 5.11 and 7.7, we addressed the eigenvalue problems associated with finding the allowed energies of an electron bound in a quantum potential well. In those treatments, we assumed the commonly used idealized profiles for the potential well, with abrupt transitions of potential between neighboring regions of constant potential. This was done to be able to write down known solution forms for the time-independent Schrödinger equation

$$\frac{d^2\psi(x)}{dx^2} = -\frac{2m}{\hbar^2}\left[E - V(x)\right]\psi(x) \tag{13.34}$$

within the regions, and join the piecewise stationary wave function solutions together with nonlinear conditions on the arbitrary constants to assemble a normalizable solution over all space coordinates. But, as discussed in Section 13.5, abrupt potential transitions are not physically realistic. We continue in this section to develop techniques to find numerical solutions of the Schrödinger equation directly, thereby allowing more realistic potential profiles to be analyzed. Initially we will develop our solution strategies with the ideal problem of the rectangular potential wells solved in Section 5.11, and then extend them to more physically realistic potential profiles.

For a solution to the Schrödinger equation to be physically meaningful, it must not only satisfy the differential trajectory imposed by the equation originating from initial conditions, but it must also conform to the requirement of a normalizable wave function over all space. This requirement ensures that the probability of the particle's existence somewhere in space may be set to the required value of one with an arbitrary multiplicative constant. These conditions define a boundary value problem. In Section 6.6 we solved the boundary value problems associated with the Laplace equation. However, the estimates of the derivatives used there were of only the second order, and a level of truncation error was seen in the plot of Figure 6.9 that may be undesirable in many applications. We will instead approach the boundary value problems in this section with the fourth-order Runge–Kutta algorithm within the strategy of the co-called "shooting" method, which can achieve significantly lower truncation error. Moreover, we will see that the shooting method also works more naturally with the eigenvalue problem inherent in seeking multiple allowed energies of quantum mechanical scenarios. For simplicity, we continue to treat these problems in only one dimension.

In Section 13.1, we solved nonlinear differential equations that describe a ballistic trajectory under the influence of air drag. One of the example programs developed was `ballistic_range_solve.c`, which used the numerical root finding tools from Chapter 5 to solve for the ball launch parameter that landed a ball at a desired target range. To land on the target, the root-finding algorithm converged on the root of a function that predicted the target error distance for a particular candidate launch angle. Consider that with this numerical process we are essentially simulating the experience of a golfer learning to land a ball in a golf course hole by adjusting the launch momentum and angle over a sequence of golf shots. The golfer is converging on an eigenvalue solution of a boundary value problem with the shooting method.

Applying this same method to the allowed energy eigenvalues of a bound electron, we assume that whatever the potential profile may be in the region of interest in which we are numerically solving the Schrödinger equation, the extremes of the x dimension represent field regions of a constant zero reference potential. The constant potential in the field regions implies that the wave function solution will be either a growing or a decaying exponential with increasing x, so that the wave function will evanescently decay to insignificant probabilities at either extreme, as shown in Figure 5.27. Since the required form for the wave function at the left end of the x range is $\psi(x) = A_1 e^{+k_1 x}$, we must "launch" the "shot" of a candidate wave function solution at the left end of the x dimension corresponding to some candidate energy level. Assuming that we assign the potential a value of zero in the field regions, and that the electron energy is less than zero so that it is bound in the intervening potential well, the wave number k_1 is determined by the candidate electron energy E through the relation

$$k_1 = \sqrt{\frac{-2mE}{\hbar^2}} \tag{13.35}$$

By using the term "launch," and invoking the image of a golf shot, we may be in danger of restricting our thinking to solving only equations of motion in which the differentiation in the system equations is with respect to time. The time-independent Schrödinger equation through which we find energy eigenvalues involves only differentiation with respect to a spatial coordinate, but we may think of initiating the sampled sequence of wave function solutions that evolve from left-hand x argument values to right-hand x values as a comparable "launching" process.

For the second-order pendulum motion solutions of Section 12.5, we applied initial conditions for both state variables, in that case the deflection angle and the angular velocity. For solving the second-order Schrödinger equation, we use the same definitions of the state variables from Equations (13.23) and (13.24), and we have the freedom of initializing the values of both $\psi(x)$ and $\frac{d\psi(x)}{dx}$ at some x coordinate value x_{init} in the field region at the left end of the geometry. In particular, we use this freedom to select an initial ratio of $\psi(x)$ to $\frac{d\psi(x)}{dx}$ that will support only a growing exponential of the form $\psi(x) = A_1 e^{+k_1 x}$ as the solution evolves at increasing x coordinates through the field region. Since finding allowed energy levels requires only normalizable wave functions, not necessarily actually normalized waves, we can set the constant A_1 arbitrarily to keep the range of wave function values reasonable within the floating point numerical range. Choosing $A_1 = 1$ arbitrarily, if we initialize the solution $\psi(x_{\text{init}}) = 1$ and $\frac{d\psi(x_{\text{init}})}{dx} = k_1$ for a candidate energy level at the initial x value in the left-side field region, we will necessarily launch a candidate wave function solution with only the growing exponential term, a wave function that will at least be normalizable on its left side.

In our first look at solutions, we will look at abrupt potential regions in order to compare against solutions from Section 5.11, but we are working toward accommodating any arbitrary potential profile that forms a bounding well or wells. Since we will evolve solutions at increasing x coordinates governed by the Schrödinger equation, including any physically feasible variation in the potential $V(x)$ we can code in our C functions, we eliminate any need to stitch together piecewise analytic solutions within the wells. What we will require, however, is a method for quantifying the error associated with any candidate wave function solution between the trajectory of the solution in the right-side field region and a normalizable solution, so that a root-finding algorithm may be used to converge on an eigenvalue of energy in the same manner that was used for solving the golf shot in Section 13.1. To appreciate graphically the task of the shooting method, see the four candidate solutions shown in Figure 13.34. Since we are starting with a rectangular potential well, and we have solved for the eigenvalues of energy for this potential in Section 5.11, we can trace candidate wave function solutions that we know are for electron energies slightly above and below the lowest two allowed energies. Although we have carefully launched normalizable solutions on the left side, we can see that for the very small energy deviations in each of these candidate trajectories, the solution on the right-side field region is contaminated with a growing exponential component of either sign that yields an unnormalizable wave function overall.

To guide the convergence of the shooting method onto an energy eigenvalue, how do we form a numerical "target error function," similar in concept to the range error function in the golf ball ballistics problem of Section 13.1? For the ball launching problem, we could form a meaningful measurement of the target error with a simple distance metric. What metric can we define that measures the "distance" between a candidate wave function trajectory and the target of a normalizable wave function, so that we may find the root at which that "distance" passes through zero? As the candidate wave function solution evolves through x, the Runge–Kutta algorithm will, of course, trace the values of all state variables, in this case, $\psi(x)$ and $\frac{d\psi(x)}{dx}$. Consider forming a test function as

$$f(E) = \psi(x_{\text{final}}) + \frac{1}{k_3}\frac{d\psi(x_{\text{final}})}{dx} \tag{13.36}$$

where x_{final} is the final x value in the grid of coordinates over which the Schrödinger equation is being solved. The motivation for this formulation is that we know the only trajectories of general solutions that the Schrödinger equation will support in the field region on the right side are a linear combination of the form

$$\psi(x) = A_3 e^{+k_3 x} + B_3 e^{-k_3 x} \tag{13.37}$$

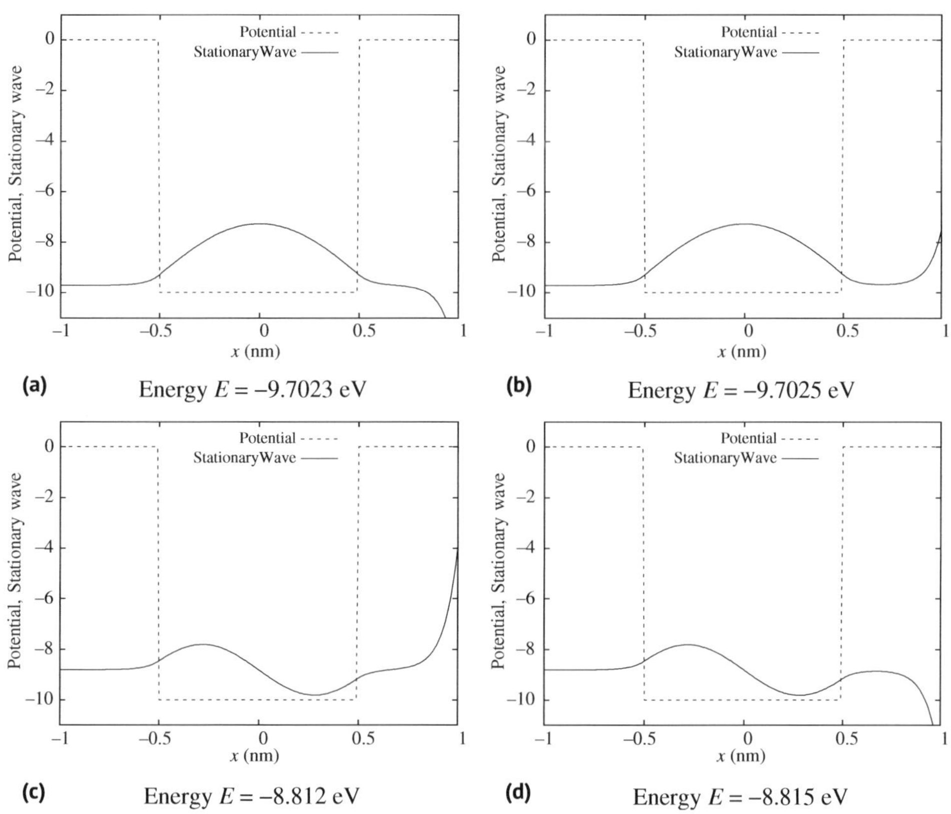

Figure 13.34 Unnormalizable quantum stationary wave candidates with energy level errors.

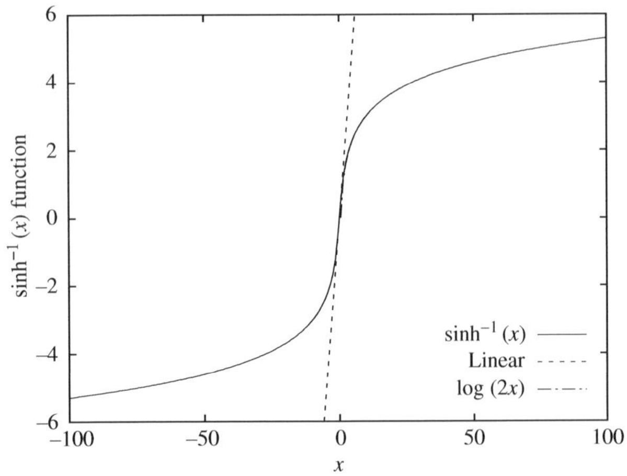

Figure 13.35 Arc hyperbolic sine numerical compression function.

where $k_3 = k_1$ is the required wave number in the field region determined by the particle's energy level by Equation (13.35). Applying the test function of Equation (13.36) to the function form of Equation (13.37) we see the decaying exponential term canceling out and leaving only

$$f(E) = 2A_3 e^{+k_3 x_{\text{final}}} \tag{13.38}$$

This function of energy must go to zero if, for some arbitrary value of x_{final}, the coefficient $A_3 = 0$, which implies that the form of the wave function at the right end of the x dimension is of solely the required decaying exponential. Using this function directly can, however, lead to numerical difficulties, as the exponential nature of the wave function in the right-side field region seen in the plots of Figure 13.34 can lead to very large numbers if a trial energy level encountered in the root convergence process is far from one of the allowed levels. We can apply a compression function to compress the orders of magnitude of such numbers for plotting and visualization. The inverse hyperbolic sine function, named `asinh()` in the C library, is a convenient function for this purpose. For x argument values in the vicinity of 0, this function is approximately linear, but for $|x| >> 1$ the function approaches a logarithmic dependence, as shown in the plot of Figure 13.35. Note that this function is applied *solely* for numerical range limitation and visualization, and has *no physical meaning* at all for the solutions. So a target error function more suitable for plotting is

$$f(E) = \sinh^{-1}\left(\psi(x_{\text{final}}) + \frac{1}{k_3}\frac{d\psi(x_{\text{final}})}{dx}\right) \tag{13.39}$$

Consider the following excerpt from the full program `quantum_rectwell_target_step.c` from the Student Companion website. The structure `QWELLWAVE` holds the physical parameters of particle mass and energy, as well as a pointer to another structure `QWELLPOT` that describes the potential profile of the well. These parameters are needed to solve the Schrödinger equation (13.34). The C function `quantum_well_target()` calculates the target error "distance" with a candidate electron energy as an argument. In the process of computing this error, the solver routine `rk4()` is called to trace out a candidate wave function solution from the left to the right end of the potential well space, after suitable initial conditions are launched. The derivatives needed by the Schrödinger equation following Equation (13.24) are computed by the `quantum_well_wave()` function, which is passed to the `rk4()` solver. This function assumes that another C function `quantum_well_potential()` has been defined in the program that calculates $V(x)$, the electron potential as a function of x. After the right end of x has been reached in the wave function solution, the target error value is calculated from Equation (13.39) and returned to the calling program. The main program steps this value over energy levels ranging from the bottom to the top of the potential well using the `step.c` tool from Chapter 4 so that we may visualize the target error function and the task of finding its roots, which are the allowed electron energies.

```
typedef struct qwell_wave_param {
  double mass;  // particle mass in kg
  double energy;  // particle energy in J
  QWELLPOT *qwpp;  // pointer to quantum well potential parameters
} QWELLWAVE;
```

```
void quantum_well_wave(double x,ARRAY1D *y,void *param_ptr,
                       ARRAY1D *yp) {
  QWELLWAVE *p;

  p = param_ptr;
  yp->elem[0] = y->elem[1];
  yp->elem[1] = (2.0 / (PLANCK_REDUCED * PLANCK_REDUCED)) * p->mass *
                (quantum_well_potential(x,p->qwpp) - p->energy) *
                y->elem[0];
}

double quantum_well_target(double energy,void *param_ptr) {
  double k,init_state[2],error;
  GRID xgrid;
  ARRAY1D init = {.elem = init_state,.n_elems = 2};
  ARRAY2D wtable;
  QWELLWAVE *p;

  p = param_ptr;
  p->energy = energy;
  k = sqrt((-2.0 / (PLANCK_REDUCED * PLANCK_REDUCED)) *
          p->mass * energy);
  init.elem[0] = 1.0;
  init.elem[1] = k;
  xgrid.init = -p->qwpp->width;
  xgrid.final = p->qwpp->width;
  xgrid.inc = p->qwpp->width * 0.01;
  alloc_array2d(&wtable,2,NUM_GRID_PTS(xgrid));
  rk4(quantum_well_wave,&xgrid,p,&init,&wtable);
  error = asinh(wtable.elem[0][wtable.n_cols - 1] +
                wtable.elem[1][wtable.n_cols - 1] / k);
  free_array2d(&wtable);
  return(error);
}
```

The student may study the program code to understand how a variety of potential profiles are specified on the command line with coded characters, and pass the profile parameters on to the `quantum_well_potential()` function. We may generate a plot of the target error function first for the rectangular potential well by compiling and running the main program with

```
$ gcc -Wall -Wshadow -O2 -o quantum_rectwell_target_step quantum_rectwell_target_step.c
    ↪ CPLibrary/step.c CPLibrary/diffeq.c CPLibrary/array.c -lm
$ ./quantum_rectwell_target_step
./quantum_rectwell_target_step <width in nm> <depth in eV> <mass * c^2 in Mev>
$ ./quantum_rectwell_target_step 1 10 0.5110 >| quantum_rectwell_target_step.dat
```

Plotting the stepped function values generates the graphics shown in Figure 13.36.

Each energy value that makes this target error function pass through zero is an energy level that launches a wave function solution that both begins and ends in an exponential form consistent with a normalizable wave function with physical meaning. Since the profile of this error function appears well behaved, we can now simply use the `step_roots.c` tool from Section 5.10, as was done for the ballistics problem, to converge on these six roots to find the allowed electron energy levels. The main program that performs this search for these energy levels is `quantum_rectwell_target_roots.c`, and is compiled and executed as below. The command line arguments specify the same rectangular potential well with a 10-eV depth and 1-nm width as analyzed in Section 5.11, and, as expected, the levels agree closely with those found with the analytical approach in that section.

```
$ gcc -Wall -Wshadow -O2 -o quantum_rectwell_target_roots quantum_rectwell_target_roots.c
    ↪ CPLibrary/step_roots.c CPLibrary/root.c CPLibrary/diffeq.c CPLibrary/step.c
    ↪ CPLibrary/array.c -lm
$ ./quantum_rectwell_target_roots
./quantum_rectwell_target_roots <width in nm> <depth in eV> <mass * c^2 in Mev>
$ ./quantum_rectwell_target_roots 1 10 0.5110 >| quantum_rectwell_target_roots.dat
6 roots found with step_regula_falsi:
-9.7023877 -8.8136836 -7.3479838 -5.3365951 -2.8546909 -0.22960619
```

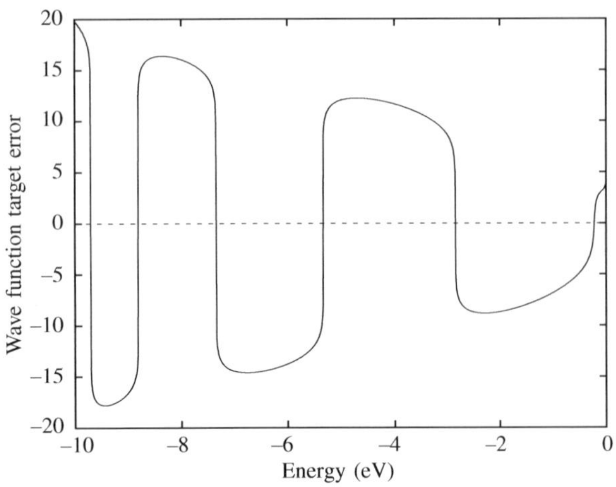

Figure 13.36 Wave function target error function for a rectangular potential well.

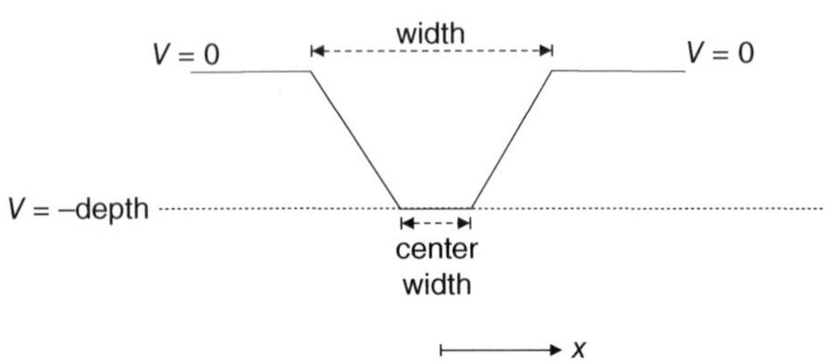

Figure 13.37 Quantum potential well with a trapezoidal profile.

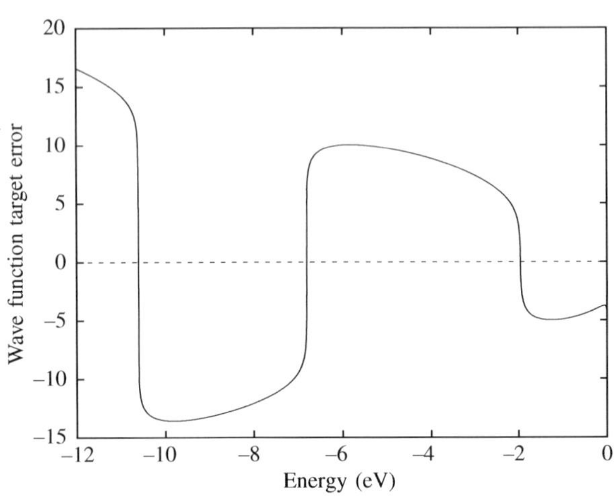

Figure 13.38 Wave function target error function for a trapezoidal potential well.

With the confidence that our numerical approach agrees with an analytic solution, where such a solution is possible, we are now able to tackle the problem of more realistic potential wells. This analysis may be accomplished simply by writing the C function `quantum_well_potential()` above so that any potential profile of interest may be calculated as a function of x. As a first step, let's consider relaxing the need for an infinitesimal region of infinite force by modeling a trapezoidal potential well with linearly graded boundaries, as shown in Figure 13.37. We'll solve for the stationary electron states for a trapezoidal well with 0.6-nm width at the top but only 0.3-nm width at the bottom, and a depth of 12 eV. When we plot the target error function in Figure 13.38, we see that this narrower well supports only three allowed electron energy levels.

Running the root-finding program for this trapezoidal profile finds the energy levels associated with each energy level.

```
$ ./quantum_trapwell_target_roots 0.6 12 0.5110 0.3 >| quantum_trapwell_target_roots.dat
3 roots found with step_regula_falsi:
-10.608807 -6.8075893 -1.9564476
```

Inspection of the program `quantum_trapwell_target_roots.c` will also show that after the allowed energy levels have been found by converging on multiple roots of the target error function, the program then uses these allowed energy levels to launch wave function solutions and generate the sampled wave solutions in the output stream. The allowed wave functions are separated in the output stream so that `gnuplot` may find and plot them as separate data sets. These curves with both odd and even parity solutions are shown in Figure 13.39 along with the potential profile. Following the usual custom, the vertical level corresponding to zero wave function value is aligned to the corresponding energy level to visualize the increasing energy states.

The flexibility to solve for allowed wave functions for any potential profile also opens the possibility of treating profiles with more than one bounding well. We have already examined the case of two adjacent wells in Section 7.7 in which we gained some physical insights into the chemical covalent bond. Treating that problem assuming idealized abrupt well boundaries required solving a very cumbersome nonlinear system of equations in order to stitch together piecewise solutions. But now with a numerical approach to solve the Schrödinger equation directly, we may arrange any number of wells of any shape, and rely on the differential equation to guide a continuous trajectory of the wave function solution through the wells. So now let us examine the case shown in Figure 13.40 with a potential of some number of quantum potential wells arranged in a regular array. We may define the geometry of the array with the parameters width, period, and depth as shown and the count of wells. The center well of the array is located with its center at the coordinate $x = 0$, and additional wells may be added symmetrically on either side of the center for an odd number of wells. Physically, we can conceive starting from a single potential well and gradually building up a discrete approximation to a one-dimensional crystal lattice of atomic ionic sites. We are treating only the case of a one-dimensional "crystal" here for simplicity, and we are considering the case of only one bound electron within the structure; but even with this simplified treatment we can gain interesting insights into physical behavior. The coding tasks for expanding the example programs for a single well to accommodate an array of wells are left to Exercise 13.8.

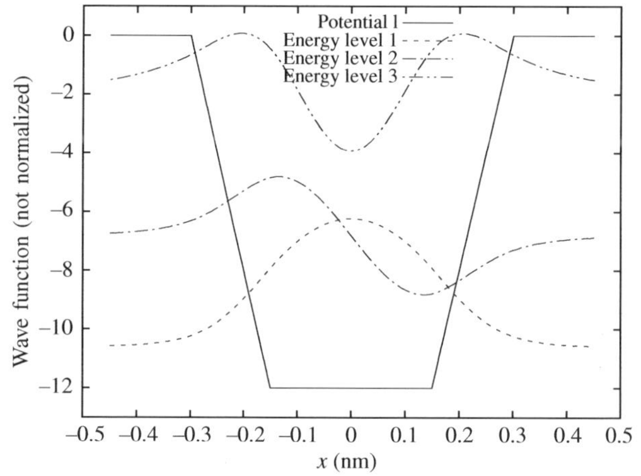

Figure 13.39 Wave function solution for three allowed electron energy states in a trapezoidal potential well.

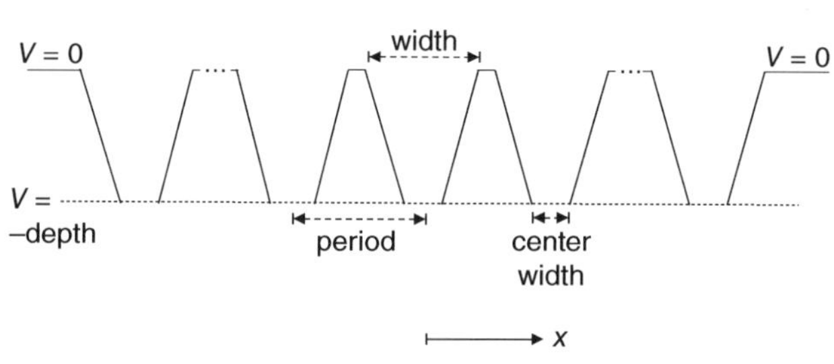

Figure 13.40 Potential of an array of quantum potential wells with repeated trapezoidal profiles.

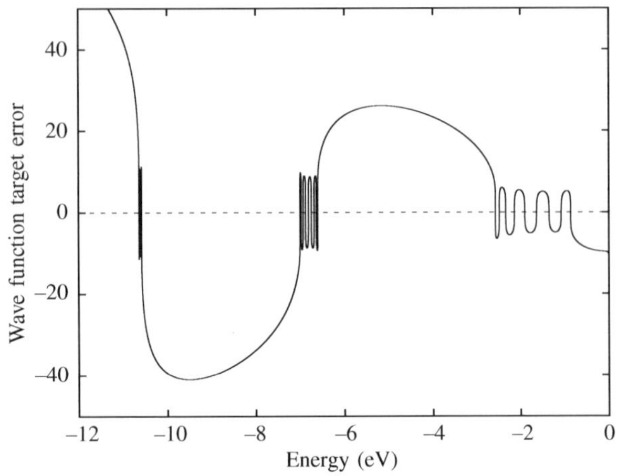

Figure 13.41 Target error function for an array of nine trapezoidal potential wells.

As an example case of a small scale array, see the plot of the target error function for an array of nine triangular potential wells in Figure 13.41. This curve was generated by an expanded version of the program `quantum_trapwell_target_step.c` that is able to accommodate an array of wells. The simulation was run with the case of trapezoidal wells with the same shape as those for Figure 13.38, with a width of 0.6 nm, a well center width of 0.3 nm, and a depth of 12 eV, but now with nine wells arranged in a one-dimensional lattice with a periodic spacing of 0.65 nm. We see in the plot of the target error function that its roots, which will be the allowed energies for an electron, are clustered into regions surrounding the location of the roots of the single well in Figure 13.38. We are able to visualize the formation of the structure of discrete energy bands, with "forbidden zones" of energy between the bands, that is, at the heart of solid state physics. Electrical conduction in metals and the nonlinear transistor behavior in semiconductors stem from this energy structure.

Note that in each root cluster there is one root in this serpentine function contour for every potential well in the array. Running an expanded version of the program `quantum_trapwell_target_roots.c` that will accommodate an array of wells, we see the numerical values of the allowed energy levels clustered about the single-well values of −10.608 807, −6.807 589 3, and −1.956 447 6 eV above

```
$ ./quantum_trapwell_array_target_roots
./quantum_trapwell_array_target_roots <width in nm> <depth in eV> <period in nm> <number of wells>
    ↪ <mass * c^2 in Mev> <center width>
$ ./quantum_trapwell_array_target_roots 0.6 12 0.65 9 0.5110 0.3
27 roots found for 9 wells
9 -10.6384664
9 -10.6342431
9 -10.6276339
9 -10.6192513
9 -10.6098878
9 -10.6004494
9 -10.5918706
9 -10.5850191
9 -10.5806
9 -7.00751257
9 -6.98019562
9 -6.93689816
9 -6.88094159
9 -6.81691456
9 -6.75054806
9 -6.68843248
9 -6.63745325
9 -6.60386481
9 -2.57598701
9 -2.48383066
9 -2.33537148
```

```
9 -2.13761215
9 -1.89956559
9 -1.63265928
9 -1.35278025
9 -1.08554311
9 -0.876263489
```

The original and expanded versions of the program `quantum_trapwell_target_roots.c` will generate the allowed wave function solutions corresponding to each of the located energy levels, as seen in Figure 13.39. We can now plot these allowed wave functions for our example case of an array of nine potential wells in Figures 13.42 and 13.43 that are clustered in the lowest of the three energy bands. The solutions alternate between even and odd parities, and after being "launched" on the left side, all make their required "landing" as a normalizable decaying exponential on the right side. We can trace through the wells and see the various combinations of oscillatory and exponential dependence expected when the electron energy is above or below the electron potential, respectively. These allowed wave functions are reminiscent of those plotted

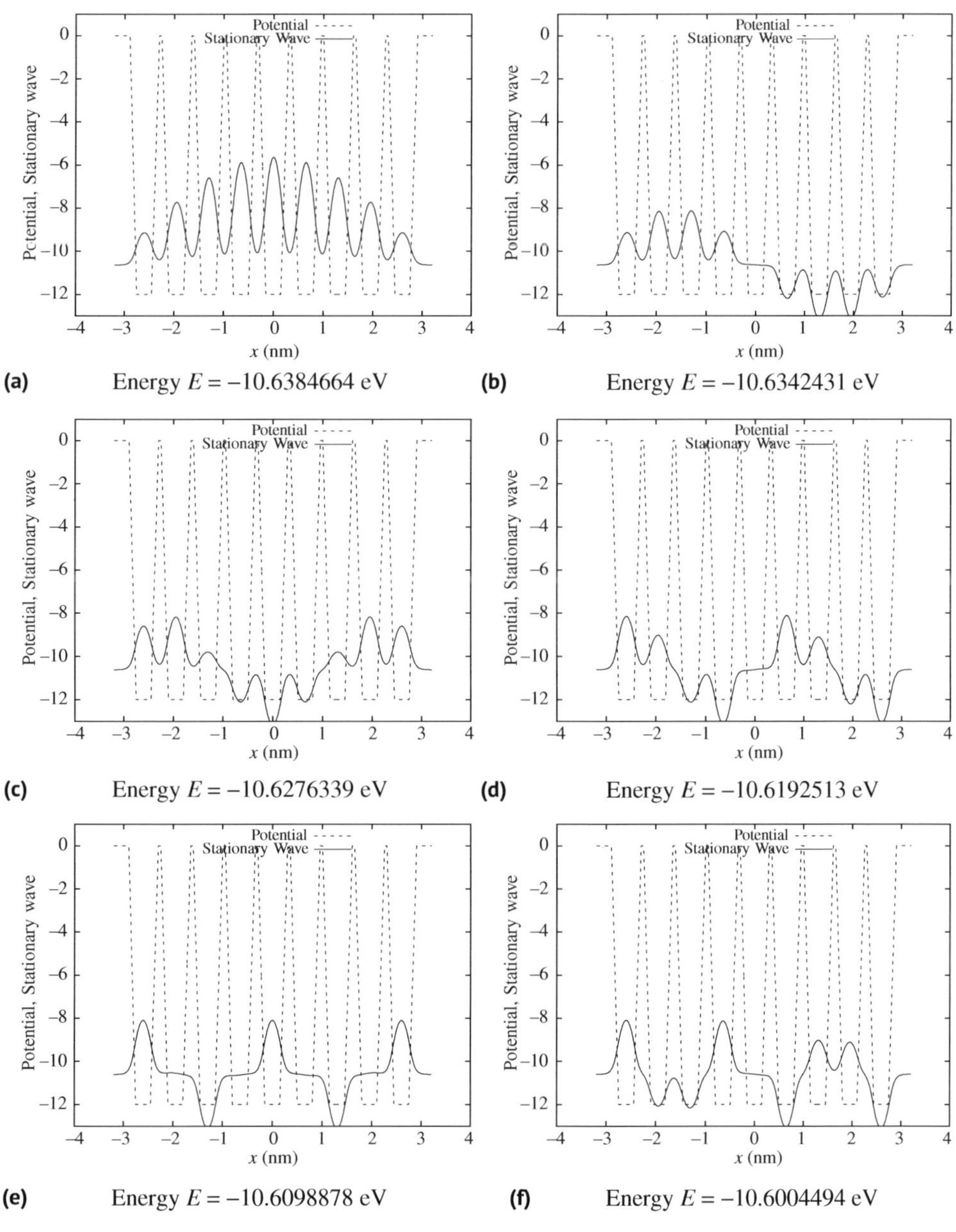

Figure 13.42 Lowest energy cluster of stationary wave function solutions for an array of nine trapezoidal wells.

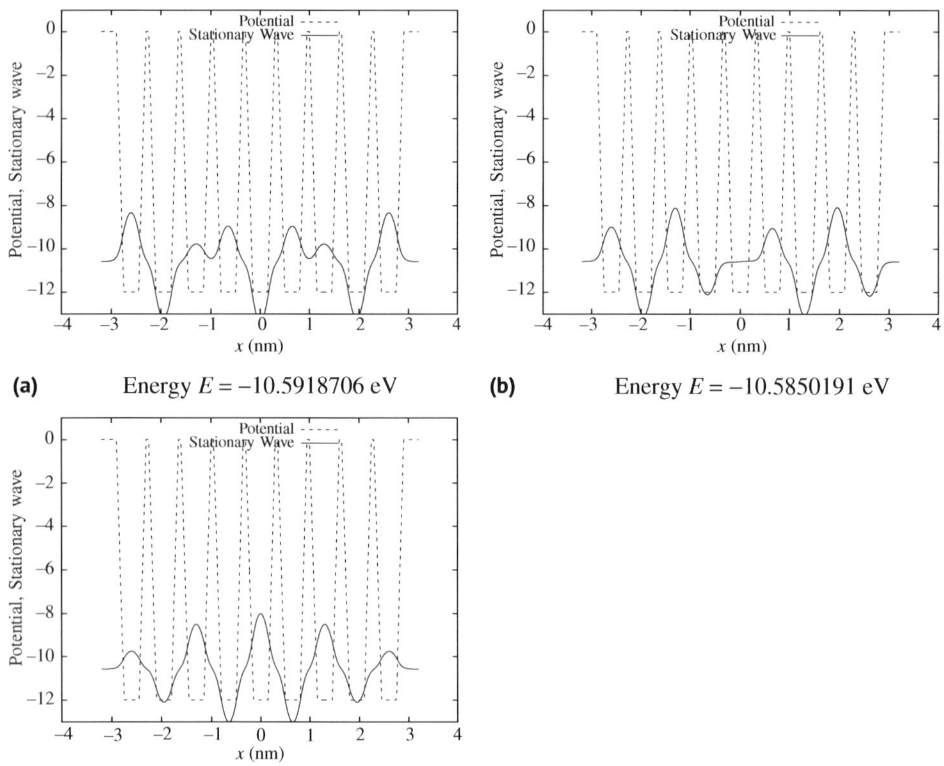

(a) Energy $E = -10.5918706$ eV

(b) Energy $E = -10.5850191$ eV

(c) Energy $E = -10.5806$ eV

Figure 13.43 Lowest energy cluster of stationary wave function solutions for an array of nine trapezoidal wells (continued).

in Figures 7.9 through 7.12 and for the solutions to Exercise 7.4 for the case of two rectangular wells, but all now developed through a direct and continuous solution of the Schrödinger equation.

Finally, the plot in Figure 13.44 is the result of rerunning the expanded `quantum_trapwell_target_roots.c` program multiple times and plotting each of the allowed energies at a corresponding point. The plot shows the allowed energy levels starting with three wells and adding a well on each side, up to 21 wells. (The stack of points for nine wells are those we have seen above.) With this summary plot we are able to see the process of gradually developing a one-dimensional approximation to a crystal structure with a regular array of potential wells corresponding to each atomic site. As each pair of wells is added to build up the crystal, a pair of available energy levels is added in each energy band, but with no states filling into the intervening forbidden regions. At this graphic resolution, the density of roots for 21 wells appears continuous. For an actual physical crystal, with a number of wells on the order of perhaps 10^{20} or 10^{30} atomic sites, the energy bands would be populated with an essentially continuous range of allowed energy levels. In Section 4.9, we approximated the construction of quantum wave packets that described a single particle in motion, subject to probabilistic uncertainties about its position and momentum. As we see in Figure 13.44, each energy band gradually fills up with allowed energy levels and approaches a continuum of energies; we are approaching the environment of a free electron, able to be described by a wave packet of continuous constituent energies, although within only the discrete energy bands. This limit is also seen using the analytic tool of the Kronig–Penny model [5, Chap. 5; 23, Chap. 10; 32, Chap. 7; 33, Chap. 1], but our numerical approach allows us to visualize the gradual transition from a single potential well to an approximation of a one-dimensional crystal structure.

The graph in Figure 13.44 was generated by running the solution program `quantum_trapwell_array_target_roots.c` to Exercise 13.8 several times for the various potential well counts shown on the horizontal axis. As the number of potential wells grows, the roots of the target error function need even finer resolution searching than those shown in Figure 13.41, and a very large number of fine steps are required to cover the energy range. Returning to the remarks at the end of Section 13.4 on computational efficiency, timing this collection of the solution code runs with the Linux `time` command reports that a total of almost two minutes of processor time on an example of typical contemporary hardware are required. From the simple benchmarking exercises in Section 1.3 showing the substantial execution time penalties involved in coding intense numerical tasks in an interpreted language such as Python relative to optimized C code, we might expect to have to wait on the order of two hours for the accumulated computation results with an interpreted programming language.

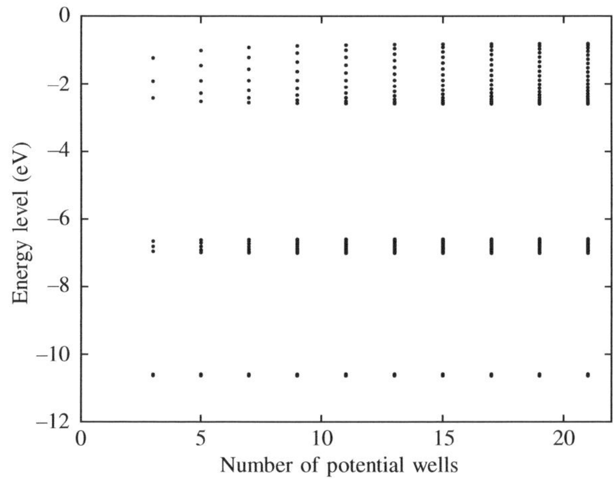

Figure 13.44 Allowed energy levels for an electron in an array of potential wells.

13.7 Exercises

13.1 Use the program `gravitation.c` from Section 13.2 to simulate the gravitational motion of the Earth orbiting around our Sun. Consider three cases of different values of the orbital eccentricity, the first being the true case in our solar system of an eccentricity of essentially zero and a radius of 1.5×10^{11} m. Then consider two fictional cases of values of 0.5 and 0.9 for the eccentricity, keeping the semi-major axis at 1.5×10^{11} m in both cases. (In comparison, the orbit of Mars exhibits an eccentricity of almost 0.1.)

(a) Evaluate Equation (5.15) to find the expected orbital period.

(b) Make use of the `orbit_step.c` program code from Section 5.6 to describe the three orbits in terms of Keplerian elements. Inspect its output file with the Linux pager tool `less` and obtain from some line the appropriate initial state variable conditions for the differential gravitational equations of motion. Use values of position and velocity at orbital locations other than perihelion and aphelion for your initial conditions. Use the orbital period calculated above to run the simulation for a few periods. Because highly elliptical orbits can involve a wide range of velocities over the orbital period, select a time step to provide approximately 1000 steps per period.

(c) For each of the three eccentricity cases, plot the two gravitational solutions (one from `orbit_step.c` and the other from `gravitation.c`) together on the same plot. In this way we may numerically reconnect Kepler's orbital laws with Newton's law of gravitation.

13.2 Expand the `collision2.c` program making use of the surface spring model to a new program `poolbreak.c` that simulates the pool break shot similar to that shown in Figure 13.12. The program must position 15 initially stationary balls in an array shaped as an equilateral triangle and a sixteenth ball incident from the left as the cue ball. Include the spring confining boundary walls around the area. In arbitrary units, set the size of the confining region to be 15 units in the x direction and 10 units in the y direction. The center of the single ball on the left side of the array that will be the first to be struck by the cue ball is at coordinates (0,0), and the centers of the balls in the array are spaced 0.2 units apart. Hint: Make the ball diameters very slightly smaller than the distance between triangular lattice points so that no spring surface force is active at the initial condition. Use a ball mass of 1 unit and a surface spring constant of 100 units for both the balls and the confining walls. The initial velocity of the cue ball is 2 velocity units solely in the positive x direction, and starting from a position with an x coordinate of -5 units. Offset the initial y coordinate of the cue ball vertically from 0 by a small distance so that momentum is imparted by the cue ball collision at some diagonal heading to scatter the balls asymmetrically. (The trajectories in Figure 13.12 resulted from an initial cue ball y coordinate of -0.05 units, but choose your own trajectory.) Include the calculation of the center of mass across all the balls to create a plot resembling Figure 13.13.

13.3 Expand your solution program to Exercise 13.2 using the animation coding techniques of Sections 4.13 and 6.9 to write a new program `poolbreak_gpanim.c` that generates animated graphics of the break shot in Figure 13.12.

13.4 Using the `efieldlines.c` program of Section 13.4 as a starting point, write a new program `magfieldlines.c` to draw magnetic field lines in the y-z plane due to a collection of current loops as shown in Figures 13.16 and 13.17.

All the current loops are assumed to be oriented parallel to the x-y plane and at regularly spaced z coordinates. As with the `magfieldvectors.c` program in Section 10.7, the `magfieldlines.c` program should generate two data sets in its output stream to allow the plots to be generated. The first data set consists of a series of point coordinates that represent the cross-sections of the circular coils as they pierce through the y-z plane. These coordinates can be plotted with solid points in the graphical output, and give a visual reference around which field vectors can be seen to curl. The second output data set lists the curve coordinates of the field lines to be plotted in line mode. The program should accept a grid of y points on the command line. With a z coordinate of zero, this set of (y, z) coordinates specify the points from where the curve solution will initiate a field line. With a positive current flow in the conductor loops, the solver will take steps of the RK2 algorithm and follow the field lines in the positive z direction, and then curl around into the negative z region. The program should follow the curve until it returns to cross back into the positive z region, at which point it will be in the vicinity of the starting point. Since it is possible for a symmetrical geometry that some field lines would extend to unbounded distances, a maximum travel distance in search of the starting point must also be accepted on the command line.

(a) Model the magnetic field from a single circular loop of radius 1 m carrying a current of 1 A. The current loop lies in the x–y plane with a z coordinate of 0. Specify a y grid of starting points to span the interior of the loop, but avoid the center at (0,0) from where the field line would extend straight upward indefinitely.

(b) Model the magnetic field from a collection of 10 circular loops of radius 1 m, each carrying a current of 1 A, as an approximate model for a 10-turn solenoid forming a helical current path. The total height of the array of coils should be 2 m, to 1 m above and below the x–y plane. Hint: Since the solenoid geometry concentrates the magnetic field inside the array, making the field outside the solenoid small, the field lines at a distance will be very widely separated, and you will need to specify a long maximum search distance for some of the field lines. As discussed in Section 13.4, you may expect this program to require a fairly long execution time.

13.5 Modify the `quantum_linbarrier_packet.c` program from Section 13.5 to write a new program `quantum_quadbarrier_packet.c` for the scattering against a quantum potential step with a quadratically shaped potential. Graphically compare the progress in the rebound direction relative to the abrupt and linearly graded potential steps in Figures 13.20f and 13.25f.

13.6 Modify the `quantum_trapwell_target_step.c` and `quantum_trapwell_target_roots.c` programs from Section 13.6 to write the new programs `quantum_quadwell_target_step.c` and `quantum_quadwell_target_roots.c`. The new programs must step the quantum well target error function and locate its roots for the case of a potential well with a limited quadratic profile, as shown in Figure 13.45. The constant a in the expression for the potential profile is set as shown for computational convenience. This profile would correspond to the case of a quantum force "spring constant" of $2a$ over a realistically limited range of displacement coordinate x.

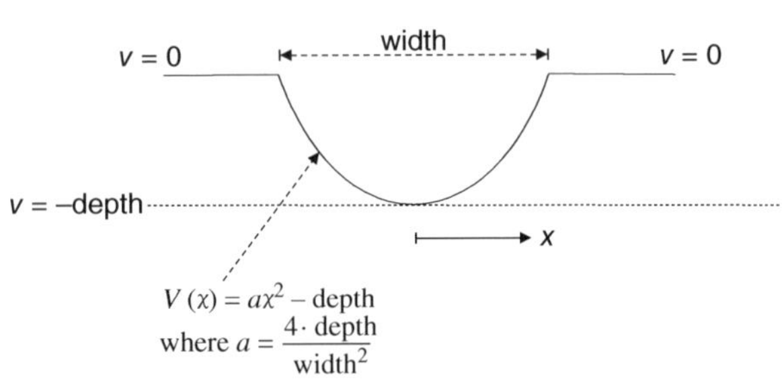

Figure 13.45 Potential well with a quadratic profile, forming a quantum harmonic oscillator of finite extent.

The theoretical energy levels of an idealized quantum harmonic oscillator for a potential that maintains an unbounded quadratic profile over all x can be found analytically. These ideal quantized levels are at [2, Chap. 7]

$$E_j = (j + \frac{1}{2})h\nu$$

where ν is the oscillation frequency of a classical spring-mass oscillator with spring constant $2a$ and mass m

$$\nu = \frac{1}{2\pi}\sqrt{\frac{2a}{m}}$$

Use your programs `quantum_quadwell_target_step.c` and `quantum_quadwell_target_roots.c` to model the more realistic oscillator potential in Figure 13.45. Plot the target error function for an electron bound in a potential well of depth 20 eV and width 1 nm and visualize the approximately linear distribution of roots over energy. Include in the `quantum_quadwell_target_roots.c` a calculation of each theoretical energy level from the expressions above and compare with each numerically found root. Plot the allowed wave functions as approximations to the Hermite polynomials known for the ideal case.

13.7 Modify the `quantum_trapwell_target_step.c` program from Section 13.6 to write a new program `quantum_2trapwell_target_step.c` to analyze the target error function for the case of two symmetrical and adjacent trapezoidal potential wells, as shown in Figure 13.46. Consider trapezoidal wells with 0.6-nm width at the top but only 0.3-nm width at the bottom and a depth of 12 eV. From the development in Section 13.6 and the run of the `quantum_trapwell_target_roots.c` program we expect that isolated potential wells with this profile would support three electron energy levels at approximately −10.6, −6.81, and −1.96 eV. By plotting the target error function versus energy for the case of a spacing between the tops of the wells of 0.1 nm that graphically shows the roots, demonstrate energy level splitting between even and odd parity states that is a more realistic demonstration of the one-dimensional covalent bond, similar to the solution for ideal rectangular wells in Section 7.7. Hint: You will need quite fine steps along the energy axis to graphically show the root at the lowest energy level.

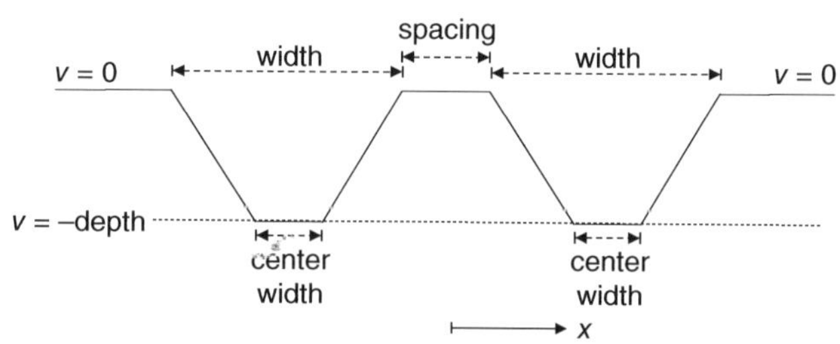

Figure 13.46 Two adjacent quantum potential wells with trapezoidal profiles.

13.8 Enhance the code for the quantum well target error function stepping and root finding in the programs `quantum_trapwell_target_step.c` and `quantum_trapwell_target_roots.c` from Section 13.6 into the new source code files `quantum_trapwell_array_target_step.c` and `quantum_trapwell_array_target_roots.c` so that the new programs accommodate regular arrays of trapezoidal potential wells as shown in Figure 13.40. Build your array with a potential well top width of 0.6 nm, a well center width of 0.3 nm, and a depth of 12 eV. Simulate a crystalline model with 21 wells arranged in a one-dimensional lattice with a periodic spacing of 0.65 nm. Generate a plot of the target error function versus electron energy and a list of the energy roots of the target error function.

14

Implicit Methods for Ordinary Differential Equations

In this chapter, we continue developing solution strategies for ordinary differential equations. We have seen several solutions for such systems of equations in Chapters 12 and 13, but the problems we tackled in those two chapters have been carefully selected to avoid a limitation of explicit solution algorithms that is important to understand. We will start this chapter with an example physics problem that illuminates the limits of the numerical stability of explicit algorithms.

14.1 Algorithms: Explicit Algorithm Instability

We can begin to see the general issue of limited regions of stability by returning to the reference differential equation (14.1) that we examined earlier in Section 12.2.

$$\frac{dy}{dx} = -ay \tag{14.1}$$

If the first-order Euler algorithm from Section 12.2.1 samples a solution following Equation (12.7), then for this solution to be bounded for increasing step index j, the step size h must satisfy the condition

$$|1 - ah| < 1 \tag{14.2}$$

Using the negative sign of the absolute value, this inequality translates to a requirement that $h < \frac{2}{a}$. We see that the maximum step size that will ensure that the numerical algorithm results in at least a convergent sequence, with no consideration of truncation error, is dependent on the constant a governing the differential equation.

For the Heun's or Improved Polygon algorithms from Sections 12.2.2 and 12.2.3, the sampled solution to the reference differential equation will follow Equation (12.10). We may derive a similar requirement for this sequence to remain bounded for increasing step index j as

$$\left|1 - ah + \frac{a^2h^2}{2}\right| < 1 \tag{14.3}$$

This constraint again implies that $h < \frac{2}{a}$ must be observed for a stable, bounded solution as the solution sequence evolves with increasing index j. Generally, for a simple differential equation such as Equation (14.1), the step size will be chosen to be much smaller than this constraint to limit truncation error to reasonable levels, and stability is not an issue. For higher-order algorithms the stability constraint is more restrictive to small step sizes [34, Chap. 13].

For systems of differential equations of higher order than the single first-order Equation (14.1) the stability requirements on the step size are more difficult to derive algebraically. We can, however, gain some further insight on stability limits by treating an example higher-order physical system numerically. We derived the normal modes of oscillation for the system of coupled spring-mass oscillators analytically in Section 6.5. But solving the equations of motion for this system numerically with the explicit methods of Chapter 12 provides an important visualization of the issue of limited regions of stability. Moreover, these simulations will also serve as an introduction to our numerical study of wave propagation in Section 14.5.

We consider again the coupled spring-mass oscillator illustrated in Figure 14.1. The two masses are identical, as are the three springs with a linear spring constant k.

Computational Physics Using C: Efficient Programming with Ease, First Edition. John W. Fattaruso. © 2026 John Wiley & Sons, Inc.
All rights reserved, including rights for text and data mining and training of artificial intelligence technologies or similar technologies.
Published 2026 by John Wiley & Sons, Inc. Companion Website: http://www.wiley.com/go/CompPhysUsingC

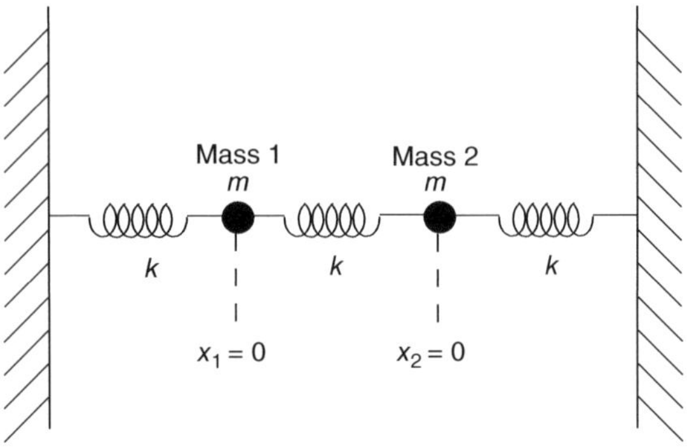

Figure 14.1 A simple coupled oscillator with two masses.

We define the square of a reference frequency as $\omega_0^2 = \frac{k}{m}$ and define x_1 and x_2 as the displacement coordinates from the equilibrium positions of the two masses. We can then write the equations of motion for the two masses from Newtonian laws as Equation (14.4).

$$\begin{aligned}\frac{d^2x_1}{dt^2} &= -2\omega_0^2 x_1 + \omega_0^2 x_2\\ \frac{d^2x_2}{dt^2} &= \omega_0^2 x_1 - 2\omega_0^2 x_2\end{aligned} \tag{14.4}$$

Since we have a system of two coupled second-order differential equations, following the practice in Chapters 12 and 13, we may define four state variables as the displacements of two masses x_1 and x_2 and their linear velocities v_1 and v_2.

$$\begin{aligned} y_0 &= x_1 \qquad & y_1 &= \frac{dy_0}{dt} = v_1\\ y_2 &= x_2 \qquad & y_3 &= \frac{dy_2}{dt} = v_2\end{aligned} \tag{14.5}$$

With these four state variables, the equations of motion (14.4) can be prepared as a system for numerical solution with the Runge–Kutta method as

$$\begin{aligned} \frac{dy_0}{dt} &= y_1 \qquad & \frac{dy_1}{dt} &= -2\omega_0^2 y_0 + \omega_0^2 y_2\\ \frac{dy_2}{dt} &= y_3 \qquad & \frac{dy_3}{dt} &= \omega_0^2 y_0 - 2\omega_0^2 y_2\end{aligned} \tag{14.6}$$

The program `coupled_osc2.c` from the Student Companion website is an example solution to modeling the motion of the two masses directly from the differential equations. The parameter structure and C function to calculate the system derivatives from Equation (14.6) to be passed to the `rk4()` solver tool are

```
typedef struct osc_param {
  double w0;   // Reference angular frequency
} OSC;

void coupled_osc(double x,ARRAY1D *y,void *param_ptr,ARRAY1D *yp) {
  OSC *p;

  p = param_ptr;
  yp->elem[0]  = y->elem[1];
  yp->elem[1]  = -2.0 * p->w0 * p->w0 * y->elem[0] +
                 p->w0 * p->w0 * y->elem[2];
  yp->elem[2]  = y->elem[3];
  yp->elem[3]  = p->w0 * p->w0 *
                 y->elem[0] - 2.0 * p->w0 * p->w0 * y->elem[2];
}
```

In this case, the `OSC` structure is needed to store only one physical parameter, the reference frequency ω_0. The initial conditions for the displacements and velocities of the two masses are taken from the command line of the main program coupled_osc2.c. Our first two simulations are with initial conditions that respectively excite the two normal modes of oscillation that were derived in Section 6.5. We will assume that the velocities of both masses are zero when they are released at time $t = 0$, but the initial displacements of the masses are either +1, −1 units for mode 1 or +1, +1 units for mode 2. Also on the command line we specify a reference frequency of 1 Hz, or a reference angular frequency $\omega_0 = 2\pi$. As derived in Section 6.5, the physical frequency of each normal mode corresponding to its respective eigenvalue λ_i is $\omega_i = \sqrt{\lambda_i}\omega_0$. Since the eigenvalues are $\lambda_1 = 1$ and $\lambda_2 = 3$, we expect that the physical frequencies of the two modes will be 1 and $\sqrt{3} \approx 1.732$ Hz. By counting the number of cycles in a 10-s simulation, we can see that the results shown in Figures 14.2 and 14.3, with the two curves in Figure 14.2 being coincident, are consistent with these two frequencies.

The motion for any other set of initial displacements for the masses will evolve for times $t > 0$ as a linear combination of these two normal modes. An example is shown in Figure 14.4, where the two masses were initially at displacements +1 and −0.5 units. With these initial displacements that do not match either of the individual normal modes, we must have energy in both normal modes, and we see a complicated motion evolve that results from the coupling between the two masses.

To complete our comparison with the eigenvalue derivations of Section 6.5, let's now extend this simulation to the case of a coupled oscillator with three masses and four identical springs shown in Figure 14.5. With the square of the reference frequency again defined as $\omega_0^2 = \frac{k}{m}$, the equations of motion for this system are

$$\begin{aligned}\frac{d^2x_1}{dt^2} &= -2\omega_0^2x_1 + \omega_0^2x_2\\ \frac{d^2x_2}{dt^2} &= \omega_0^2x_1 - 2\omega_0^2x_2 + \omega_0^2x_3\\ \frac{d^2x_3}{dt^2} &= \omega_0^2x_2 - 2\omega_0^2x_3\end{aligned} \tag{14.7}$$

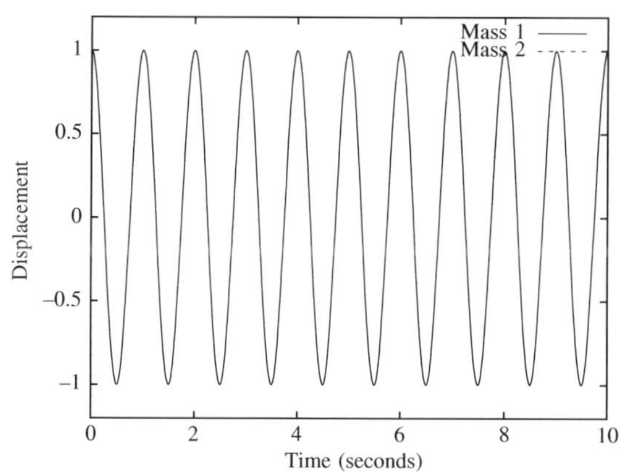

Figure 14.2 Coupled oscillator with two masses, $\omega_0 = 2\pi$, normal mode 1 with initial displacements +1 and +1, time step 0.01 s.

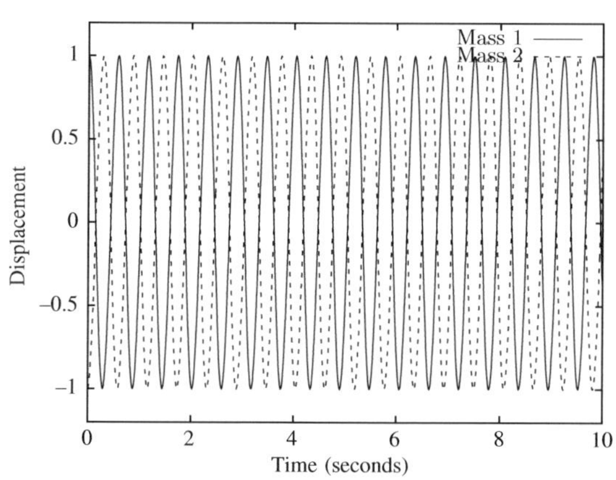

Figure 14.3 Coupled oscillator with two masses, $\omega_0 = 2\pi$, normal mode 2 with initial displacements +1 and −1, time step 0.01 s.

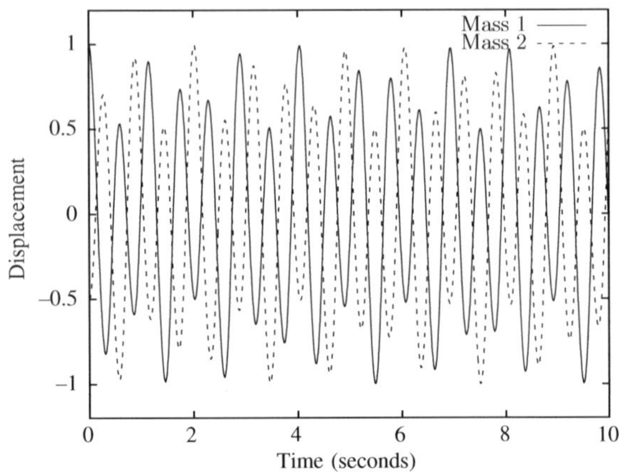

Figure 14.4 Coupled oscillator with two masses, $\omega_0 = 2\pi$, mixed mode with initial displacements +1 and −0.5, time step 0.01 s.

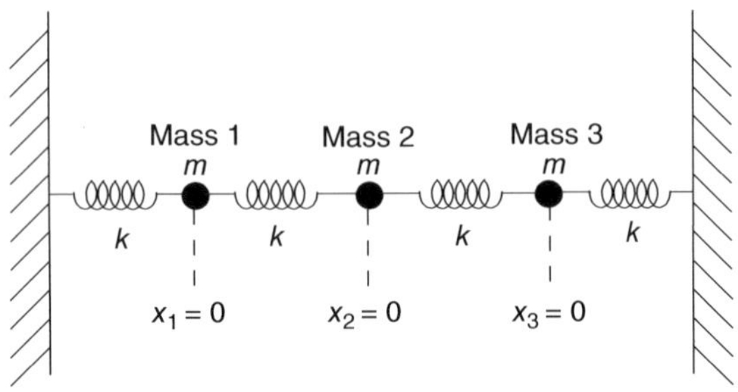

Figure 14.5 A simple coupled oscillator with three masses.

We now require six state variables to characterize the motion, namely, the displacements and velocities of the three masses in Equation (14.8).

$$\begin{aligned} y_0 &= x_1 \qquad & y_1 &= \frac{dy_0}{dt} = v_1 \\ y_2 &= x_2 \qquad & y_3 &= \frac{dy_2}{dt} = v_2 \\ y_4 &= x_3 \qquad & y_5 &= \frac{dy_4}{dt} = v_3 \end{aligned} \tag{14.8}$$

This extends the system of coupled first-order equations for a numerical solution with the Runge–Kutta method to

$$\begin{aligned} \frac{dy_0}{dt} &= y_1 \qquad & \frac{dy_1}{dt} &= -2\omega_0^2 y_0 + \omega_0^2 y_2 \\ \frac{dy_2}{dt} &= y_3 \qquad & \frac{dy_3}{dt} &= \omega_0^2 y_0 - 2\omega_0^2 y_2 + \omega_0^2 y_4 \\ \frac{dy_4}{dt} &= y_5 \qquad & \frac{dy_5}{dt} &= \omega_0^2 y_2 - 2\omega_0^2 y_4 \end{aligned} \tag{14.9}$$

The solution program `coupled_osc3.c` is also available on the Student Companion website, and the expanded parameter structure and vector field function are

```
typedef struct osc_param {
  double w0;  // Reference angular frequency
} OSC;

void coupled_osc(double x,ARRAY1D *y,void *param_ptr,ARRAY1D *yp) {
  OSC *p;

  p = param_ptr;
```

```
  yp->elem[0]  = y->elem[1];
  yp->elem[1]  = -2.0 * p->w0 * p->w0 * y->elem[0] +
                 p->w0 * p->w0 * y->elem[2];
  yp->elem[2]  = y->elem[3];
  yp->elem[3]  = p->w0 * p->w0 * y->elem[0] -
                 2.0 * p->w0 * p->w0 * y->elem[2] +
                 p->w0 * p->w0 * y->elem[4];
  yp->elem[4]  = y->elem[5];
  yp->elem[5]  = p->w0 * p->w0 * y->elem[2] -
                 2.0 * p->w0 * p->w0 * y->elem[4];
}
```

With this program, we can plot three normal modes of oscillation in Figures 14.6–14.8 for a reference frequency of 1 Hz, or $\omega_0 = 2\pi$. The eigenvalues for these three modes found in Section 6.5 are $\lambda_1 = 2$, $\lambda_2 = 2 - \sqrt{2}$, and $\lambda_3 = 2 + \sqrt{2}$. Since the physical frequency for each normal mode is again $\omega_i = \sqrt{\lambda_i}\omega_0$, we expect oscillation frequencies of $\sqrt{2} \approx 1.414$, $\sqrt{2-\sqrt{2}} \approx 0.765$, and $\sqrt{2+\sqrt{2}} \approx 1.848$ Hz.

As in the case of the two-mass oscillator, we can initialize the displacements of the masses with some arbitrary positions that result in a complicated mixed-mode motion. For example, initial displacements of -1, 1, and 0.5 units result in the motion plotted in Figure 14.9. Note that this plot matches the motion plotted in Figure 6.3 that was derived from the normal mode eigenvalues with the same initial conditions.

As a verification that the Runge–Kutta algorithm with a step size of 0.01 s allows us to follow the correct dynamic solution within an acceptable truncation error tolerance, we may use the method of plotting of the total energy of the system per

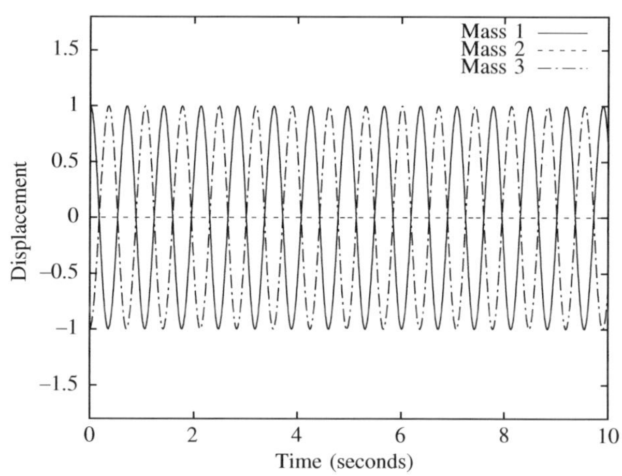

Figure 14.6 Coupled oscillator with three masses, $\omega_0 = 2\pi$, normal mode 1 with initial displacements +1, 0, and −1, time step 0.01 s.

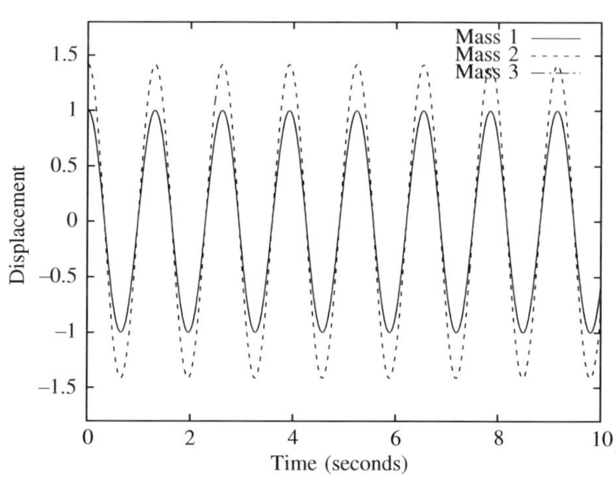

Figure 14.7 Coupled oscillator with three masses, $\omega_0 = 2\pi$, normal mode 2 with initial displacements 1, $\sqrt{2}$, and 1, time step 0.01 s.

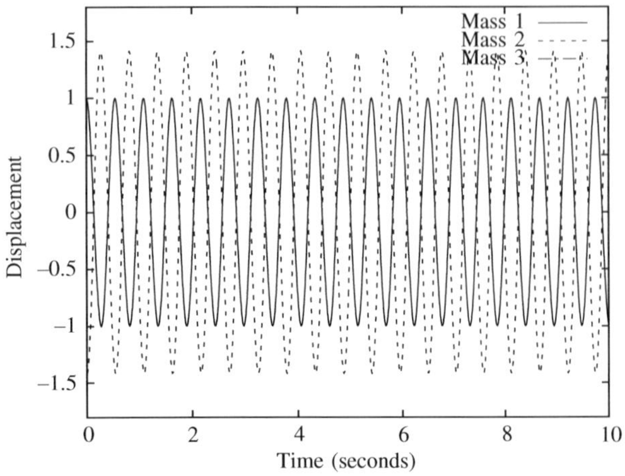

Figure 14.8 Coupled oscillator with three masses, $\omega_0 = 2\pi$, normal mode 3 with initial displacements 1, $-\sqrt{2}$, and 1, time step 0.01 s.

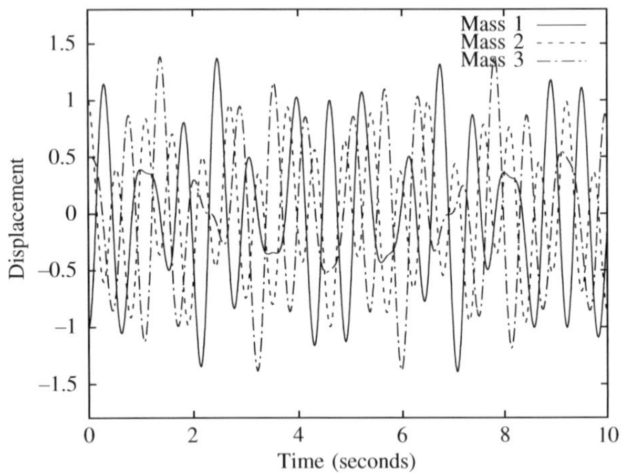

Figure 14.9 Coupled oscillator with three masses, $\omega_0 = 2\pi$, mixed modes with initial displacements −1, 1, and 0.5, time step 0.01 s.

unit mass that we used in Sections 12.5 and 12.9. This method tests the conformance of the numerical solution to the physical principle of conservation of energy. Equation (14.10) expresses the total energy for this system, with the first three terms being the kinetic energy of the three masses, and the last four terms being the potential energy dependent on the compression or expansion of the four springs.

$$\frac{E}{m} = \frac{v_1^2}{2} + \frac{v_2^2}{2} + \frac{v_3^2}{2} + \frac{\omega_0^2 x_1^2}{2} + \frac{\omega_0^2 (x_2 - x_1)^2}{2} + \frac{\omega_0^2 (x_3 - x_2)^2}{2} + \frac{\omega_0^2 x_3^2}{2} \tag{14.10}$$

We have the displacements x_i and the velocities v_i in this expression available at each time step as the values of the state variables. With a step size of 0.01 s, we can see from the span of the vertical axis in Figure 14.10 that the energy conservation error is less than 30 ppm. More accuracy may be purchased with a finer time step.

Besides providing another study of numerical solutions of equations of motion, these coupled oscillator systems can illuminate the limits of stability of explicit algorithms such as the fourth-order Runge–Kutta method. Let us now return to the two-mass oscillator and examine the case where the center coupling spring is allowed to have a different spring constant than the springs attached to the anchor walls, as illustrated in Figure 14.11. Letting the square of the two different reference frequencies be $\omega_w^2 = \frac{k_w}{m}$ and $\omega_c^2 = \frac{k_c}{m}$, the eigenvalue problem from Section 6.5 becomes

$$\det \begin{bmatrix} \frac{\omega_w^2}{\omega_c^2} + 1 - \lambda & -1 \\ -1 & \frac{\omega_w^2}{\omega_c^2} + 1 - \lambda \end{bmatrix} = 0 \tag{14.11}$$

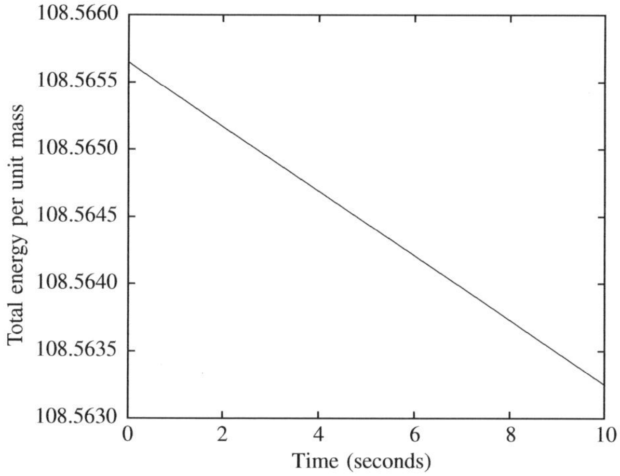

Figure 14.10 Total energy versus time of coupled oscillator with three masses, $\omega_0 = 2\pi$, mixed modes with initial displacements −1, 1, and 0.5.

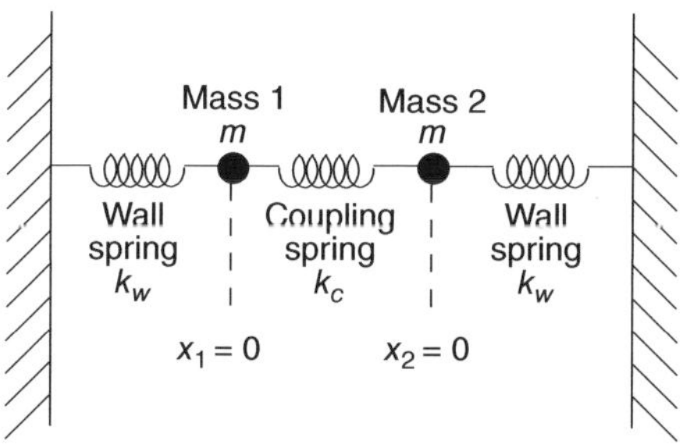

Figure 14.11 A simple coupled oscillator with two identical masses and different spring constants.

We may find the two allowed eigenvalues λ_1 and λ_2 as the roots of the characteristic polynomial

$$\left(\frac{\omega_w^2}{\omega_c^2} - \lambda\right)\left(\frac{\omega_w^2}{\omega_c^2} + 2 - \lambda\right) = 0 \tag{14.12}$$

The two roots are the eigenvalues $\lambda_1 = \frac{\omega_w^2}{\omega_c^2}$ and $\lambda_2 = \frac{\omega_w^2}{\omega_c^2} + 2$. These correspond to the physical frequencies of the two normal modes of oscillation being $\omega_1 = \left(\frac{\omega_w}{\omega_c}\right)\omega_c = \omega_w$ and $\omega_2 = \left(\sqrt{\frac{\omega_w^2}{\omega_c^2} + 2}\right)\omega_c$.

The vector field functions for this system with four state variables are now

$$\begin{aligned} \frac{dy_0}{dt} &= y_1 & \frac{dy_1}{dt} &= -(\omega_w^2 + \omega_c^2)y_0 + \omega_c^2 y_2 \\ \frac{dy_2}{dt} &= y_3 & \frac{dy_3}{dt} &= \omega_c^2 y_0 - (\omega_w^2 + \omega_c^2)y_2 \end{aligned} \tag{14.13}$$

The program `coupled_osc22.c` from the Student Companion website is an example solution for this oscillator with two different reference frequencies now stored in the parameter structure. That structure and the C function to calculate the system vector field from Equation (14.13) to be passed to the `rk4()` solver tool are

```
typedef struct osc_param {
  double ww;  // Reference frequency of masses with wall spring
  double wc;  // Reference frequency of masses with coupling spring
} OSC;

void coupled_osc(double x,ARRAY1D *y,void *param_ptr,ARRAY1D *yp) {
  OSC *p;
```

```
    p = param_ptr;
    yp->elem[0] = y->elem[1];
    yp->elem[1] = -(p->ww * p->ww + p->wc * p->wc) * y->elem[0] +
                  p->wc * p->wc * y->elem[2];
    yp->elem[2] = y->elem[3];
    yp->elem[3] = -(p->ww * p->ww + p->wc * p->wc) * y->elem[2] +
                  p->wc * p->wc * y->elem[0];

    return;
}
```

Let us consider the case where the center coupling spring is significantly stiffer than the two more relaxed springs connected to the walls. For an example case of a factor of 10 ratio between the coupling and wall frequencies of 5 and 0.5 Hz, respectively, or $\omega_c = 10\pi$ and $\omega_w = \pi$, initializing the two normal modes with the same initial conditions as in Figures 14.2 and 14.3 gives the simulation results plotted in Figures 14.12 and 14.13. Note that the two curves in Figure 14.12 coincide. The physical oscillation frequency shown in these plots of the two modes are consistent with the values of 0.5 and $\sqrt{2.01} \cdot 5 \approx 7.089$ Hz from the eigenvalues derived above.

If we are interested in accurately simulating the oscillatory behavior of this system that exhibits two such widely varying physical frequencies, it would be important to choose two different time steps to follow the motion of these two normal modes. For instance, the coarse time step of 0.05 s for the slow oscillations in Figure 14.12 still provides 40 steps per cycle. This step size has been reduced by a factor of 10 to 0.005 s for the simulation of Figure 14.13.

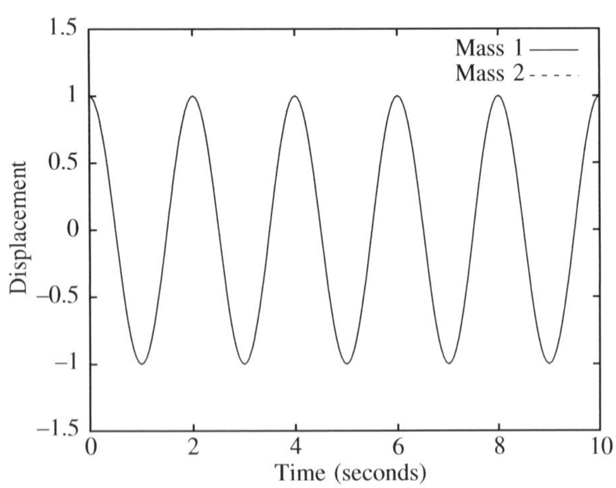

Figure 14.12 Coupled oscillator with different spring constants, $\omega_w = \pi$, $\omega_c = 10\pi$, mode 1 with initial displacements +1 and +1, time step 0.05 s.

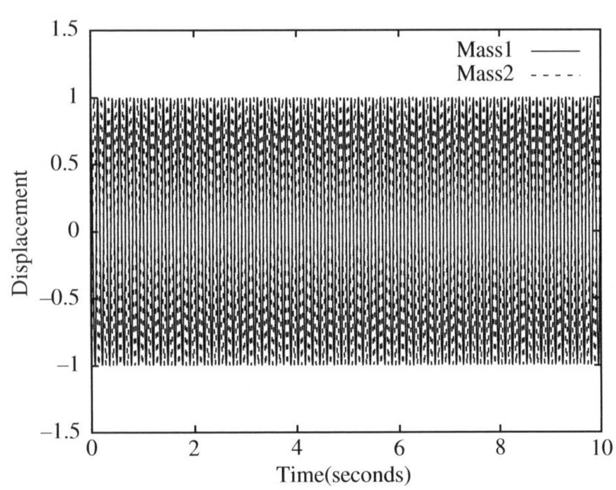

Figure 14.13 Coupled oscillator with different spring constants, $\omega_w = \pi$, $\omega_c = 10\pi$, mode 2 with initial displacements +1 and −1, time step 0.005 s.

Let us consider the case, however, that the slow oscillations of normal mode 1 are to be treated as the primary motion we are interested in capturing numerically, and that our intention is to avoid initiating any significant motion from mode 2. As a model for this sort of interest, let's rerun our simulations with the initial displacements of 1.001 and 0.999 for the two masses. These initial conditions almost completely match those of the normal mode 1, with an almost insignificant contribution from mode 2. For this case, the simulation plotted in Figure 14.14 shows that, for the resolution of the graphical image, the motion still closely resembles the slow oscillations of mode 1 in Figure 14.12. Then let's see the result of relaxing our time step size even further, to 0.1 s. The plot in Figure 14.15 of this numerical output generated by the Runge–Kutta algorithm bears little resemblance to the oscillatory behavior we expect. We notice from the span of the vertical axis that the numerical solution has "blown up" to unbounded limits as time progresses to the far right side. If we were to zoom in to the region of the plot that appears to be lying coincident with the horizontal axis at this graphics resolution, we would see exponentially growing state variable values. It should be understood that this numerical behavior has nothing to do with the physical behavior of the system we have submitted to the Runge–Kutta solver, but is instead the consequence of the step size chosen for this explicit algorithm being outside a limited region of numerical stability. These results can be thought of as the higher dimensional equivalent of the violation of the conditions for a convergent sequence we derived using Equations (14.2) and (14.3) for the one-dimensional prototype differential equation.

This simulation of a coupled spring-mass oscillator can be thought of as the prototype of some more complex system with nearly insignificant embedded eigenvalues that govern physical behavior at associated high frequencies. We can think of such eigenvalues as those associated with the stiff spring in the coupled oscillator. This mathematical configuration can

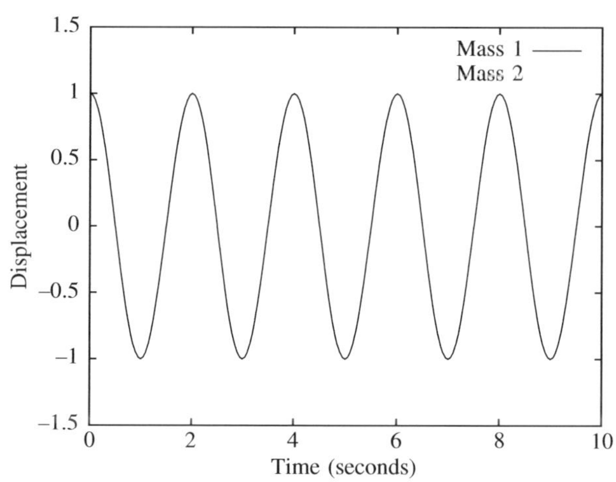

Figure 14.14 Coupled oscillator with different spring constants, $\omega_w = \pi$, $\omega_c = 10\pi$, slightly mixed modes, time step 0.05 s.

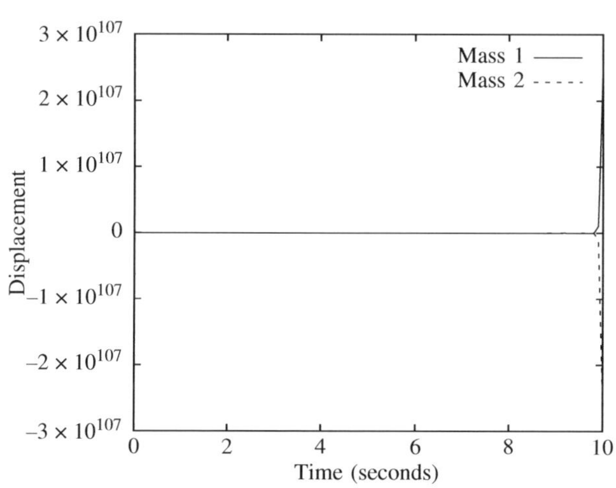

Figure 14.15 Coupled oscillator with different spring constants, $\omega_w = \pi$, $\omega_c = 10\pi$, slightly mixed modes, time step 0.1 s.

easily cause an equivalent stability condition for explicit methods to be violated when coarse step sizes are chosen to model relatively slower behavior in the interest of computational efficiency and avoiding accumulation of roundoff errors. Systems of equations with this unfortunate property are called "stiff" equations, and are encountered in many interesting physical systems, so studying alternative algorithms with wider regions of stability is important. Returning to the visualization of the "channels" in Figure 12.1 that guide the evolution of the desired solution function from its initial condition, instability results from the information of only *the prior* right-hand side function values being insufficient to constrain the solution within the prescribed channels as the next steps are projected in the solution space. To keep the solution conformed to the trajectory defined by the differential equation vector space, we must make use of the values of the right-hand side function not only at prior points but also *implicitly* at the next point we are seeking.

14.2 Algorithms: Implicit Methods for Differential Equations

Returning to the single first-order differential equation that was our starting point for the explicit algorithms in Chapter 12

$$\frac{dy}{dx} = f(x, y) \tag{14.14}$$

the general solution strategy of a stepped algorithm shown in Figure 12.2 still applies. Just as with the explicit methods detailed in Chapter 12, we endeavor to develop the solution function to the differential equation sampled at the uniform grid of values of the independent variable $x_j = x_{init} + jh$. We start by looking at a simple first-order implicit algorithm closely related to the "Forward Euler" method of Section 12.2.1.

14.2.1 First-order Adams–Moulton or Backward Euler Method

Figure 14.16 shows the method of using a first-order truncation of the Taylor series of the solution function to project a step with some slope to the next sample grid point, but involving an implicit relationship with the derivative function $f(x, y)$ in Equation (14.14). This implicit relationship is given by

$$y_{j+1} = y_j + h \cdot f(x_{j+1}, y_{j+1}) \tag{14.15}$$

This method is called the "Backward Euler" algorithm [11, Chap. 15; 34, Chap. 13], as we can think of the slope of the line used to project to the next sample point as facing "backward" from that point. The slope of the projection line will no longer be the line tangent to the solution curve at the current point (x_j, y_j) as it was with the Forward Euler algorithm, but instead will be the line tangent to the solution curve at the supposed next point. Since we are not taking into account any curvature of the solution, there is the potential of some significant truncation error, as we would expect with any simple first-order algorithm. The Backward Euler method is the first-order algorithm among a wider class of solution methods of various orders known as Adams–Moulton methods. In contrast to the Forward Euler method of Section 12.2.1, where the value of the next step y_{j+1} could be calculated with an explicit formula, with the Backward Euler method we must solve the implicit relationship in Equation (14.15) for y_{j+1}.

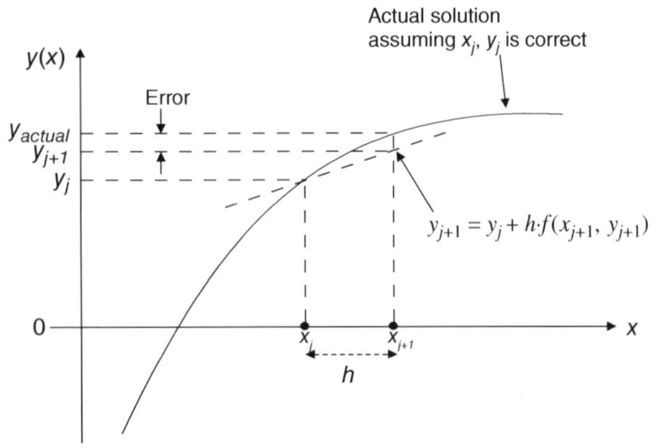

Figure 14.16 Solution step of the Backward Euler method.

To see a simple example where the solution for y_{j+1} can be accomplished easily, and also to address the issue of stability, we return to Equation (14.16), the same reference differential equation we treated in Chapter 12.

$$\frac{dy}{dx} = -ay \tag{14.16}$$

Substituting Equation (14.16) into Equation (14.15) we arrive at the implicit relationship for the next step as

$$y_{j+1} = y_j - ahy_{j+1} \tag{14.17}$$

For this simple differential equation, this relationship may easily be solved as

$$y_{j+1} = \frac{1}{1+ah} y_j \tag{14.18}$$

Therefore, beginning from the initial condition at $x_{init} = 0$, the solution that is developed by this algorithm for any value of the sample index j is

$$y_j = \frac{1}{(1+ah)^j} y_0 \tag{14.19}$$

If the step interval h is small enough, this solution will approximate the true solution $y_j = y_0 e^{-ax_j}$. In the next subsection we will examine how closely this sequence approximates the true solution in comparison to a second-order algorithm. From Equation (14.19), we can see that the requirement for stability is now

$$\frac{1}{1+ah} < 1 \tag{14.20}$$

This is satisfied with any step size h, so the step size may be chosen solely on the basis of limiting truncation error, and all solutions will be bounded.

Solving the implicit relationship of Equation (14.15) for some general first-order differential equation, either by itself or as part of a general system of equations, will likely not be the easy exercise of Equation (14.18), and for the sorts of problems addressed in this text will typically require a numerical solution by methods developed in Chapters 5, 6, and 7. But before we explore these methods, let us move on to a more practical second-order implicit algorithm for differential equations with significantly less truncation error.

14.2.2 Second-order Adams–Moulton or Trapezoidal Method

Figure 14.17 shows the method of projecting a step to the next sampled solution point with what is often called the "trapezoidal" algorithm, but is also the second-order version of the Adams–Moulton class of methods. The general idea is similar to that of trapezoidal integration from Section 10.1.1 and Heun's method in Section 12.2.2 in that a corrected slope with which we project the next step, and that takes into account the curvature of the solution, is calculated as the average of the starting and ending slopes of the projection. In this case, however, the ending slope comes from an implicit relationship with the derivative function evaluated at each point being solved for, $f(x_{j+1}, y_{j+1})$, as shown in Equation (14.21).

$$y_{j+1} = y_j + \frac{h}{2} \cdot [f(x_j, y_j) + f(x_{j+1}, y_{j+1})] \tag{14.21}$$

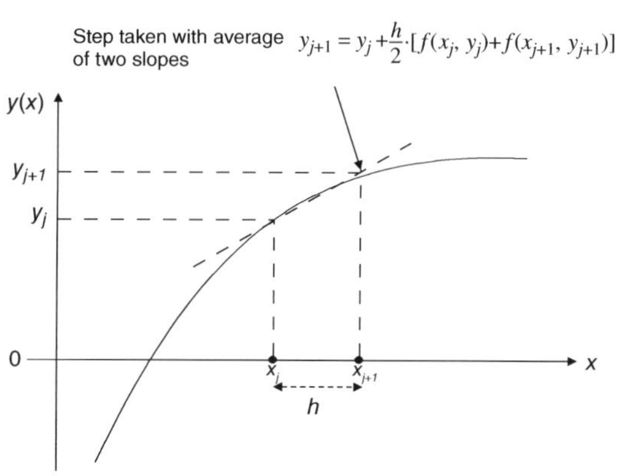

Figure 14.17 Solution step of the second-order Adams–Moulton or trapezoidal method.

As a simple example of solving this relationship, we again return to the reference differential equation (14.16). Substituting the differential equation into the second-order Adams–Moulton algorithm of Equation (14.21) gives us the implicit relationship for the next step as

$$y_{j+1} = y_j - \frac{ah}{2}y_j - \frac{ah}{2}y_{j+1} \tag{14.22}$$

This may easily be solved to give the relationship from one sample to the next of

$$y_{j+1} = y(x_{j+1}) = \left(\frac{1 - \frac{ah}{2}}{1 + \frac{ah}{2}} \right) y_j \tag{14.23}$$

This implies that the solution that is developed by this algorithm for any value of the sample index j is

$$y_j = y_0 \left(\frac{1 - \dfrac{ah}{2}}{1 + \dfrac{ah}{2}} \right)^j \tag{14.24}$$

With increasing sample index j this sequence will also remain bounded for any step size h, so, as with the Backward Euler algorithm, the step size may be chosen solely on the basis of limiting truncation error. Since we know the solution to this simple reference equation, we can easily plot the error between the first- and second-order implicit solution methods with the known solution in Figure 14.18.

Similar to the plot of Figure 12.8 for explicit methods, we see significant truncation error for the first-order Backward Euler algorithm. Whereas the Forward Euler method tends to deviate from the true solution on the outside of curves, the truncation errors of the Backward Euler method tend to lead its projections on the inside of curves. The second-order trapezoidal algorithm exhibits much smaller errors, but with a coarse step size of 0.1, its deviation is still just visible in the zoomed-in inset plot in Figure 14.18.

We can see the signature dependence on step size of the errors of the first- and second-order algorithms in Figure 14.19, which appears similar to that of the explicit methods in Figure 12.9. Reasonable errors levels can be seen with the second-order trapezoidal algorithm, at the expense of the high computation load required with fine step sizes. Higher-order implicit algorithms with smaller truncation error levels are possible, but their complexity leaves them outside the scope of this text.

In the same manner that we generalized solving a first-order differential equation to a system of equations in Section 12.3, we may generalize the trapezoidal algorithm of Equation (14.21) to a system of equations as well. We must be prepared, however, to solve the implicit relationship projecting to the next sample point also as a system of equations.

Assume that the system of first-order differential equations that we must solve can be put in the form

$$\frac{d\vec{y}}{dx} = \vec{f}(x, \vec{y}) \tag{14.25}$$

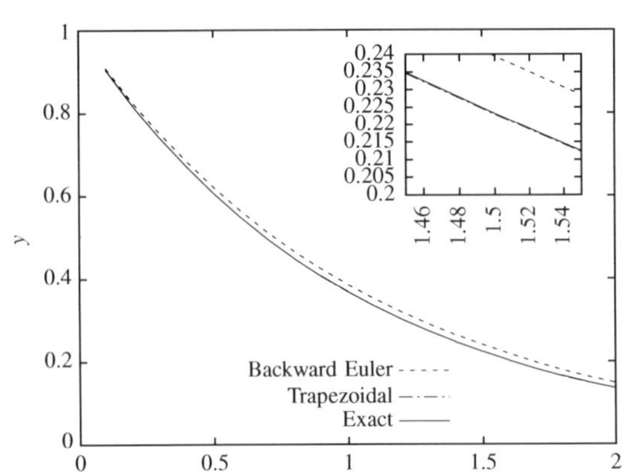

Figure 14.18 Example solution of reference differential equation with implicit methods, step size 0.1.

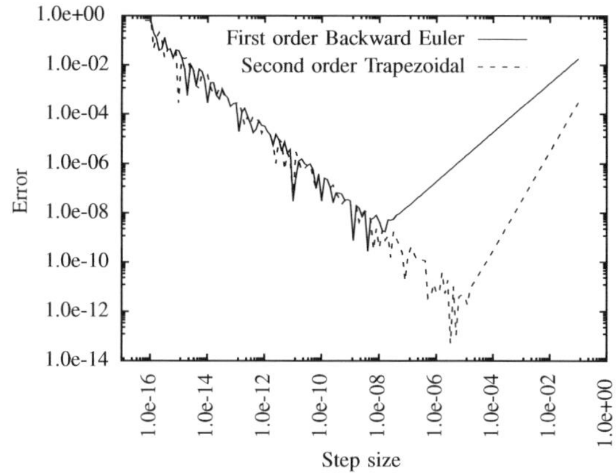

Figure 14.19 Tradeoff of roundoff and truncation error versus step size for first- and second-order differential equation implicit solution estimates.

where the components of the vector $\vec{y}$ are the n state variables of the system. Then at each sample step with index j, the state variable vector is

$$\vec{y}_j = \begin{bmatrix} y_{0,j} \\ y_{1,j} \\ y_{2,j} \\ \vdots \\ y_{n-2,j} \\ y_{n-1,j} \end{bmatrix} \tag{14.26}$$

where the first subscript is the state variable index ranging from 0 to $n-1$ and the second index is the sample index. The elements of the vector $\vec{f}(x, \vec{y})$ in Equation (14.25) are the right-hand side derivative functions of the differential equations in the system

$$\vec{f}(x, \vec{y}) = \begin{bmatrix} f_0(x, \vec{y}) \\ f_1(x, \vec{y}) \\ f_2(x, \vec{y}) \\ \vdots \\ f_{n-2}(x, \vec{y}) \\ f_{n-1}(x, \vec{y}) \end{bmatrix} \tag{14.27}$$

with the subscripts of f indexing the state equations. For this system, we extend the second-order method of projecting the state variables from a current vector to the next sample vector described by Equation (14.21) to the vector implicit relationship

$$\vec{y}_{j+1} = \vec{y}_j + \frac{h}{2} \cdot \left[\vec{f}(x_j, \vec{y}_j) + \vec{f}(x_{j+1}, \vec{y}_{j+1})\right] \tag{14.28}$$

At this point we consider two mathematical possibilities. For many systems, such as that representing the coupled spring-mass oscillators of Section 14.1, the $\vec{f}(x, \vec{y})$ functions are linear combinations of the state variables, and the implicit relationship described by Equation (14.28) may be solved by the tools for systems of linear equations in Chapter 6. However, many more systems of interest, such as n-body gravitational systems, are characterized by nonlinear functions as $\vec{f}(x, \vec{y})$, and we must apply the algorithms from Chapter 7 for finding the roots of systems of nonlinear equations.

In the linear case, we assume the system of Equation (14.25) may be cast in the form

$$\frac{d\vec{y}}{dx} = \mathbf{A}\vec{y} + \vec{v}(x) \tag{14.29}$$

where **A** is a square matrix of constant coefficients and the vector function $\vec{v}(x)$ is a "forcing function," a function of only the independent variable x, and not dependent on any state variable. In this case, the implicit trapezoidal step relationship of Equation (14.28) becomes

$$\vec{y}_{j+1} = \vec{y}_j + \frac{h}{2}\left[\mathbf{A}\vec{y}_j + \vec{v}(x_j) + \mathbf{A}\vec{y}_{j+1} + \vec{v}(x_{j+1})\right] \tag{14.30}$$

This relationship may be rearranged as

$$\left[\mathbf{I} - \frac{h}{2}\mathbf{A}\right]\vec{y}_{j+1} = \left[\mathbf{I} + \frac{h}{2}\mathbf{A}\right]\vec{y}_j + \frac{h}{2}\left[\vec{v}(x_j) + \vec{v}(x_{j+1})\right] \tag{14.31}$$

where **I** is the identity matrix. Assuming the solution has evolved from the initial conditions up through the current sample index j and we are solving for the next sample $j+1$, the entire right side of Equation (14.31) is a vector with known components, and the factor $\left[\mathbf{I} - \frac{h}{2}\mathbf{A}\right]$ on the left side is a square matrix of known constant coefficients. We can construct a system of linear equations with this expression and solve for the unknown vector of the next sample points $\vec{y}_{j+1}$ by calling a Gaussian elimination tool in the `lineq.c` library code developed in Chapter 6 at each sample step. Therefore, we can write a utility solving tool with which a calling program can numerically evolve a solution of a linear differential equation by supplying a matrix of coefficients **A** and a C function that calculates the vector forcing function $\vec{v}(x)$. From Equation (14.31) we can see that for the system of linear equations to be nonhomogeneous, we must have either a nonzero initial condition for at least one state variable or a nonzero value for at least one component of the forcing function vector $\vec{v}$.

On the other hand, in the nonlinear case, where the $\vec{f}(x, \vec{y})$ functions in Equation (14.25) are any nonlinear functions that may include the effect of forcing, we can rewrite the trapezoidal step relationship of Equation (14.28) as

$$\left[\vec{y}_{j+1} - \frac{h}{2}\vec{f}(x_{j+1}, \vec{y}_{j+1})\right] - \left[\vec{y}_j + \frac{h}{2}\vec{f}(x_j, \vec{y}_j)\right] = 0 \tag{14.32}$$

As we project the next sample vector $\vec{y}_{j+1}$, we treat the current sample vector $\vec{y}_j$ at the independent variable value x_j as a known constant. Also known are all the x values on the sampling grid $x_j = x_{init} + jx_{inc}$, including the next value $xj + 1$. So we note that the first bracketed term in Equation (14.32) is dependent on the unknown next sample vector $\vec{y}_{j+1}$, but the second bracketed term is a constant and known vector. We can define the entire left side of Equation (14.32) as a nonlinear function of $\vec{g}(\vec{y}_{j+1})$, so this relationship is essentially in the form of

$$\vec{g}(\vec{y}_{j+1}) = 0 \tag{14.33}$$

Solving this relationship for the next sample vector y_{j+1} is essentially solving for the vector root of a system of nonlinear equations, which is the same numerical exercise as the exercise that was treated in Chapter 7. Therefore, we could solve for the unknown vector of the next sample points $\vec{y}_{j+1}$ by calling the `sys_newton_raphson()` tool in the `sys_root.c` library code file at each sample step. However, that tool spends a significant amount of computation assessing whether the sequence of its iterations has converged on a root to within a given tolerance. In our application of interest now, finding samples of a differential equation solution, a single step of the simple Forward Euler method can provide a good initial guess of the root $\vec{y}_{j+1}$. Since the Newton–Raphson algorithm converges rapidly with a good initial starting point, we are likely to converge within the truncation error of the second-order trapezoidal algorithm simply by performing a fixed number of three Newton–Raphson iterations. We are required to perform this root search at each sample point in the differential equation solution, so we may save considerable computation simply by counting a fixed number of iterations at each sample. Following the derivation of Equation (7.6), our iteration formula for converging on the root of Equation (14.33) is

$$\vec{y}_{j+1,k+1} = \vec{y}_{j+1,k} - \mathbf{J}^{-1}\left(\vec{y}_{j+1,k}\right) \cdot \vec{g}\left(\vec{y}_{j+1,k}\right) \tag{14.34}$$

At this point we are on the verge of being overwhelmed by subscript notation, and must exercise some care to keep the algorithm clear. The first subscript on $\vec{y}$ in Equation (14.34) is the sample index j of the solution we are evolving, and the second is the iteration index k as we use the Newton–Raphson method to converge on the root of Equation (14.33), which will be our best estimate of the next sample, index $j+1$, at the known independent variable value x_{j+1}. Just as in Section 7.1, we begin with an initial guess of the root $\vec{y}_{j+1,0}$ and then at each iteration we successively displace the current candidate root at iteration index k to the next candidate root at iteration index $k+1$, for $k = 0, 1, 2$. That displacement is the second term on the right side of Equation (14.34), and we find it by solving the linear system with the coefficients being the elements of the Jacobian matrix and the right-hand side vector being the value $g(\vec{y}_{j+1,k})$ at the current iteration index k.

If we had only one state variable in the system, we could visualize the convergence process as being similar to that in Figure 5.1. In that plot we see the Newton–Raphson algorithm searching for a root by starting from an initial guess and then successively making adjustments in candidate values of the root horizontally along the x axis. In each iteration, the function derivative is evaluated successively at each candidate root value to guide the convergence. But for the implicit trapezoidal differential equation method, the convergence instead takes place rotated vertically along the dotted line above the x_{j+1} independent variable value in Figure 14.17 and converges on the y_{j+1} root. The Jacobian matrix of Equation (14.36) (which would consist only of a single derivative in the case of one state variable) similarly must be evaluated at each candidate root. Expanding to a vector $\vec{y}$ of more than one state variable, the visualization would be difficult, but the same sort of process applies.

Following the derivation in Section 7.1, the iteration formula of Equation (14.34) requires the Jacobian matrix $\mathbf{J}$ to be evaluated at the successive candidate values of the root y_{j+1}, which also means at the known independent variable value x_{j+1}. That Jacobian matrix is filled with the first partial derivatives of the vector function $\vec{g}(\vec{y})$ whose root is sought.

$$\mathbf{J}(\vec{y}_{j+1}) = \begin{bmatrix} \frac{\partial g_0(\vec{y}_{j+1})}{\partial y_{0,j+1}} & \frac{\partial g_0(\vec{y}_{j+1})}{\partial y_{1,j+1}} & \cdots & \frac{\partial g_0(\vec{y}_{j+1})}{\partial y_{n-2,j+1}} & \frac{\partial g_0(\vec{y}_{j+1})}{\partial y_{n-1,j+1}} \\ \frac{\partial g_1(\vec{y}_{j+1})}{\partial y_{0,j+1}} & \frac{\partial g_1(\vec{y}_{j+1})}{\partial y_{1,j+1}} & \cdots & \frac{\partial g_1(\vec{y}_{j+1})}{\partial y_{n-2,j+1}} & \frac{\partial g_1(\vec{y}_{j+1})}{\partial y_{n-1,j+1}} \\ \frac{\partial g_2(\vec{y}_{j+1})}{\partial y_{0,j+1}} & \frac{\partial g_2(\vec{y}_{j+1})}{\partial y_{1,j+1}} & \cdots & \frac{\partial g_2(\vec{y}_{j+1})}{\partial y_{n-2,j+1}} & \frac{\partial g_2(\vec{y}_{j+1})}{\partial y_{n-1,j+1}} \\ & & \vdots & & \\ \frac{\partial g_{n-2}(\vec{y}_{j+1})}{\partial y_{0,j+1}} & \frac{\partial g_{n-2}(\vec{y}_{j+1})}{\partial y_{1,j+1}} & \cdots & \frac{\partial g_{n-2}(\vec{y}_{j+1})}{\partial y_{n-2,j+1}} & \frac{\partial g_{n-2}(\vec{y}_{j+1})}{\partial y_{n-1,j+1}} \\ \frac{\partial g_{n-1}(\vec{y}_{j+1})}{\partial y_{0,j+1}} & \frac{\partial g_{n-1}(\vec{y}_{j+1})}{\partial y_{1,j+1}} & \cdots & \frac{\partial g_{n-1}(\vec{y}_{j+1})}{\partial y_{n-2,j+1}} & \frac{\partial g_{n-1}(\vec{y}_{j+1})}{\partial y_{n-1,j+1}} \end{bmatrix} \tag{14.35}$$

where the subscript on g is the vector component index i, the first subscript on y is also the vector component index i, and the second subscript on y is the sample index $j + 1$. We may evaluate this Jacobian matrix by referring back to the definition of the function $\vec{g}$ in Equation (14.32). Since the terms in the second pair of brackets are independent of $\vec{y}_{j+1}$, the elements of the Jacobian matrix are determined by the partial derivatives of only the terms in the first pair of brackets. We rewrite the Jacobian as

$$\mathbf{J}(\vec{y}_{j+1}) = \begin{bmatrix} 1-\frac{h}{2}\frac{\partial f_0}{\partial y_0} & -\frac{h}{2}\frac{\partial f_0}{\partial y_1} & \cdots & -\frac{h}{2}\frac{\partial f_0}{\partial y_{n-2}} & -\frac{h}{2}\frac{\partial f_0}{\partial y_{n-1}} \\ -\frac{h}{2}\frac{\partial f_1}{\partial y_0} & 1-\frac{h}{2}\frac{\partial f_1}{\partial y_1} & \cdots & -\frac{h}{2}\frac{\partial f_1}{\partial y_{n-2}} & -\frac{h}{2}\frac{\partial f_1}{\partial y_{n-1}} \\ -\frac{h}{2}\frac{\partial f_2}{\partial y_0} & -\frac{h}{2}\frac{\partial f_2}{\partial y_1} & \cdots & -\frac{h}{2}\frac{\partial f_2}{\partial y_{n-2}} & -\frac{h}{2}\frac{\partial f_2}{\partial y_{n-1}} \\ & & \vdots & & \\ -\frac{h}{2}\frac{\partial f_{n-2}}{\partial y_0} & -\frac{h}{2}\frac{\partial f_{n-2}}{\partial y_1} & \cdots & 1-\frac{h}{2}\frac{\partial f_{n-2}}{\partial y_{n-2}} & -\frac{h}{2}\frac{\partial f_{n-2}}{\partial y_{n-1}} \\ -\frac{h}{2}\frac{\partial f_{n-1}}{\partial y_0} & -\frac{h}{2}\frac{\partial f_{n-1}}{\partial y_1} & \cdots & -\frac{h}{2}\frac{\partial f_{n-1}}{\partial y_{n-2}} & 1-\frac{h}{2}\frac{\partial f_{n-1}}{\partial y_{n-1}} \end{bmatrix} \tag{14.36}$$

where all the partial derivatives of $\vec{f}$ will be evaluated at the current candidate root arguments $x_{j+1}, \vec{y}_{j+1,k}$ at each Newton–Raphson iteration. The Jacobian matrix of Equation (14.36) is in the form

$$\mathbf{J} = \mathbf{I} - \frac{h}{2}\mathbf{J}_f \tag{14.37}$$

where $\mathbf{J}_f$ is a Jacobian matrix composed of the partial derivatives of the system of vector functions $\vec{f}(x, \vec{y})$ in Equation (14.27). Therefore, we can write a utility solving tool with which a calling program can numerically evolve a solution of a system of nonlinear differential equations by supplying two C functions, one calculating the vector functions $\vec{f}(x, \vec{y})$ and the other calculating the matrix of partial derivatives $\mathbf{J}_f$, similar to invoking the `sys_newton_raphson()` utility described in Section 7.2.

14.3 Coding: The Implicit Differential Equation Solver Tool

The tool code for solving systems of differential equations with implicit algorithms can be found in the files `diffeq_implicit.c` and `diffeq_implicit.h`, part of the `CPLibrary` package from the Student Companion download package.

```
int am2_linear(ARRAY2D *coeff,void (*force)(double x,void *p,ARRAY1D *v),
               GRID *xgrid_ptr,void *param,ARRAY1D *init,double min,
               ARRAY2D *ytab_ptr) {
  ...
}

int am2_linear_band(ARRAY2D_BAND *coeff,void (*force)(double x,
                     void *p,ARRAY1D *v),GRID *xgrid_ptr,void *param,
                     ARRAY1D *init,double min,ARRAY2D *ytab_ptr) {
  ...
}

int am2_nonlinear(void (*func)(double x,ARRAY1D *y,void *p,
                   ARRAY1D *yp),void (*jacobian)(double x,ARRAY1D *y,
                   void *p,ARRAY2D *partial),GRID *xgrid_ptr,
                   void *param,ARRAY1D *init,double min,
                   ARRAY2D *ytab_ptr) {
  ...
}
```

The `am2_linear()` and `am2_linear_band()` functions are called to solve systems of first-order differential equations with the linear version of the second-order Adams–Moulton, or trapezoidal, method over a specified grid of x values. The trapezoidal method is described in Section 14.2.2. The `am2_linear()` routine expects the coefficients of the linear system of Equation (14.29) to be stored in a full square matrix in an instance of the `ARRAY2D` structure, and will consequently call the `gauss()` tool in the `lineq.c` package to solve the implicit relationship to find each next solution sample point. In contrast, the `am2_linear_band()` routine is used for a linear system with coefficients that lie in a band around the central matrix diagonal, and expects the coefficients to be stored in a `ARRAY2D_BAND` structure instance. To solve for the implicit relationship at each sample point, the `gauss_band()` tool is called. A pointer to the respective coefficient array type is expected as the first argument. Beyond this first argument, both the `am2_linear()` and `am2_linear_band()` functions expect the same sequence of arguments. The second argument is a pointer to a C function that computes the set of n forcing functions in Equation (14.29). The prototype of that function is `void (*func)(double x,void *p,ARRAY1D *v)`. This prototype is similar to the sort of function we passed to the `rk4()` Runge–Kutta solver in Chapter 12 to compute the vector field functions, except for the omission of an array of state variables, since the forcing functions for the linear system of differential equations of Equation (14.29) are assumed to be only dependent on the grid values of the independent variable x_j. As with other functions supplied to the tools in our library, a pointer to a set of physical parameters that may be needed for the function calculations is also passed, followed by a final argument of a pointer to an array where the function should store all the forcing values.

Returning to the argument list for `am2_linear()` and `am2_linear_band()`, those functions next expect a pointer to a grid of the x independent variable over which the solution is to be found, the parameter structure pointer that will be passed on to the forcing function, a pointer to an array of the initial conditions for the state variables y_i, a minimum value for a pivot element to test for a singular coefficient matrix, and finally a pointer to a two-dimensional array where the solver is to store the solution. The memory for the solution array in the `ARRAY2D` structure must be allocated by the calling code, just as in the case for the explicit solvers developed in Chapter 12. Just as for the solutions developed by the `rk4()` solver in `diffeq.c`, the row index i of the array indexes the equations in the system and the state variables $y_i(x)$, and the column index j of the array indexes the samples in each state variable $y_{i,j}$ over the grid of x_j values. The sample index j corresponds to the independent variable value $x = x_{init} + jx_{inc}$, and the ARRAY2D must be allocated by the calling code with sufficient size to accommodate x values up to and including x_{final}, where x_{init}, x_{final}, and x_{inc} are specified in the xgrid argument. As before, making use of the NUM_GRID_PTS() macro is a convenient way to arrive at the column dimension for allocation. The implicit routines may return a value of DIFFEQ_IMPLICIT_SINGULAR in the case of a request to solve a degenerate system of linear equations.

After am2_linear() and am2_linear_band(), the third utility for an implicit algorithm for systems of differential equations in diffeq_implicit.c is the function am2_nonlinear(). This tool must be called for systems of equations in the form of Equation (14.25) where the right-hand side functions $f_i(x, \vec{y})$ are nonlinear functions of their arguments. The arguments to the am2_nonlinear() tool provide the method for specifying both $f_i(x, \vec{y})$, the right side functions, and $\mathbf{J}_f$, the Jacobian of the right side functions, as shown in Equations (14.36) and (14.37). The first two arguments to am2_nonlinear() are pointers to functions that calculate the function values and the partial derivatives in the Jacobian matrix. Just as with similar functions passed to the sys_newton_raphson() tool of Chapter 7, these functions are given pointer arguments to the one-dimensional and two-dimensional types ARRAY1D and ARRAY2D, respectively, where the calculated values are to be written and used by the Adams–Moulton solution algorithm in am2_nonlinear(). The remaining expected arguments to am2_nonlinear() are a pointer to a grid of the x independent variable over which the solution is to be found, the parameter structure pointer that will be passed on to the right side and Jacobian functions, a pointer to an array of the initial conditions for the state variables y_i, a minimum value for a pivot element to test for a singular coefficient matrix, and finally a pointer to a two-dimensional array where the solver is to store the solution. As in the case of the am2_linear() tool, the memory for the solution array in the ARRAY2D structure must be allocated by the calling code.

The following sections provide examples and exercises demonstrating the use of all three of these solvers.

14.4 Application: Coupled Oscillators with the Implicit Solver

For an important lesson on stability with an example more complex than the simple reference problem of Equation (14.16), we now return to the coupled spring-mass oscillator we analyzed in Section 14.1. In the case of widely separated frequency eigenvalues, the instability of the explicit fourth-order Runge–Kutta method led us to the numerical breakdown of unbounded errors in Figure 14.15.

The program to simulate the coupled two-mass oscillator with the trapezoidal algorithm is in the source code coupled_osc22_implicit.c from the Student Companion website. The program must populate the coefficient matrix for the equation system and supply a function that returns the value of the forcing vector, which, in the case of the coupled oscillator with nonzero initial conditions, is zero. The excerpt from this program is

```
void zero_force(double x,void *p,ARRAY1D *v) {
  zero_array1d(v);
}

int main(int argc,char *argv[]) {
  ...
  alloc_array2d(&coupled,4,4);
  zero_array2d(&coupled);
  coupled.elem[0][1]  = 1.0;
  coupled.elem[1][0]  = -(w0w * w0w + w0c * w0c);
  coupled.elem[1][2]  = w0c * w0c;
  coupled.elem[2][3]  = 1.0;
  coupled.elem[3][0]  = w0c * w0c;
  coupled.elem[3][2]  = -(w0w * w0w + w0c * w0c);
  alloc_array2d(&ytable,4,NUM_GRID_PTS(tgrid));
// Solve the system of differential equations with the AM2 linear
// implicit algorithm
  if (am2_linear(&coupled,zero_force,&tgrid,NULL,&init,1.0e-12,
                 &ytable) == DIFFEQ_IMPLICIT_SUCCESS) {
    gen_output(&ytable,&tgrid);
  }
  ...
  exit(0);
}
```

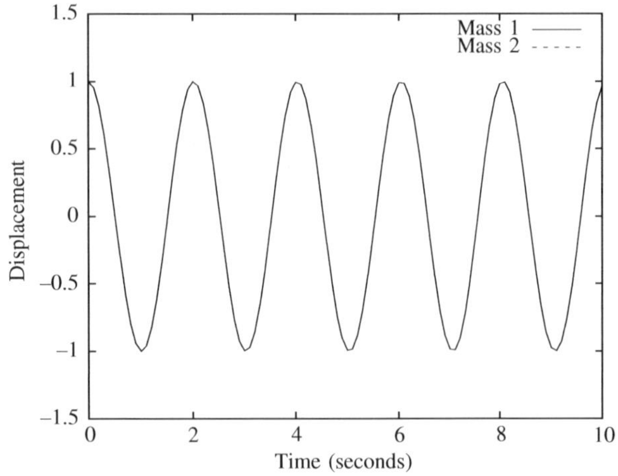

Figure 14.20 Coupled oscillator with different spring constants, $\omega_w = \pi$, $\omega_c = 10\pi$, slightly mixed modes, time step 0.1 s, implicit AM2 algorithm.

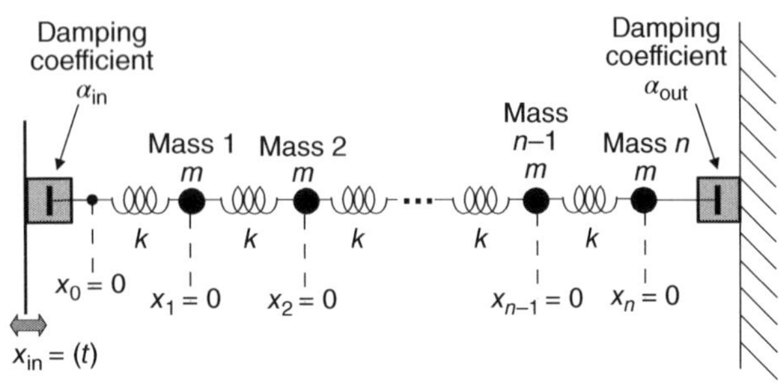

Figure 14.21 A longitudinal wave propagation line with n coupled spring-mass oscillators and damping at the input and output.

Repeating a simulation of this same system with a coarse time step of 0.1 seconds, but now with the implicit second-order Adams–Moulton, or trapezoidal, algorithm, generates the results shown in Figure 14.20. With the wider region of stability of the implicit algorithm, the fidelity of the solution algorithm has been restored, at the expense of the additional computational load required to numerically solve the implicit relationships stemming from Equation (14.21) at each sample point.

14.5 Application: Waves

If we increase the number of masses and springs in the coupled spring-mass oscillator analyzed in Section 14.1, the eigenvalues of the system spread out to encompass a very large range, as illustrated in Figure 6.7. As the frequency of the slowest eigenvalue becomes a very small fraction of the fastest, this will eventually lead to the sort of instability seen in Figure 14.15 for any practical time step, even with the case of all the springs and masses being identical. These systems must be treated with an implicit algorithm for a stable solution. Greatly increasing the number of elements, as shown in Figure 14.21, will allow us to see wave propagation behavior emerge, but these systems must be treated with an implicit algorithm for a stable solution.

We assume that the damping devices at the input and output, generally constructed with some viscous fluid, are characterized by a damping coefficient α such that a reaction damping force is exerted by the device in a direction opposite to the velocity of its connection arm v and with magnitude given by

$$F_{damp} = \alpha v \tag{14.38}$$

We may derive the equations of motion for all the masses in this system with the straightforward application of Newtonian mechanics. For mass index i with displacement from its equilibrium position x_i, where $1 \le i \le n-1$, the motion is determined by the reaction forces of the two springs connected to each side.

$$\ddot{x}_i - \frac{k}{m}(x_{i-1} - 2x_i + x_{i+1}) = 0 \tag{14.39}$$

We consider the anchor position of the left end to be the "input" source of the stimulus that launches a wave that will propagate down the spring-mass medium. We will assume that the anchor position will be driven by some mechanical means, and that its x coordinate is known to follow the forcing function $x_{in}(t)$. We track the position of the connection point between the damper device and the first spring and designate the displacement of that connection point x_0. The reaction force of the damper must be supported by the reaction force of the first spring, so we may write

$$\ddot{x}_0 - \frac{k}{\alpha_{in}}(x_1 - x_0) = \dot{x}_{in} \tag{14.40}$$

where α_{in} is the damping coefficient of the input end damper. Note that only the time derivative of $x_{in}(t)$ has any physical effect on the wave medium. At the right end, which we consider to be the "output" of the transmitted wave, the displacement of mass n is x_n, and the motion is determined by the reaction forces of both the last spring and the damper device.

$$\ddot{x}_n - \frac{k}{m}(x_{n-1} - x_n) + \frac{\alpha_{out}}{m}\dot{x}_n = 0 \tag{14.41}$$

where α_{out} is the damping coefficient of the output end damper. To apply the linear version of the second-order Adams–Moulton algorithm, as detailed in Section 14.2.2, we need to bundle all these equations of motion together with an appropriate definition of a vector of state variables $\vec{y}$ and a forcing function that represents the input displacement x_{in}, into the form of Equation (14.42)

$$\frac{d\vec{y}}{dt} = \mathbf{A}\vec{y} + \vec{v}(t) \tag{14.42}$$

The first state variable y_0 is the displacement x_0 of the connection point between the input damper and the first spring. State variables y_1 and y_2 are the displacement and velocity x_1 and $\dot{x}_1$ of mass 1, respectively. This pattern continues, with state variables y_{2i-i} and y_{2i} being the displacement and velocity x_i and $\dot{x}_i$ of mass i, respectively. Finally, the last two state variables y_{2n-i} and y_{2n} are the displacement and velocity x_n and $\dot{x}_n$ of mass n, respectively. This totals to $2n + 1$ state variable components in the state vector $\vec{y}$ for n masses.

We may develop the system of state equations extending the manner of Equation (14.6) for the two-mass coupled oscillator. The pair of state equations associated with mass 1 that contribute to the system are $\frac{dy_{2i-i}}{dt} = y_{2i}$ and $\frac{dy_{2i}}{dt} = \frac{F_i}{m}$, where F_i is the net force on mass i due to either two connected springs or to a spring and the output damper in the case of mass n. With these definitions the full coefficient matrix $\mathbf{A}$ we would pass to the `am2_linear()` solver routine is

$$\begin{bmatrix}
-\frac{k}{\alpha_{in}} & \frac{k}{\alpha_{in}} & 0 & 0 & 0 & 0 & 0 & 0 & \cdots & 0 & 0 & 0 & 0 & 0 & 0 \\
0 & 0 & 1 & 0 & 0 & 0 & 0 & 0 & \cdots & 0 & 0 & 0 & 0 & 0 & 0 \\
\frac{k}{m} & -2\frac{k}{m} & 0 & \frac{k}{m} & 0 & 0 & 0 & 0 & \cdots & 0 & 0 & 0 & 0 & 0 & 0 \\
0 & 0 & 0 & 0 & 1 & 0 & 0 & 0 & \cdots & 0 & 0 & 0 & 0 & 0 & 0 \\
0 & \frac{k}{m} & 0 & -2\frac{k}{m} & 0 & \frac{k}{m} & 0 & 0 & \cdots & 0 & 0 & 0 & 0 & 0 & 0 \\
0 & 0 & 0 & 0 & 0 & 0 & 1 & 0 & \cdots & 0 & 0 & 0 & 0 & 0 & 0 \\
0 & 0 & 0 & \frac{k}{m} & 0 & -2\frac{k}{m} & 0 & \frac{k}{m} & \cdots & 0 & 0 & 0 & 0 & 0 & 0 \\
 & & & & & & & & \vdots & & & & & & \\
0 & 0 & 0 & 0 & 0 & 0 & 0 & 0 & \cdots & 0 & 0 & 0 & 1 & 0 & 0 \\
0 & 0 & 0 & 0 & 0 & 0 & 0 & 0 & \cdots & \frac{k}{m} & 0 & -2\frac{k}{m} & 0 & \frac{k}{m} & 0 \\
0 & 0 & 0 & 0 & 0 & 0 & 0 & 0 & \cdots & 0 & 0 & 0 & 0 & 0 & 1 \\
0 & 0 & 0 & 0 & 0 & 0 & 0 & 0 & \cdots & 0 & 0 & \frac{k}{m} & 0 & -\frac{k}{m} & -\frac{\alpha_{out}}{m}
\end{bmatrix} \tag{14.43}$$

However, we can see this coefficient matrix is fairly sparsely populated, and the benefit of greatly increased computational efficiency may be gained by using the banded matrix version of the linear equation solver with a half band width of 3. The alert reader may be alarmed that this coefficient matrix has zeroes all down the center diagonal except at the very ends. We learned in Section 6.1 that the Gaussian elimination used for solving linear system depended on nonzero elements down the diagonal as the denominator of elimination fractions. However, we must keep in mind that this matrix of coefficients **A** will not be used directly in the trapezoidal algorithm for Gaussian elimination. In Equation (14.31), we see that the actual coefficient matrix defining the linear system to be solved for the next solution sample point is $\mathbf{I} - \frac{h}{2}\mathbf{A}$. In general, the "1" elements down the diagonal of the identity matrix will form a dominant central diagonal (in terms of absolute value) relative to the other elements that are multiplied by a small step size h, and, therefore, we may rely on good numerical stability in the Gaussian elimination algorithm.

To complete the mathematical model of Equation (14.42) we finally require the $2n + 1$ dimensional forcing vector on the right side, defined as

$$\vec{v}(t) = \begin{bmatrix} \dot{x}_{in} \\ 0 \\ 0 \\ \vdots \\ 0 \\ 0 \end{bmatrix} \tag{14.44}$$

where $\dot{x}_{in}$ is the time derivative of the mechanical forcing position of the anchor on the left side.

We can visualize the continuous limit of this simulation with discrete masses and springs by comparing Equation (14.39) with Equation (2.11) for a discrete second-order estimate of the second derivative. Imagine the limit where the number of segments n becomes large, and the mass m and spring constant k become the mass and stiffness in a segment length η in the spatial dimension along the line. We can see that Equation (14.39) is essentially related to an estimate of the second spatial derivative of the displacement

$$x(\xi - \eta) - 2x(\xi) + x(\xi + \eta) \approx \eta^2 \frac{\partial^2 x(\xi)}{\partial \xi^2} \tag{14.45}$$

where ξ is the general spatial coordinate in the longitudinal direction of the mass displacements x and η is the spatial step size over one spring-mass segment. Identifying the constant c^2 as $\eta^2 \frac{k}{m}$, in the limit of large n Equation (14.39) becomes

$$\frac{\partial^2 x}{\partial t^2} = c^2 \frac{\partial^2 x}{\partial \xi^2} \tag{14.46}$$

This is exactly the form of the longitudinal wave equation [9, Chap. 7; 10, Chap. 12; 13, Chap. 9; 30, Chap. 3] with a propagation speed of

$$c = \eta \sqrt{\frac{k}{m}} \tag{14.47}$$

Therefore, we identify $\sqrt{\frac{k}{m}}$ as the reciprocal of the propagation time through a single spring-mass element spanning a step of η in distance. The step size η for the spatial second derivative is what will determine the truncation error for the right side of Equation (14.46).

In any wave propagating medium, there is a characteristic "impedance" of the propagation mechanism that characterizes how the energy in two complementary forms alternate as the wave propagates. In the spring-mass line, the longitudinal kinetic energy of the masses alternates with the potential energy in the springs, and the characteristic impedance is given by

$$Z = \frac{1}{\sqrt{km}} \tag{14.48}$$

For the condition where all the energy in the wave launched at the input end is to be absorbed by the terminating damping device at the output end of the medium, the characteristic impedance of the medium must match the reciprocal of the damping coefficient, or $\frac{1}{\alpha}$. Otherwise, if $\sqrt{km} \neq \alpha$, there is a mismatch between the characteristic impedance and the output damping, and part of the energy of an incident wave will be reflected back into a wave propagating in the opposite direction when it reaches the output end of the medium. This behavior will be demonstrated in the simulations of this model.

The equations of motion and wave propagation above, although derived for a medium supporting longitudinal displacements, are essentially in the same form as those governing transverse waves, such as on a vibrating string, or torsional waves propagating on a rotational medium. Moreover, by making the substitutions shown in Table 14.1 the same equations that govern these mechanical longitudinal waves also govern electric waves propagating down a sequence of segments consisting of a series inductor and a shunt capacitor to ground. These components formed the resonant circuit of the electronic oscillator in Section 12.10. In that case, the ends of the wave medium are terminated by electrical resistances.

The source code file `wave1.c` may be downloaded from the Student Companion website. Note in the source code that we prepare the coefficients in the matrix of Equation (14.43) in the banded matrix structure `ARRAY2D_BAND` with a half band width of 3 and call the `am2_linear_band()` utility in `diffeq_implicit.c` tool to solve the linear system of Equation (14.31). This choice takes advantage of the banded structure of (14.43) to save an enormous amount of computation. The `wave1.c` program may be compiled with the commands

```
$ gcc -Wall -Wshadow -O2 -o wave1 wave1.c CPLibrary/diffeq_implicit.c CPLibrary/lineq.c
    ↪ CPLibrary/array.c -lm
$ ./wave1
./wave1 <tfinal> <tinc> <num segments> <springk> <mass> <alphai> <alphao> <input code> <input
    ↪ period>
```

This modeling program essentially provides a numerical wave propagation laboratory where we can perform experiments to visualize all the classical wave phenomena. If we specify a fairly large number of mass elements, we are able to visualize the behavior of a physical system somewhere in the boundary region between discrete oscillator segments and a continuous medium supporting wave propagation.

Two types of input forcing function dependent on time may be chosen in the code: a single pulse with a "raised cosine" profile or a continuous sinusoidal excitation. A normalized raised cosine pulse, plotted in Figure 14.22, follows the expression

$$x_{in}(t) = \begin{cases} 0 & t < 0 \\ \dfrac{1 - \cos(2\pi t/T)}{2} & 0 < t < T \\ 0 & T < t \end{cases} \tag{14.49}$$

for a time period T. The name comes from the formulation of "raising" a cosine function with a constant so that its minima are aligned with a zero baseline. This function is important in signal processing as a pulse shape with both a value and derivative that are continuous with a baseline zero level outside the single time period and with a spectral content that is highly concentrated at the fundamental pulse period.

For the continuous sine wave excitation option the forced displacement on the left side follows

$$x_{in}(t) = \sin(2\pi t/T) \tag{14.50}$$

Following Equation (14.44), the `wave_force()` function in the `wave1.c` program provides the time derivative of either Equation (14.49) or (14.50) for the linear trapezoidal algorithm.

Table 14.1 Equivalence between a spring-mass mechanical wave medium and an inductor–capacitor electric wave medium.

Spring-mass	**Inductor–capacitor**
Mass m	Capacitance C
Stiffness k	Reciprocal inductance $\frac{1}{L}$
Damping α	Conductance $\frac{1}{R}$
Impedance $\frac{1}{\sqrt{km}}$	Impedance $\sqrt{\frac{L}{C}}$

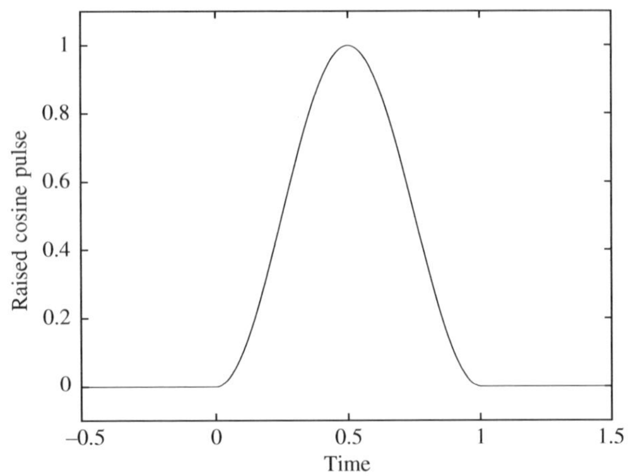

Figure 14.22 Raised cosine profile with period $T = 1$ for the single input pulse.

For a first example, the following dialog will simulate a nominal case of a sequence of 100 spring-mass elements with a characteristic impedance $\frac{1}{\sqrt{km}} = 1$, and with a raised cosine pulse applied as the displacement of the left-hand input wall. With the output damping coefficient $\alpha_{out} = 1$ the impedance of the wave medium is matched to that of the output wall.

```
$ ./wave1 180 0.1 100 1 1 1 1 pulse 20 >| wave1.dat
```

Figure 14.23 shows the energy of the initial pulse introduced on the left side, the wave propagating down the line to the right, and the pulse energy absorbed in the termination damping on the right-hand side.
The results of the simulation are close to ideal wave behavior following the wave equation (14.46). Some residual behavior can be seen in these plots, as a line consisting of 100 discrete resonant segments is still only an approximation to a fully continuous transmission line with infinitesimal elements, and we see the artifacts of the resulting spatial truncation error.

Next, we both increase the spring constant k by a factor of two and decrease the mass m by a factor of two, which, according to Equation (14.47), should double the propagation speed $\sqrt{\frac{k}{m}}$ but keep the characteristic impedance $\frac{1}{\sqrt{km}}$ the same relative to the nominal case.

```
$ ./wave1 90 0.1 100 2 0.5 1 1 pulse 20 >| wave1.dat
```

Note that the time sample values in the resulting plots of Figure 14.24 are half that of Figure 14.23, showing the doubling of propagation velocity.

Next we repeat the simulation with the original propagation speed but for a mismatch of the damping coefficient at the output end. This case is for very light output damping with a coefficient of $\alpha_{out} = 0.01$, allowing almost free displacement at the right end with very little energy loss.

```
$ ./wave1 180 0.1 100 1 1 1 0.01 pulse 20 >| wave1.dat
```

Since we now have a very significant mismatch between the characteristic impedance of the line and the damping, this case generates a reflection of the incident wave, with the reflected wave having the same positive longitudinal displacement direction as the incident wave, shown in Figure 14.25. The small following echo propagating back to the right after the main body of the pulse is again due to the finite number of mass elements.

Next we repeat the simulation but for a very rigid damping coefficient at the output end of $\alpha_{out} = 100$.

```
$ ./wave1 180 0.1 100 1 1 1 100 pulse 20 >| wave1.dat
```

This case, plotted in Figure 14.26, generates another sort of reflection of the incident wave, this time with longitudinal displacements in the negative direction.

Finally, we keep the large mismatch at the output end with a damping coefficient of $\alpha_{out} = 100$ but change the input mode from a single pulse to a continuous sine wave excitation with a period of 20 units.

```
$ ./wave1 200 0.1 80 1 1 1 100 sine 20 >| wave1.dat
```

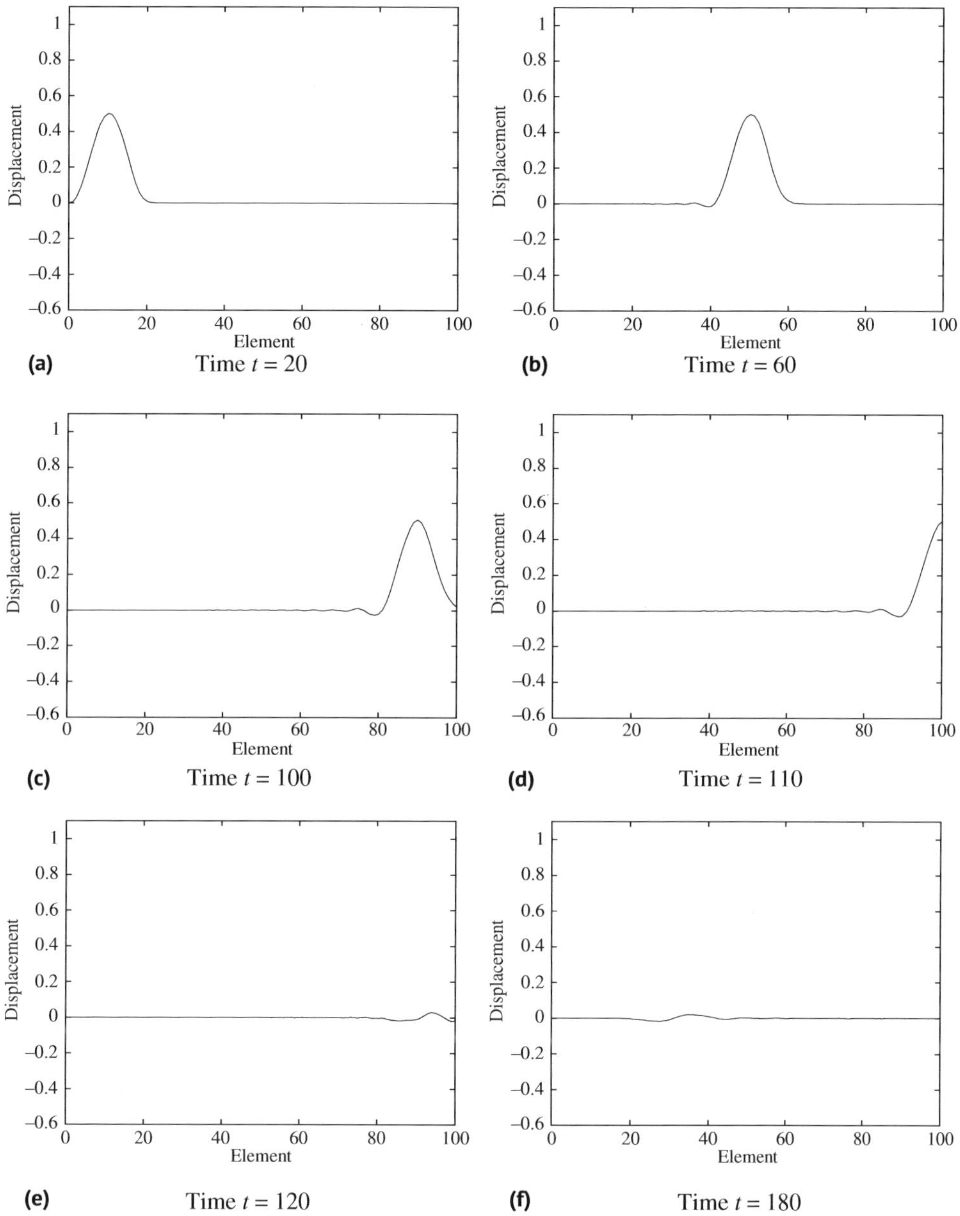

Figure 14.23 Simulated pulse propagation through 100 spring-mass elements to a matched termination impedance of $\alpha_{out} = 1$.

This case allows us to visualize the process of developing a standing wave on the transmission line. We can see in the first two frames of Figure 14.27 the initial sine wave entering and propagating from the left to the right, and then being reflected back on subsequent incident sine wave cycles. The last four frames show a complete cycle of the standing wave that has developed during a span of later sample times. We know from Equation (14.47) that the wave propagation speed c for this medium is $\sqrt{\frac{k}{m}}$ in units of line segments. Since the wavelength through any wave medium must be $\lambda = cT$ for a signal period of T, our observed wavelength of 20 segments in Figure 14.27 is consistent with the simulation parameters of $k = 1$, $m = 1$, and $T = 20$. This standing wave simulation arrives at patterns similar to those plotted in Figure 6.5 from studying the eigenvalues of coupled spring-mass oscillators. Since the output termination with a damping coefficient of 100 is not a perfectly rigid termination, there is a small minimum wave pattern visible.

The wave motion in all these simulations is most vividly appreciated by applying the animation techniques of Sections 4.13 and 6.9. Modifying the `wave1.c` program with a version of a `gen_output_gpanim()` function to generate animated graphical frames is available as Exercise 14.3 for the student.

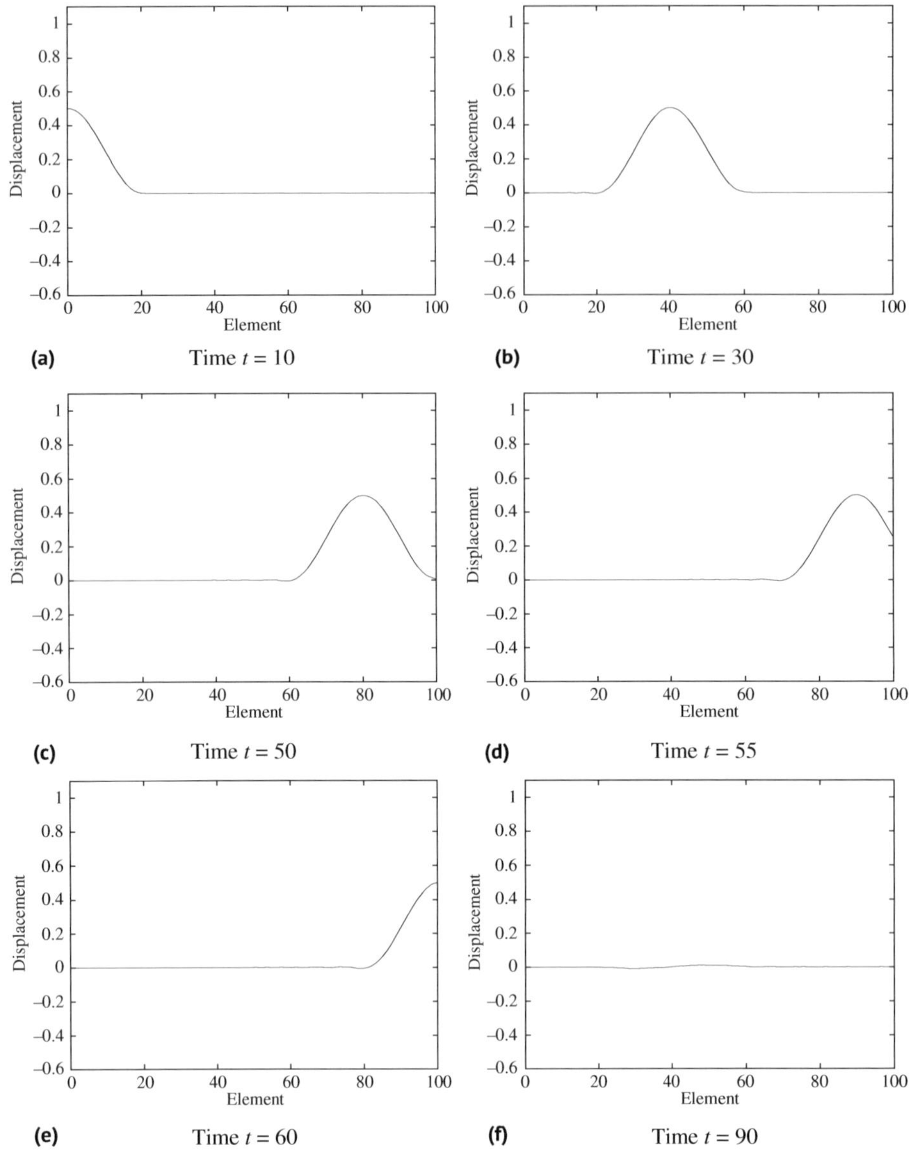

Figure 14.24 Simulated pulse at twice the propagation speed through 100 spring-mass elements to a matched termination impedance of $\alpha_{out} = 1$.

Now consider the scenario of two different transmission lines with different characteristic impedances connected in sequence carrying a mechanical pulse. The physical arrangement is still that pictured in Figure 14.21 but with a change in the mass *m* and the spring constant *k* half way through the sequence of *n* segments. The forcing input is applied to left side of the line, and the output termination is connected to damp the motion at the right side of the line. While the propagation speed $\sqrt{\frac{k}{m}} = 1$ everywhere, the characteristic impedances $\frac{1}{\sqrt{km}}$ of the two portions of the line are different by a factor of four, and this difference presents a sharp discontinuity of impedance to a pulse as it propagates from left to right. This mismatch at the discontinuity point causes both a reflection of the incident wave back to the left, as well as a continuing transmission to the right toward the termination. The coding exercise for this case, which may be most easily accomplished by copying the `wave1.c` source code file into a file `wave2.c` and expanding it, is available as Exercise 14.2 for the student. Figure 14.28 shows the simulated results for the case of the left half of the line having higher impedance than the right half, that is, $Z_1 = 2$ and $Z_2 = 0.5$. The total length of the line has been doubled relative to Figure 14.23 to 200 segments to allow both the reflected and transmitted waves to be followed.

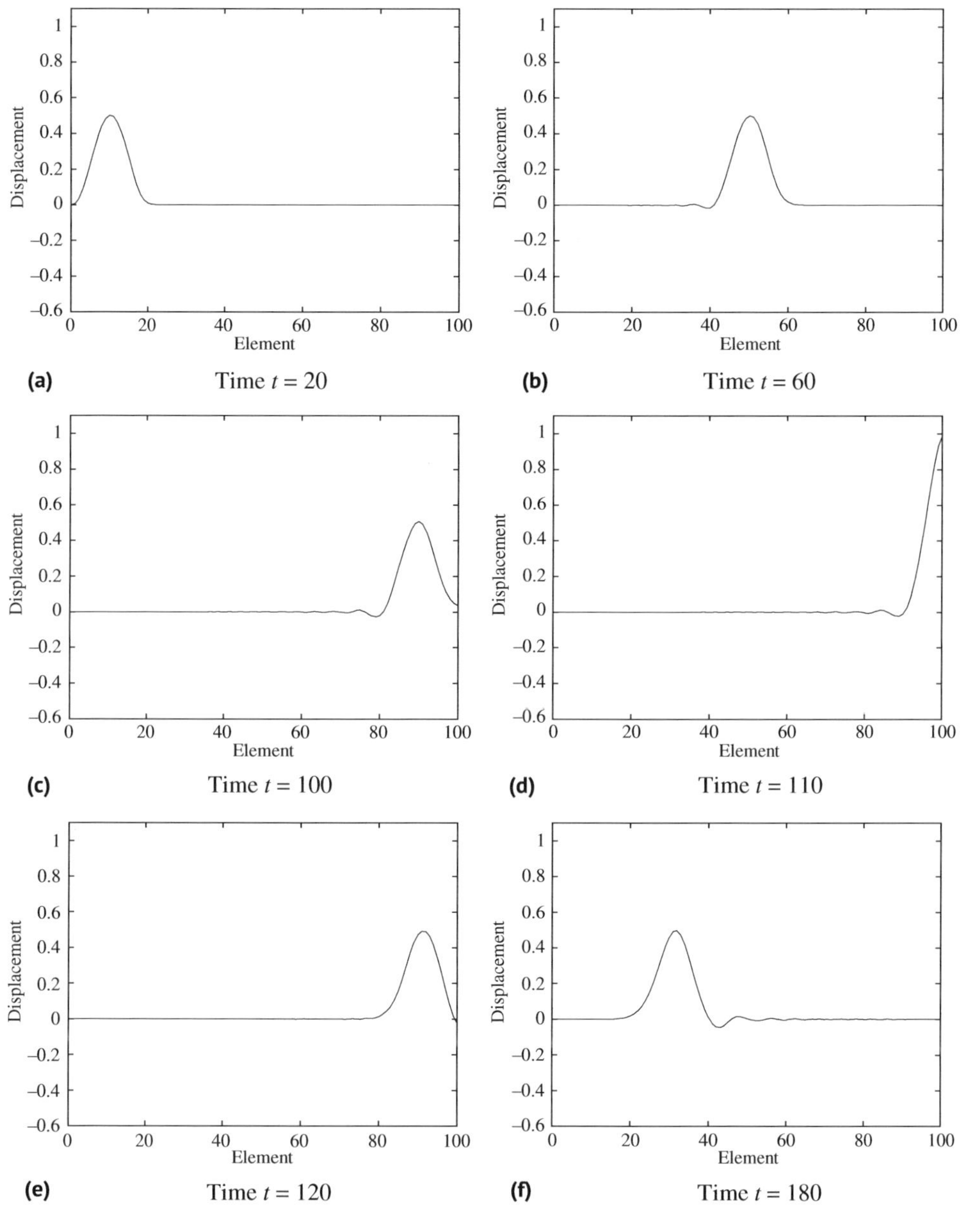

Figure 14.25 Simulated pulse propagation through 100 spring-mass elements to an unmatched termination impedance of $\alpha_{out} = 0.01$.

Figure 14.29 shows the behavior with the left half of the line having lower impedance than the right half, that is, $Z_1 = 0.5$ and $Z_2 = 2$. Note that the polarity of the reflected pulse is reversed compared with Figure 14.28.

14.6 Application: The Large-angle Pendulum with the Implicit Method

Using the nonlinear implicit Adams–Moulton algorithm detailed in Section 14.2.2 is somewhat more involved than the more straightforward explicit Runge–Kutta method in Section 12.2.4. So it will be instructional to return to the familiar problem of the damped large angle pendulum of Section 12.5, but now solved with the implicit method of this chapter, as a relatively simple example before tackling general n-body gravitational systems in Section 14.7.

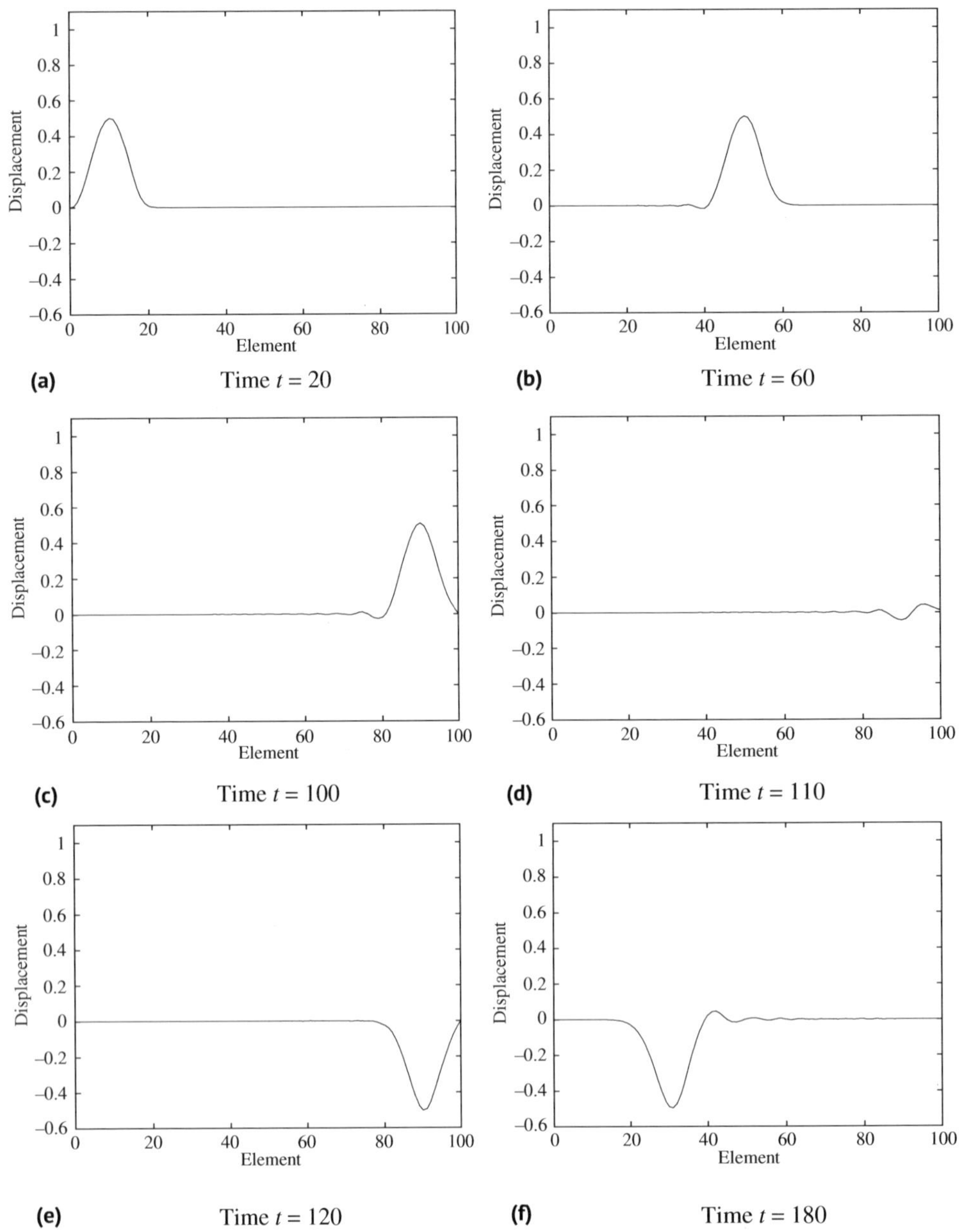

Figure 14.26 Simulated pulse propagation through 100 spring-mass elements to an unmatched termination impedance of $\alpha_{out} = 100$.

In Section 12.5, we defined two state variables as the deflection angle and angular velocity and then derived the system of first-order differential equations to submit to a numerical solver as

$$\begin{aligned}\frac{dy_0}{dt} &= y_1 \\ \frac{dy_1}{dt} &= -\frac{g}{l}\sin(y_0) - 2\zeta\sqrt{\frac{g}{l}}y_1\end{aligned} \tag{14.51}$$

where l is the pendulum length, g is the gravitational acceleration, and ζ is the equivalent damping factor for small angle oscillations. We must supply to the nonlinear implicit trapezoidal algorithm not only a C function that will compute the two functions on the right sides of Equation (14.51), which are the same functions required by the explicit Runge–Kutta solver, but now also a second C function that will compute the matrix of the four partial derivatives $\mathbf{J}_f$ from these two functions.

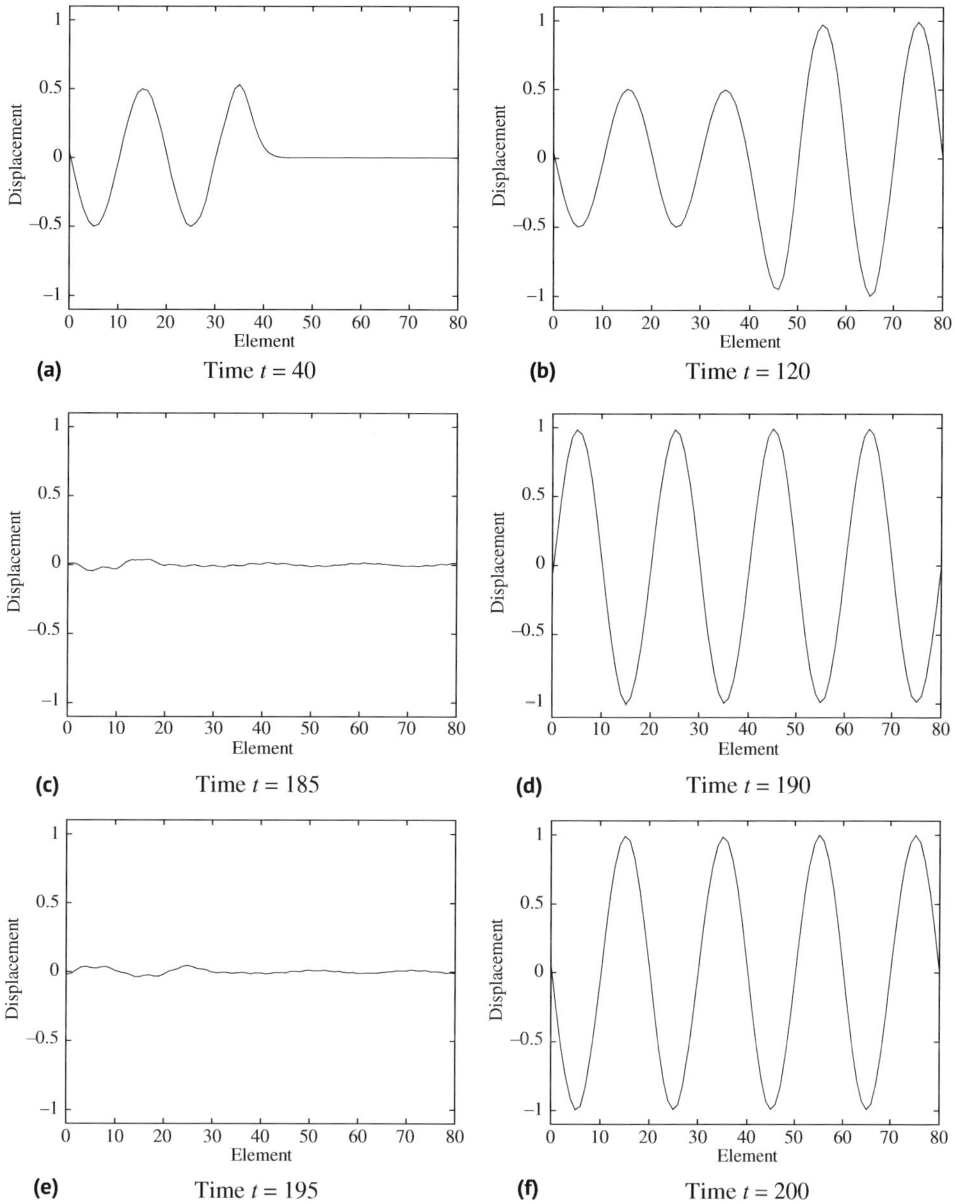

Figure 14.27 Simulated standing wave developing through 80 spring-mass elements with an unmatched termination impedance of $\alpha_{out} = 100$.

These are easily derived from Equation (14.51) as

$$\mathbf{J}_f = \begin{bmatrix} 0 & 1 \\ -\frac{g}{l}\cos(y_0) & -2\zeta\sqrt{\frac{g}{l}} \end{bmatrix} \tag{14.52}$$

We observe that in the case of small damping this $\mathbf{J}_f$ matrix has weak entries along the central diagonal, but the same comments made about the matrix in Equation (14.43) for the linear case apply here – namely that the actual Jacobian matrix used in the linear system of the iteration Equation (14.34) follows Equation (14.37), and the resulting diagonal elements will be dominated, in an absolute value sense, by the "1" elements of the identity matrix. Therefore, we can expect satisfactory numerical results from the Gaussian elimination process for solving the linear system. The C code in the source file `pendulum_implicit.c` from the Student Download website contains the following lines that set up the parameter structure and two required C functions as

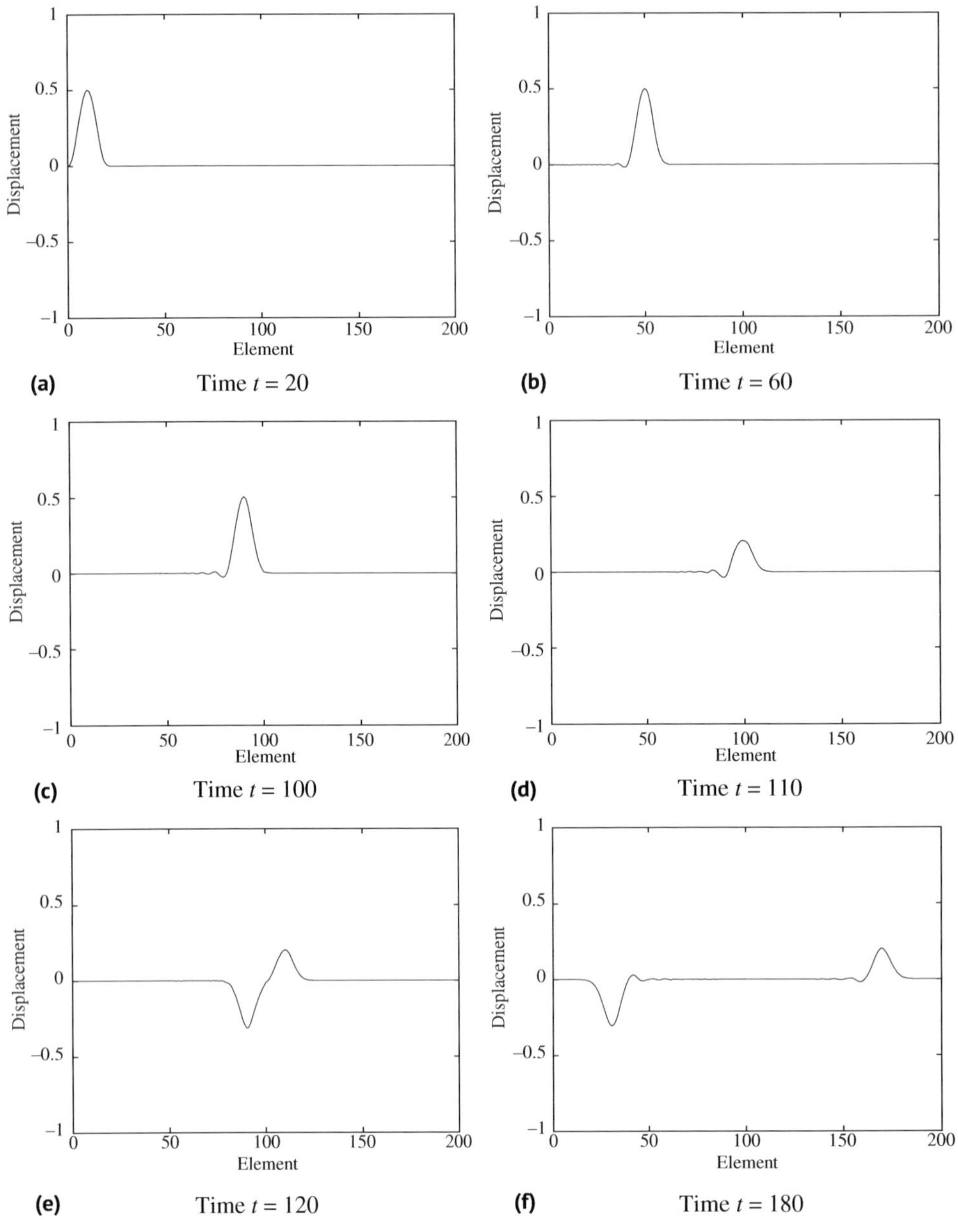

Figure 14.28 Simulated pulse propagation through 200 spring-mass elements and impedance discontinuity $Z_1 = 2, Z_2 = 0.5$.

```
typedef struct pend_param {
  double length;  // pendulum length in meters
  double zeta;  // damping coefficient in 1/sec
} PEND;

void pendulum(double x,ARRAY1D *y,void *param_ptr,ARRAY1D *yp) {
  double omegasq,omega;
  PEND *p;
  p = param_ptr;
  omegasq = GRAVITATIONAL_ACCEL/p->length;
  omega = sqrt(omegasq);
  yp->elem[0] = y->elem[1];
  yp->elem[1] = -omegasq * sin(y->elem[0]) -
              2.0 * p->zeta * omega * y->elem[1];
}
```

```
void pendulum_jacobian(double x,ARRAY1D *y,void *param_ptr,
                       ARRAY2D *partial) {
  double omegasq,omega;
  PEND *p;

  p = param_ptr;
  omegasq = GRAVITATIONAL_ACCEL/p->length;
  omega = sqrt(omegasq);
  partial->elem[0][0] = 0.0;
  partial->elem[0][1] = 1.0;
  partial->elem[1][0] = -omegasq * cos(y->elem[0]);
  partial->elem[1][1] = -2.0 * p->zeta * omega;
}
```

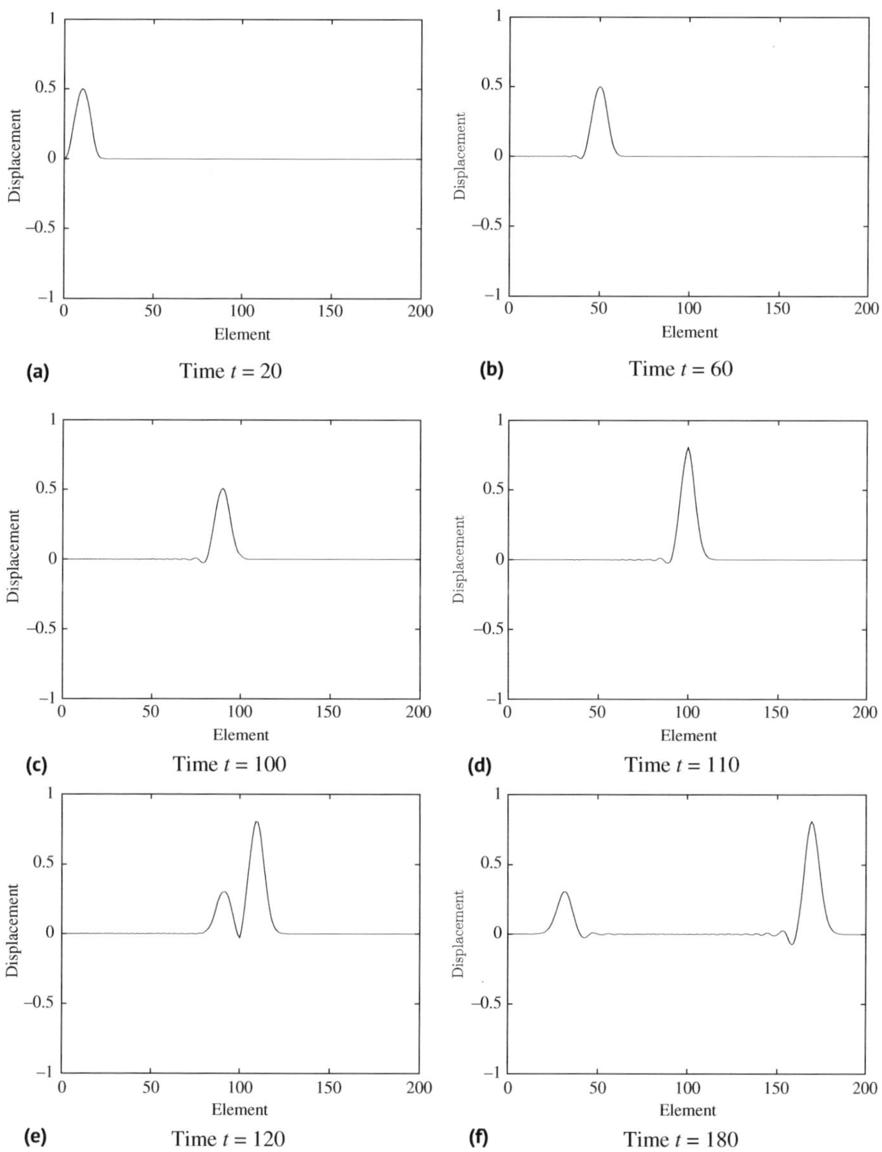

Figure 14.29 Simulated pulse propagation through 200 spring-mass elements and impedance discontinuity $Z_1 = 0.5, Z_2 = 2$.

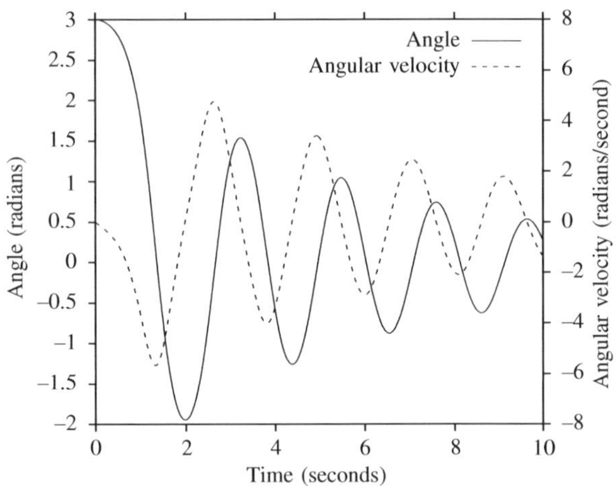

Figure 14.30 Pendulum with initial angle of 3 rad, damping coefficient 0.05, solved with the implicit AM2 method.

The rest of the program extracts the same physical parameters from its command line as used in `pendulum.c` and calls the solver tool with

```
if (am2_nonlinear(pendulum,pendulum_jacobian,&tgrid,&param,&init,
    1.0e-12,&ytable) == DIFFEQ_IMPLICIT_SUCCESS) {
  gen_output(&ytable,&tgrid);
}
```

The program is compiled and run with the dialog

```
$ gcc -Wall -Wshadow -O2 -o pendulum_implicit pendulum_implicit.c CPLibrary/diffeq_implicit.c
    ↪ CPLibrary/lineq.c CPLibrary/array.c -lm
$ ./pendulum_implicit
./pendulum_implicit <length> <zeta> <init y0> <init y1> <tfinal> <tinc>
$ ./pendulum_implicit 1 0.05 3 0 10 0.01 >| pendulum.dat
```

The same plotting procedure from Section 12.5 shows the solution in Figure 14.30. This compares with Figure 12.17 as an identical solution, as expected.

14.7 Application: *n*-body Gravitational Systems Revisited

With the numerical example of the large-angle pendulum in Section 14.6 to draw from, we are now ready to return to the *n*-body gravitation problems we first treated in Section 13.2. In that section we steered clear of gravitational scenarios that would lead to very close encounters between the bodies. In that situation, the gravitational forces following the Newtonian inverse square law can grow to be very large, causing very rapid accelerations. Although mathematically we are now discussing a nonlinear system, situations where the velocities of the bodies can vary over a large range are the counterpart to linear systems with a very wide range of eigenvalues. Either situation leads to instability with the explicit algorithms of Chapter 12.

We begin by deriving the gravitational properties of the example stable binary star system pictured in Figure 14.31, consisting of two identical stars of the same mass M as our Sun. We assume that each star is in a circular orbit about the center of mass, at the center of the circle. Using simple Newtonian gravitational laws to equate the gravitational force on each star with the centripetal force required to maintain each star in its circular orbit, we set

$$F = \frac{Gm^2}{4R^2} = \frac{mv^2}{R} = \frac{4\pi^2 Rm}{T^2} \tag{14.53}$$

From this we can solve for the square of the orbital period

$$P^2 = \frac{16\pi^2 R^3}{Gm} \tag{14.54}$$

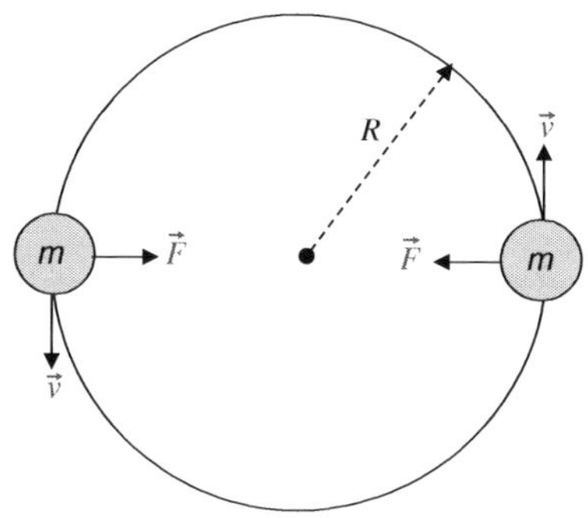

Figure 14.31 A symmetrical binary star gravitational system.

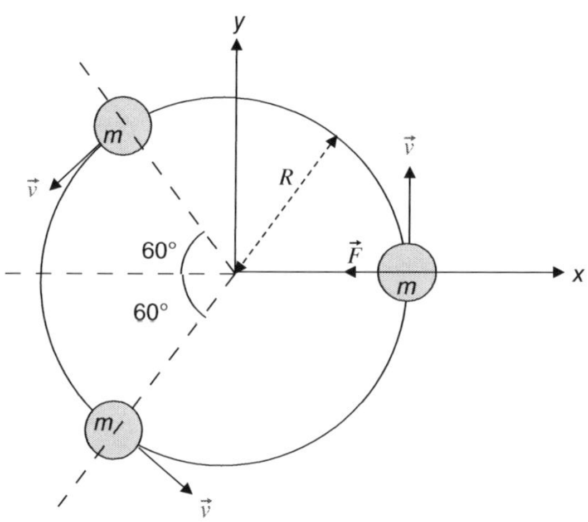

Figure 14.32 Initial configuration of an unstable symmetrical trinary star gravitational system.

If two identical stars may be configured in a stable orbit about their center of mass, is a similar stable configuration with three identical stars possible? Let us start our analysis using the implicit methods of this chapter and consider a hypothetical trinary star system with a symmetrical initial condition shown in Figure 14.32. We follow the same approach used for the binary system leading to Equation (14.54) and assume an initial configuration as shown with one star along the x axis of the coordinate system and two other stars located on the circular orbit path at 60° angles relative to the x axis. Some trigonometric analysis leads to the modified relationship

$$F = \frac{Gm^2}{\sqrt{3R^2}} = \frac{mv^2}{R} = \frac{4\pi^2 Rm}{T^2} \tag{14.55}$$

and a revised square of the orbital period of

$$P^2 = \frac{4\sqrt{3}\pi^2 R^3}{Gm} \tag{14.56}$$

We can take some example values to assign to our hypothetical system, with three stars identical to our Sun with mass 1.9891×10^{30} kg and an orbital period of one year. From Equation (14.56) we find a circular radius of 1.2460141×10^{11} m and an orbital velocity of 24807.726 m/s. From these figures we can derive initial vector positions and vector velocities of the three bodies and initialize the simulation state variables.

The C code in the source file `gravitation_implicit.c` from the Student Companion website contains the following lines that set up the parameter structure and the required C functions to calculate the differential equation derivative functions as

```
typedef struct grav_param {
  ARRAY1D mass;
  ARRAY1D inc;
} GRAV;

void gravitation(double x,ARRAY1D *y,void *param_ptr,ARRAY1D *yp) {
  int i_body,j_body,n_bodies,i,j;
  double dsq;
  GRAV *p;

  p = param_ptr;
  n_bodies = y->n_elems / 4;
  i_body = 0;
  while (i_body < n_bodies) {
    i = 4 * i_body;
    j_body = 0;
    yp->elem[i] = y->elem[i+2];
    yp->elem[i+1] = y->elem[i+3];
    yp->elem[i+2] = 0.0;
    yp->elem[i+3] = 0.0;
    while (j_body < n_bodies) {
      if (j_body != i_body) {
        j = 4 * j_body;
        dsq = (y->elem[j]-y->elem[i])*(y->elem[j]-y->elem[i]) +
              (y->elem[j+1]-y->elem[i+1])*(y->elem[j+1]-y->elem[i+1]);
        yp->elem[i+2] += p->mass.elem[j_body] *
                         (y->elem[j]-y->elem[i]) / (dsq * sqrt(dsq));
        yp->elem[i+3] += p->mass.elem[j_body] *
                         (y->elem[j+1]-y->elem[i+1]) / (dsq * sqrt(dsq));
      }
      j_body++;
    }
    yp->elem[i+2] *= GRAVITY;
    yp->elem[i+3] *= GRAVITY;
    i_body++;
  }
  return;
}
```

Analytically deriving the partial derivatives to fill in the Jacobian matrix would be a daunting task, so we surrender to the expediency of including a C function to numerically calculate the Jacobian matrix $\mathbf{J}_f$ borrowed from Section 7.6. This Jacobian function requires two additional parameters from the program command line, the discrete increment values for the positions and velocities of the bodies over which numerical derivative estimates are calculated.

The program is compiled and run with the dialog

```
$ gcc -Wall -Wshadow -O2 -o gravitation_implicit gravitation_implicit.c
    ↪ CPLibrary/diffeq_implicit.c CPLibrary/lineq.c CPLibrary/array.c -lm
$ ./gravitation_implicit
./gravitation_implicit <tfinal> <tinc> <num bodies> <x1,y1,vx1,vy1,m1> <x2,y2,vx2,vy2,m2> ...
    ↪ <rinc> <vinc>
$ ./gravitation_implicit 1e9 1e5 3 1.2460141e+11,0,0,24807.726,1.99e30
    ↪ -6.2300704e+10,1.0790798e+11,-21484.121,-12403.863,1.99e30
    ↪ -6.2300704e+10,-1.0790798e+11,21484.121,-12403.863,1.99e30 1e9 1e2 >| gravitation.dat
```

Figure 14.33 shows the results of the simulation. Note that as the simulation proceeds, the inevitable floating point round-off errors will contribute a tiny bit of perturbation from the ideal initial conditions. We can think of these numerical perturbations as modeling the physical influence of variations in the star mass distribution, or the influence of some nearby gravitational object in the actual physical system. From the simulation results we see that the system is physically unstable, resulting in two of the stars being ejected in one direction as a tightly orbiting binary couple, and the third star ejected in the other direction at a faster velocity, since the total momentum of the system would, of course, be conserved.

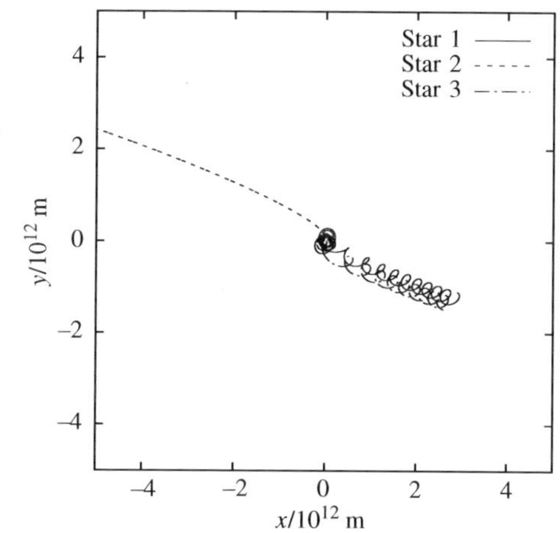

Figure 14.33 Simulation of an unstable trinary star system.

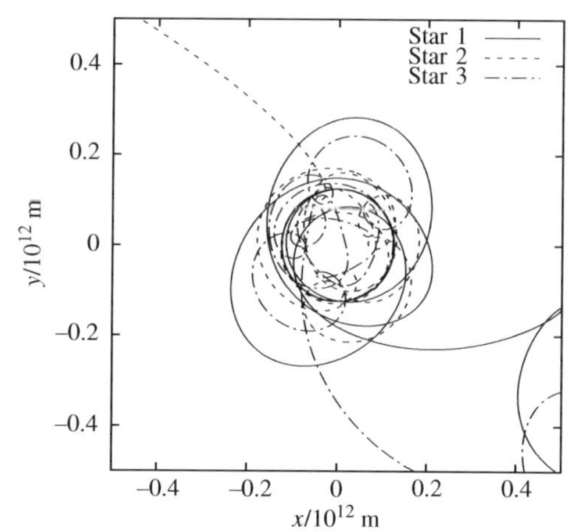

Figure 14.34 Simulation of an unstable trinary star system – zoomed-in view.

Figure 14.34 shows a zoomed-in view of the chaotic and intense interaction around the original orbit radius. These intense whiplash dynamics would result in instability of the numerical solution if this system were to be simulated with an explicit differential equation solution algorithm.

Exercise 14.4 gives the student an opportunity to explore two limited scenarios where a stable gravitational relationship among three stars is possible. You may refer to the animated simulations shown in the video files `gravitation_implicit_gpanim_lgsun-smsun.gif`, `gravitation_implicit_gpanim_lgsun-lgsun-smsun.gif`, and `gravitation_implicit_gpanim_lgsun-smsun-smsun.gif`, available for download from the Student Companion website, for a visualization of the three scenarios explored in that exercise.

14.8 Application: Magnetic Mirrors

In Section 12.12, we solved for the spiral paths that charged particles follow as they are deflected by regions of a magnetic field. In that section, we assumed that the particle initially would be incident from a region of zero field and then abruptly enter and leave a region of a uniform magnetic field. This is a rather idealized scenario, so in this section we revisit this same general problem with more realistic magnetic fields generated by the circular loops of electric current we have treated in prior chapters.

In Section 10.7, we plotted vectors that followed the direction of the local magnetic field due to two circular loops of electric current in the arrangement of a Helmholtz coil. That field was calculated with a numerical integration of the

Biot–Savart law analyzed in Section 10.6. As shown in Section 4.4, in the nominal Helmholtz coil arrangement, the distance D from the x-y plane to the planes of each of the current loops must be half the radius R of the coils for optimum uniformity of the magnetic field magnitude at the center of the geometry. Consider now the contours of the magnetic field when the distance D is extended by a factor of four from the nominal Helmholtz geometry, out to twice the coil radius. For example, let us examine the geometry with $R = 1$ m and $D = 2$ m. We can see the resulting curvature of the field vectors in Figure 14.35, where the field magnitude is coded in the plot vector lengths. See the image file `magfieldvectors_n2high_color.png` in the Student Companion website for the same field vector plot with the field magnitude coded in the plot vector colors.

Then in Section 13.4, we developed algorithms to plot classic field lines that followed the direction of a magnetic field generated by circular loops of electric current. Using the program code that is the solution to Exercise 13.4 for the extended Helmholtz coil, again with $R = 1$ m and $D = 2$ m, we can visualize the field lines in Figure 14.36.

Of course, Figure 14.36 is only a two-dimensional plot in one plane of a cylindrically symmetric arrangement, and the actual physical fields enclose the region between the coils in three dimensions. This configuration is referred to as a "magnetic bottle." In Section 12.12, we saw a simple simulation of a charged particle entering and leaving an idealized region of a uniform magnetic field, and traced the trajectory of the particle over a helical path that encircled and followed the magnetic field lines. What sort of trajectory would we expect from a particle injected into the central region of the fields in Figure 14.36? We would expect the particle to follow the field lines upward or downward toward one of the coils into a

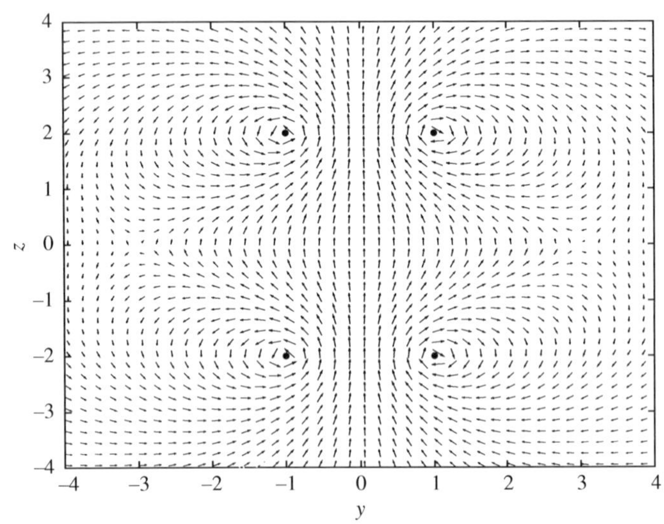

Figure 14.35 Magnetic field vectors around an extended Helmholtz coil, radius 1 m and distance 2 m, coded in length.

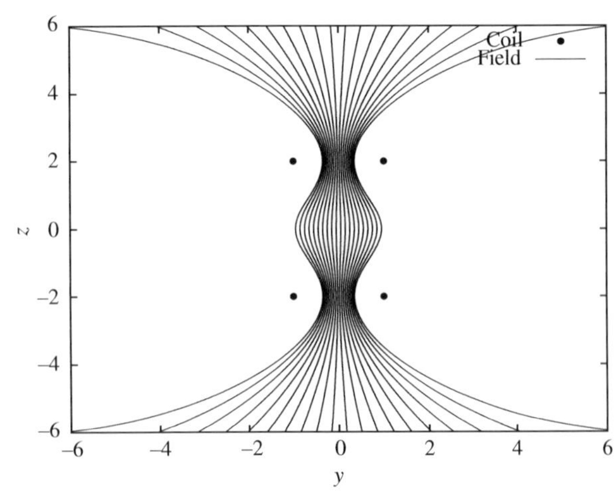

Figure 14.36 Field lines through an extended Helmholtz coil, radius $R = 1$ m and distance $D = 2$ m.

region where the field lines converge together, indicating a high field strength. In one of the curious consequences of the dynamic laws of motion in this system, when the local field reaches a concentration at a certain level, the particle will lose its velocity along the field line; any further motion of the particle in its original direction cannot be supported and the direction of travel suddenly reverses. This reflection of motion is referred to as a "magnetic mirror." This sudden acceleration can be captured by an implicit method solving the equations of motion, but could easily be the cause of numerical instability using an explicit method. Figure 14.37 shows an example simulation of a charged particle injected into a magnetic bottle. The particle travels upward and downward, continually spiralling around the field lines, and reversing direction at the two magnetic mirrors near each current loop. This arrangement of confining charged particles in a magnetic bottle is commonly used in high-temperature plasma experiments. The coding task for this simulation is available as Exercise 14.5 for the student.

Another example simulation geometry we can explore gives us insight into a natural phenomenon on a planetary scale. Perhaps the most visible manifestation of the properties of the Earth's ionosphere is the display known as the aurora borealis, photographed in Figure 14.38.

This visually stunning fluorescence is the result of electrically charged particles streaming from our Sun interacting with gas molecules in our atmosphere. Why are these particles primarily funneled into a circular region at very high Earth latitudes, into a region known as the "aurora oval"? To visualize a very basic model of the magnetic field of the earth that is responsible for this collection, we can make a simple approximation that our planet's magnetic field is generated by the equivalent of a single circular loop of electric current deep under the equator. We will approximate the resulting field to be similar to that of a simple magnetic dipole, with field lines shaped in the contours illustrated in Figure 13.16.

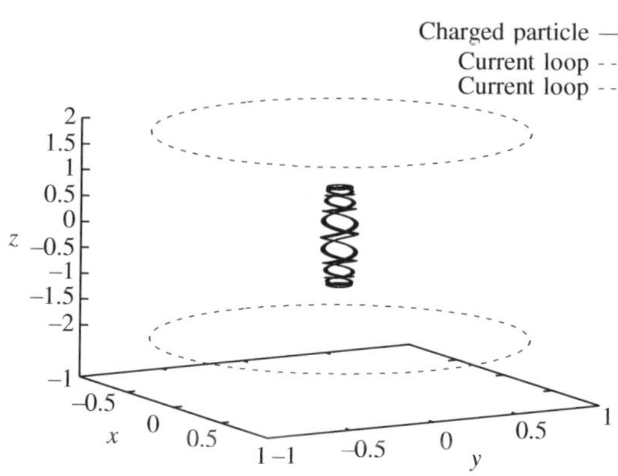

Figure 14.37 Simulation of charged particle trajectories in a magnetic bottle.

Figure 14.38 The aurora borealis photographed in Coldfoot, Alaska.

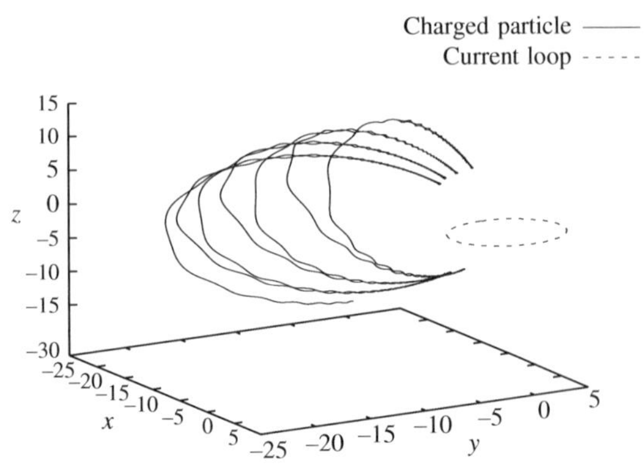

Figure 14.39 Simulation of charged particle trajectories in the ionosphere.

Just as in Section 12.12, we can then model the equations of motion of an incident charged particle traveling into the region of this magnetic field. For our exploration of the geometric principles of the dynamics of the charged particles, we can be content simply to use some convenient numerical choices of electric current, particle mass and charge, and velocity. With this very basic model, a simulation of the motion of incident particles is shown in Figure 14.39. We can again see trajectories taking the form of tight helical paths circulating around the magnetic field lines of the dipole. Similar to the behavior in the magnetic bottle, when the particles approach the regions of the magnetic poles where the field lines converge, the continuation of their forward velocity can no longer be supported mathematically, and the particles reverse course at the magnetic mirror points. This reflection results in the particles oscillating back and forth from pole to pole at high altitudes, forming the ionosphere. The altitude near the poles where their velocity reverses, as seen by the tips of the plotted trajectories, is dependent on the incident energy of the particles. Particularly energetic particles will plunge deep into the atmosphere around the poles, reacting with atmospheric gas molecules and generating light of characteristic colors. This very simplified model is insufficient in scope to capture the distortion of the contours of the magnetic field by the effective electric current of the particles themselves, but the general form of the mechanics is apparent. The coding task for this simulation is available for the student as Exercise 14.6.

14.9 Exercises

14.1 Following the downloaded source code file `coupled_osc22_implicit.c` from Section 14.1, write a new program `coupled_oscN_implicit.c` to simulate the motion of a coupled spring-mass oscillator with an arbitrary number of masses specified on the command line. After an argument specifying the number of masses, additional arguments specifying all initial displacements must follow. Assume that all springs are identical.

- **(a)** Run a simulation for the case of three masses with the initial conditions of Figure 14.9 and compare with the results in the figure.
- **(b)** Run a simulation for the case of nine masses and the initial conditions 1, 0, 1, 0, 1, 0, 1, 0, 1. Your results should demonstrate fluctuations over a wide frequency range that would tend toward numerical instability with an explicit algorithm as the number of masses is increased. Compare the simulation results with that of Exercise 6.2.

14.2 Starting with the `wave1.c` program from Section 14.5, write a new program `wave2.c` that joins two wave media with different characteristic impedances, and thereby forms a longer wave medium line with an abrupt transition at the halfway point. Provide perfect impedance matching at the input and output and simulate a raised cosine pulse traveling down the transmission line toward the impedance transition. First consider the case that the left half of the line has a higher impedance than the right half, that is, $Z_1 = 2$ and $Z_2 = 0.5$. Follow this simulation with the case that the left half of the line has a lower impedance than the right half, that is, $Z_1 = 0.5$ and $Z_2 = 2$. Plot the results to show both a reflected and transmitted wave component at the transition, similar to the plots shown in Figures 14.28 and 14.29, and note the wave polarities.

14.3 Starting with the `wave1.c` program, write a program `wave1_gpanim.c` using the animation coding techniques of Sections 4.13 and 6.9 to generate animated graphics of wave reflections and standing waves. Use the example cases presented in Figures 14.23 through 14.27.

14.4 In Section 14.7 we simulated an unstable trinary star system where all three stars were of identical masses. Are there trinary star systems with different mass distributions that evolve with stable orbits? In general, the presence of a relatively large dominant mass can stabilize these systems. We will explore here first an unsymmetrical binary system and then two special cases where three stars can have a stable gravitational relationship. You may refer to the animated simulations shown in the video files `gravitation_implicit_gpanim_lgsun-smsun.gif`, `gravitation_implicit_gpanim_lgsun-lgsun-smsun.gif`, and `gravitation_implicit_gpanim_lgsun-smsun-smsun.gif`, available for download from the Student Companion website, for a visualization of the three scenarios simulated in the following cases.

(a) In Section 14.7 we derived the relationship between the orbital period and radius of a symmetrical binary star system. To prepare for analyzing two candidate trinary star systems, first we consider the unsymmetrical binary system shown in Figure 14.40 where a large star of mass m_{lg} and a small star of mass $m_{sm} < m_{lg}$ are bound in a mutual orbit. The larger mass circles with a radius R_{lg}, and the smaller mass orbits in a circle of radius $R_{sm} > R_{lg}$. Assume that the relationship between the masses is $m_{lg} = \alpha m_{sm}$, with $\alpha > 1$. For a mutual orbit period P, leveraging the derivation in Section 14.7, show that

$$R_{sm} = \alpha R_{lg}$$

$$P^2 = \frac{4\pi^2 R_{lg}^3 (1+\alpha)^2}{G m_{sm}}$$

Note that for the case of $\alpha = 1$ the expression for P^2 reduces to that of Equation 14.54.

(b) Simulate and plot the two trajectories in an unsymmetrical binary star system with the `gravitation_implicit.c` program. Assume that the orbital period P is one year, that the smaller star has a mass equal to that of our Sun, but that the larger star is 10 times more massive. Establish the initial conditions as shown in 14.40 with the origin of the coordinate system at the center of both circular orbits and the initial positions of both stars along the x axis. Run your simulation for approximately 10 orbital periods, and use a time step giving about 300 steps per period.

(c) Now construct one candidate trinary system evolved from the binary system simulated in part (b). Split the small star into two stars, each of half the mass of our Sun, as illustrated in Figure 14.41. Use the relationships in Section 14.7 to find the radius and velocities in this local binary system for a binary period of 30 days. Displace the two small stars each in $\pm y$ directions by the local binary radius. Assign initial velocity vectors that maintain the y components at the original velocity from the long-range binary relationships in part (b), but x components from the local binary relationship, as shown in the figure. Simulate the system with the `gravitation_implicit.c` program and plot the three trajectories.

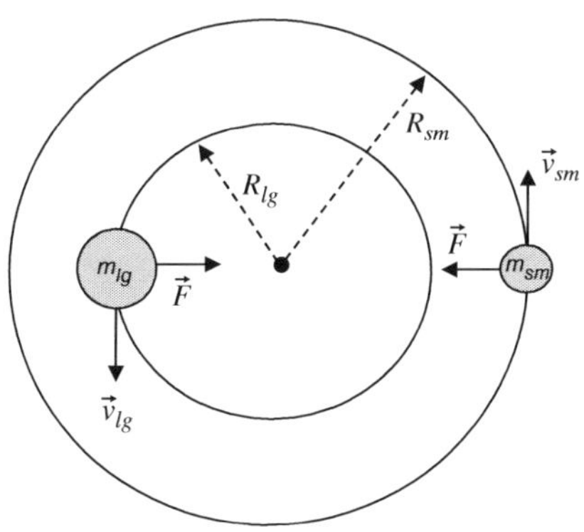

Figure 14.40 An unsymmetrical binary star gravitational system.

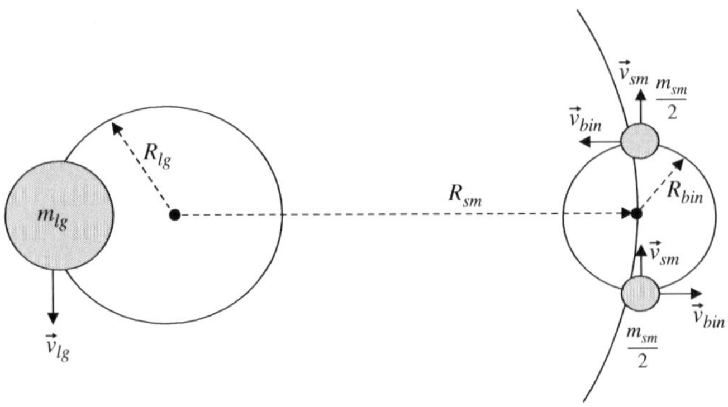

Figure 14.41 A stable trinary star gravitational system with a dominant large star and a binary system of small stars.

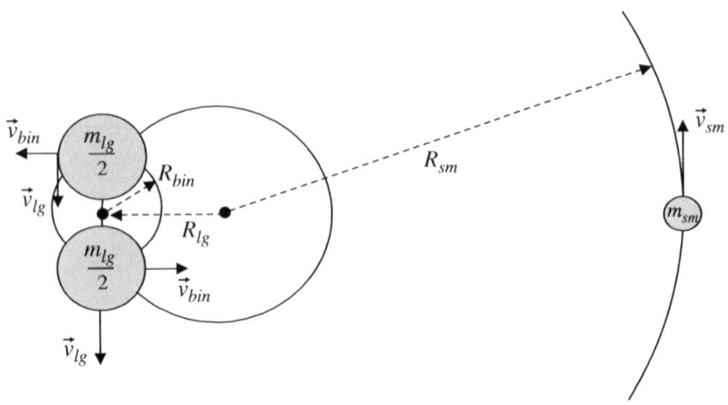

Figure 14.42 A stable trinary star gravitational system with a dominant binary system of large stars and a small star.

(d) In a similar manner, now simulate the trinary system shown in Figure 14.42. The large star from part (b) has now been split into two stars, each five times as massive as our Sun. Similar to part (c), displace the two large stars in $\pm y$ directions, and initialize the radius and velocities of the local binary system for a short period of three days. Simulate this system with the `gravitation_implicit.c` program and plot the three trajectories. Hint: The local binary system of the two massive stars rotates quite rapidly, so a shorter time step relative to that used in parts (b) and (c) will be required.

14.5 Write a program `magbottle_implicit.c` to simulate the system of a charged particle confined in the magnetic bottle of Figure 14.37 using the `am2_nonlinear()` nonlinear implicit solver routine. You may include a function to calculate the Jacobian matrix numerically. Use the example physical parameters of the coil radius $R = 1$ meter, a distance between the horizontal coils $D = 2$ meters, a current in both coils $I = 1$ A, a particle mass $m = 10^{-9}$ kg, and a charge $q = 1$ C. Initialize the charged particle at a location of $(0, 0.1, 0)$ with a velocity vector of $(10, 0, 10)$. Besides a table of the (x, y, z) components of the particle's position as a function of time, the program should include two extra data sets in the output stream to allow plotting of the two circular current paths for geometric reference. Run the simulation for 10 seconds.

14.6 Write a program `aurora_implicit.c` to simulate the system of a charged particle oscillating in the field of a magnetic dipole of Figure 14.39 using the `am2_nonlinear()` nonlinear implicit solver routine. You may include a function to calculate the Jacobian matrix numerically. Use the example physical parameters of the coil radius $R = 5$ m, a current in the coil $I = 100$ A, a particle mass $m = 10^{-7}$ kg, and a charge $q = 1$ C. Initialize the charged particle at a location of $(-20, 0, 10)$ with a velocity vector of $(0.5, 0, 0)$. Besides a table of the (x, y, z) components of the particle's position as a function of time, the program should include an extra data set in the output stream to allow plotting of the circular current path for geometric reference. Run the simulation for 1000 s.

Bibliography

1 Robert Martin Eisberg and Robert Resnick. *Quantum Physics of Atoms, Molecules, Solids, Nuclei and Particles*. Wiley, New York, NY, second edition, 1985.

2 Donald Rapp. *Quantum Mechanics*. Holt, Rinehart and Winston, New York, NY, 1971.

3 David Park. *Introduction to the Quantum Theory*. McGraw-Hill, New York, NY, second edition, 1974.

4 Claude Cohen-Tannoudji, Bernard Diu, and Franck Laloë. *Quantum Mechanics*. Wiley, New York, NY, first edition, 1977.

5 David K. Ferry and D. Ronald Fannin. *Physical Electronics*. Addison-Wesley, Reading, MA, 1971.

6 S. Ramo, J.R. Whinnery, and T. Van Duzer. *Fields and Waves in Communication Electronics*. Wiley, 1965.

7 John D. Kraus. *Electromagnetics*. McGraw-Hill, New York, NY, 1953.

8 John R. Reitz, Frederick J. Milford, and Robert W. Christy. *Foundations of Electromagnetic Theory*. Addison-Wesley, Reading, MA, third edition, 1980.

9 Brice Carnahan, H.A. Luther, and James O. Wilkes. *Applied Numerical Methods*. Wiley, New York, NY, 1969.

10 E. Ward Cheney and David R. Kincaid. *Numerical Mathematics and Computing*. Brooks Cole, Boston, MA, seventh edition, 2013.

11 William H. Press, Saul A. Teukolsky, William T. Vetterling, and Brian P. Flannery. *Numerical Recipes in C*. Cambridge University Press, Cambridge, USA, second edition, 1992.

12 Steven C. Chapra and Raymond P. Canale. *Numerical Methods for Engineers*. McGraw-Hill, New York, NY, 1988.

13 Curtis Gerald and Patrick Wheatley. *Applied Numerical Analysis*. Addison-Wesley, Reading, MA, third edition, 1985.

14 Richard Heathfield, Lawrence Kirby, et al. *C Unleashed*. Sams, Indianapolis, USA, 2000.

15 John David Jackson. *Classical Electrodynamics*. Wiley, New York, NY, second edition, 1975.

16 René Descartes. *Discourse on the Method*. 1637.

17 David Halliday, Robert Resnick, and Jearl Walker. *Fundamentals of Physics*. Wiley, New York, NY, sixth edition, 2001.

18 Randall D. Knight. *Physics for Scientists and Engineers*. Addison-Wesley, Reading, MA, 2004.

19 Douglas C. Giancoli. *Physics for Scientists and Engineers*. Prentice-Hall, Englewood Cliffs, NJ, third edition, 2000.

20 Raymond A. Serway and Robert J. Beichner. *Fundamentals of Physics*. Harcourt College Publishers, Philadephia, PA, fifth edition, 2000.

21 Marvin L. De Jong. *Introduction to Computational Physics*. Addison-Wesley, Reading, MA, 1991.

22 Adam M. Dziewonski and Don L. Anderson. Preliminary reference earth model. *Physics of the Earth and Planetary Interiors*, 25: 297–356, 1981.

23 Paul A. Tipler and Ralph A. Llewellyn. *Modern Physics*. W. H. Freeman, New York, NY, fourth edition, 2004.

24 Raymond A. Serway, Clement J. Moses, and Curt A. Moyer. *Modern Physics*. Saunders College Publishing, Philadelphia, PA, 1989.

25 Marshall L. Burns. *Modern Physics for Science and Engineering*. Harcourt Brace Jovanowich, San Diego, CA, 1988.

26 Anthony Ralston and Herbert S. Wilf. *Mathematical Methods for Digital Computers*. Wiley, New York, NY, 1961.

27 John R. Taylor. *Classical Mechanics*. University Science Books, Sausalito, California, 2005.

28 David Morin. *Introduction to Classical Mechanics: With Problems and Solutions*. Cambridge University Press, 2008.

29 Grant R. Fowles. *Analytical Mechanics*. Saunders College Publishing, Philadelphia, PA, fourth edition, 1986.

Computational Physics Using C: Efficient Programming with Ease, First Edition. John W. Fattaruso. © 2026 John Wiley & Sons, Inc.
All rights reserved, including rights for text and data mining and training of artificial intelligence technologies or similar technologies.
Published 2026 by John Wiley & Sons, Inc. Companion Website: http://www.wiley.com/go/CompPhysUsingC

30 Lawrence Kinsler and Austin Frey. *Fundamentals of Acoustics*. Wiley, New York, NY, second edition, 1963.

31 John A. Nelder and Roger Mead. A simplex method for function minimization. *Computer Journal*, 7: 308–313, 1965.

32 Charles Kittel. *Introduction to Solid State Physics*. Wiley, New York, NY, fifth edition, 1976.

33 S.M. Sze. *Physics of Semiconductor Devices*. Wiley, New York, NY, second edition, 1981.

34 Leon O. Chua and Pen-Min Lin. *Computer-Aided Analysis of Electronic Circuits: Algorithms and Computational Techniques*. Prentice-Hall, Englewood Cliffs, NJ, 1975.

Index

Computational Physics Using C: Efficient Programming with Ease, First Edition. John W. Fattaruso.

© 2026 John Wiley & Sons, Inc.
All rights reserved, including rights for text and data mining and training of artificial intelligence technologies or similar technologies.

Published 2026 by John Wiley & Sons, Inc. Companion Website: http://www.wiley.com/go/CompPhysUsingC